AF812037

Environmental Science and Engineering

Series Editors

Ulrich Förstner, Buchholz, Germany

Wim H. Rulkens, Department of Environmental Technology, Wageningen, The Netherlands

The ultimate goal of this series is to contribute to the protection of our environment, which calls for both profound research and the ongoing development of solutions and measurements by experts in the field. Accordingly, the series promotes not only a deeper understanding of environmental processes and the evaluation of management strategies, but also design and technology aimed at improving environmental quality. Books focusing on the former are published in the subseries Environmental Science, those focusing on the latter in the subseries Environmental Engineering.

Aravind Jeyaseelan · Kamaraj Murugasen
Editors

Sustainable Environmental Remediation: Avenues in Nano and Biotechnology

Springer

Editors
Aravind Jeyaseelan
Department of Biotechnology
Saveetha School of Engineering
Saveetha University
Chennai, Tamil Nadu, India

Kamaraj Murugasen
Department of Biotechnology
SRM Institute of Science and Technology
Ramapuram Campus
Chennai, Tamil Nadu, India

ISSN 1863-5520 ISSN 1863-5539 (electronic)
Environmental Science and Engineering
ISBN 978-3-031-78482-8 ISBN 978-3-031-78483-5 (eBook)
https://doi.org/10.1007/978-3-031-78483-5

© The Editor(s) (if applicable) and The Author(s), under exclusive license to Springer Nature Switzerland AG 2025

This work is subject to copyright. All rights are solely and exclusively licensed by the Publisher, whether the whole or part of the material is concerned, specifically the rights of translation, reprinting, reuse of illustrations, recitation, broadcasting, reproduction on microfilms or in any other physical way, and transmission or information storage and retrieval, electronic adaptation, computer software, or by similar or dissimilar methodology now known or hereafter developed.
The use of general descriptive names, registered names, trademarks, service marks, etc. in this publication does not imply, even in the absence of a specific statement, that such names are exempt from the relevant protective laws and regulations and therefore free for general use.
The publisher, the authors and the editors are safe to assume that the advice and information in this book are believed to be true and accurate at the date of publication. Neither the publisher nor the authors or the editors give a warranty, expressed or implied, with respect to the material contained herein or for any errors or omissions that may have been made. The publisher remains neutral with regard to jurisdictional claims in published maps and institutional affiliations.

This Springer imprint is published by the registered company Springer Nature Switzerland AG
The registered company address is: Gewerbestrasse 11, 6330 Cham, Switzerland

If disposing of this product, please recycle the paper.

Contents

Chapter 1
The Role of Actinomycetes in a Cleaner Environment

Veilumuth Pattapulavar and John Godwin Christopher

Abstract Actinomycetes, a diverse group of bacteria, have emerged as invaluable tools in environmental remediation due to their versatile metabolic capabilities and enzymatic diversity. This chapter delves into the pivotal role of specific actinomycete species as effective environmental cleaners across various domains. Highlighting their proficiency in addressing diverse pollutants, including heavy metals, hydrocarbons, synthetic dyes, and soil contaminants, this chapter elucidates how certain species excel in specific remediation tasks. For instance, *Streptomyces* spp. demonstrate remarkable efficacy in heavy metal immobilization and detoxification, contributing to ecosystem restoration. Moreover, *Thermobifida fusca* BCRC 19214 exhibits exceptional ability to metabolize hydrocarbons, making it instrumental in combating petroleum-based pollution in aquatic and terrestrial environments. The species *Streptomyces krainskii* SUK-5 plays a crucial role in soil bioremediation, enhancing soil fertility and ecosystem health. By shedding light on these specific species and their environmental remediation potential, this chapter underscores the importance of harnessing actinomycetes as effective agents for sustainable environmental cleanup initiatives.

1.1 Introduction

Pollution, in various forms, poses a significant threat to our planet's ecosystems, human health, and the prospects of future generations. From the contamination of air, water, and soil to the proliferation of plastic waste, pollution exacts a heavy toll on both the natural world and human society. Understanding its multifaceted effects is crucial for implementing effective strategies to mitigate its harmful consequences. The three major impacts on the ecosystem are the loss of biodiversity, destruction of habitat, and alteration of the ecosystem.

V. Pattapulavar · J. G. Christopher (✉)
Vellore Institute of Technology, Vellore, Tamil Nadu 632014, India
e-mail: godwinj@vit.ac.in

© The Author(s), under exclusive license to Springer Nature Switzerland AG 2025

A. Jeyaseelan and K. Murugasen (eds.), *Sustainable Environmental Remediation: Avenues in Nano and Biotechnology*, Environmental Science and Engineering,
https://doi.org/10.1007/978-3-031-78483-5_1

Pollution disrupts ecosystems by poisoning or suffocating plants, animals, and microorganisms. Chemical pollutants, such as pesticides and industrial waste, contaminate water bodies and soil, leading to the decline of species populations and the disruption of food chains. Pollution contributes to the loss of habitat, threatens the survival of countless species and reduces the resilience of ecosystems to withstand environmental changes. Altered Ecosystem Dynamics: Excessive pollution alters ecological processes, such as nutrient cycling and water purification, destabilizes ecosystems and diminishes their capacity to provide essential services such as clean air and water (Manisalidis et al. 2020).

Air pollution, primarily caused by vehicle emissions, industrial activities, and biomass burning, is a leading cause of respiratory illnesses such as asthma, bronchitis, and lung cancer. Fine particulate matter and toxic gases penetrate deep into the lungs, exacerbating existing health conditions and posing serious risks, especially to vulnerable populations (Ghorani-Azam et al. 2016). Contaminated water sources, tainted by pollutants such as heavy metals, pathogens, and chemicals from agricultural runoff, industrial discharge, and improper waste disposal, contribute to the spread of waterborne diseases such as cholera, typhoid, and dysentery (Lin et al. 2022). Prolonged exposure to environmental pollutants is associated with a range of long-term health effects, including cardiovascular diseases, neurological disorders, and developmental abnormalities. Children, pregnant women, and elderly people are particularly susceptible to the adverse effects of pollution on health.

Pollution, particularly greenhouse gas emissions from burning fossil fuels, drives climate change, leading to rising global temperatures, extreme weather events, and sea level rise. The consequences of unchecked climate change, such as food insecurity, water scarcity, and mass displacement, threaten the well-being of future generations. Pollution contributes to the depletion of natural resources, including clean water, arable land, and biodiversity, jeopardizing the ability of future generations to meet their needs and maintain a high quality of life. The economic costs of pollution, including healthcare expenses, environmental cleanup efforts, and loss of productivity, impose a heavy burden on societies and future generations, diverting resources away from investments in sustainable development and infrastructure (Shivanna 2022).

Therefore, a clean environment is essential for safeguarding human health and well-being, reducing the incidence of pollution-related diseases and promoting a greater quality of life for present and future generations. Maintaining a clean environment preserves biodiversity and ecosystem services, ensuring the resilience of natural systems to withstand environmental challenges and supporting essential functions such as pollination, nutrient cycling, and climate regulation. A clean environment is fundamental to achieving sustainable development goals, including poverty alleviation, food security, and equitable access to resources. By promoting eco-friendly practices and technologies, we can create a more prosperous and resilient future for all (Malhi et al. 2020).

Pollution poses a grave threat to ecosystems, human health, and the prospects of future generations. To address this pressing challenge, concerted efforts are needed to mitigate pollution sources, promote sustainable practices, and protect the environment for the benefit of all life on Earth. By prioritizing the importance of a clean environment and taking decisive action to combat pollution, we can build a healthier, more resilient planet for generations to come.

Bioremediation is a multifaceted approach that utilizes biological agents to detoxify, degrade, or eliminate pollutants from the environment. The concept of bioremediation gained momentum in the late twentieth century as a response to the growing concern over anthropogenic pollution and its detrimental effects on ecosystems and human health.

Bioremediation can be defined as "The use of living organisms, primarily microorganisms, to degrade environmental contaminants into less toxic or nontoxic forms" (Robles-González et al. 2008). This process relies on the metabolic activities of microorganisms to break down pollutants into simpler compounds through enzymatic reactions, ultimately rendering them harmless or less harmful to the environment.

The importance of environmental cleanup cannot be overstated in the face of escalating pollution levels worldwide. Anthropogenic activities such as industrial processes, agricultural practices, and improper waste disposal have led to the release of a myriad of pollutants into the air, water, and soil. These pollutants include organic compounds, heavy metals, pesticides, and petroleum hydrocarbons, among others, which pose significant risks to human health, wildlife, and ecosystem integrity (Abubakar et al. 2022).

Environmental contamination has far-reaching consequences, including the disruption of ecosystem functioning, loss of biodiversity, and contamination of food and water resources. Additionally, pollutants can accumulate in the environment over time, leading to persistent contamination and long-term environmental degradation. Bioremediation offers a sustainable and cost-effective approach to address these challenges by harnessing natural processes to clean up contaminated sites and restore environmental quality (Adla et al. 2022).

Microbial bioremediation involves the use of microorganisms, including bacteria, fungi, algae, and protozoa, to degrade or transform pollutants in the environment. Microorganisms possess diverse metabolic capabilities, allowing them to utilize a wide range of organic and inorganic compounds as energy and carbon sources. In bioremediation, microorganisms play key roles in detoxifying pollutants through various mechanisms, including enzymatic degradation, metabolic transformation, and immobilization (Haritash and Kaushik 2009).

Microbial bioremediation can be categorized into different strategies based on the metabolic pathways involved, including aerobic bioremediation, anaerobic bioremediation, and phytoremediation. Aerobic bioremediation relies on the presence of oxygen to support microbial growth and metabolism, making it suitable for degrading organic contaminants such as petroleum hydrocarbons and solvents. In contrast, anaerobic bioremediation occurs in the absence of oxygen and is effective for treating pollutants such as chlorinated solvents and nitroaromatic compounds (Megharaj et al. 2011). Phytoremediation utilizes plants and their associated microorganisms to uptake, metabolize, or sequester contaminants from the soil, water, or air, offering a sustainable and aesthetically pleasing remediation approach (Singh and Pant 2023).

Actinomycetes, a group of gram-positive filamentous bacteria belonging to the phylum Actinobacteria, have emerged as key players in bioremediation due to their diverse metabolic capabilities and wide distribution in various environmental habitats. Actinomycetes are known for producing a vast array of antibiotics but also produce extracellular enzymes, secondary metabolites, and bioactive compounds, enabling them to degrade complex organic molecules and detoxify pollutants (Devanshi et al. 2021).

Actinomycetes play crucial roles in bioremediation processes, including the degradation of petroleum hydrocarbons, polycyclic aromatic hydrocarbons (PAHs), pesticides, and heavy metals (Jagannathan et al. 2021). These microorganisms produce a variety of enzymes involved in pollutant degradation, including oxidoreductases, hydrolases, and lyases, which catalyze the breakdown of pollutants into simpler, less toxic compounds. Furthermore, actinomycetes have been shown to possess mechanisms for metal sequestration, precipitation, and transformation, contributing to the remediation of metal-contaminated sites (Cimermanova et al. 2021).

Based on the above, this book chapter focuses on actinomycetes as a valuable resource for bioremediation applications, offering a natural and sustainable approach to environmental cleanup. Their versatile metabolic capabilities and adaptive traits make them suitable for degrading a wide range of pollutants and restoring contaminated environments to their natural state.

1.2 Actinomycetes: Nature's Bioengineers

Actinomycetes, a fascinating group of filamentous bacteria, play pivotal roles in various ecological processes, including bioremediation. Actinomycetes, derived from the Greek words "actis", meaning rays, and "mykes", meaning fungi, are a diverse group of gram-positive bacteria characterized by their filamentous growth morphology. These bacteria are ubiquitous in nature and inhabit a wide range of environments, including soil, water, air, and even extreme habitats such as hot springs, deep-sea vents, and polar regions. Actinomycetes exhibit diverse metabolic capabilities, contributing to their significance in various ecological processes, including decomposition, nutrient cycling, and bioremediation (Mincer et al. 2002).

1.2.1 Taxonomy and Diversity

Taxonomically, actinomycetes belong to the phylum Actinobacteria, which comprises a vast array of bacteria with high G+C content in their DNA. Actinomycetes are further classified into multiple families, including *Streptomycetaceae*,

Micromonosporaceae, *Nocardiaceae*, and *Actinomycetaceae*, among others. The genus Streptomyces is perhaps the most well-known and extensively studied group of actinomycetes and is renowned for its prolific production of bioactive secondary metabolites with diverse pharmacological and industrial applications (Bérdy 2005).

The diversity of actinomycetes is staggering, with an estimated 10,000–15,000 species described to date and many more awaiting discovery. These bacteria exhibit a wide range of morphological, physiological, and ecological traits, reflecting their adaptation to diverse environmental niches and functional roles in ecosystems. Actinomycetes display remarkable metabolic versatility, allowing them to utilize various carbon and energy sources, including simple sugars, complex carbohydrates, and organic acids (Barka et al. 2015).

1.2.2 Unique Features

Actinomycetes possess several unique features that distinguish them from other bacteria. One of the most striking characteristics of actinomycetes is their filamentous growth morphology, characterized by long, branching hyphae that form extensive mycelial networks in soil and other substrates. This filamentous growth enables actinomycetes to explore and exploit their surroundings more efficiently, facilitating nutrient acquisition, interspecies interactions, and biofilm formation (Barka et al. 2015).

Moreover, actinomycetes produce a wide array of extracellular enzymes, including proteases, cellulases, lipases, and chitinases, which play crucial roles in organic matter degradation and nutrient cycling. These enzymes enable actinomycetes to decompose complex organic molecules and utilize them as energy and carbon sources, contributing to soil fertility and ecosystem functioning (Goodfellow et al. 1988).

1.2.3 Adaptations for Survival in Diverse Environments

Actinomycetes have developed numerous adaptations to survive and thrive in diverse environmental conditions. These adaptations include the production of osmolytes, heat shock proteins, and cell wall components that confer resistance to environmental stressors such as desiccation, high salinity, and extreme temperatures. Actinomycetes also engage in various symbiotic and mutualistic interactions with plants, animals, and other microorganisms, allowing them to colonize and persist in challenging environments (Ventura et al. 2007).

Furthermore, actinomycetes are prolific producers of secondary metabolites, including antibiotics, antifungals, anticancer agents, and immunosuppressants, which serve as chemical defenses against competitors and predators. These bioactive compounds have significant ecological implications and are of great interest for biotechnological and pharmaceutical applications (Bérdy 2005). Actinomycetes exhibit a myriad of unique features and adaptations that allow them to act as bioengineers, shaping microbial communities, nutrient cycles, and ecosystem dynamics in diverse environments.

1.3 Biodegradation Mechanisms of Actinomycetes

Biodegradation, the process by which microorganisms breakdown complex organic compounds into simpler, less harmful substances, lies at the heart of bioremediation efforts. In this chapter, we delve into the fascinating world of biodegradation mechanisms employed by actinomycetes, shedding light on the biodegradation processes, the enzymatic pathways involved in pollutant degradation, the metabolic versatility of actinomycetes, and the factors influencing biodegradation rates.

1.3.1 Biodegradation Processes

Biodegradation is a complex biochemical process that involves the enzymatic breakdown of pollutants by microorganisms. Due to their diverse metabolic capabilities, Actinomycetes play crucial roles in biodegradation processes, contributing to the removal of various pollutants from the environment. Biodegradation can occur through different pathways, including aerobic degradation in the presence of oxygen and anaerobic degradation in the absence of oxygen. These pathways may involve sequential or parallel enzymatic reactions, leading to the degradation of pollutants into simpler compounds that can be utilized as carbon and energy sources by microorganisms (Fig. 1.1).

1.3.2 Enzymatic Pathways Involved in Pollutant Degradation

Actinomycetes employ a wide array of extracellular enzymes to degrade diverse pollutants, including hydrocarbons, aromatic compounds, pesticides, and industrial chemicals. These enzymes catalyze specific biochemical reactions, leading to the cleavage of chemical bonds and the conversion of complex molecules into simpler intermediates. Some of the key enzymatic pathways (Table 1.1) involved in pollutant degradation by actinomycetes include the following:

1. Oxidoreductases: Enzymes such as cytochrome P450 monooxygenases, dehydrogenases, and oxygenases catalyze oxidation–reduction reactions, facilitating the breakdown of organic compounds by introducing oxygen atoms or removing hydrogen atoms from substrate molecules (Kharwar et al. 2014).

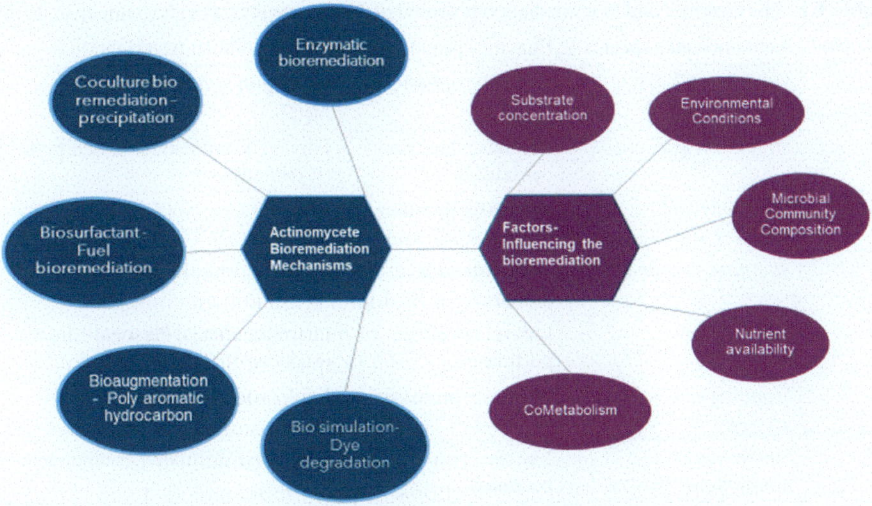

Fig. 1.1 Actinomycete biodegradation mechanisms and their influencing factors

2. Hydrolases: Actinomycetes produce a variety of hydrolases, including esterases, lipases, proteases, and glycosidases, which hydrolyze ester, lipid, protein, and carbohydrate bonds, respectively, in organic pollutants. These enzymes play crucial roles in the degradation of complex organic molecules into smaller, more manageable substrates (Mukhtar et al. 2017). Lyases: Lyases are enzymes that catalyze the cleavage of chemical bonds without the addition of water. Actinomycetes produce lyases such as decarboxylases and deaminases, which mediate the removal of functional groups from organic compounds, leading to their degradation into simpler compounds (Mukhtar et al. 2017).

1.3.3 Metabolic Versatility

One of the key strengths of actinomycetes in biodegradation is their metabolic versatility, allowing them to utilize a wide range of organic and inorganic compounds as carbon and energy sources. Actinomycetes possess diverse metabolic pathways, including glycolysis, the tricarboxylic acid (TCA) cycle, and various biosynthetic pathways, enabling them to metabolize complex pollutants through catabolic and anabolic reactions (Luz et al. 2004).

Actinomycetes can adapt their metabolic machinery in response to changing environmental conditions, such as changes in substrate availability, pH, temperature, and redox potential. This metabolic flexibility enables actinomycetes to thrive in diverse habitats and play crucial roles in the degradation of pollutants under different environmental scenarios (Barka et al. 2015).

Table 1.1 The enzymes and bioremediation potential of Actinomycetes species (Saini et al. 2015)

S. No	Actinomycetes species	Enzymes produced	Bioremediation material
1	*Streptomyces griseus*	Protease, lipase, amylase	Petroleum hydrocarbons, heavy metals
2	*Streptomyces coelicolor*	Cellulase, laccase, peroxidase	Polycyclic aromatic hydrocarbons (PAHs), dyes
3	*Streptomyces venezuelae*	Xylanase, ligninase, pectinase	Pesticides, herbicides
4	*Streptomyces albus*	Chitinase, urease, protease	Cyanides, chlorinated solvents
5	*Streptomyces avermitilis*	Lipase, cellulase, pectinase	Pharmaceuticals, Personal care products
6	*Streptomyces griseoflavus*	Amylase, protease, cellulase	Explosives, heavy metals
7	*Streptomyces antibioticus*	Laccase, peroxidase, pectinase	Endocrine-disrupting compounds, VOCs
8	*Streptomyces rimosus*	Chitinase, keratinase, amylase	Polyethylene, polystyrene
9	*Actinomadura madurae*	Urease, lipase, cellulase	Petroleum-contaminated soil, radioactive waste
10	*Actinomadura peucetia*	Xylanase, protease, ligninase	Heavy metals, cyanobacterial toxins
11	*Micromonospora echinospora*	Protease, lipase, amylase	Petroleum hydrocarbons, PAHs
12	*Micromonospora purpureochromogenes*	Cellulase, laccase, peroxidase	Dyes, explosives
13	*Actinoplanes missouriensis*	Ligninase, peroxidase, amylase	Pesticides, herbicides
14	*Actinoplanes utahensis*	Pectinase, urease, protease	Pharmaceuticals, Endocrine disruptors
15	*Actinoplanes philippinensis*	Cellulase, laccase, amylase	Polyethylene, polystyrene
16	*Nocardia brasiliensis*	Lipase, protease, cellulase	Petroleum hydrocarbons, PAHs
17	*Nocardia asteroides*	Amylase, chitinase, laccase	Cyanides, chlorinated solvents
18	*Nocardia farcinica*	Cellulase, xylanase, pectinase	Explosives, heavy metals
19	*Nocardia nova*	Laccase, peroxidase, amylase	Dyes, pesticides
20	*Actinopolyspora alba*	Protease, lipase, amylase	Petroleum-contaminated soil, radioactive waste

1.3.4 Factors Influencing Biodegradation

Several factors influence the rates of biodegradation mediated by the species of actinomycetes

1. Substrate Concentration: The availability of pollutants in the environment directly impacts biodegradation rates. Higher substrate concentrations may initially stimulate microbial activity but can lead to substrate inhibition at excessive concentrations.
2. Environmental Conditions: Factors such as temperature, pH, moisture content, and oxygen availability profoundly influence microbial activity and biodegradation rates. Optimal environmental conditions promote microbial growth and enzymatic activity, enhancing biodegradation efficiency.
3. Microbial Community Composition: The composition and diversity of microbial communities in the environment play a crucial role in biodegradation processes. Synergistic interactions between different microbial species may enhance pollutant degradation rates, whereas competition for resources or inhibitory interactions may hinder biodegradation efficiency.
4. Nutrient Availability: The adequate availability of essential nutrients, such as carbon, nitrogen, phosphorus, and trace elements, is critical for microbial growth and metabolism. Limiting nutrient availability can constrain microbial activity and biodegradation rates, necessitating supplementation with nutrients to enhance bioremediation efficiency (Atlas and Philp n.d.).
5. Cometabolism: Actinomycetes may exhibit cometabolic degradation of pollutants, whereby the degradation of primary substrates is coupled with the degradation of cosubstrates or intermediate metabolites. Cometabolic pathways can enhance the degradation of recalcitrant pollutants and broaden the substrate range of actinomycetes (Megharaj et al. 2011).

Actinomycetes employ a diverse array of enzymatic pathways and metabolic strategies to degrade pollutants in the environment. Understanding the mechanisms underlying biodegradation processes and the factors influencing biodegradation rates is crucial for optimizing bioremediation strategies and harnessing the full potential of actinomycetes in environmental cleanup efforts. Table 1.2 shows different chemical compounds were degraded by actinomycetes species.

1.4 Applications of Actinomycetes in Bioremediation

With their diverse metabolic capabilities and enzymatic machinery, Actinomycetes have emerged as valuable tools for bioremediation, offering sustainable solutions to environmental contamination. In this chapter, we explore the applications of actinomycetes in various bioremediation projects, including soil remediation, water and wastewater treatment, and the remediation of hydrocarbon-contaminated sites, through successful case studies and innovative approaches.

Table 1.2 Actinobacteria with degradation potential

S.No.	Strain	Compound degradation	Source	References
1	*Arthrobacter* sp.	α-Endosulfan, β–endosulfan	Soil from different agricultural fields contaminated with pesticides, India	Kumar et al. (2008)
2	*Arthrobacter* sp. strain DAT1	Atrazine	Atrazine-contaminated agricultural soil in Hebei Province, China	Wang and Xie (2012)
3	*Frankia alni* ACN14a	Atrazine	Not reported	Rehan et al. (2014)
4	*Gordonia* sp. JAAS1	Chlorpyrifos	Paddy field soil exposed to continuous chlorpyrifos applications in Tamil Nadu, India	Abraham et al. (2013)
5	*Kocuria* sp. CL2	Pentachlorophenol	Sludge of pulp and paper mill	Karn et al. (2011)
6	*Kocuria kristinae*	Chlorpyrifos	Chlorpyrifos-contaminated agricultural soil in India	Hamsavathani et al. (2017)
7	*Rhodococcus* sp. BCH2	Atrazine	Soil from grape farm with long-term atrazine treatment near Sangli, India	Kolekar et al. (2014)
8	*Streptomyces aureus* HP-S-01	Deltamethrin, cypermethrin	Activated sludge from a pyrethroid manufacturing wastewater treatment system in Zhongshan, China	Chen et al. (2011)
9	*Nocardioides* sp. strain PD653	Pentachloronitrobenzene, hexachlorobenzene, pentachloropheno	Soil from an agricultural field contaminated with pentachloronitrobenzene in Japan	Takagi et al. (2009)
10	*Streptomyces* sp. strains A2, A5, A11	Lindane	Organochlorine pesticide-contaminated soil in Santiago del Estero, Argentina	Fuentes et al. (2011)

1.4.1 Bioremediation of Heavy Metals Using Acinomycetes

Industrialization has brought about a surge in the release of heavy metals into the environment, surpassing the natural capacity of geochemical processes to handle them. This influx poses a significant challenge to environmental sustainability, particularly impacting microbial communities Fig. 1.2. Heavy metal stress has emerged as a pivotal factor shaping the structure and function of these microbial populations (Hemme et al. 2010).

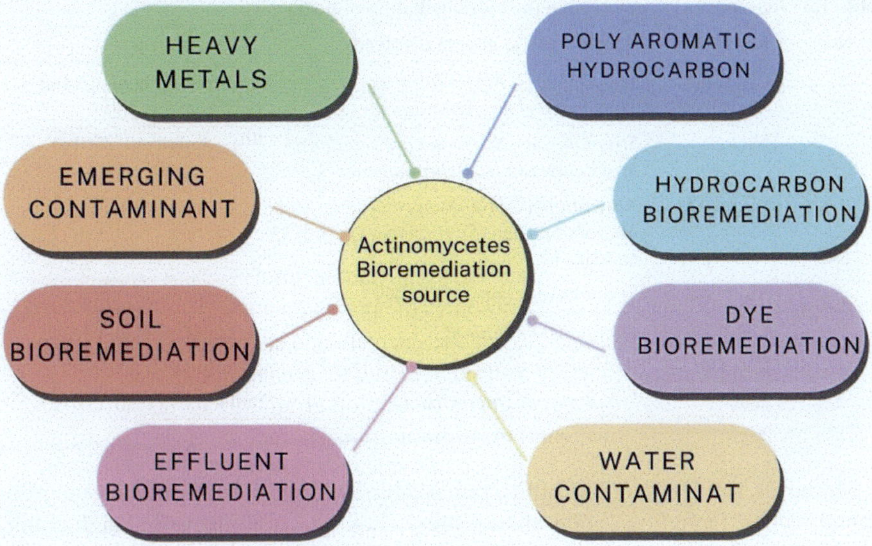

Fig. 1.2 Actinomycetes as a promising source of bioremediation of organic and inorganic contaminants

However, due to this environmental turmoil, microorganisms have emerged as potent agents for remediation. They possess inherent capabilities to mitigate heavy metal pollution, offering a glimmer of hope for environmental restoration. In polluted sites, the native microbial communities undergo a transformative journey, adapting to the presence of pollutants. Through genetic modifications and regulatory adjustments, these microbes have evolved strategies for bioremediation, enhancing their tolerance and efficacy in combating heavy metal contamination (Grace Liu et al. 2011).

The removal of heavy metals by microorganisms occurs through diverse mechanisms, highlighting the versatility of these microbial agents. Mobilization and immobilization stand out as prominent strategies employed by microorganisms in this endeavor. Mobilization involves the active extraction of heavy metals from their surroundings, rendering them more accessible for subsequent remediation processes. Conversely, immobilization entails the sequestration of heavy metals, effectively trapping them and preventing their further dissemination in the environment (Hemme et al. 2010).

Actinomycetes, a group of diverse and resilient bacteria, play a crucial role in this bioremediation saga. Due to their metabolic versatility and adaptive properties, actinomycetes have emerged as frontline soldiers in the battle against heavy metal pollution. These bacteria possess unique enzymatic machinery capable of catalyzing the transformation and degradation of heavy metals, offering an effective means of detoxification (Grace Liu et al. 2011).

Table 1.3 Actinobacteria involved in the bioremediation of heavy metals

S. No	Metal	Actinobacteria genera involved	References
1	Chromium	*Streptomyces, Amycolatopsis, Microbacterium, Arthrobacter*	Benimeli et al. (2006)
2	Nickel	*Streptomyces, Streptomyces seoulensis, Streptomyces viridochromogenes*	Fashola et al. (2016)
3	Cadmium	*Nocardiopsis, Streptomyces, Microbacterium, Streptomyces* spp. K33 and HL-12	Lin et al. (2012)
4	Mercury	*Streptomyces* spp. VITSVK9	Sanjenbam et al. (2012)
5	Lead	*Streptomyces fradiae*	Timková et al. (2018)
6	Copper	*Amycolatopsis tucumanensis* DSM 45259	Polti et al. (2014)
7	Zinc	*Streptomyces zinciresistens*	Timková et al. (2018)

Moreover, actinomycetes exhibit a remarkable ability to colonize diverse environmental niches, including contaminated sites. Their ability to thrive in harsh conditions and metabolize a wide array of organic and inorganic compounds makes them indispensable allies in the quest for a cleaner environment. By harnessing the bioremediation potential of actinomycetes, we can unlock new avenues for environmental restoration and ushers in a sustainable future (aburase 2016). The role of actinomycetes in bioremediation is promising because of the environmental challenges posed by heavy metal pollution. Through their adaptive mechanisms and enzymatic arsenal, these microbial warriors stand poised to lead the charge toward a cleaner and healthier environment. Table 1.3 shows the different actinobacteria involved in the bioremediation of heavy metals.

1.4.2 Soil Remediation Using Actinomycetes

Actinomycetes play crucial roles in soil remediation by degrading organic pollutants, immobilizing heavy metals, and promoting soil health and fertility. Soil contaminated with petroleum hydrocarbons, pesticides, heavy metals, and other pollutants can be effectively remediated using actinomycete-based bioremediation strategies.

One of the key mechanisms by which actinomycetes facilitate soil remediation is through the production of extracellular enzymes, such as lipases, proteases, and cellulases, which catalyze the breakdown of organic contaminants into simpler, more manageable compounds. Additionally, actinomycetes can secrete biosurfactants that enhance the solubility and bioavailability of hydrophobic pollutants, facilitating their degradation by microbial consortia.

1.4.3 Water and Wastewater Treatment Applications

Actinomycetes play vital roles in water and wastewater treatment processes, contributing to the removal of organic pollutants, nutrients, and pathogens from contaminated water sources. Actinomycetes can degrade a wide range of organic pollutants, including pharmaceuticals, pesticides, dyes, and industrial chemicals, through enzymatic degradation pathways (Bérdy 2005).

In wastewater treatment plants, actinomycetes are involved in the biodegradation of organic matter, nitrogen, and phosphorus compounds, leading to the purification of wastewater before discharge into the environment. Moreover, actinomycetes produce antimicrobial compounds that inhibit the growth of pathogenic bacteria and fungi, contributing to the disinfection of water and safeguarding public health (Agunbiade et al. 2019).

1.4.4 Bioremediation of Hydrocarbon Contaminated Cites

Hydrocarbon contamination is a widespread environmental problem that arises from industrial activities, oil spills, and improper waste disposal practices. Actinomycetes have demonstrated remarkable abilities to degrade various hydrocarbons, including alkanes, alkenes, and aromatic compounds, making them valuable agents for the bioremediation of hydrocarbon-contaminated sites (Chandra et al. 2013).

Innovative bioremediation approaches, such as bioaugmentation and biostimulation, harness the metabolic potential of indigenous or exogenous actinomycetes to accelerate the degradation of hydrocarbons in contaminated environments (Haritash and Kaushik 2009). These approaches have been successfully implemented in numerous field trials and pilot-scale projects, resulting in the restoration of contaminated sites and the mitigation of environmental risks associated with hydrocarbon pollution. The Table 1.4 shows the approaches used by actinomycetes for the bioremediation of hydrocarbons.

The applications of actinomycetes in bioremediation are diverse and multifaceted, ranging from soil remediation to water and wastewater treatment and the cleanup of hydrocarbon-contaminated sites. Successful case studies and innovative approaches highlight the effectiveness of actinomycetes in addressing environmental contamination and restoring ecosystems to their natural state. Continued research and technological advancements in actinomycete-based bioremediation hold promise for sustainable environmental cleanup and the preservation of ecosystem health.

Table 1.4 Approaches used by actinomycetes for the bioremediation of hydrocarbons

S.No	Bioremediation	Actinomycetes in involving bioremediation
1	Pure culture petroleum hydrocarbon degradation	Actinobacteria genera like *Actinomyces, Streptomyces, Rhodococcus, Gordonia, Micrococcus*, and *Nocardia* exhibit significant biodegradation potential for petroleum hydrocarbons Actinobacteria secrete biosurfactants, enhancing the bioavailability of pollutants and facilitating their degradation Comparative analysis reveals *Gordonia's* efficiency in degrading aliphatic hydrocarbons, while *Rhodococcus* mediates petroleum hydrocarbon degradation effectively
2	Mixed cultures (actinomycete and bacteria) Polycyclic Aromatic Hydrocarbons (PAHs)	Studies involving mixed cultures of Actinobacteria (*Rhodococcus, Arthrobacter, Gordonia*) and *Pseudomonas* strains show synergistic activity in degrading hydrocarbons Cometabolism of various PAHs, including naphthalene, phenanthrene, and fluoranthene, observed under aerobic conditions by *Rhodococcus* spp., suggests enhanced biodegradation efficiency through further study
3	Polycyclic Aromatic Hydrocarbons (PAHs)	PAHs, recognized as toxic pollutants, pose environmental risks due to natural and anthropogenic sources Biological approaches, particularly involving Actinobacteria genera like *Streptomyces, Rhodococcus, Microbacterium,* and *Arthrobacter*, offer economical and safe means for PAH removal from the environment Actinobacteria, such as *Streptomyces rochei* and *Mycobacterium* spp., demonstrate efficient degradation of PAHs, including those with 3–4 rings, through biosurfactant production and modified ortho-cleavage pathways

1.4.5 Bioremediation of Synthetic Dye

Synthetic dyes, which are extensively utilized in the textile industry, contribute substantially to wastewater generation during dyeing processes. The discharge of colored effluents into rivers and lakes is a primary cause of reduced dissolved oxygen levels, leading to anoxic conditions and acute toxicity to aquatic flora and fauna. Moreover, the presence of these colored effluents hampers photosynthesis by impeding light dispersion in water, posing a major environmental concern. Despite these efforts, synthetic dyes persist in wastewater treatment plants due to their recalcitrant nature.

Azo dyes, particularly water-soluble reactive compounds, represent a versatile class of synthetic dyes widely used across various industries. When left untreated in water bodies, azo dyes contribute to environmental pollution and pose risks of toxicity, carcinogenicity, and mutagenicity. Microorganisms, including isolated actinomycetes, demonstrate promising capabilities for degrading azo dyes. Through acclimatization to different dye concentrations and consortium development, actinomycetes exhibit concentration-dependent degradation, facilitated by enzymes such as lignin peroxidase, laccase, and tyrosinase. Additionally, indigenous *Streptomyces* spp. optimize conditions for degrading azo blue and azo orange dyes via metabolic pathways, demonstrating their potential in sustainable bioremediation efforts.

Actinomycetes, such as *Thermobifida fusca* BCRC 19214 and *Streptomyces* spp., play pivotal roles in the decolorization of synthetic dyes. Through the production of laccase enzymes, thermophilic actinomycetes oxidize dye intermediates, enhancing biodegradability without altering dye properties. Furthermore, specific structural modifications induced by actinomycetes in sulfonated azo dyes contribute to enhanced biodegradability, thereby reducing environmental persistence (Chen et al. 2013).

Reactive dyes, a prominent type of azo dye, pose significant challenges to environmental sustainability due to their adverse effects on water quality and ecosystem health. Actinomycete strains demonstrate efficacy in decolorizing effluents containing reactive dyes such as anthraquinone, phthalocyanine, and azo dyes, offering a viable solution for wastewater treatment.

The actinomycete *Streptomyces krainskii* SUK-5 potently degrades and decolorizes the textile azo dye Reactive Blue-59, underscoring the importance of enzymatic pathways such as lignin peroxidase and NADH-DCIP reductase in degradation processes. Actinomycetes have emerged as valuable assets for addressing the challenges posed by synthetic dye pollution in aquatic environments (Gurav et al. 2008). Their diverse enzymatic machinery and metabolic versatility enable efficient degradation and decolorization of various synthetic dyes, thereby mitigating environmental pollution and safeguarding ecosystem health. Harnessing the bioremediation potential of actinomycetes holds promise for achieving cleaner and healthier water bodies, contributing to a more sustainable future (Table 1.5).

1.5 Challenges and Limitations in Actinomycete-Mediated Bioremediation

Actinomycetes have shown tremendous potential in bioremediation, offering sustainable solutions for environmental remediation. However, the application of actinomycetes in bioremediation is not without challenges and limitations. The various factors that hinder the effectiveness of actinomycete-mediated bioremediation include environmental factors affecting actinomycete activity, challenges in scaling up bioremediation processes, ecological implications and safety concerns, and regulatory considerations for bioremediation projects.

1.5.1 Environmental Factors Affecting

Actinomycetes are highly sensitive to environmental conditions, and their activity and efficacy in bioremediation can be influenced by various factors. Temperature,

Table 1.5 Actinomycetes for dye clearance in aquatic environments

S. No	Synthetic dye	Actinomycete genus	Degradation/Decolorization process	References
1	Reactive yellow biodegradation	Consortium of actinomycetes	Biodegradation of Reactive Yellow dye by acclimatized actinomycetes consortium, dependence on dye concentration and actinomycetes growth, involvement of lignin peroxidase, laccase, and tyrosinase enzymes	Bagewadi (2011)
2	Azo blue and azo orange dyes	*Streptomyces* spp.	Optimization of conditions for metabolic pathways degradation with responsible enzymes	Jai et al. (2014)
3	Monosulfonated mono azo dye derivatives of azobenzene	*Streptomyces* spp.	Decolorization of dyes, enhancing biodegradability without affecting dye properties, change in dye structure observed, decolorization of Orange I	Pasti-Grigsby et al. (1992)
4	Sulfonated azo dyes	*Streptomyces chromofuscus*	Mineralization of modified dyes, enhanced biodegradability due to 5 specific structural changes in the azo dye molecules	Paszczynski et al. (1992)
5	Reactive blue-59	*Streptomyces krainskii* SUK-5	Degradation and decolorization of reactive blue-59, induction of lignin peroxidase, NADH-DCIP reductase, and MR reductase enzymes	Gurav et al. (2008)

pH, moisture content, oxygen availability, and nutrient availability are among the key environmental factors that affect actinomycete metabolism and growth. Suboptimal environmental conditions can inhibit actinomycete activity and limit its capacity to degrade pollutants efficiently (Pandey et al. 2009).

Furthermore, the presence of inhibitory substances, such as heavy metals, pesticides, and recalcitrant pollutants, in contaminated environments can adversely affect actinomycete activity and bioremediation efficacy. These environmental stressors may disrupt microbial metabolism, enzyme function, and cell viability, hampering the degradation of pollutants and prolonging the remediation process (Gkorezis et al. 2016).

1.5.2 Challenges in Scaling up Bioremediation Processes

While laboratory-scale studies have demonstrated the efficacy of actinomycetes in bioremediation, scaling up these processes to field or industrial scales presents numerous challenges. The successful implementation of bioremediation strategies requires careful consideration of factors such as site heterogeneity, pollutant distribution, microbial ecology, and process optimization (Mishra et al. 2023).

One of the primary challenges in scaling up bioremediation processes is achieving uniform distribution and adequate contact between actinomycetes and contaminants in large-scale treatment systems. Poor mixing, limited mass transfer, and heterogeneous substrate distribution can create microenvironments with suboptimal conditions for microbial growth and pollutant degradation, leading to uneven remediation and incomplete cleanup (Huang et al. 2017).

1.5.3 Ecological Implications and Safety Concerns

Although actinomycetes offer promising bioremediation solutions, their introduction into the environment may have unintended ecological consequences. Concerns regarding the potential risks associated with the release of genetically modified or nonnative actinomycetes into natural ecosystems have raised questions about the safety and sustainability of bioremediation approaches (Farda et al. 2022).

The ecological implications of actinomycete-mediated bioremediation include the potential for unintended impacts on native microbial communities, soil and water quality, and ecosystem functioning. Competition for resources, displacement of indigenous species, and alteration of microbial diversity and community structure are among the ecological risks associated with the introduction of exogenous actinomycetes into environmental settings.

1.5.4 Regulatory Considerations for Bioremediation Projects

Regulatory frameworks governing bioremediation projects vary widely across different jurisdictions and may pose significant challenges for project implementation and approval. Regulatory considerations for bioremediation projects encompass a range of issues, including environmental permitting, risk assessment, monitoring and reporting requirements, and stakeholder engagement (Mohee and Mudhoo 2012).

Environmental agencies and regulatory bodies play crucial roles in overseeing bioremediation activities, ensuring compliance with relevant laws, regulations, and standards. Permitting processes may involve comprehensive environmental impact assessments, risk analyses, and public consultations to assess potential environmental risks, mitigate adverse impacts, and secure community acceptance of bioremediation projects.

Despite the promising potential of actinomycetes in bioremediation, several challenges and limitations must be addressed to realize their full benefits. Environmental factors, scaling-up challenges, ecological implications, and regulatory considerations present complex hurdles that require interdisciplinary approaches and stakeholder collaboration.

Addressing these challenges will require continued research and innovation in bioremediation science, technology, and policy development. Integrating advanced analytical techniques, computational modeling, and risk assessment tools can enhance our understanding of actinomycete-mediated bioremediation processes and inform sustainable environmental management practices.

1.6 Advances in Actinomycete Bioremediation

Actinomycetes have long been recognized for their remarkable ability to degrade a wide range of environmental pollutants, making them valuable assets in bioremediation efforts. In this chapter, we explore recent advances in actinomycete bioremediation, focusing on genetic engineering approaches for enhancing biodegradation capabilities, biotechnological innovations in actinomycete-based bioremediation, the integration of omics technologies for understanding microbial communities, and future prospects and emerging trends in actinomycete bioremediation.

1.6.1 Genetic Engineering Approaches

Genetic engineering holds immense potential for enhancing the biodegradation capabilities of actinomycetes, enabling the development of tailored strains with improved pollutant-degrading enzymes and metabolic pathways. Through targeted genetic modifications, researchers can optimize the expression of key genes involved in pollutant degradation, enhance enzyme activities, and introduce novel metabolic pathways to expand the substrate range of actinomycetes (Tramper and Zhu 2011).

For example, the introduction of foreign genes encoding xenobiotic-degrading enzymes, such as cytochrome P450 monooxygenases and dioxygenases, into actinomycete genomes can enhance their ability to metabolize recalcitrant pollutants, including polycyclic aromatic hydrocarbons (PAHs), chlorinated compounds, and pesticides. Additionally, metabolic engineering strategies can be employed to optimize carbon and energy utilization pathways, improving the efficiency and specificity of pollutant degradation by actinomycetes (Peng et al. 2008).

1.6.2 Biotechnological Innovations

Recent years have witnessed significant advancements in biotechnological tools and techniques for harnessing the bioremediation potential of actinomycetes. High-throughput screening methods, such as metagenomic and metatranscriptomic analyses, enable the rapid identification and characterization of novel actinomycete strains and enzymes with desirable biodegradation capabilities (Kim 2021).

Moreover, the development of synthetic biology platforms allows researchers to engineer custom-designed actinomycete strains with tailored metabolic pathways and regulatory circuits for targeted pollutant degradation (Li et al. 2019). Synthetic biology approaches offer unprecedented control over microbial behavior and functionality, paving the way for the design of next-generation bioremediation systems with enhanced efficiency, specificity, and environmental compatibility.

1.6.3 Integration of Omics Technologies for Understanding Microbial Communities

Omics technologies, including genomics, proteomics, metabolomics, and transcriptomics, have revolutionized our understanding of microbial communities and their interactions in bioremediation processes. By characterizing the genetic, metabolic, and functional profiles of actinomycetes and associated microorganisms, omics approaches provide valuable insights into the mechanisms underlying pollutant degradation, microbial community dynamics, and ecosystem responses to environmental perturbations (Mishra et al. 2021).

Furthermore, advances in bioinformatics and computational biology have enabled the integration of omics data to construct predictive models of microbial behavior and ecosystem functioning, facilitating the design and optimization of bioremediation strategies. Through multidimensional omics analyses, researchers can elucidate complex microbial networks, identify key genes and pathways driving biodegradation processes, and identify novel biomarkers for monitoring the progress of environmental remediation.

1.6.4 Future Prospects and Emerging Trends

Overall, actinomycete bioremediation holds great promise for addressing emerging environmental challenges, including the remediation of emerging pollutants, such as pharmaceuticals, personal care products, and microplastics, and the restoration of degraded ecosystems. Advances in synthetic biology, metabolic engineering, and omics technologies are expected to drive innovations in actinomycete bioremediation, enabling the development of tailored solutions for specific contaminants and environmental settings.

Moreover, the integration of interdisciplinary approaches, including nanotechnology, bioinformatics, and machine learning, will further enhance the efficiency, selectivity, and sustainability of actinomycete-based bioremediation processes (Sharma et al. 2022). By harnessing the collective power of biotechnology, informatics, and environmental science, actinomycete bioremediation is poised to play a central role in shaping the future of environmental sustainability and ecosystem health. Recent advancements in actinomycete bioremediation represent significant strides toward the development of innovative and sustainable solutions for environmental cleanup. Genetic engineering, biotechnological innovations, omics technologies, and future prospects offer unprecedented opportunities to expand the scope and efficacy of actinomycete-mediated bioremediation, paving the way for a cleaner and healthier environment.

1.7 Environmental Impact Assessment and Monitoring

Environmental impact assessment (EIA) and monitoring play crucial roles in evaluating the effectiveness, sustainability, and ecological implications of bioremediation strategies. We investigated methods for assessing the efficacy of bioremediation strategies, monitoring techniques for tracking pollutant degradation, evaluating ecological restoration and ecosystem resilience, and long-term monitoring and sustainability considerations.

1.7.1 Methods for Assessing the Efficacy of Bioremediation Strategies

Assessing the efficacy of bioremediation strategies involves determining the extent to which pollutants are degraded or removed from contaminated environments. Various methods are employed to measure pollutant concentrations, assess microbial activity, and evaluate changes in environmental parameters before, during, and after bioremediation treatments.

Quantitative analyses, such as gas chromatography–mass spectrometry (GC–MS) and high-performance liquid chromatography (HPLC), are commonly used to quantify pollutant concentrations in soil, water, and air samples (Islam et al. 2019). Biomarker assays, enzymatic activity assays, and molecular techniques, such as polymerase chain reaction (PCR) and next-generation sequencing (NGS), provide insights into microbial community dynamics, metabolic pathways, and pollutant degradation rates (Ojo 2023).

Additionally, geospatial mapping, remote sensing, and geographic information systems (GIS) enable the visualization and spatial analysis of contaminated sites, facilitating the identification of hotspots and monitoring of remediation progress. Integrating multiple lines of evidence from chemical, biological, and geospatial analyses allows for comprehensive assessments of bioremediation efficacy and environmental impact (Sharma et al. 2022).

1.7.2 Monitoring Techniques for Tracking Pollutant Degradation

Monitoring techniques are essential for tracking pollutant degradation, assessing contaminant fate and transport, and guiding remediation efforts. Continuous monitoring of key environmental parameters, including temperature, pH, dissolved oxygen, and nutrient concentrations, provides real-time data on microbial activity and biodegradation rates (Yadav 2018).

In situ monitoring tools, such as gas sensors, biosensors, and microbial fuel cells (MFCs), offer rapid and cost-effective methods for assessing microbial activity and pollutant concentrations in contaminated environments. Additionally, stable isotope probing (SIP) techniques enable the tracking of isotopically labeled pollutants and the identification of microbial taxa responsible for their degradation.

Furthermore, passive sampling devices, such as diffusive gradients in thin films (DGTs) and passive samplers, can be deployed to measure pollutant fluxes and bioavailability over time, providing insights into contaminant mobility and bioaccumulation in environmental matrices (Sousa et al. 2018). By combining complementary monitoring techniques, researchers can obtain comprehensive datasets for assessing bioremediation performance and informing remediation strategies.

1.7.3 Evaluation of Ecological Restoration and Ecosystem Resilience

Beyond pollutant removal, the success of bioremediation efforts hinges on the restoration of ecological functions and ecosystem resilience in contaminated environments. Ecological assessment tools, such as ecological indices, biodiversity surveys, and ecosystem service assessments, are used to evaluate the recovery of native flora and fauna, soil quality, and ecosystem services following remediation (Samson and Knopf 1996).

Functional gene analyses, stable isotope analysis, and microbial community profiling provide insights into the resilience of microbial communities and their functional diversity in response to environmental perturbations (Li et al. 2019; Zhang et al. 2019). Furthermore, ecosystem modeling and ecological network analysis offer predictive insights into ecosystem dynamics, trophic interactions, and feedback mechanisms driving ecosystem recovery (Holling 2001).

Long-term monitoring studies are essential for assessing the sustainability and long-term efficacy of bioremediation treatments, as well as detecting potential rebound effects and secondary pollution risks. By evaluating ecological indicators, ecosystem services, and socioeconomic factors over extended timeframes, researchers can identify trends, detect emerging threats, and adapt management strategies to ensure the continued health and resilience of remediated ecosystems.

1.7.4 Long-Term Monitoring and Sustainability Considerations

Long-term monitoring is critical for assessing the sustainability of bioremediation practices and mitigating potential risks associated with residual contamination, ecosystem disturbance, and secondary impacts. Sustainable remediation frameworks, such as the Triad Approach and adaptive management strategies, emphasize the importance of iterative assessment, stakeholder engagement, and continuous improvement in environmental management (NRC 2019).

Moreover, incorporating socioeconomic factors, stakeholder perceptions, and community engagement in monitoring and decision-making processes is essential for promoting social equity, transparency, and accountability in bioremediation projects. Community-based monitoring programs, citizen science initiatives, and participatory approaches empower local communities to contribute to environmental monitoring, decision-making, and governance processes, fostering collaboration and trust among stakeholders.

Environmental impact assessment and monitoring are integral components of bioremediation projects, providing essential insights into the efficacy, sustainability, and ecological implications of remediation efforts. By employing a multidisciplinary approach and integrating advanced monitoring techniques, researchers can assess bioremediation performance, track pollutant degradation, evaluate ecosystem recovery, and ensure the long-term sustainability of remediated environments.

1.8 Studies in Actinomycete Bioremediation

Actinomycetes have emerged as powerful tools in the field of bioremediation, offering innovative solutions for cleaning contaminated environments. Here, we present comprehensive case studies that showcase diverse bioremediation applications, success stories, and lessons learned from real-world projects. We also explore the importance of cross-disciplinary collaboration and community engagement initiatives in driving successful actinomycete bioremediation projects.

1.8.1 Case Studies

1.8.1.1 Case Study 1. Soil Bioremediation

In a contaminated industrial site in the United States, soil remediation was undertaken to address the accumulation of heavy metals and organic pollutants resulting from decades of industrial activities. A consortium of indigenous actinomycetes was isolated from the site and augmented with organic amendments to enhance pollutant degradation. Over the course of two years, significant reductions in soil metal concentrations and organic pollutant levels were observed, accompanied by improvements in soil microbial diversity and ecosystem functions. This case study highlights the efficacy of indigenous actinomycetes in soil remediation and the importance of site-specific bioremediation approaches (Alori et al. 2022).

1.8.1.2 Case Study 2. Water and Wastewater Treatment

In a municipal wastewater treatment plant in India, actinomycetes were utilized to enhance the removal of organic pollutants and nutrients from wastewater effluents. A bioaugmentation strategy involving the introduction of selected actinomycete strains into activated sludge systems led to improvements in pollutant removal efficiency and effluent quality. By harnessing the synergistic interactions between actinomycetes and indigenous microbial communities, treatment plants achieve significant reductions in biochemical oxygen demand (BOD), chemical oxygen demand (COD), and nutrient concentrations, thereby meeting regulatory discharge standards and protecting receiving water bodies.

1.8.1.3 Case Study 3. Hydrocarbon Contaminated Sites

In an oil spill-affected coastal region in Brazil, actinomycetes were employed in the bioremediation of hydrocarbon-contaminated soils and sediments. A combination of bioaugmentation and biostimulation techniques was used to enhance the degradation of petroleum hydrocarbons by indigenous actinomycete populations. Through regular monitoring and optimization of environmental conditions, including nutrient availability and oxygen levels, substantial reductions in total petroleum hydrocarbon (TPH) concentrations were achieved within six months of treatment. This case study demonstrated the effectiveness of tailored bioremediation strategies in addressing complex hydrocarbon contamination scenarios.

1.8.2 Success Stories and Lessons

These case studies exemplify the success stories and lessons learned from real-world actinomycete bioremediation projects. The key factors contributing to project success include the selection of appropriate actinomycete strains, the optimization of environmental conditions, and the integration of complementary remediation techniques. Cross-disciplinary collaboration between microbiologists, environmental engineers, and stakeholders is essential for developing holistic bioremediation strategies that consider site-specific challenges and regulatory requirements.

The lessons learned from these case studies underscore the importance of long-term monitoring, adaptive management, and community engagement in ensuring the sustainability and effectiveness of bioremediation efforts. Continuous assessments of remediation progress, feedback mechanisms, and knowledge sharing among project stakeholders are critical for addressing emerging challenges, optimizing remediation strategies, and building trust and confidence in bioremediation technologies (Chen et al. 2020).

1.8.3 Cross-Disciplinary Collaborations and Community Engagement Initiatives

Cross-disciplinary collaboration and community engagement initiatives are integral to the success of actinomycete bioremediation projects. By involving diverse stakeholders, including local communities, regulatory agencies, industry partners, and academic institutions, projects can benefit from interdisciplinary expertise, local knowledge, and stakeholder perspectives. Community engagement activities, such as public consultations, outreach events, and participatory monitoring programs, foster transparency, trust, and shared responsibility in environmental decision-making.

Moreover, cross-disciplinary collaboration facilitates the transfer of knowledge and technology between scientific disciplines, accelerating innovation and problem solving in bioremediation research and practice. By fostering collaborative networks and partnerships, researchers can leverage synergies, pool resources, and address complex environmental challenges more effectively.

Case studies in actinomycete bioremediation offer valuable insights into the practical applications, challenges, and opportunities of using these versatile microorganisms for environmental cleanup. Success stories from diverse bioremediation projects demonstrate the efficacy of actinomycetes in remediation scenarios ranging from soil and water treatment to hydrocarbon-contaminated site remediation. The lessons learned highlight the importance of adaptive management, stakeholder engagement, and interdisciplinary collaboration in achieving sustainable and effective bioremediation outcomes.

1.9 Collaborative Approaches for Sustainable Bioremediation

Bioremediation, the use of biological agents to degrade or remove environmental pollutants, has emerged as a promising approach for addressing diverse contamination challenges. The importance of collaborative approaches in bioremediation research, including interdisciplinary collaboration, industry-academia partnerships, stakeholder engagement, and global cooperation. By fostering synergies between diverse stakeholders, collaborative efforts can drive innovation, enhance technology transfer, promote environmental stewardship, and address transboundary pollution issues.

1.9.1 Importance of Interdisciplinary Collaboration in Bioremediation Research

Bioremediation research is inherently interdisciplinary, drawing on expertise from fields such as microbiology, environmental science, engineering, chemistry, and ecology. Interdisciplinary collaboration enables researchers to leverage diverse perspectives, methodologies, and approaches to address complex environmental challenges.

For example, microbiologists have contributed expertise in microbial ecology, physiology, and genetics to understanding the metabolic capabilities of biodegrading microorganisms, while environmental engineers have designed innovative bioremediation technologies and strategies for pollutant removal. Chemists play a crucial role in characterizing pollutants and developing analytical methods for monitoring bioremediation progress, while ecologists assess the ecological impacts of remediation activities on soil and aquatic ecosystems.

1.9.2 Industry-Academia Partnerships for Technology Transfer and Innovation

Collaboration between academia and industry is essential for translating bioremediation research into practical applications and commercial technologies. Industry-academia partnerships facilitate technology transfer, knowledge exchange, and the development of market-ready solutions for environmental cleanup.

Industry partners provide valuable insights into market needs, regulatory requirements, and scalability considerations, guiding academic researchers in the development of cost-effective, scalable, and environmentally sustainable bioremediation technologies. In return, academia offers cutting-edge research expertise, innovation capabilities, and access to state-of-the-art facilities for proof-of-concept studies and technology validation.

1.9.3 Engaging Stakeholders and Local Communities in Environmental Stewardship

Stakeholder engagement and community involvement are integral to the success and acceptance of bioremediation projects. Engaging stakeholders, including regulatory agencies, industry stakeholders, environmental NGOs, and local communities, foster transparency, trust, and collaborative decision-making in environmental management.

Community engagement initiatives, such as public consultations, citizen science projects, and participatory monitoring programs, empower local communities to contribute to environmental stewardship efforts, share local knowledge, and codesign remediation strategies that align with their needs and priorities. By fostering social acceptance and ownership of remediation projects, stakeholder engagement enhances project sustainability and long-term success.

1.9.4 Global Cooperation for Addressing Transboundary Pollution Issues

Environmental pollution knows no boundaries, and addressing transboundary pollution requires coordinated action at the regional, national, and international levels. Global cooperation and partnerships are essential for sharing knowledge, resources, and best practices and addressing cross-border pollution challenges.

International collaborations facilitate the exchange of expertise, capacity-building initiatives, and joint research projects aimed at addressing common environmental concerns, such as marine pollution, air quality degradation, and cross-border contamination of water resources. Multilateral agreements, such as the Stockholm Convention on Persistent Organic Pollutants and the Minamata Convention on Mercury, provide frameworks for international cooperation and policy coordination in addressing global pollution issues.

Collaborative approaches are vital for advancing sustainable bioremediation solutions and addressing complex environmental challenges. Interdisciplinary collaboration fosters innovation, industry-academia partnerships drive technology transfer and commercialization, stakeholder engagement promotes social acceptance and ownership, and global cooperation enhances collective action and policy alignment.

By harnessing the collective expertise, resources, and goodwill of diverse stakeholders, collaborative efforts can catalyze transformative change in environmental management, leading to cleaner, healthier, and more resilient ecosystems for present and future generations.

1.10 Future Directions and Concluding Remarks

In conclusion, our exploration of actinomycete bioremediation is crucial because of the key findings, insights gained, and the future trajectory of this field. The key points discussed throughout the book includes identifying opportunities for further research and development, envisioning a sustainable future powered by actinomycete bioremediation, and issuing a call for collective efforts toward a cleaner environment. Throughout this chapter, we have delved into the remarkable capabilities of actinomycetes in bioremediation, exploring their diverse metabolic pathways, unique adaptations, and applications in environmental cleanup. We have examined the importance of bioremediation in addressing pollution challenges, the role of actinomycetes in degrading various contaminants, and the collaborative approaches necessary for successful remediation projects.

Key findings include the versatility of actinomycetes in degrading pollutants such as hydrocarbons, heavy metals, and pesticides, their ability to thrive in diverse environmental conditions, and their potential to be harnessed for sustainable remediation strategies. We have also highlighted the importance of interdisciplinary collaboration, industry-academia partnerships, stakeholder engagement, and global cooperation in advancing bioremediation research and practice.

1.10.1 Opportunities for Further Research and Development

Although significant progress has been made in the field of actinomycete bioremediation, several opportunities for further research and development exist. Future studies could focus on elucidating the molecular mechanisms underlying pollutant degradation by actinomycetes, exploring the potential of synthetic biology and genetic engineering to enhance biodegradation capabilities, and optimizing bioremediation strategies for specific contaminants and environmental settings.

Additionally, there is a need for comprehensive assessments of the ecological impacts of bioremediation activities, including long-term monitoring studies, ecosystem modeling, and socioeconomic analyses. Collaborative research initiatives that bring together experts from different disciplines can foster innovation, address knowledge gaps, and drive the development of sustainable bioremediation technologies.

1.10.2 Vision for a Sustainable Future Powered by Actinomycete Bioremediation

In the future, actinomycete bioremediation holds immense potential to contribute to a sustainable future by mitigating environmental pollution, restoring ecosystem health, and promoting human well-being. In the future, we envision that actinomycetes will play a central role in cleaning contaminated sites, protecting natural resources, and safeguarding biodiversity. Actinomycete-based bioremediation technologies could be deployed on a global scale to address a wide range of pollution challenges, from industrial waste sites and oil spills to urban brownfields and agricultural runoff. By harnessing the power of nature's cleanup crew, we can create cleaner, healthier environments for current and future generations, paving the way for a more sustainable and resilient planet.

1.10.3 Call to Action for Collective Efforts Toward a Cleaner Environment

As we embark on this journey toward a cleaner environment powered by actinomycete bioremediation, it is imperative that we act collectively to address environmental pollution and its impacts. This call for action extends to scientists, policymakers, industry leaders, community members, and citizens worldwide. Scientists are called upon to conduct rigorous research, develop innovative technologies, and share knowledge to advance the field of bioremediation. Policymakers are urged to enact robust environmental regulations, support sustainable remediation initiatives, and invest in clean technologies. Industry leaders are encouraged to adopt green practices, embrace corporate responsibility, and collaborate with academia and government agencies. Community members and citizens are invited to participate in environmental stewardship efforts, advocate for clean air and water, and support local bioremediation projects. By working together across sectors and borders, we can overcome pollution challenges, protect our planet's natural resources, and build a healthier, more sustainable future for all.

1.11 Conclusion

In conclusion, actinomycete bioremediation represents a powerful and sustainable approach for addressing environmental pollution. By harnessing the unique capabilities of these microorganisms, we can transform contaminated sites into thriving ecosystems, mitigate pollution impacts, and create a cleaner, healthier world.

As we embark on the journey ahead, let us remain committed to collaborative action, scientific innovation, and environmental stewardship. Together, we can harness the power of nature to heal our planet and ensure a brighter future for generations to come.

References

Abraham J, Shanker A, Silambarasan S (2013) Role of Gordonia sp JAAS1 in biodegradation of chlorpyrifos and its hydrolyzing metabolite 3,5,6-trichloro-2-pyridinol. Lett Appl Microbiol 57(6):510–516. https://doi.org/10.1111/LAM.12141

Abubakar IR, Maniruzzaman KM, Dano UL, AlShihri FS, AlShammari MS, Ahmed SMS, Al-Gehlani WAG, Alrawaf TI (2022) Environmental sustainability impacts of solid waste management practices in the global south. Int J Environ Res Public Health 19(19):12717. https://doi.org/10.3390/IJERPH191912717

Adla K, Dejan K, Neira D, Dragana Š (2022) Degradation of ecosystems and loss of ecosystem services. In: One health: integrated approach to 21st century challenges to health, pp 281–327. https://doi.org/10.1016/B978-0-12-822794-7.00008-3

Agunbiade MO, Pohl C, Heerden EV, Oyekola O, Ashafa A (2019) Evaluation of fresh water actinomycete bioflocculant and its biotechnological applications in wastewaters treatment and removal of heavy metals. Int J Environ Res Public Health 16(18). https://doi.org/10.3390/IJERPH16183337

Alori ET, Gabasawa AI, Elenwo CE, Agbeyegbe OO (2022) Bioremediation techniques as affected by limiting factors in soil environment. Front Soil Sci 2:937186. https://doi.org/10.3389/FSOIL.2022.937186/BIBTEX

Atlas RM, Philp J (n.d.) Bioremediation: applied microbial solutions for real-world environmental cleanup, p 366. https://www.wiley.com/en-br/Bioremediation%3A+Applied+Microbial+Solutions+for+Real+World+Environmental+Cleanup-p-9781683671978. Accessed 17 April, 2024

Bagewadi ZK (2011) Biodegradation of industrially important textile dyes by actinomycetes isolated from activated sludge. https://www.researchgate.net/publication/266419347

Barka EA, Vatsa P, Sanchez L, Gaveau-Vaillant N, Jacquard C, Klenk H-P, Clément C, Ouhdouch Y, van Wezel GP (2015) Taxonomy, physiology, and natural products of actinobacteria. Microbiol Mol Biol Rev 80(1):1–43. https://doi.org/10.1128/MMBR.00019-15

Benimeli CS, Castro GR, Chaile AP, Amoroso MJ (2006) Lindane removal induction by Streptomyces sp. M7. J Basic Microbiol 46(5):348–357. https://doi.org/10.1002/JOBM.200510131

Bérdy J (2005) Bioactive microbial metabolites. J Antibiot 58(1):1–26. https://doi.org/10.1038/ja.2005.1

Chandra S, Sharma R, Singh K, Sharma A (2013) Application of bioremediation technology in the environment contaminated with petroleum hydrocarbon. Ann Microbiol 63(2):417–431. https://doi.org/10.1007/S13213-012-0543-3/TABLES/5

Chen S, Lai K, Li Y, Hu M, Zhang Y, Zeng Y (2011) Biodegradation of deltamethrin and its hydrolysis product 3-phenoxybenzaldehyde by a newly isolated Streptomyces aureus strain HP-S-01. Appl Microbiol Biotechnol 90(4):1471–1483. https://doi.org/10.1007/S00253-011-3136-3

Chen CY, Huang YC, Wei CM, Meng M, Liu WH, Yang CH (2013) Properties of the newly isolated extracellular thermo-alkali-stable laccase from thermophilic actinomycetes, Thermobifida fusca and its application in dye intermediates oxidation. AMB Express 3(1):1–9. https://doi.org/10.1186/2191-0855-3-49/TABLES/6

Chen C, Zhang X, Chen J, Chen F, Li J, Chen Y, Hou H, Shi F (2020) Assessment of site contaminated soil remediation based on an input output life cycle assessment. J Cleaner Prod 263:121422. https://doi.org/10.1016/J.JCLEPRO.2020.121422

Cimermanova M, Pristas P, Piknova M (2021) Biodiversity of actinomycetes from heavy metal contaminated technosols. Microorganisms 9(8). https://doi.org/10.3390/MICROORGANISMS9081635/S1

Devanshi S, Shah KR, Arora S, Saxena S, Devanshi S, Shah KR, Arora S, Saxena S (2021) Actinomycetes as an environmental scrubber. In: Crude oil—new technologies and recent approaches. https://doi.org/10.5772/INTECHOPEN.99187

Farda B, Djebaili R, Vaccarelli I, Del Gallo M, Pellegrini M (2022) Actinomycetes from caves: an overview of their diversity, biotechnological properties, and insights for their use in soil environments. Microorganisms 10(2). https://doi.org/10.3390/MICROORGANISMS10020453

Fashola MO, Ngole-Jeme VM, Babalola OO (2016) Heavy metal pollution from gold mines: environmental effects and bacterial strategies for resistance. Int J Environ Res Public Health 13(11). https://doi.org/10.3390/IJERPH13111047

Fuentes MS, Sáez JM, Benimeli CS, Amoroso MJ (2011) Lindane biodegradation by defined consortia of indigenous Streptomyces strains. Water Air Soil Pollut 222(1–4):217–231. https://doi.org/10.1007/S11270-011-0818-5/METRICS

Ghorani-Azam A, Riahi-Zanjani B, Balali-Mood M (2016) Effects of air pollution on human health and practical measures for prevention in Iran. J Res Med Sciences Off J Isfahan Univ Med Sci 21(5). https://doi.org/10.4103/1735-1995.189646

Gkorezis P, Daghio M, Franzetti A, Van Hamme JD, Sillen W, Vangronsveld J (2016) The interaction between plants and bacteria in the remediation of petroleum hydrocarbons: an environmental perspective. Front Microbiol 7(NOV):214884. https://doi.org/10.3389/FMICB.2016.01836/BIBTEX

Goodfellow M, Williams ST, Stanley T, Mordarski M (1988) Actinomycetes in biotechnology. Academic Press. http://www.sciencedirect.com:5070/book/9780122896736/actinomycetes-in-biotechnology

Grace Liu PW, Chang TC, Whang LM, Kao CH, Pan PT, Cheng SS (2011) Bioremediation of petroleum hydrocarbon contaminated soil: effects of strategies and microbial community shift. Int Biodeterior Biodegradation 65(8):1119–1127. https://doi.org/10.1016/J.IBIOD.2011.09.002

Gurav PN, Deshmukh AM, Govindwar SP (2008) Degradation of textile dye reactive navy-blue Rx (Reactive blue-59) by an isolated Actinomycete Streptomyces krainskii SUK-5. Malays J Microbiol 4(2):1–5

Hamsavathani V, Aysha OS, Ajith AR (2017) Isolation and identification of chlorpyrifos degrading bacteria from agricultural soil. Int J Adv Res 5(5):1209–1221. https://doi.org/10.21474/IJAR01/4234

Haritash AK, Kaushik CP (2009) Biodegradation aspects of Polycyclic Aromatic Hydrocarbons (PAHs): a review. J Hazard Mater 169(1–3):1–15. https://doi.org/10.1016/J.JHAZMAT.2009.03.137

Hemme CL, Deng Y, Gentry TJ, Fields MW, Wu L, Barua S, Barry K, Tringe SG, Watson DB, He Z, Hazen TC, Tiedje JM, Rubin EM, Zhou J (2010) Metagenomic insights into evolution of a heavy metal-contaminated groundwater microbial community. ISME J 4(5):660–672. https://doi.org/10.1038/ISMEJ.2009.154

Holling CS (2001) Understanding the complexity of economic, ecological, and social systems. Ecosystems 4(5):390–405. https://doi.org/10.1007/S10021-001-0101-5/METRICS

Huang X, Stärz M, Gohl K, Knorr G, Lohmann G (2017) Impact of Weddell Sea shelf progradation on Antarctic bottom water formation during the Miocene. Paleoceanography 32(3):304–317. https://doi.org/10.1002/2016PA002987

Islam MS, Islam MZ, Islam MS (2019) Discriminate and indiscriminate use of amoxicillin antibiotic and detection of its residue in poultry edible tissue by thin layer chromatography (TLC) method. Asian-Australas J Food Saf Secur 3(2):96–102. https://doi.org/10.3329/AAJFSS.V3I2.55932

Jagannathan SV, Manemann EM, Rowe SE, Callender MC, Soto W (2021) Marine actinomycetes, new sources of biotechnological products. Mar Drugs 19(7). https://doi.org/10.3390/MD19070365

Jai HP, Pillai S, Girish K, Agsar D (2014) Isolation, characterization and screening of actinomycetes from textile industry effluent for dye degradation. Int J Curr Microbiol App Sci 3(11):105–115. http://www.ijcmas.com

Karn SK, Chakrabarti SK, Reddy MS (2011) Degradation of pentachlorophenol by Kocuria sp. CL2 isolated from secondary sludge of pulp and paper mill. Biodegradation 22(1):63–69. https://doi.org/10.1007/S10532-010-9376-6

Kharwar RN, Upadhyay RS, Dubey NK, Raghuwanshi R (2014) Microbial diversity and biotechnology in food security, pp 1–610. https://doi.org/10.1007/978-81-322-1801-2/COVER

Kim ES (2021) Recent advances of actinomycetes. Biomolecules 11(2):1–3. https://doi.org/10.3390/BIOM11020134

Kolekar PD, Phugare SS, Jadhav JP (2014) Biodegradation of atrazine by Rhodococcus sp. BCH2 to N-isopropylammelide with subsequent assessment of toxicity of biodegraded metabolites. Environ Sci Pollut Res 21(3):2334–2345. https://doi.org/10.1007/S11356-013-2151-6

Kumar M, Lakshmi CV, Khanna S (2008) Biodegradation and bioremediation of endosulfan contaminated soil. Biores Technol 99(8):3116–3122. https://doi.org/10.1016/J.BIORTECH.2007.05.057

Li L, Liu X, Jiang W, Lu Y (2019) Recent advances in synthetic biology approaches to optimize production of bioactive natural products in Actinobacteria. Front Microbiol 10:495220. https://doi.org/10.3389/FMICB.2019.02467/BIBTEX

Lin L, Yang H, Xu X (2022) Effects of water pollution on human health and disease heterogeneity: a review. Front Environ Sci 10:880246. https://doi.org/10.3389/FENVS.2022.880246/BIBTEX

Lin Y, Wang X, Wang B, Mohamad O, Wei G (2012) Bioaccumulation characterization of zinc and cadmium by Streptomyces zinciresistens, a novel actinomycete. Ecotoxicol Environ Saf 77:7–17. https://doi.org/10.1016/J.ECOENV.2011.09.016

Luz AP, Pellizari VH, Whyte LG, Greer CW (2004) A survey of indigenous microbial hydrocarbon degradation genes in soils from Antarctica and Brazil. Can J Microbiol 50(5):323–333. https://doi.org/10.1139/W04-008

Malhi Y, Franklin J, Seddon N, Solan M, Turner MG, Field, CB, Knowlton N (2020) Climate change and ecosystems: threats, opportunities and solutions. Philos Trans R Soc B 375(1794). https://doi.org/10.1098/RSTB.2019.0104

Manisalidis I, Stavropoulou E, Stavropoulos A, Bezirtzoglou E (2020) Environmental and health impacts of air pollution: a review. Front Public Health 8:505570. https://doi.org/10.3389/FPUBH.2020.00014/BIBTEX

Megharaj M, Ramakrishnan B, Venkateswarlu K, Sethunathan N, Naidu R (2011) Bioremediation approaches for organic pollutants: a critical perspective. Environ Int 37(8):1362–1375. https://doi.org/10.1016/J.ENVINT.2011.06.003

Mincer TJ, Jensen PR, Kauffman CA, Fenical W (2002) Widespread and persistent populations of a major new marine actinomycete taxon in ocean sediments. Appl Environ Microbiol 68(10):5005. https://doi.org/10.1128/AEM.68.10.5005-5011.2002

Mishra PN, Jain S, Bore T, Chang I, Kwon YM, Wang Y, Dash HR, Kumar A, Tiwari S, Jiang N, Das SK, Scheuermann A (2023) Biological perspectives in geotechnics: application and monitoring. J Rock Mech Geotech Eng. https://doi.org/10.1016/J.JRMGE.2023.10.007

Mishra S, Lin Z, Pang S, Zhang W, Bhatt P, Chen S (2021) Recent advanced technologies for the characterization of xenobiotic-degrading microorganisms and microbial communities. Front Bioeng Biotechnol 9:632059. https://doi.org/10.3389/FBIOE.2021.632059/BIBTEX

Mohee R, Mudhoo A (2012) Bioremediation and sustainability: research and applications, p 407. https://www.wiley.com/en-us/Bioremediation+and+Sustainability%3A+Research+and+Applications-p-9781118371251

Mukhtar S, Zaheer A, Aiysha D, Abdulla Malik K, Mehnaz S (2017). Actinomycetes: a source of industrially important enzymes. J Proteomics Bioinform 10(12). https://doi.org/10.4172/JPB.1000456

Natural Resource Commission (2019) By—Group Annual report NRC Group—Issuu. https://issuu.com/nrcgroup/docs/nrc_group_annual_report_2019_-_web. Accessed 4 Dec 2024

Ojo AO (2023) An overview of lignocellulose and its biotechnological importance in high-value product production. Fermentation 9(11):990. https://doi.org/10.3390/FERMENTATION9110990

Pandey J, Chauhan A, Jain RK (2009) Integrative approaches for assessing the ecological sustainability of in situ bioremediation. FEMS Microbiol Rev 33(2):324–375. https://doi.org/10.1111/J.1574-6976.2008.00133.X

Pasti-Grigsby MB, Paszczynski A, Goszczynski S, Crawford DL, Crawford RL (1992) Influence of aromatic substitution patterns on azo dye degradability by Streptomyces spp. and Phanerochaete chrysosporium. Appl Environ Microbiol 58(11):3605–3613. https://doi.org/10.1128/AEM.58.11.3605-3613.1992

Paszczynski A, Pasti-Grigsby MB, Goszczynski S, Crawford RL, Crawford DL (1992) Mineralization of sulfonated azo dyes and sulfanilic acid by Phanerochaete chrysosporium and Streptomyces chromofuscus. Appl Environ Microbiol 58(11):3598. https://doi.org/10.1128/AEM.58.11.3598-3604.1992

Peng RH, Xiong AS, Xue Y, Fu XY, Gao F, Zhao W, Tian YS, Yao QH (2008) Microbial biodegradation of polyaromatic hydrocarbons. FEMS Microbiol Rev 32(6):927–955. https://doi.org/10.1111/J.1574-6976.2008.00127.X

Polti MA, Aparicio JD, Benimeli CS, Amoroso MJ (2014) Simultaneous bioremediation of Cr (VI) and lindane in soil by actinobacteria. Int Biodeterior Biodegrad 88:48–55. https://doi.org/10.1016/J.IBIOD.2013.12.004

Rehan M, Kluge M, Fränzle S, Kellner H, Ullrich R, Hofrichter M (2014) Degradation of atrazine by Frankia alni ACN14a: gene regulation, dealkylation, and dechlorination. Appl Microbiol Biotechnol 98(13):6125–6135. https://doi.org/10.1007/S00253-014-5665-Z

Robles-González IV, Fava F, Poggi-Varaldo HM (2008) A review on slurry bioreactors for bioremediation of soils and sediments. Microb Cell Fact 7(5):5. https://doi.org/10.1186/1475-2859-7-5

Saini A, Aggarwal NK, Sharma A, Yadav A (2015) Actinomycetes: a source of Lignocellulolytic enzymes. Enzym Res. https://doi.org/10.1155/2015/279381

Samson FB, Knopf FL (1996) Ecosystem management. https://doi.org/10.1007/978-1-4612-4018-1

Sanjenbam P, Saurav K, Kannabiran K (2012) Biosorption of mercury and lead by aqueous Streptomyces VITSVK9 sp. isolated from marine sediments from the bay of Bengal, India. Front Chem Sci Eng 6(2):198–202. https://doi.org/10.1007/S11705-012-1285-2/METRICS

Sharma P, Singh SP, Iqbal HMN, Tong YW (2022) Omics approaches in bioremediation of environmental contaminants: an integrated approach for environmental safety and sustainability. Environ Res 211:113102. https://doi.org/10.1016/J.ENVRES.2022.113102

Shivanna KR (2022) Climate change and its impact on biodiversity and human welfare. Proc Indian Natl Sci Acad Part A Phys Sci 88(2):160. https://doi.org/10.1007/S43538-022-00073-6

Singh H, Pant G (2023) Phytoremediation: low input-based ecological approach for sustainable environment. Appl Water Sci 13(3):1–8. https://doi.org/10.1007/S13201-023-01898-2/TABLES/6

Sousa JCG, Ribeiro AR, Barbosa MO, Pereira MFR, Silva AMT (2018) A review on environmental monitoring of water organic pollutants identified by EU guidelines. J Hazard Mater 344:146–162. https://doi.org/10.1016/J.JHAZMAT.2017.09.058

Takagi K, Iwasaki A, Kamei I, Satsuma K, Yoshioka Y, Harada N (2009) Aerobic mineralization of hexachlorobenzene by newly isolated pentachloronitrobenzene-degrading Nocardioides sp. strain PD653. Appl Environ Microbiol 75(13):4452–4458. https://doi.org/10.1128/AEM.02329-08

Timková I, Sedláková-Kaduková J, Pristaš P (2018) Biosorption and bioaccumulation abilities of actinomycetes/streptomycetes isolated from metal contaminated sites. Separations 5(4):54. https://doi.org/10.3390/SEPARATIONS5040054

Tramper J, Zhu Y (2011) Modern biotechnology Panacea or new Pandora's box? Mod Biotechnol. https://doi.org/10.3920/978-90-8686-725-7

Ventura M, Canchaya C, Tauch A, Chandra G, Fitzgerald GF, Chater KF, van Sinderen D (2007) Genomics of Actinobacteria: tracing the evolutionary history of an ancient phylum. Microbiol Mol Biol Rev: MMBR 71(3), 495–548. https://doi.org/10.1128/MMBR.00005-07

Wang Q, Xie S (2012) Isolation and characterization of a high-efficiency soil atrazine-degrading Arthrobacter sp. strain. Int Biodeterior Biodegrad 71:61–66. https://doi.org/10.1016/J.IBIOD.2012.04.005

Yadav MK (2018) Role of biofilms in environment pollution and control. Microb Biotechnol 1:377–398. https://doi.org/10.1007/978-981-10-6847-8_16

Zhang JJ, Tang X, Moore BS (2019) Genetic platforms for heterologous expression of microbial natural products. Nat Prod Rep 36(9):1313–1332. https://doi.org/10.1039/C9NP00025A

Chapter 2
Harmonizing Nature and Innovation: Sustainable Environmental Solutions Through Nano-biotech Synergy

H. Samson Prince, Nazarene Marylene Nicky Macarius, Sanket Chandrashekhar Walkikar, Shubankar Das, Anand Prem Rajan, and Kanagavel Deepankumar

Abstract As traditional solutions fall short in mitigating the complex issues of pollution, resource depletion, and climate change, there is a growing imperative to develop sustainable approaches that work in harmony with nature. This chapter elucidates the concept of nano-biotech synergy and its applications in various environmental domains, including water purification, air quality improvement, soil remediation, biodiversity conservation, and waste management. Through a comprehensive analysis of emerging trends, scaling-up opportunities, collaborative partnerships, and interdisciplinary approaches, the chapter outlines a roadmap for future research and innovation in the field. Additionally, it offers policy recommendations aimed at fostering responsible development and deployment of nano-biotechnological solutions. By bridging the gap between cutting-edge science and environmental stewardship, this chapter advocates for a holistic approach to sustainability that leverages the synergistic potential of nature-inspired innovation.

Keywords Sustainable · Nanotechnology · Waste management · Remediation

2.1 Introduction

An interdisciplinary field known as "nanobiotechnology" combines the enormous potential of biotechnology with the knowledge and instruments of nanotechnology. One billionth of a meter, or the nanoscale, is the scale at which biological systems are manipulated to create materials, devices, and systems that, because of their small size, have unique qualities and uses. The use of nanotechnology in the biological sciences

H. Samson Prince · N. M. N. Macarius · S. Chandrashekhar Walkikar · S. Das · A. P. Rajan · K. Deepankumar (✉)
School of Biosciences and Technology, Vellore Institute of Technology, Vellore, Tamil Nadu 632014, India
e-mail: deepankumar.k@vit.ac.in

© The Author(s), under exclusive license to Springer Nature Switzerland AG 2025
A. Jeyaseelan and K. Murugasen (eds.), *Sustainable Environmental Remediation: Avenues in Nano and Biotechnology*, Environmental Science and Engineering,
https://doi.org/10.1007/978-3-031-78483-5_2

is known as nanobiotechnology. It entails studying biological systems and creating new medical technologies using nanoscale materials and methods. This covers the development of nanoscale systems, tools, and materials for illness detection, management, and prevention. New instruments for biological research and medical applications are made possible by the special qualities of nanoscale materials, such as their small size, high reactivity, and high surface area (Kapiel 2023; Omran 2020).

Nanobiotechnology covers a wide range of important topics. Nanomedicine is the use of nanotechnology in medicine, such as imaging agents, drug delivery systems, and diagnostic tools. Biosensors: creating nanoscale instruments with the ability to identify biological substances, infections, or environmental factors. Environmental Applications: Monitoring and cleaning the environment with nanobiotechnology. Agriculture: Improving nutrient delivery systems, protecting crops better, and creating innovative pest control techniques. Biosensors, biochips, and other nanoelectronic devices can be created by combining biological systems with electronics and computing. Material science is the study of creating new materials with improved qualities for a range of uses, such as biocompatible materials for prosthetics and implants.

Because nanobiotechnology is multidisciplinary, it can have an impact on a variety of industries, including pharmaceuticals, healthcare, agriculture, and environmental sciences. There is constant development in this field as new technologies and applications are created. Significant progress has already been made in the fields of therapy, drug delivery, contrast agents, diagnostic tools, and analytical tools as a result of the fusion of nanomaterials and biology (Shoseyov and Levy 2008). Research in these areas is still ongoing. Nano biotechnology has a wide range of applications, and it is always changing due to new discoveries and developments in technology. The combination of biology and nanotechnology provides creative answers to difficult problems in many different industries. Nanobiotechnology has enormous potential to influence science and technology in a variety of domains as it develops.

2.1.1 Historical Perspective on the Convergence of Nanotechnology and Biotechnology

A major turning point in the history of science and technology has been reached by the convergence of nanotechnology and biotechnology, which has given rise to a new multidisciplinary field called nanobiotechnology. This convergence has been sparked by the discovery that powerful tools for comprehending and modifying biological systems can be obtained through the manipulation of matter at the nanoscale.

Early Foundations and Conceptualization: Physicist Richard Feynman's seminal lecture "There's Plenty of Room at the Bottom" in 1959 envisions the potential for atom and molecular manipulation to create new materials and devices (Bhushan 2016). This is where the conceptual roots of nanotechnology can be found. This concept established the foundation for the advancement of nanotechnology.

2.1.1.1 Instrumentation Advancements

During the 1980s, there were notable developments in instrumentation that made it possible for scientists to view and work with individual atoms and molecules. The instruments required to operate at the nanoscale were made possible by the 1981 invention of the scanning tunneling microscope (STM) and the atomic force microscope (AFM) that followed (Bayda et al. 2019).

2.1.1.2 Finding of Fullerenes

In 1985, a new form of carbon comprising 60 carbon atoms arranged in a spherical structure was discovered. This finding revealed the special characteristics of materials at the nanoscale and sparked increased interest in nanotechnology (Bayda et al. 2019).

2.1.1.3 Creation Engines

In 1986, The concept of nanotechnology gained popularity when K. Eric Drexler published "Engines of Creation," a book that described the possible uses of molecular nanotechnology, including in the medical field (Bayda et al. 2019).

2.1.1.4 Emergence of Nanobiotechnology

The term "nanobiotechnology" was first applied to the intersection of nanotechnology and biology. It emphasizes the application of nanoscale materials and techniques to biological problems, symbolizing the convergence of engineering and life science (Choi and Montemagno 2016).

2.1.1.5 National Nanotechnology Initiative

Established in 2000, the National Nanotechnology Initiative (NNI) has been a major source of funding and support for the field of nanotechnology research and development in the United States. It formalized the federal commitment to this endeavor (Bhushan 2016).

2.1.1.6 Synergistic Leveraging

The combination of biotechnology and nanotechnology has resulted in nanobiotechnology, which uses the fundamental control of physical, chemical, and biological processes at the nanoscale to develop new technologies and methods for addressing environmental and health concerns (Choi and Montemagno 2016).

2.1.1.7 Current Trends and Applications

The contemporary patterns and uses of nanobiotechnology are extensive, comprising targeted drug delivery mechanisms, diagnostic tools, as well as the innovation of novel materials for tissue engineering and regenerative medicine. Research in this field persists in expanding, investigating the complete capabilities of this integration (Bhushan 2016).

In essence, the historical viewpoint regarding the intersection of nanotechnology and biotechnology underscores a progression of forward-thinking concepts, technological progressions, and interdisciplinary partnerships that have resulted in the emergence of a dynamic and swiftly developing domain with the capacity to transform numerous facets of our society.

2.1.2 Sustainable Approaches in Environmental Management

Sustainable strategies in environmental management are crucial in tackling the increasing issues of pollution, depletion of resources, and ecological instability. The integration of nanotechnology and biotechnology, known as nanobiotechnology, presents novel methods for promoting environmental sustainability.

2.1.2.1 Addressing Water Scarcity and Contamination

Water scarcity is a pressing concern that is worsened by the pollution caused by industrialization and urbanization. Conventional techniques for treating water, such as chemical precipitation and membrane filtration, often prove inadequate due to problems like secondary contamination and energy inefficiency. However, nanobiotechnology offers a potential solution by utilizing microbial-assisted green nanomaterials (NMs) for the remediation of wastewater. These NMs possess a large surface area, enabling them to efficiently eliminate various water pollutants (Malik et al. 2022).

2.1.2.2 Eco-friendly Remediation Techniques

Nanobiotechnology facilitates the advancement of environmentally friendly remediation strategies that are both less harmful and more economical in contrast to traditional approaches. For example, the utilization of microbial cells, magnetotactic bacteria, and integrated bioreactors can effectively eliminate pollutants without resulting in additional harm to the environment (Malik et al. 2022).

2.1.2.3 Sustainable Crop Production

Nanobiotechnology in agriculture plays a crucial role in enhancing sustainable crop production through the enhancement of soil quality and plant health. The utilization of nano-bioremediation techniques aids in the restoration of deteriorated soils, guaranteeing that forthcoming agricultural demands are fulfilled while safeguarding the environment (Malik et al. 2022).

2.1.2.4 Resource Recovery and Pollution Monitoring

Nanobiotechnology plays a crucial role in enabling the extraction of resources from wastewater and the creation of sensors based on bio-nanomaterials for the surveillance of air and water contamination. This dual functionality aids in waste reduction and pollution prevention through the implementation of real-time monitoring and control measures (Dwivedi and Dwivedi 2022).

2.1.2.5 Energy-Efficient Industrial Processes

The production of nanomaterials through biological approaches proves to be a sustainable and eco-friendly method, as it has the capability to harness renewable energy sources. This shift away from non-renewable energy sources not only decreases the environmental impact of industrial procedures but also enhances energy efficiency (Shahcheraghi et al. 2022).

2.1.2.6 Integration with Other Technologies

Nanobiotechnology has the potential to be combined with various other technologies, such as the Internet of Things (IoT), in order to develop environmentally friendly nanotechnology solutions. This amalgamation has the capability to enhance the effectiveness of environmental management systems by making them more adaptable and responsive to the ever-changing environmental conditions (Dwivedi and Dwivedi 2022).

2.1.2.7 Life-Cycle Assessment

Nanobiotechnology facilitates a thorough evaluation of environmental contaminants throughout their life cycle. This aids in comprehending the enduring consequences of these pollutants and the efficacy of remedial approaches, guaranteeing the sustainability of interventions in the long run (Dwivedi and Dwivedi 2022).

2.2 The Nano-biotech Interface

The exploration of the relationship between nanotechnology and biotechnology presents an intriguing pursuit that combines the precision and capabilities of manipulating materials at the nanoscale with the intricate and dynamic systems of living organisms. This interdisciplinary method, referred to as nanobiotechnology, capitalizes on the advantages of both disciplines to develop groundbreaking solutions for a diverse array of applications. Nanobiotechnology is characterized as the utilization of nanotechnology within the realm of life sciences, encompassing the examination, creation, and control of biological and biochemical systems and devices at the molecular level. The fusion of nanotechnology and biotechnology holds the promise of transforming numerous industries, such as medicine, agriculture, environmental science, and manufacturing.

In the field of medicine, nanobiotechnology plays a crucial role in enhancing drug delivery systems. By engineering nanoparticles to target specific cells or tissues, the side effects of treatments can be reduced while increasing their effectiveness. For instance, nano-sized carriers can transport drugs directly to cancer cells, thereby minimizing harm to healthy cells and tissues. In the industrial sector, enzymes and other biological molecules are being immobilized on nanomaterials to improve their stability and activity. This advancement has significant implications for biofuel production, as nanobiotechnology can enhance conversion processes. Agriculture also benefits from nanobiotechnology with the development of nano-fertilizers and nano-pesticides, which can decrease the need for chemicals and reduce environmental impact. Additionally, nanosensors can be utilized to monitor soil health and crop status in real-time. Furthermore, nanobiotechnology shows promise in environmental remediation by using nanomaterials to eliminate pollutants in water and soil, transforming harmful substances into less toxic forms (Shahcheraghi et al. 2022).

The intersection of nanotechnology and biotechnology is part of a larger trend of converging technologies, which also encompasses information technology and cognitive sciences (NBIC) (Choi and Montemagno 2016). This convergence is anticipated to have a significant impact on shaping a sustainable future, tackling issues such as climate change and renewable energy sources. An intriguing aspect of nanobiotechnology involves the synthesis of nanoparticles utilizing plant extracts and essential oils. This method combines the antimicrobial properties of plant-based compounds with the distinctive characteristics of nanomaterials, potentially leading to the development of new antimicrobial agents (Antunes Filho et al. 2023). The collaboration between nanotechnology and biotechnology is based on the precise control of chemical, physical, and biological processes. At the nanoscale level, there exists a clear synergy between biological and nanoscale entities, reinforced by their proximity in size to the fundamental building blocks of matter (Díaz et al. 2019). The term biointerface denotes the interaction between biological systems and nanomaterials. Nanobiotechnology seeks to leverage advancements in nanotechnology to enhance biotechnological applications, while bionanotechnology utilizes natural or biomimetic systems to fabricate unique nanoscale structures (Goodsell 2004; Mirkin and Niemeyer 2007; Niemeyer and Mirkin 2004). The synergy between nanotechnology and biotechnology exemplifies the potential of interdisciplinary research. By merging the capabilities of these two domains, researchers and engineers can develop innovative technologies and solutions to address some of the most critical challenges of our era. As this field progresses, it holds the promise of significant advancements that could bring numerous benefits to humanity.

2.2.1 Key Materials and Methods

The field of nano-biotech interface is continuously progressing, with its primary objective being the study of the interaction between nanoscale materials and biological systems. The utilization of key materials and methodologies at this interface plays a pivotal role in the advancement of applications in the domains of medicine, environmental science, and technology.

2.2.1.1 Key Materials

Two-Dimensional Nanomaterials (2D NMs) encompass a range of materials including graphene, transition metal carbides/nitrides (MXenes), transition metal dichalcogenides (TMDs), and others. These materials possess a flat, sheet-like structure that enables efficient drug loading, precise chemical modifications, and the creation of hierarchical structures (Wang et al. 2024). Nanoparticles (NPs), such as titanium dioxide nanoparticles, have been extensively studied for their interactions with biological molecules. These investigations have provided valuable insights into the bio-nano interface (Rouse et al. 2021). Bio-Nanomaterial Interfaces are formed by combining biological molecules such as proteins, DNAs, and lipid membranes with engineered nanomaterials like SiO_2, TiO_2, CNT, and graphene (Carrara 2018).

2.2.1.2 Methods

- **First Principles Analysis**: This encompasses the anticipation of interactions occurring at the interface between bio and nano materials through the utilization of computational materials modeling. It entails the assessment of properties exhibited by the nano-bio interface in the presence of water, as well as the parameterization of the atomistic force field for the material (Rouse et al. 2021).
- **Synthesis and Manufacturing**: The development of protocols for the synthesis, manufacturing, and construction of interfaces involving bio-nanomaterials is of utmost importance in the creation of functional materials tailored for specific applications (Bergese and Hamad-Schifferli 2013).
- **Characterization Procedures**: These procedures encompass the examination and analysis of interfaces involving bio-nanomaterials, which plays a crucial role in

comprehending their properties and interactions with biological systems (Bergese and Hamad-Schifferli 2013).

2.2.1.3 Applications

The potential of 2D NMs and their nano-bio interfacial interactions in revolutionizing cancer treatment strategies is currently being investigated (Wang et al. 2024). The development of advanced biosensors and nanomedicine applications heavily relies on the interactions between nanomaterials and biological molecules (Carrara 2018). The convergence of nanotechnology and biotechnology at the nano-biotech interface offers great opportunities for innovation. Through a comprehensive understanding and effective utilization of essential materials and techniques, researchers can create groundbreaking solutions that have the potential to greatly influence diverse fields, ranging from healthcare to environmental sustainability.

2.2.2 Successful Nano-biotech Applications in Environmental Technology

Nano-biotechnology represents a hopeful area of study that combines the complexities of nanotechnology with the fundamentals of biotechnology in order to develop novel approaches for addressing environmental issues.

2.2.2.1 Nano-biotechnology for Sustainable Future

The role of nano-biotechnology in shaping a sustainable future is explored in a research paper published in the journal 3 Biotech. The study sheds light on the utilization of nano-sized materials such as nanoparticles, nanotubes, and nanofibers across different domains, with a specific focus on environmental technology. The article underscores the significance of integrating nano-based strategies to augment conventional biotechnological techniques, particularly in the realms of water and soil remediation, as well as enzymatic processes. By harnessing the potential of nano-biotechnology, we can pave the way for a more sustainable and eco-friendly future (Shahcheraghi et al. 2022).

2.2.2.2 Nanotechnology in Environmental Remediation

The article in Vegetos provides a comprehensive examination of how nanotechnology can be utilized in environmental cleanup efforts. Specifically, it highlights the environmentally friendly technique of synthesizing nanoparticles without the use of harmful substances, offering a cost-effective and sustainable solution. The review delves into the utilization of biogenic materials in creating nanoparticles and discusses their various environmental applications, including reducing pollution, treating water, and degrading dyes (Das et al. 2019).

2.2.2.3 Environmental Biotechnology in India

Springer presents a publication detailing various case studies related to environmental monitoring, restoration of environmental quality, resource recovery, and the replacement of non-renewable resources with renewable ones in the context of India. The paper highlights the utilization of environmental biotechnology in activities such as oil spill remediation, recovery of hydrocarbons from oily sludges, and the production of biosurfactants from distillery spentwash (Khanna 1996).

2.2.2.4 Agricultural Biotechnology Case Studies

The FAO has published a document that includes case studies in agricultural biotechnology, which also have implications for environmental technology. These studies demonstrate the technical constraints tackled through biotechnologies to reduce impediments to productivity in challenging biophysical environments (Tyczewska et al. 2023).

2.2.2.5 Green Nanotechnology Applications

An excerpt from a publication found on Taylor & Francis delves into the practical implementations of green nanotechnology. This innovative approach transforms the utilization of eco-friendly and easily degradable materials, playing a crucial role in upholding ecological equilibrium and lessening environmental impacts (Jain and Kejariwal 2022).

The presented instances serve as tangible examples of the wide-ranging and influential uses of nano-biotechnology in tackling environmental challenges, highlighting its capacity to make substantial contributions towards a sustainable tomorrow.

2.3 Nano-biotech for Environmental Monitoring

2.3.1 Detection of Pollutants and Pathogens

The field of nanosensors has brought about a significant transformation in the identification of pollutants and pathogens, presenting advanced capabilities such as heightened sensitivity, specificity, and swift response durations. This cutting-edge technology has revolutionized the way we detect and monitor these harmful substances, providing us with more accurate and efficient methods.

2.3.1.1 DNA-Nanosensors for Environmental Monitoring

The detection of environmental pollutants has been the subject of a study published in Current Pollution Reports. The study focuses on the utilization of DNA-nanosensors, which employ DNA as a recognition element to identify a wide range of pollutants such as pathogens, antibiotics, pesticides, heavy metals, and other toxic substances. The review emphasizes the progress made in the design of DNA nanosensors and the potential for developing nanosensors to detect previously unexplored pollutants (Kumar and Guleria 2020).

2.3.1.2 Nanomaterial-Enabled Sensors for Environmental Contaminants

An article was recently published in The Journal of Nanobiotechnology focusing on the utilization of nanomaterial-enabled sensors to detect environmental contaminants. The research delves into the practical application of these sensors in identifying pesticides, heavy metals, and pathogens. These sensors are capable of detecting contaminants at extremely low concentrations, ranging from nanomolar to sub-picomolar levels, and are specifically engineered for on-site detection without the need for costly laboratory equipment (Willner and Vikesland 2018).

2.3.1.3 Nanosensors for Water Quality Monitoring

The scientists at the Indian Institute of Technology Madras are dedicated to the development of nanomaterial-based “nanosensors” designed for the monitoring of water quality. These innovative sensors have the ability to detect organic pollutants, pharmaceuticals, heavy metals, pathogens, and pesticides with high sensitivity, selectivity, speed, efficiency, and can be used for on-site detection, all of which pose a threat to various ecosystems (Vikesland 2018).

2.3.1.4 Development of Environmental Nanosensors

One of the chapters in Springer provides an in-depth analysis of the progress made in the field of environmental nanosensors. These nanosensors have been designed to detect a wide range of substances including pathogens, toxic gases, organic chemicals, heavy metals, and pesticides. The chapter explores the different types of sensor materials that are utilized in the fabrication of nanosensors, including metal-based, metal-oxide-based, carbon-based, and polymer-based nanomaterials. This comprehensive discussion sheds light on the advancements and potential applications of environmental nanosensors (Chakraborty et al. 2021).

2.3.2 Bioassays and Biomarkers for Environmental Surveillance

Bioassays and biomarkers play a crucial role in environmental monitoring by offering valuable information regarding the existence and effects of contaminants on ecosystems. These tools are indispensable for understanding the health of the environment and assessing the potential risks posed by various pollutants.

2.3.2.1 Bioassays for Environmental Surveillance

1. Bioassays serve as experimental systems in which living organisms are subjected to environmental samples in order to evaluate the toxicity and general well-being of ecosystems. These assessments can detect variations in mortality, growth, or reproductive capabilities of the organisms under study, offering a comprehensive perspective on the environmental consequences of pollutants. Bioassays are especially valuable for identifying the impacts of intricate combinations of contaminants that may not be entirely elucidated through chemical evaluations alone (Hagger et al. 2006).

2.3.2.2 Biomarkers as Indicators of Environmental Stress

Biomarkers are biological indicators at the sub-individual or individual level that signal exposure to environmental stressors. These indicators can serve as functional assessments of exposure to stressors manifested at different tiers, such as molecular, cellular, physiological, or behavioral. Biomarkers play a crucial role in the timely identification of environmental alterations prior to the occurrence of substantial ecological harm (Calisi 2023).

2.3.2.3 Integrated Environmental Risk Assessment

The integration of biomarkers into environmental risk assessment frameworks is gaining traction. The European Commission's Water Framework Directive (WFD) emphasizes the need for an integrated approach that incorporates both chemical and ecological objectives. Biomarkers can be incorporated within this framework to provide a weight-of-evidence approach for environmental management (Hagger et al. 2006).

2.3.2.4 Bioindicators and Biomarkers in Aquatic Ecotoxicology

1. Within the field of aquatic ecotoxicology, the combination of bioindicators and biomarkers is utilized to evaluate the potential ecological harm caused by chemical contaminants. The analysis of biomarkers in bioindicator species has proven to be a crucial method for monitoring the environment, assessing risks, and documenting the restoration of aquatic ecosystems. These tools play a significant role in identifying the physiological impacts of new pollutants, even when present in minimal amounts or within intricate combinations (Calisi 2023).

2.3.2.5 Effect-Based Methods for Monitoring Strategies

Effect-based methods, including bioassays and biomarkers, play a vital role in the implementation of monitoring strategies. These methods rely on living organisms as indicators of environmental quality, enabling a direct assessment of the impact of pollutants on the biosphere. The significance of this approach lies in its capacity to accurately reflect the actual consequences of environmental contaminants (Girotti et al. 2020).

2.3.2.6 Biomarkers in Regulatory Legislation

Although biomarkers are widely used in laboratory and field studies, their integration into regulatory legislation for environmental risk assessment has been restricted. Nevertheless, as environmental protection strategies become more comprehensive, biomarkers are now being recognized as valuable instruments for environmental management. They provide a more comprehensive approach to evaluating the impacts of environmental samples and wastes on ecosystems compared to solely relying on chemical-based monitoring (Milinkovitch et al. 2019).

These sources offer a thorough examination of the present condition and progressions in bioassays and biomarkers for environmental surveillance, highlighting their capacity to greatly enhance environmental monitoring and assessment of ecosystem health.

2.3.3 *Data Analytics and IoT Integration for Real-Time Monitoring*

The field of real-time environmental monitoring, which combines data analytics and IoT (Internet of Things) integration, is undergoing rapid advancements. This emerging field utilizes the capabilities of data science and interconnected devices to effectively observe and control environmental conditions.

2.3.3.1 IoT-Based Environmental Monitoring Systems

Environmental monitoring systems utilizing IoT technology encompass a network of sensors and devices designed to gather data pertaining to a range of environmental factors, including temperature, humidity, air quality, and pollutant concentrations. By offering instantaneous updates on environmental conditions, these systems facilitate timely reactions to potential threats and the preservation of ecological equilibrium (Hassan et al. 2020).

2.3.3.2 Real-Time Monitoring of Contaminants

The incorporation of Internet of Things (IoT) technology in environmental monitoring enables the continuous monitoring of pollutants in the air, water, and soil. This plays a vital role in promoting sustainable development and safeguarding natural resources. IoT sensors have the capability to identify various environmental factors and transmit the collected data to cloud platforms for evaluation, enabling prompt responses when certain limits are surpassed (Pamula et al. 2022).

2.3.3.3 IoT and Machine Learning for Environmental Insights

The integration of machine learning algorithms with IoT technology has the potential to enhance the predictive capabilities of environmental monitoring systems. By combining these two technologies, it becomes possible to generate more accurate forecasts of environmental changes. This, in turn, enables better decision-making when it comes to environmental protection and resource management (Charan et al.).

2.3.3.4 Smart Environmental Monitoring and Analytics

The utilization of IoT for environmental monitoring is effectively demonstrated by the SEMAR (Smart Environmental Monitoring and Analytics in Real-Time) platform. This platform seamlessly combines sensor data collection, display, and analysis into a unified system, offering a holistic approach for swift implementation of IoT application systems (Panduman et al. 2024).

2.3.3.5 Challenges and Frameworks

Although IoT-based environmental monitoring systems provide various advantages, they also pose certain difficulties, including the need to ensure reliability and cost-effectiveness. To overcome these challenges, it is crucial to establish a comprehensive framework that integrates sensing, network, and visualization layers. By doing so, stakeholders can effectively analyze and interpret environmental data trends (Pamula et al. 2022).

To summarize, the integration of data analytics and IoT for real-time environmental monitoring signifies a notable progression in environmental surveillance. This integration holds the promise of improving the identification, examination, and control of environmental conditions, thereby benefiting ecosystems and public health.

2.4 Nanoparticle Production from Waste Management

Nanobiotechnology is transforming waste management through the provision of effective solutions. Nanoparticles (NPs) increase reactivity and selectivity, streamlining waste treatment processes and cutting down on expenses. The fusion of nanotechnology with biological science displays potential in wastewater treatment, utilizing different nanomaterials and methods. Nanobiotechnology facilitates the repurposing of valuable elements from various wastes, like metals and phosphorous, via microbial processes, providing sustainable options to conventional methods. Progress in nano bioremediation entails the utilization of nanomaterials to address industrial water pollution and persistent compounds, tackling environmental issues and advocating for Circular Economy models. These advancements underscore the promise of nanobiotechnology in reshaping waste management practices towards a more sustainable and effective future.

The use of nanotechnology in waste treatment and recycling processes has shown promising advancements in recent years (Ahmed et al. 2022). These innovations have the potential to revolutionize waste management practices, improve resource efficiency, and minimize environmental impacts (Shrivastava 2018). However, there are still several challenges that need to be addressed, including the scalability and cost-effectiveness of nanomaterials, as well as concerns regarding their potential toxicity and environmental fate. With proper research and development, nanomaterials have the potential to play a crucial role in waste treatment and recycling processes, leading to more sustainable and efficient waste management practices. Nanomaterials have emerged as a promising solution for waste treatment and recycling processes, offering potential advancements in resource efficiency and reducing environmental impacts. Furthermore, their unique properties and capabilities allow for the development of innovative technologies that can address the challenges faced in traditional waste management methods. Recent advancements in nanomaterials for waste treatment and recycling processes have opened new possibilities for sustainable environmental solutions. Nanomaterials such as carbon nanotubes, graphene, and metal–organic frameworks have shown great potential in improving the efficiency of waste treatment and recycling processes.

Carbon nanotubes, for instance, have been used in water purification processes due to their high surface area and adsorption capabilities. These nanomaterials can remove pollutants and heavy metals from wastewater, making them a valuable tool in waste treatment (Nasrollahzadeh et al. 2021). Graphene, with its excellent conductivity and high strength, has been utilized in the development of nanocomposite materials for efficient waste separation and recycling. Its unique properties enable the creation of filters and membranes that can efficiently separate different types of waste materials, contributing to a more effective recycling process (Tiwari et al. 2020). Metal–organic frameworks have also gained attention for their potential in capturing and storing gases, which can be beneficial in the treatment and recycling of industrial emissions and waste gases. These nanomaterials offer a high surface area and tunable porosity, allowing for selective gas adsorption and separation (Gandara-Loe et al. 2021).

The utilization of these advanced nanomaterials in waste treatment and recycling processes marks a significant step towards sustainable environmental practices. Continued research and development in the field of nanotechnology will further enhance the capabilities of these materials, addressing the challenges and maximizing the opportunities for more efficient and environmentally friendly waste management.

2.4.1 Biogenic Synthesis of Nanoparticles

Microorganisms can create a variety of unusual nanostructures. Scientists are increasingly interested in using microorganisms to create nanostructures for diverse uses. Bacteria and fungi can synthesise inorganic compounds through biologically mediated and induced mechanisms (Moitra et al. 2020). Controlling biological synthesis allows for the formation of desirable nanostructures with specific geometries and composition. Although nanoparticle physicochemical synthesis is accurate, biological synthesis still lacks control over particle geometry and process scalability (Ghosh and Webster 2021). Biologically induced synthesis enables scientists to create organic nanoparticles from common metal precursors (Grasso et al. 2019), with diverse compositions available. Microorganisms like bacteria, fungus, and algae are increasingly employed to produce nanosized particles. Various microorganisms create nanoparticles through various reactions with metal precursors. Bacteria and fungi can produce both intracellularly and extracellularly, with distinct routes across different microorganisms. Intracellular synthesis uses the cell wall to transport metal ions, with positive ions interacting with the negative wall. Enzymes in the cells convert these ions to metal nanoparticles. The process begins with the nucleation of HAuCl ions, resulting in the formation of nanoclusters via an electrostatic contact.

2.4.1.1 Silver

Silver nanoparticles (AgNPs) have been intensively explored for their antibacterial and antioxidant characteristics, particularly in the biomedical field (Ameen et al. 2020). Smaller particles have stronger antibacterial effects against certain diseases.

2.4.1.2 Gold

Gold nanoparticles (AuNPs), another highly investigated metal, have a wide range of potential in cancer therapy, medication transport, and other biomedical applications such as antibacterial (Hooda et al. 2020). Nanobubbles containing gold nanoparticles can be targeted to a tumorous location and explode when exposed to increased heat from a laser or infrared radiation source, according to research (Kalimuthu et al. 2020). These nanoparticles infiltrate cancer cells and reduce their growth rate. AuNPs are also used for a variety of applications, including Raman spectroscopy and imaging (Siddiqi and Husen 2017).

2.4.1.3 Metal Oxides

Metal oxides have also been studied, including zinc oxide (ZnO), copper oxide (CuO), magnesium oxide (MgO), titanium dioxide (TiO_2), aluminium (Al_2O_3), iron oxide (Fe_3O_4), and many others. Because of its magnetic characteristics, iron oxide has recently emerged as a promising nanomaterial. Iron oxide (Fe_3O_4) nanoparticles are employed in a variety of applications, including biomedicine, catalysis, and environmental safety. The most popular applications for such nanoparticles include targeted medication delivery, biosensors, MRI, cancer diagnostics, and tissue engineering (Moreira et al. 2018). Tin oxide (SnO_2) and tungsten (tri)oxide (WO_3) have unique electrical characteristics that vary with nanoparticle size and are utilised as gas sensors (Liu et al. 2020). Titanium dioxide is used in the optical and solar energy sectors because of its electrical conductivity (Bai et al. 2014).

2.4.1.4 Bacteria

Bacteria are freely available in the environment, can be grown quickly, and can adapt to a variety of environments, making them ideal candidates for nanoparticle synthesis. Temperature and oxygenation in bacterial cultures can be easily controlled. It is crucial to manage such circumstances so that different sized nanoparticles can be synthesised for various applications (Poreddy et al. 2015). The biochemical principles driving the microbial creation of nanoparticles have been partially established and are currently being investigated for a better understanding. Microbes use many processes for nanoparticle formation, including changes in solubility and toxicity, bioadsorption, bioaccumulation, metal precipitation, and efflux systems (Liu and Catchmark 2019). These processes serve as microbial resistance strategies for enzyme-mediated cellular detoxification of inorganic particles. The key enzymes involved include oxidoreductases, such as nitrate reductase and sulphite reductase, as well as cellular transporters. Certain bacteria have demonstrated the ability to synthesise organic nanoparticles. Aerobic acetic bacteria, such as Gluconacetobacter, have been utilised to produce cellulose nanocomposites (Peigneux et al. 2016). Bacterial cellulose is a three-dimensional network of cellulose nanofibrils that exhibits superior qualities to nanocrystalline and nanofibrillated cellulose. The primary applications for bacterially manufactured nanocellulose are biomedical, such as antibacterial agents, drug delivery systems, and biosensors. Microbes can also produce nanomineral crystals, magnetic materials, and oxide metals. Magnetotactic bacteria excel in constructing magnetosomes capable of producing magnetic radicals (Yan et al. 2017). Nanoparticles can be controlled in terms of geometry and composition. Magnetosomes are organically coated nanocrystals of iron oxide and iron sulphide produced by both magnetotactic and non-magnetotactic bacteria. The magnetic characteristics are determined by bacterial species, and the organic layer is formed by the bacterial phospholipid bilayer membrane (Boucher et al. 2017). Magnetosomes are used in cancer therapy, molecular imaging, and as a biosensor for toxicity assessments (Tsekhmistrenko et al. 2020). Bacteria can synthesise nanoparticles either extracellularly or intracellularly. However, there are limitations to the bacterial creation of nanoparticles. The primary problems are nanoparticle purification and geometric control due to a lack of complete understanding of mechanisms (Mughal et al. 2021). It is difficult to manage particle form and size, manufacture monodispersed particles, and achieve large-scale capabilities. As a result, these technologies require additional research to better our understanding of the mechanisms before they may be considered for practical applications.

2.4.1.5 Fungi

Nanoparticle production utilising fungi has advantages over bacteria, therefore study in this area has grown in popularity over the last decade. Mycelium has several advantages, including the ability to scale up and downstream the process, as well as being economically and environmentally benign, allowing it to cover a broad surface (Syed et al. 2012). Fungi include enzymes in their cytoplasm and cell walls that convert metal ions into nanoparticles. The metal ions have a positive charge, which attracts the fungi and initiates biosynthesis. Metal ions can also trigger certain proteins, which hydrolyze the ions. Fungi are capable of secreting large amounts of protein, which increases nanoparticle synthesis (Rana et al. 2020). Studies on the manufacture of silver nanoparticles using fungi are becoming more prevalent, and these particles can be used in electronics, antimicrobials, antioxidants, and other medicinal applications. There are more studies on the microbial synthesis of gold nanoparticles, but fewer on the use of fungi in this process. Silver and metal oxides have been the subject of numerous research. Using the fungus *Cladosporium cladosporioides*, which is derived from the seaweed *Sargassum wightii*, Joshi et al. synthesised gold nanoparticles (Bhargava et al. 2016). After looking into the mechanism of nanoparticle formation, the study concluded that phenolic chemicals and NADPH-dependent reductase enzymes were in charge of the process. The *Fusarium oxysporum* fungus is used in a number of experiments, including numerous studies on AgNPs, to synthesise nanoparticles. AgNPs (10–15 nm in size) were synthesised by (Pham et al. 2019) utilising the fungus *Fusarium oxysporum*, and they showed encouraging antibacterial properties against *Pseudomonas aeruginosa* and *Escherichia coli*. Other nanoparticles made using *F. oxysporum* are magnesium oxide, copper, and gold (Abdel-Aziz et al. 2020; Chhipa 2019; Naimi-Shamel et al. 2019). Fungi create intra- and extracellular nanoparticles, just like other microorganisms. The organism produces many nanoparticles that are smaller than those that are formed outside of it. The nucleation of the particles inside the species is connected to the size limit. Additionally, intracellular creation has a lot of positive benefits. The removal of depleted metals from the environment, such as copper and platinum, is necessary to reverse the consequences of pollution. There have been reports of fungus-Fe_3O_4 bionanocomposites for managing radioactive waste (Hu et al. 2019). The wood-growing fungus *Penicillium shijie* provided the nano Fe_3O_4 particles. The SEM results showed that a thin layer of Fe_3O_4 had been consistently applied to the particles to beautify the fungus' surface. The FTIR analysis demonstrated that the iron oxide particles were chemically linked to the fungus's surface. They discovered that fungus—Fe_3O_4 sorption was unaffected by ion strengths in the presence of ions, suggesting that its surface complexion—which was thought to effect sorption—was in fact insignificant.

2.4.1.6 Algae

Algae are single or multicellular, autotrophic, aquatic photosynthetic creatures that belong to the kingdom Protista. These organisms lack the multicellular features found in plants, making them eukaryotic. Algae range widely in size from tiny microalgae to enormous macroalgae. The majority of aquatic life uses them as a food source and they are well recognised for creating oxygen. One common type of macroalgae that has been used as a source of protein, vitamins, and minerals in diet is seaweed. Microalgae are thought to be green entry points for the creation of nanoparticles because they have the ability to change metal ions into pliable forms

(Venkatesan et al. 2014). The presence of active chemicals in the cell walls of algae is thought to be responsible for their capacity to synthesise nanoparticles. There have been many published investigations on the use of algae in the synthesis of gold and silver. It has been demonstrated that *Tetraselmis kochinensis* may create intracellular gold nanoparticles (Shankar et al. 2016). In addition to seaweed, microalgae such as diatoms (*N. atomus* and *D. gallica*) can be used to synthesise AuNPs and Au-silica bionanocomposites (Kerry et al. 2019). Compared to other microorganisms like bacteria and fungi, algae is equally important for the synthesis of nanoparticles. As a result, an environmentally benign method for the algae-mediated biosynthesis of nanoparticles has recently been investigated. Even though algae provide a green way to synthesise nanoparticles, there are still issues that need to be resolved before this process can be regarded as long-term sustainable.

2.4.2 Trash: Repository of Nanomaterial

2.4.2.1 Nanomaterial from Kitchen Waste

Our kitchen produces a large amount of biodegradable green garbage every day. The main ingredients are the peels from fruits and vegetables. These leftover peels are loaded with metabolites that could lead to nanoconversion. Peels from the fruit known by the common name "longkong," *Lansium domesticum*, contain a variety of metabolites, including organic substances (methyl ester and lanternic acid) and terpenoids (triterpene glycosides and lanternic acid), which cap the reduction of metal ions to stable nanoparticles. Strong reducing potential is possessed by these minerals (Shankar et al. 2014). Similarly, polyphenols such as gallic and ellagic acids, as well as condensed and hydrolyzable tannins, are abundant in pomegranate peels (Goudarzi et al. 2016). Mango peels contain polysaccharides, lignins, flavonoids, hemicelluloses, and pectins (Yang and Li 2013); orange peels contain limonene, citral, neohesperidin, naringin, rutin, rhamnose, eriocitrin, and vitamin C (Rao et al. 2017). Other peels include papayas, which are rich in metabolites like chymopapain and papain (Kokila et al. 2016). According to Bankar et al. (2010), banana peels are packed with polymers such as lignins, hemicelluloses, and pectins, whereas the rind of watermelon is made up of minerals including cellulose, citrulline, pectins, proteins, and carotenoids (Patra et al. 2016).

2.4.2.2 Nanomaterial from Poultry and Slaughters Waste

Nobody in the world can dispute their love of eating; we all drool over delicious cuisine and have the option of choosing between plant-based and animal-based foods. Nonetheless, the goal is to raise awareness of the tonnes of food processing waste produced worldwide. This section validates the exorcism of poultry and abattoir waste. Both edible and inedible components are included in animal food (choice-based). Slaughterhouse and poultry refuse such as hatchery waste, sawdust, wood shavings, straws, peanuts, feathers, entrails and the organs of killed birds, as well as processing wastewater, bio-solids, blood, hides and bones, are produced and processed, and the majority of these materials are made of both organic and inorganic nutrients. The effects lessen the significance of water, soil, and air (Williams 2013).

2.4.2.3 Nanomaterials from Combustible Waste

In its literal sense, combustion is the phenomenon in which light and heat are released by burning caused by the reaction of a reductant and an oxidant. Combustible materials are characterised as those that can burn, and similarly, combustible trash is described. Based on where it came from, combustible waste can be any type. There are two different kinds of combustible waste: (a) natural (NCW) and (b) synthetic (SCW). Fallen leaves, wastepaper, wood, bagasse, and other materials are included in NCW, whereas diapers, plastic bottles, cigarette butts, ceramics, porcelain, rubber, leather, clothing, and other materials are included in SCW. Many researchers have worked to turn specific flammable trash into products with added value, including nanoparticles.

The production of graphene using inexpensive progenitor sources such as bagasse, wood, leaves, fruit waste, bones, cow dung, and newspapers is described by (Akhavan et al. 2014) in their work. They tried to overcome the drawback of utilising especially pure graphite as the raw material for production. They have selected sources with a good carbon content. Their synthesised graphene sheets were exclusive of the starting material, and their purity was maintained in a manner akin to that of the graphene sheets produced using Hummers' approach.

Another team of researchers created graphene quantum dots using waste neem leaves as a carbon source. The dried neem leaves were subjected to pyrolysis, yielding a black powder, and subsequently ground in a ball mill to produce a fine powder. After diluting the precursor with water, the mixture was poured through a filter to separate the larger particles and the unreacted carbon. Strong fluorescence was displayed by the filtrate, making it resemble graphene quantum dots (Suryawanshi et al. 2014). Like this, Gong et al. (2014), Rajarao et al. (2014), Sharma et al. (2014) used pyrolysis to harness the potential of solid waste plastic, wood wool, tyre rubber, waste polypropylene, and silicon carbide NPs to produce graphene flakes, graphene crystals, and carbide nano-onions, respectively. The process underlying the production of nanoparticles (NPs) from agricultural waste is mostly unknown in research studies. In their study, Malhotra et al. (2014) examined the process by which rice bran, an agricultural waste, is biomineralized into gold nanoparticles. They discovered that the bioconversion of gold salt to gold metal is attributed to ferulic acid, the function molecule. They also kept an eye on how the bioconversion process changed in temperature and concluded that it was exothermic.

2.4.2.4 Nanomaterials from Pharmaceutical Waste

The combined result of unused or wasted chemicals, contaminated drugs, and expired medications or items is known as pharmaceutical waste. Since there hasn't been much research done in this field, pharmaceutical corporations may find it more appealing to fabricate nanomaterials as a means of recycling their leftover medications. Only the functional portion of the ingested pharmaceuticals is metabolised by our biological system, while the non-metabolized portion is excreted into the environment, according to Kadam et al. (2016). Lakes and other naturally occurring water sources are the last destination for excrement. According to their documentation, India's health centres produce almost 40,000 tonnes of garbage per day.

Jha and Prasad (2012) made it possible to recycle leftover medication to create metal (gold), oxide (zirconia oxide), and chalcogenide (cadmium sulphide) nanoparticles. Pharmaceutical businesses, who have adopted green protocols to achieve nanosynthesis, may find this suggestion beneficial in recycling trash. The quinolone group of the antibiotic norfloxacin and the combination of tinidazole were used to create the nanomaterials. They used 4 months-old Norflox-TZ that had been first dissolved in sterile distilled water to prepare the reaction broth. After adding more HCl to completely dissolve the tablet, the mixture was heated to 40 °C over a steam bath. For the creation of NPs, the prepared broth was used. They discussed how tinidazole and norfloxacin's chemical structures support redox reactions.

2.5 Water Purification and Remediation

Industrial and agricultural operation leads to water contamination that negatively impacts quality of life and environment (Reddy et al. 2015). Harmful industrial wastewater contaminants removal is big threat for the society. Metal ions are the most hazardous inorganic pollutants found in water and wastewater. Organic contaminants include phenols, polychlorinated biphenyls and dyes (Pan and Kurumada 2008; Zhang et al. 2015). A significant pollutant called chromium (Cr) while its existence in two states oxidation states C(III) and Cr(IV). With Cr(IV) being more cancerous and toxic in nature (Costa 2003; Wu et al. 2013). The extensive manufacturing and use of Bisphenol A propane (BPA) in industry endangers human health and aquatic ecosystems (Im and Löffler 2016; Yang et al. 2016). To counteract these effects, modern water purification techniques such as filtration, chemical treatment, and biological processes are critical to maintaining freshwater supplies and protecting human health.

2.5.1 Nano Filtration and Biofiltration Technologies

Biofiltration is a water treatment method that involves passing water through a filter containing activated charcoal and biologically active sand to remove contaminants.

Conventional water treatment processes, such as precipitation, coagulation, flocculation, and sand filtration, successfully remove internal poisons by physically removing cells, but are ineffective for extracellular toxins resulting from algal lysis (Acero et al. 2005).

Biofiltration was used to degrade microcystin-LR (MC-LR) by naturally occurring bacteria. The process involved filling a biofiltration column with dechlorinated water, packing material consisting of activated charcoal and biologically active sand, and bubbling air into the setup to facilitate the growth of aerobic bacteria and prevent the growth of anaerobic bacteria. The presence of naturally occurring bacteria was determined using heterotrophic plate counts, which showed a viable number of bacteria was present. Additionally, a 10-ppb MC-LR solution was left in the column for 8 days and then filtered out, with samples collected daily. The activated carbon used for biofiltration was virgin (Eleuterio and Batista 2010), and the amount of MC-LR adsorbed to it was not differentiated from the amount that was biodegraded (Eke et al. 2018).

Nanofiltration is a membrane filtration method that uses membranes with fewer holes than ultrafiltration to extract microscopic particles and molecules from water. Nanofiltration membranes were examined for their ability to remove algal toxins, particularly microcystin-LR (MC-LR), from water at basic pH levels (Eke et al. 2018).

Tannic acid (TA) is a naturally occurring chemical compound found in aqueous solutions. According to reports, it presents a serious threat to aquatic life. Disinfecting water with TA may create toxic byproducts that disrupt internal operations and pose health hazards (Jung et al. 2015). Study by Beck et al., investigated the purification of TA using various carbon-based materials (PAN-based and lignin-based CNFs). Lignin-based CNFs had a greater clearance capacity for TA (220 mg g^{-1}) than PAN-based CNFs (12 m^2 g^{-1}). The elimination capacity was 13.4 times that of the other substance. CNFs without prospective purification materials may be considered as possible purification materials (Beck et al. 2017).

Carbon nanofibers (CNFs) are a potential material for water purification because of their simplicity of manufacturing, chemical resilience, low cost, and environmental friendliness. CNFs, with their one-dimensional structure, high aspect ratio, and mechanical flexibility, show great potential for water filtration. Their hexagonal carbon configuration enables bonding with four others, resulting in a high aspect ratio (Valenkov et al. 2011).

2.5.2 *Photocatalytic Degradation of Contaminants*

Photo catalysis is a green water treatment process that uses light to turn hazardous contaminants into nontoxic ones (Ravelli et al. 2009). The conversion of photon energy in chemical energy takes place. It employs photon energy to create a transitory state, releasing electron–hole pairs and creating molecules in the form of products (Haw et al. 2016). Photocatalysis is a low-impact technology that effectively

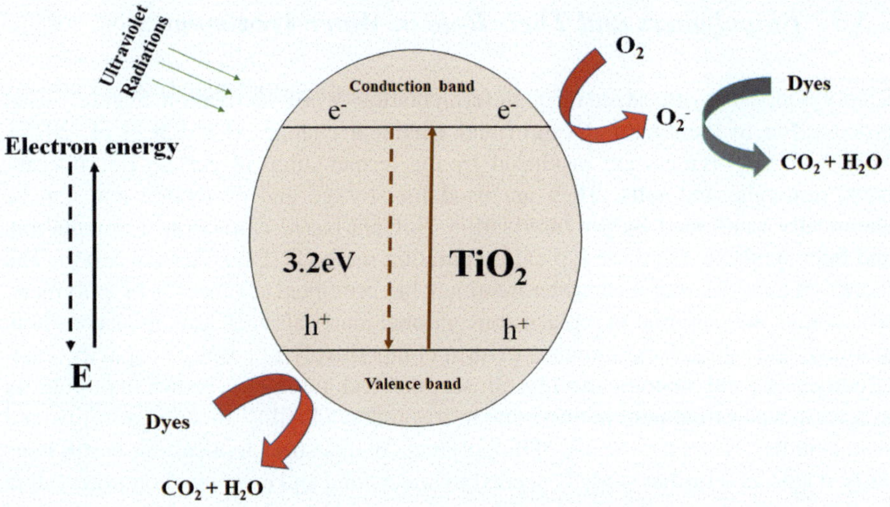

Fig. 2.1 Reaction mechanism schematics for photocatalysis by TiO_2

removes organic contaminants and non-biodegradable chemicals from wastewater. This approach successfully removes organic pollutants from water and wastewater, including herbicides, humic acid, insecticides, medicines, and halophenols (Alem et al. 2009; Diaz-Angulo et al. 2020; Habibpanah et al. 2011; Zeltner and Tompkins 2005).

TiO_2-based hetero structured materials use photocatalytic capabilities to effectively remove contaminants from wastewater as shown in (Fig 2.1). Examples are TiO_2/Au-WO_3 and TiO_2/g-C_3N_4, which have shown considerable improvements in organic pollutant degradation. These materials have many channels for mass and electron transport, which increases their potential to adsorb and enrich contaminants. They also aid in the movement of photo excited charge carriers, resulting in effective pollutant breakdown in wastewater. As a result, TiO_2-based hetero structured materials have potential in environmental cleanup operations (Madkhali et al. 2023).

Various materials have been employed as catalysts or adsorbents for the removal of pollutants, such as nanoscale zero-valent iron (nZVI) (Pang et al. 2018), MXene (Gu et al. 2018), layered double hydroxide (LDH) (Zhang et al. 2020) and metal–organic frameworks (MOFs), etc. there are some limitation which can be shown by such materials for removing pollutants from aqeous environment, they can be aggregated in environment and high synthesis cost and low solubility.

2.5.3 Biopolymers and Their Role in Water Treatment

Biopolymers are synthesized through fermentation or polymerization, a widely used process due to their natural origin and chemical changes (Cardon et al. 2017). Microbial biopolymers are produced by the fermentation of certain microorganisms, minerals, and salts. They are used for storage and protection and can be chemically reinforced by polymerization. Solvent-based extraction, fermentation, and bulk synthesis are three typical preparation methods (Díez-Pascual 2019). The increased concern over wastewater treatment has prompted advances in biopolymers, which have the potential to replace conventional materials such as activated carbon, alumina, and silica. Biopolymers perform competitively in terms of capacity, cost-effectiveness, and biocompatibility. Biosorption technology has grown in popularity as a wastewater treatment method due to its biodegradability, biocompatibility, and sustainability. However, the use of biopolymers in industrial applications is still in its early stages, and further study is needed to understand and examine their potential in drinking and wastewater treatment (Elgarahy et al. 2023). There are different types of biopolymers used for the wastewater treatment, such as chitosan, cellulose, alginate, pectin, gelatin, etc.

Biopolymers serve an important role in a variety of applications, including wastewater treatment. Biopolymers are natural polymers generated from renewable natural sources or biosynthesized by living organisms. Exopolysaccharides, polyhydroxyalkanoates, and polypeptides are examples of compounds with a wide range of physical, chemical, and biological characteristics. These biopolymers are non-toxic, ecologically benign, and biodegradable, making them attractive options for water treatment (Awe et al. 2017).

In wastewater treatment, biopolymers are commonly used as coagulants, flocculants, adsorbents, membranes, and photocatalytic agents to remove heavy metals, dyes, natural organic matter, and other water contaminants. They have outstanding mechanical, physical, chemical, and biological properties, including renewability, environmental compatibility, antibacterial activity, biocompatibility, and biodegradability (Khademian et al. 2020). Their structure contains reactive functional groups such as amide, carbonyl, carboxyl, and hydroxyl, which makes them suitable for wastewater treatment.

Biopolymers may engineer to have certain qualities, increasing their effectiveness in meeting various wastewater treatment demands. Their sustainable manufacture helps to minimize energy usage and environmental effect, which aligns with global sustainability objectives. Furthermore, biopolymers promote biological compatibility and may be used with other treatment methods to provide diverse solutions to wastewater concerns. Scaling up biopolymer production not only boosts economic growth, but it also stimulates community participation in sustainable practices, which leads to better water quality and ecosystem health.

2.6 Energy Harvesting and Storage

Nanotechnology has transformed energy harvesting by developing novel methods for catching, storing, and transforming energy. Using materials smaller than 100 nm, it has increased the efficiency of solar energy systems, resulting in improved solar collectors, fuel cells, photocatalysis, and solar photovoltaic technologies. Nanomaterials can improve incident radiation by nine times and solar collector efficiency by ten percent when compared to traditional flat plate collectors. Nanowires and carbon nanotubes have showed promise in enhancing solar energy conversion and storage due to their distinct features (Abdin et al. 2013).

2.6.1 Nanostructured Materials for Energy Conversion and Storage

Advanced material development has evolved away from single materials and toward integrating several components to form hybrid composites. This method improves matrices' characteristics by introducing reinforcing elements such as fibers. Recently, interest has switched to nanoscale materials, notably nanoparticle-based composites, which provide considerable property gains (Ji et al. 2018; Mallakpour and Nezamzadeh Ezhieh 2018; Sahoo et al. 2018; Sharma et al. 2019). These composites have been applied in engineering and healthcare, indicating their promise in a variety of sectors.

Research conducted by Andrew et al., has focused on the synthesis and characterization of magnetoelectric multi-ferroic materials, which are uncommon single-phase ferromagnetic and ferroelectric materials. This resulted in the invention of multiferroic composite materials. However, the features of magnetoelectric materials have not been completely expressed, highlighting the need for more study to optimize the material's capabilities. Understanding the structure–property connection at the nanoscale is critical for improving the qualities and potential applications of nanoparticle-based composite materials (Andrew et al. 2014).

Study by Kumar et al., highlights the importance of nanostructured materials as an effective alternative energy storage option, which include metal-based, inorganic, coordination polymers, and carbon nanomaterials. They cover how nanostructured materials are used in Li-ion batteries, supercapacitors, and solar cells, as well as the benefits and drawbacks of putting them into polymer composites. Recent research has focused on the synthesis and characterization of magnetoelectric multi-ferroic materials, which are uncommon single-phase ferromagnetic and ferroelectric materials. This resulted in the invention of multiferroic composite materials. However, the features of magnetoelectric materials have not been completely expressed, highlighting the need for more study to optimize the material's capabilities. Understanding the structure–property connection at the nanoscale is critical for improving the qualities and potential applications of nanoparticle-based composite materials (Kumar et al. 2017).

A Copper oxide and polyaniline (CuO/PANI) nanocomposite was created through in-situ polymerization, characterized using cyclic voltammetry and electrochemical impedance spectroscopy. The material showed a 75% capacitance retention rate and specific capacitance, significantly higher than pure CuO nanoparticles at 30% and 76 F/g, respectively (Gholivand et al. 2015).

Photovoltaic (PV) technology transforms solar radiation into electrical energy by exposing it to a semiconductor, which increases conductivity. PV technology is classified into two types: crystalline silicon-based cells and thin film technologies that use semiconductor materials such as cadmium telluride, amorphous silicon, and copper indium gallium dieline. Both kinds are available on the market (Sagadevan 2013).

Electrocatalytic technologies are crucial for long-term energy sustainability because they enable hydrogen production through water splitting. This approach allows hydrogen-based fuels to be combined with other renewable sources including wind, solar, and biomass while maintaining high energy conversion efficiency (Ouchi and Henzie 2017).

Advanced materials, particularly nanostructured and composite materials, have transformed a variety of sectors, including energy production, conversion, and storage. These materials provide qualities such as high thermal conductivity, strength-to-weight ratio, and electrical conductivity, which improve energy conversion tactics. Supercapacitors and solar cells have seen performance improvements. However, difficulties such as high prices and unavailability persist. Ruthenium and palladium are being researched as possibilities for cost-effective and long-lasting solutions (Ibrahim et al. 2020).

2.6.2 Bio-inspired Energy Systems

The development of sophisticated materials, notably nanostructured and composite materials, has transformed several sectors, including energy. These materials possess features that conventional materials lack, making them critical for strategic breakthroughs in energy generation, conversion, and storage. Nanostructured materials, such as gold, silver, platinum, and silicon, have considerably improved energy conversion and storage technologies due to their outstanding capabilities. Advances in nanostructured materials have enhanced the performance of supercapacitors and solar cells, increasing specific capacitance and energy conversion efficiency (Jamei and Vrcelj 2021). In the search for efficient hydrogen fuel production via electrocatalytic water splitting, nanostructured materials like platinum encounter hurdles such as high cost and scarcity. However, current research is looking into alternatives such as ruthenium (Ru) and palladium (Pd), which show promise for cost-effective and long-lasting solutions. Continued work in this subject has the potential to reveal novel materials with improved capabilities for energy generation, conversion, and storage, resulting in additional improvements in sustainable energy technology.

Advanced materials, particularly nanostructured and composite materials, have transformed a variety of sectors, including energy production, conversion, and storage. These materials provide qualities such as high thermal conductivity, strength-to-weight ratio, and electrical conductivity, which improve energy conversion tactics. Supercapacitors and solar cells have seen performance improvements. However, difficulties such as high prices and unavailability persist. Ruthenium and palladium are being researched as possibilities for cost-effective and long-lasting solutions (Regassa et al. 2023).

Bio-inspired structural engineering design faces severe problems in developing sustainable energy conversion technologies due to a variety of design, execution, and material constraints. The kingfisher beak-inspired bullet train nose design decreases noise and energy consumption, making it more environmentally friendly (Boaretto et al. 2021). Bio-inspired design ideas for sustainable energy structures are motivated by the need to reduce environmental effect while increasing energy production and consumption. Researchers want to create bio-inspired wind turbine designs (Guchhait and Sarkar 2023), solar panels (Zähr et al. 2010), and energy storage systems by pulling inspiration from natural materials such as Nenuphar (Water Lily) and Sunflower. Bio-inspired design has several opportunities, including increased energy efficiency, infinite renewable energy sources, and nature as a source of creativity. However, issues like as complexity, biocompatibility, material sustainability, and dynamic adaptability persist.

2.6.3 Integration of Nano-biotech in Renewable Energy Infrastructures

Though environmentally friendly, renewable energy sources including solar, wind, geothermal, ocean thermal, and tidal are being used at a much lower rate than energy derived from nonrenewable sources. example, in the USA in 2005, the amount of energy produced from renewable sources (such as biomass and hydroelectric power) was only 7%; on the other hand, 8 and 85% of the energy came from nuclear power and fossil fuels, respectively (Serrano et al. 2009). Incorporating nanobiotechnology into renewable energy infrastructures can lead to considerable reductions in energy consumption and greenhouse gas emissions while also offering promising advances in clean energy production and storage (Alvarez and Cervantes 2011). In order to improve the effectiveness and environmental friendliness of renewable energy technologies including hydro, wind, solar, geothermal, and biomass, this integration makes use of nanoparticles (NP) and nanomaterials (Serrano et al. 2009). By using the special qualities of nanomaterials, such as their larger surface area and higher reactivity, scientists want to raise the effectiveness and performance of renewable energy sources, such as solar panels.

Furthermore, it was observed that multi-layer graphene nanoribbon (NP) networks can quickly respond and recover to detect hydrogen at parts-per-million concentration levels at room temperature (Johnson et al. 2010). Furthermore, TiO_2 nanoparticles' photocatalytic activity since better water treatment and hydrogen production result from the use of TiO_2 NP because it increases the surface area available for light adsorption. As a result, the photocatalytic activity described in reactions can be significantly enhanced by the large carbon nanotube network (Ni et al. 2007).

Utilization of nanomaterials in a variety of processes, including adsorption, photocatalysis, and filtration, is the mechanism of integration (Alvarez and Cervantes 2011). For example, azo dyes and cadmium have been adsorbed using carbon nanoparticles and multi-walled carbon nanotubes, respectively (Mishra et al. 2010). Fe-carbon nanotube/TiO_2 composites have shown improved photocatalytic activity for the breakdown of contaminants such as (Zhao and Zhu 2006). Moreover, effective benzene, toluene, ethylbenzene, and p-xylene (BTEX) adsorption has been demonstrated by carbon nanotubes oxidized with NaOCl (Su et al. 2010).

The applications of nanobiotechnology in renewable energy infrastructures are extensive and include improved hydrogen detection, enhanced water treatment, and increased efficiency of renewable energy technologies. Anyway still, there are limitations to consider, such as the cost associated with the production and utilization of nanomaterials, as well as the potential environmental and health impacts of nanoparticles (Alvarez and Cervantes 2011; Johnson et al. 2010). Furthermore, the long-term fate and transport of nanoparticles in the environment are not fully understood, and their ecological and human health impacts require further research and assessment.

So, Nanobiotechnology has a wide range of applications in renewable energy infrastructures, such as better water treatment, better hydrogen detection and higher renewable energy technology efficiency. But there are restrictions to consider, like the price of producing and using nanomaterials and the possible negative effects of nanoparticles on the environment and human health. Furthermore, more investigation and evaluation are needed to completely understand the long-term destiny and transit of nanoparticles in the environment, as well as the implications for human and ecological health. Therefore, by using nanoparticles in a variety of processes, the incorporation of nanobiotechnology into renewable energy infrastructures presents a substantial opportunity to advance clean energy production and storage. To guarantee this integration's sustainable and responsible application, it is crucial to consider its applications and restrictions.

2.7 Agricultural Enhancement Through Nano-biotech

Nanotechnology is an emerging field used in various sectors that including agriculture. Nanotechnology helps in improving the plant yield as well as to improve soil quality as well. Through the advantage of distinct characteristics of nanomaterials scientists are investigating innovative solutions like fertilizers, pesticides, and their delivery system so as to get plants their nutrition in adequate amount.

2.7.1 Nanopesticides and Nanofertilizers for Crop Improvement

The application of nanofertilizers and nanopesticides has enabled regions with adverse climate elevations to compete with nations with ideal environmental conditions in food production, therefore revolutionizing crop growth. The extensive increase in agricultural output has been made possible by the application of agrochemicals such as pesticides and fertilizers. Since agrochemicals are supposed to be more efficient, pose less harm to non-target organisms, and prevent target organisms from developing resistance, the application of nanotechnology in agrochemicals has grown over the past ten years. It is necessary to do a thorough risk evaluation of these products due to worries over the possible genotoxic effects of nano pesticides, particularly in human cells. 2,4-dichlorophenoxyacetic acid (2,4-D) and Dichlorodiphenyltrichloroethane (DTT) were the two principal representatives of the first generation of insecticides. All parties benefitted from their successful introduction into the world market, and the second generation of pesticides of which urease, organophosphates, and carbamates were the most notable example was created between 1945 and 1955 (Paz-Trejo et al. 2023).

2.7.2 Nanopesticides

Nanoparticles are at such small level of organization and can be useful for plant condition and soil health monitoring sensors. Most of the Nano pesticides do not have size exactly about the range of 1–100 nm but this 'Nano' term used because some of its content remains in the nano range (Kah and Hofmann 2014). Based on the incredible possibilities that nanotechnology offers, metallic nanoparticles such as silica, titanium oxide, or aluminum oxide are making their way into a variety of applications, including building materials. Research findings indicate that these minuscule particles have remarkable properties that facilitate their smooth integration into biological systems, resistance to external stimuli, ability to modify their size, and ability to communicate with biological entities. This unveils a new way to interpret their behavior and their effects, illuminating the complex relationship between nanotechnology and biological systems (Kumar and Dhawan 2013).

In commercial agriculture, pesticide application is a continuous process that entails screening a large number of pesticides each year (Resh and Cardé 2009). Only 0.1% of pesticides used are effective against the target pests; the rest 99.9% harm the environment, threatening human health and food chains (Carriger et al. 2006). Pesticide resistance in weeds, insects, and illnesses is another consequence of pesticide use on non-target species (Rai and Ingle 2012). Though there are bio pesticides which overcomes such limitations but their use is halted due to slow efficacy against environment dependent pesticides. In addition to that us of nano pesticides would be a better option for the sustainable release of active ingredients and has potential to reduce these limitations by other conventional pesticide (Chhipa 2017). The use of nanopesticides overcomes the implementations of synthetic chemicals and their harmful effect on the environment (Kah et al. 2019). Implementation of nanopesticides would be cost effective and conserves water as well as energy, resulting in higher yield of crops and less labor cost as well (Ragaei and Sabry 2014).

Study by Suresh Kumar et al., there is significantly increase in utilization of permethrin by nanoformulations *Aedes aegypti* compared to conventional methods, in addition harmful effects on plants and sol bacteria were reduced with help of nanoparticles implications (Suresh Kumar et al. 2013). polyethylene glycol based Nanoformulations on carbofuran and acephate has shown lower toxicity on the non-targeted organisms when compared with conventional acephate. The efficiency is due to slower active ingredient release not by nano formulations (Pankaj et al. 2012; Pradhan et al. 2013a).

On the other hand, study demonstrated that nanopesticides can cause Geno toxicity in the environment. Since the nanotechnology in new in its phase it requires toxicological research focuses on acute effects, aquatic repercussions, and environmental impact. Some of the examples like increase in decrease in fertility and increase mortality of daphnia (Oberdörster et al. 2006). Reduction in activity of glutathione enzyme by Titanium dioxide (TiO_2) in rainbow trout, simultaneously affects the phytoplankton combining with sunlight (Federici et al. 2007; Miller et al. 2012).

2.7.3 *Nanofertilizers*

Nanofertilizers increase crop quality and yield, enhance nutrient efficiency, support sustainable agriculture, and reduce production costs. Because crop varieties have evolved to respond to fertilizer, there has been a significant increase in the usage of chemical fertilizers in agriculture. Nevertheless, the inefficiency of these fertilizers, which is reflected in losses due to volatilization and leaching, boosts production costs and pollutes the environment (FAO 2017). Most of the nitrogen gets lost in environment and affect the environment (DeRosa et al. 2010).

According to the study, growth rate (32%) and seed yield (20%) of the Soya bean (*Glycin max* L.) after applications of phosphate nanofertilizers compared with convention fertilizer (Liu and Lal 2014). In addition to that, nanocomposites of hydroxyapatite and urea controlled release of nitrogen and lower NH_3 volatilization continued availability of phosphate following 4 weeks of incubation (Giroto et al. 2017). Usually the nanofertilizers are slow releasing the active ingredient and working for better efficacy of plant yield but it should release active ingredient whenever is necessary and wherever required. Researchers are finding the way to achieve this by one of the method based on plant signaling which reduced the leaching of the fertilizer and conversion of it into gaseous forms (DeRosa et al. 2010). It stimulates ethylene production in plant roots and acidification of rhizosphere which help to sustained release of phosphorous and potassium in the soil (Rop et al. 2019).

Materials with nanostructures, such as zeolite, hydroxyapatite, chitosan, clay minerals, and poly acrylic acid, are used to make fertilizers for soil and/or foliar application. The high surface area of hydroxyapatite and its strong interactions with urea are the reason for the slow release of nitrogen from urea (Kottegoda et al. 2011). Outlined the use of superparamagnetic iron oxide to raise the chlorophyll content of soybeans. Applying 0.5 g L^{-1} solution of Fe nanoparticles to the leaves of cowpea (*Vigna unguiculata* (L. Walp.)) resulted in a 10% increase in chlorophyll content, which was evaluated in comparison to the same solution of conventional forms of Fe (Karami et al. 2014).

Thus increasing global demand of food is achieved by increase in crop yield by using biofertilizers, pesticides, synthetic fertilizers, etc. ether the production of such fertilizers has been complicated process or excess release in the environment caused harmful effect. For such time Nano particles (NPs) make effective way for increasing the crop yield.

2.7.4 Genetic Engineering for Stress-Resistant Plant Varieties

The next major problem with which the plants growth is concerned is pathogenic fungi, bacteria, virus causing diseases to the crops. Conventional methods of breeding do not stand out for presenting the resistant genes in gene pool. So genetic engineering mainly focuses on incorporation of genes resistant against several pathogens. For examples, resistance against fungi and bacteria Genes such as chitinase osmotin, glucanase, and defensin are being incorporated into diverse horticulture crops worldwide (Antony Ceasar and Ignacimuthu 2012).

Crops are designed and genetically modified to improve absorption and nutritional value. Plant breeding techniques can boost the micronutrient content of specific cereals, legumes, and tuber crops. As a policy, bio fortification via breeding has several advantages: it contributes to the production of staple foods that are consumed daily while also aggressively catering to the poor. Again, these evolved types may be employed throughout time, cutting costs, and making it more sustainable.

Abiotic stresses such as heat, dryness, and salt have a substantial influence on horticultural crop yield and production. Conventional plant breeding has failed to solve these difficulties because several genes influence qualities at a quantitative trait locus (Parmar et al. 2017). Genetic manipulation of genes encoding stress-protective chemicals in plants and creatures can strengthen fragile genotypes against stresses. This allows researchers to tap into shared stress-response systems across species, enabling the production of more resilient crops. This method not only strengthens agricultural systems but also has the potential to solve food security challenges in the face of changing climates and environmental stressors (Parmar et al. 2017).

Bt. (*Cry*) gene isolated from *Bacillus thuringiensis* (a soil bacteria), it has shown effective control against the lepidopteran insects in number of crops. First insect resistance was found in Tomato using *Bt.* gene in 1987. Genetically modified *Bt.* tomato showcased immunity towards *Tobacco budworm* and *Tobacco cutworm* (Fischhoff et al. 1987). CRISPR/Cas9 technology has revolutionized genome editing in plants, enabling the creation of resistance to various infections (Ali et al. 2015; Baltes et al. 2015). As there is no necessary transgene in the gene editing of plants hence they don't require stringent conditions for regulation and can be easily accepted by consumers (Jones 2015).

2.7.5 *Precision Agriculture Leveraging Nano-biotech Tools*

The science of enhancing agricultural yields and supporting management choices using advanced sensor and analytic technologies is known as precision agriculture, or PA (Fig. 2.2). The globe has embraced PA, a novel idea that promises to boost output, cut down on labor costs, and guarantee efficient control of irrigation and fertilizer systems. It can be achieved by nanotechnology through various processes includes:

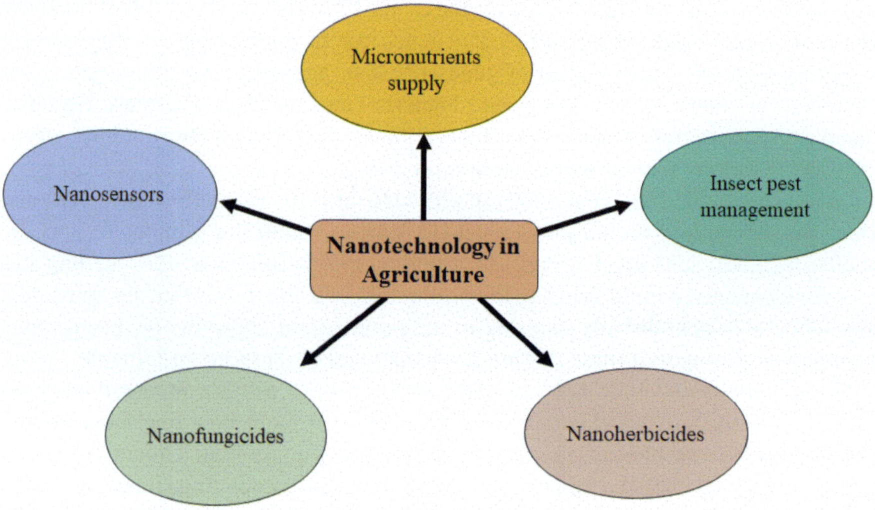

Fig. 2.2 Schematic representation of nanotechnology in agriculture

2.7.5.1 Insect Pest Management

Synthetic pesticides have been changing the way of agriculture production and its nature. It also contributing to the resistance of pest against the insecticides. Nanoparticles have been used in modern agriculture. Nanoparticles coated with polyethylene glycol enhanced insecticidal activity of garlic essential oil against *Tribolium castaneum* (red flour beetle) (Goswami et al. 2010). Nanostructured alumina has shown great insecticidal activity against *Sitophilus oryzae* and *Rhyzoperttha dominica* have proven to be more cheap and reliable than any other commercial insecticides (Stadler et al. 2010).

2.7.5.2 Micronutrient Supply

Secondary nutrients such as sulfur, calcium, and magnesium are required in significant quantities, but micronutrients such as iron, boron, copper, manganese, zinc, molybdenum, nickel, chlorine, and molybdenum are required in tiny numbers for crop development, similar to nitrogen, phosphorus, and potassium (Elemike et al. 2019). The availability of micronutrients for plants can be enhanced by nanotechnology. Spraying of nanonutrient composition on plants or in order to improve root absorption and soil health (Peteu et al. 2010). According to the study by V. Kumar et al., MWCNT composites has shown sustained released of zinc (Zn) and promotes the germination of onion seeds since it has implicated more the telophases in onion root tips mitotic cell division (Kumar et al. 2018).

Chitosan nanoparticles loaded with 1-napthlyacetic acid (a growth hormone) tested at different pH and temperature for slow release. Results were significantly drawn that such formulations make slow and sustained released of agrochemicals like hormones (Tao et al. 2012). Manganese nanoparticles have been shown to improve mung bean (*Vigna radiata*) growth and photosynthesis (Pradhan et al. 2013).

2.7.5.3 Nanofungicides

A commercially available fungicide Trifloxystrobin 25% + Tebuconazole 50% was converted to a nanoform and causes hyphal abnormality, hyphal lysis and abnormality in sclerotial formation in *Macrophomina phaeolina.* It is effective as compared to the conventional fungicides or the fungicide in bulk (Kumar et al. 2016). Maize treated with nanosilica (20–40 nm) was more resistant to phytopathogens such as *Fusarium oxysporum* and *Aspergillus niger* than bulk silica. Maize leaf extracts included more phenolic compounds and less stress-responsive enzymes that protect against fungi. Maize had increased disease resistance and expression of total phenols, phenylalanine ammonia lyase, peroxidase, and polyphenol oxidase (Suriyaprabha et al. 2014).

Study revealed that the silica nanoparticle size and shape also affect the killing effect on the organism, particles with triangular shape have higher effect of killing fungi than spherical or rod shaped particles. Zinc (Zn) and Sulphur (S) nanoparticles has shown great fungicidal activity against *Aspergillus niger* and *Fusarium oxysporum* (Patra et al. 2012; Roy Choudhury et al. 2011).

2.7.5.4 Nanosensors

Crops to be grown without disease, and increased productivity has become demand of ever-increasing human population. To support such actions advancement in technology is needed in agriculture. Various biotic and abiotic stress on plant affects their growth and may jeopardize it. Hence detection of stress condition even before their occurrence may help to increase the crop yield as well as to prevent from diseases. Nanosensors can detect environmental stress condition even at a nanoscale. Nanosensors helps to determine the humidity of soil, nutrients requirement, pesticide residues and plant pathogen.

Agricultural output has increased by smart sensors which gives reliable information for farmers to improve precision agriculture. Nanosensors are the devices that uses water, nutrients and agrochemical for detection of the environmental condition (Rai et al. 2012). Bionanosensors provide quick and precise detection of bacteria and viruses, improving food safety for customers (Otles and Yalcin 2010). Zn–Se quantum dot-immobilized acetylcholinesterase was used to detect organophosphate pesticides on a graphene-chitosan nanocomposite modified electrode. The biosensor detects organophosphate pesticides by using methyl parathion as an enzyme inhibitor, with a detection limit of 0.2 $nmol/dm^3$ (Dong et al. 2013).

2.7.5.5 Nanoherbicides

Weeds are among the largest jeopardy to the agriculture, consumes the nutrients which generally can be utilized by the crop plants for their growth. Thus, to implicate proper methods for removal of the weed has become an essential step. To remove weeds from the crop field various synthetic herbicides have been used but thus they can accumulate in the environment and causes soil pollution. Conversely, Nanoherbicides plays a crucial role in removing the weeds from crops in efficient way without disturbing the environment (Pérez-de-Luque and Rubiales 2009).

Hybrid materials have ability to incorporate advantage of two or more materials into a single structure (Gao et al. 2020). These multifunctional nanomaterials exhibit different size, properties, morphologies and chemical composition (Aich et al. 2014; Ananikov 2019). Apart from this hybrid Nanoherbicides possesses stimuli-responsive attributes, traceability, and better targetability (Chen et al. 2018; Li et al. 2021; Zhao et al. 2020). After passing through various biological barriers Nanoherbicides enters into the plant cell it causes the electrolyte leakage in the target species because of cell disruption.

2.8 Ethical, Legal, and Social Implications (ELSI)

Nanotechnology, heralded as a transformative technology for the twenty-first century, has substantial benefits but requires cautious implementation owing to its largely unknown adverse consequences. Despite its widespread uses in a variety of industries, including agriculture, public awareness of nanotechnology remains low, underlining the need for more education and knowledge of its possible consequences. Many people don't know about the nanotechnology and its implication in field of agriculture.

In the lack of perfect knowledge, people frequently rely on heuristics such as trust to assess the dangers and advantages of new technology (Coles and Frewer 2013). As a result, when people have more faith in institutions, researchers, and business, particularly based on legislative history, they are more likely to believe claims about the safety of nanotechnologies, including those used in agriculture. Furthermore, public opinion of nanotechnology is impacted by how the media depicts technological challenges.

Combining all the factors that influences the perception and awareness among the people about nanotechnology in agriculture there are some ethical issues which raised. Ethical issues arise regarding nanotechnology in agriculture, including potential risks to human health and the environment, labeling requirements, funding, public awareness, and regulation. If not addressed, negative perception and reluctance may arise, like GM food stuffs in the European Union.

2.8.1 Regulatory Frameworks Governing Nano-biotech Applications

Ethical discussions in this area can be structured around equity, privacy, security, environment, and metaphysical considerations, enabling thorough analysis of the multifaceted dimensions.

2.8.1.1 Equity

The worldwide inequality between rich and developing nations is a major concern, and nanotechnology offers promising answers for poverty reduction, energy, water, health, and biodiversity protection. However, significant implementation costs frequently impede wider adoption, worsened by the wealth divide. Political will and investment capability are critical, but debt complicates things. Nanotechnology advancements provide greener energy, safer medicine delivery, and improved healthcare, but they require equal access and deliberate investments.

2.8.1.2 Security

The proliferation of surveillance technology, such as "invisible" microphones and cameras, has raised worries about privacy violations and security dangers. These technologies improve law enforcement and national security monitoring, but they also open up chances for malevolent actors to use them, perhaps leading to bio-terrorism (Rathjen and Shahbodaghi 2021), techno-terrorism, or nano terrorism (Pandey 2020). To overcome these dangers, both the government and the private sector must provide open control and monitoring of military nanotechnology research in both defensive and offensive capacities. Monitoring, ownership, and control of invisible things raises ethical and legal concerns. Interdisciplinary collaboration among politicians, technologists, ethicists, and legal specialists is critical.

2.8.1.3 Environment

The development of nano photovoltaic panels presents a potential alternative to traditional power generation, but it also raises worries about the environmental and health consequences of creating nanomaterials such as fullerenes and carbon nanotubes. The disposal of these materials after their mechanical life cycle raises concerns about their possible impact on ecosystems and human health. The use of nanoparticles into medical equipment increases safety and has long-term ramifications. Risk assessment for nanomaterials smaller than 100 nm is critical because they can enter biological systems via inhalation, skin absorption, or biological barriers (Jacobstein 2006; Stark 2011). Existing rules may not effectively address the unique properties and possible risks of nanoscale particles.

2.8.1.4 Privacy

Advances in monitoring technology have generated worries about privacy invasion. The use of "invisible" microphones and cameras raises the potential of unauthorized monitoring. Wireless monitoring in healthcare improves mobility while also introducing weaknesses that might lead to privacy breaches. The integration of wireless communication and physiological sensors in e-health systems opens up new possibilities, but also necessitates careful consideration of privacy and security concerns. Balancing advantages and privacy is critical for ethical and responsible technology adoption.

2.8.1.5 Metaphysical

The inclusion of implants into the human body raises ethical concerns about human values and autonomy, particularly for nanobots that can make independent decisions and communicate. This raises concerns of control and responsibility. Societal mistrust of these breakthroughs highlights the need for a nuanced approach to nano ethics. Study emphasizes the moral implications of rising human capabilities while warning against excessive technological advancement (Bensaude-Vincent 2010). A proactive engagement between research and society is necessary, with interdisciplinary collaboration and the precautionary principle driving the convergence of bio-nano-info-neuro technologies. Ethicbots, an Europen Union initiative, seeks to address techno-ethical issues, whereas Rodotà and Capurro emphasize the need of maintaining individual liberty and independence when examining the ethical implications of ICT implants (Rodotà and Capurro 2005).

2.8.2 Ethical Considerations in the Development and Deployment of Nano-biotech

Ethical considerations surrounding the advancement and deployment of nano-biotechnology are multifaceted and require careful examination. Key among these concerns is the imperative to prioritize safety, necessitating rigorous risk assessment procedures and robust regulatory oversight to safeguard human health and environmental integrity. Moreover, issues of equity must be addressed to ensure that the benefits of nano-biotech are accessible to all, irrespective of socioeconomic status, thus mitigating the risk of exacerbating existing disparities. Additionally, the blurring of boundaries between human and machine in nano-biotech applications prompts reflection on autonomy and privacy, underscoring the importance of upholding individual rights and informed consent. Environmental impact assessments are critical to understand and mitigate potential ecological consequences, particularly regarding the release of nanoparticles. Furthermore, the dual-use nature of nano-biotech necessitates measures to prevent misuse and promote responsible innovation, while transparency and accountability mechanisms are vital for fostering public trust and engagement. Anticipating and addressing long-term societal and cultural implications are essential to ensure that nano -biotechnology contributes positively to human well-being and societal advancement.

2.9 Future Directions

2.9.1 Emerging Trends in Nano-biotech Synergy

2.9.1.1 Growth of Multifunctional Nanomaterials

Predicted increase in the development of nanomaterials with multiple functionalities, allowing for diverse applications in environmental remediation and sustainability.

2.9.1.2 Integration of Artificial Intelligence (AI)

Utilization of AI algorithms for designing nanobiotechnological solutions, enhancing efficiency and effectiveness in addressing environmental challenges.

2.9.1.3 Advancements in Bioremediation Techniques

Continued research into genetically engineered microorganisms and nanobiosensors for targeted and efficient bioremediation of contaminated environments.

2.9.1.4 Nanotechnology for Climate Change Mitigation

Exploration of nanomaterials for carbon capture and utilization, as well as novel approaches for enhancing renewable energy production and storage.

2.9.2 Potential for Scaling up Solutions

2.9.2.1 Industrial Adoption of Nano-biotech Solutions

Increased investment and adoption of nano-biotechnological solutions by industries for sustainable manufacturing processes and waste management.

2.9.2.2 Infrastructure Development

Integration of nano-biotech solutions into existing infrastructure for water treatment plants, air purification systems, and waste management facilities.

2.9.2.3 Global Collaboration and Knowledge Sharing

Establishment of international collaborations and platforms for sharing research findings, best practices, and technological innovations in nano-biotechnology for environmental sustainability.

2.9.3 Collaborative Partnerships and Interdisciplinary Approaches

2.9.3.1 Cross-Sector Collaboration

Partnerships between academia, industry, government agencies, and non-profit organizations to leverage expertise and resources for addressing complex environmental challenges.

2.9.3.2 Interdisciplinary Research Initiatives

Promotion of interdisciplinary research programs integrating nanotechnology, biotechnology, environmental science, engineering, and social sciences to develop holistic and sustainable solutions.

2.9.3.3 Public–Private Partnerships

Collaboration between public and private sectors to accelerate the translation of research findings into real-world applications and policy frameworks for environmental conservation and sustainability.

2.9.4 Policy Recommendations

2.9.4.1 Regulatory Frameworks for Nano-biotech Applications

Development of robust regulatory frameworks to ensure the safe and responsible use of nanomaterials and biotechnologies in environmental applications, while addressing potential risks and ethical concerns.

2.9.4.2 Incentivizing Sustainable Innovation

Implementation of incentives, grants, and funding schemes to encourage research and development in nano-biotechnology for environmental sustainability, prioritizing projects with high social and environmental impact.

2.9.4.3 Education and Outreach Initiatives

Integration of nano-biotechnology and environmental sustainability into educational curricula and public outreach programs to raise awareness, foster scientific literacy, and promote community engagement in environmental conservation efforts.

2.10 Conclusion

In conclusion, "Harmonizing Nature and Innovation: Sustainable Environmental Solutions through Nano-Biotech Synergy" underscores the critical importance of integrating nature-inspired innovation into our quest for environmental sustainability. The chapter has illuminated the transformative potential of nano-biotechnology in addressing a myriad of environmental challenges, from pollution abatement to resource conservation. By harnessing the synergistic power of nanomaterials and biotechnological processes, we stand poised to revolutionize the way we interact with and protect our natural world.

Through a nuanced exploration of emerging trends, scaling-up opportunities, collaborative partnerships, and interdisciplinary approaches, we have outlined a clear path forward for advancing the field of nano-biotech synergy. From industrial adoption to global collaboration, the opportunities for impactful change are vast and promising. However, this journey is not without its challenges. Regulatory frameworks must be strengthened, ethical considerations must be addressed, and equitable access to sustainable solutions must be ensured.

As we navigate these complexities, it is imperative that we remain steadfast in our commitment to responsible innovation and environmental stewardship. By prioritizing the development and deployment of nano-biotechnological solutions, supported by robust policies and collaborative efforts, we can forge a more sustainable future for generations to come. In essence, "Harmonizing Nature and Innovation" serves as a rallying cry for collective action, urging us to harness the ingenuity of human creativity while honoring the wisdom of the natural world. Through this harmonious integration, we have the opportunity to not only mitigate environmental degradation but also to cultivate a flourishing ecosystem where innovation thrives in tandem with nature's resilience. Let us seize this opportunity with determination, empathy, and a steadfast commitment to building a brighter, more sustainable tomorrow.

References

Abdel-Aziz MM, Emam TM, Elsherbiny EA (2020) Bioactivity of magnesium oxide nanoparticles synthesized from cell filtrate of endobacterium Burkholderia rinojensis against Fusarium oxysporum. Mater Sci Eng C 109:110617

Abdin Z, Alim MA, Saidur R, Islam MR, Rashmi W, Mekhilef S et al (2013) Solar energy harvesting with the application of nanotechnology. Renew Sustain Energy Rev 26:837–852

Acero JL, Rodriguez E, Meriluoto J (2005) Kinetics of reactions between chlorine and the cyanobacterial toxins microcystins. Water Res 39(8):1628–1638

Ahmed SF, Mofijur M, Rafa N, Chowdhury AT, Chowdhury S, Nahrin M et al (2022) Green approaches in synthesising nanomaterials for environmental nanobioremediation: technological advancements, applications, benefits and challenges. Environ Res 204:111967

Aich N, Plazas-Tuttle J, Lead JR, Saleh NB (2014) A critical review of nanohybrids: synthesis, applications and environmental implications. Environ Chem 11(6):609–623

Akhavan O, Bijanzad K, Mirsepah A (2014) Synthesis of graphene from natural and industrial carbonaceous wastes. RSC Adv 4(39):20441–20448

Alem A, Sarpoolaky H, Keshmiri M (2009) Titania ultrafiltration membrane: preparation, characterization and photocatalytic activity. J Eur Ceram Soc 29(4):629–635

Ali Z, Abulfaraj A, Idris A, Ali S, Tashkandi M, Mahfouz MM (2015) CRISPR/Cas9-mediated viral interference in plants. Genome Biol 16(1):238. https://doi.org/10.1186/s13059-015-0799-6

Alvarez LH, Cervantes FJ (2011) (Bio)nanotechnologies to enhance environmental quality and energy production. J Chem Technol Biotechnol 86(11):1354–1363. https://doi.org/10.1002/jctb.2697

Ameen F, AlYahya S, Govarthanan M, ALjahdali N, Al-Enazi N, Alsamhary K et al (2020) Soil bacteria Cupriavidus sp. mediates the extracellular synthesis of antibacterial silver nanoparticles. J Mol Struct 1202:127233

Ananikov VP (2019) Organic–inorganic hybrid nanomaterials. Nanomaterials 9(9):1197, MDPI. [Online]. https://www.mdpi.com/2079-4991/9/9/1197. Accessed 26 Apr 2024

Andrew JS, Starr JD, Budi MA (2014) Prospects for nanostructured multiferroic composite materials. Scr Mater 74:38–43

Antony Ceasar S, Ignacimuthu S (2012) Genetic engineering of crop plants for fungal resistance: role of antifungal genes. Biotechnol Lett 34(6):995–1002. https://doi.org/10.1007/s10529-012-0871-1

Antunes Filho S, Dos Santos MS, Dos Santos OAL, Backx BP, Soran M-L, Opriş O et al (2023) Biosynthesis of nanoparticles using plant extracts and essential oils. Molecules 28(7):3060

Awe OW, Zhao Y, Nzihou A, Minh DP, Lyczko N (2017) A review of biogas utilisation, purification and upgrading technologies. Waste Biomass Valor 8:267–283

Bai Y, Mora-Sero I, De Angelis F, Bisquert J, Wang P (2014) Titanium dioxide nanomaterials for photovoltaic applications. Chem Rev 114(19):10095–10130

Baltes NJ et al (2015) Conferring resistance to geminiviruses with the CRISPR–Cas prokaryotic immune system. Nat Plants 1(10):1–4

Bankar A, Joshi B, Kumar AR, Zinjarde SJC, Physicochemical SA, Aspects E (2010) Banana peel extract mediated novel route for the synthesis of silver nanoparticles. Colloids Surf A Physicochem Eng Aspects 368(1–3):58–63

Bayda S, Adeel M, Tuccinardi T, Cordani M, Rizzolio F (2019) The history of nanoscience and nanotechnology: from chemical–physical applications to nanomedicine. Molecules 25(1):112

Beck RJ, Zhao Y, Fong H, Menkhaus TJ (2017) Electrospun lignin carbon nanofiber membranes with large pores for highly efficient adsorptive water treatment applications. J Water Process Eng 16:240–248

Bensaude-Vincent B (2010) Opening the field of nanoethics. HYLE Int J Phil Chem 16:1–2

Bergese P, Hamad-Schifferli K (2013) Nanomaterial interfaces in biology, vol 1025. Springer

Bhargava A, Jain N, Khan MA, Pareek V, Dilip RV, Panwar J (2016) Utilizing metal tolerance potential of soil fungus for efficient synthesis of gold nanoparticles with superior catalytic activity for degradation of rhodamine B. J Environ Manag 183:22–32

Bhushan BJ (2016) Introduction to nanotechnology: history, status, and importance of nanoscience and nanotechnology education. In: Global perspectives of nanoscience and engineering education, pp 1–31

Boaretto J, Fotouhi M, Tende E, Aver GF, Marcon VRR, Cordeiro GL et al (2021) Biomimetics and composite materials toward efficient mobility: a review. J Compos Sci 5(1):22
Boucher M, Geffroy F, Prévéral S, Bellanger L, Selingue E, Adryanczyk-Perrier G et al (2017) Genetically tailored magnetosomes used as MRI probe for molecular imaging of brain tumor. Biomaterials 121:167–178
Calisi A (2023) Integrating bioindicators and biomarkers in aquatic ecotoxicology: an overview. Appl Sci 13(21):11920
Cardon L, Ragaert K, De Santis R, Gloria A (2017) Design and fabrication methods for biocomposites. In: Biomedical composites. Elsevier, pp 17–36
Carrara S (2018) Integrated Bio/Nano/CMOS interfaces for electrochemical molecular sensing. IEEJ Trans Electr Electron Eng 13(11):1534–1539
Carriger JF, Rand GM, Gardinali PR, Perry WB, Tompkins MS, Fernandez AM (2006) Pesticides of potential ecological concern in sediment from south florida canals: an ecological risk prioritization for aquatic arthropods. Soil Sediment Contam Int J 15(1):21–45. https://doi.org/10.1080/15320380500363095
Chakraborty U, Kaur G, Chaudhary GR (2021) Development of environmental nanosensors for detection monitoring and assessment. In: New frontiers of nanomaterials in environmental science, pp 91–143
Charan D, Mounika M, Pravalika V, Changala R, Gummadi A Iot driven machine learning platform for real time environmental monitoring
Chen C et al (2018) Fabrication of light-responsively controlled-release herbicide using a nanocomposite. Chem Eng J 349:101–110
Chhipa H (2017) Nanofertilizers and nanopesticides for agriculture. Environ Chem Lett 15(1):15–22. https://doi.org/10.1007/s10311-016-0600-4
Chhipa H (2019) Mycosynthesis of nanoparticles for smart agricultural practice: a green and eco-friendly approach. In: Green synthesis, characterization and applications of nanoparticles. Elsevier, pp 87–109
Choi H-J, Montemagno CD (2016) Convergence of nanotechnology and biotechnology. In: Handbook of science and technology convergence. Springer, pp 253–277
Coles D, Frewer LJ (2013) Nanotechnology applied to European food production—a review of ethical and regulatory issues. Trends Food Sci Technol 34(1):32–43
Costa M (2003) Potential hazards of hexavalent chromate in our drinking water. Toxicol Appl Pharmacol 188(1):1–5
Das A, Kamle M, Bharti A, Kumar PJV (2019) Nanotechnology and it's applications in environmental remediation: an overview. Vegetos 32:227–237
DeRosa MC, Monreal C, Schnitzer M, Walsh R, Sultan Y (2010) Nanotechnology in fertilizers. Nat Nanotechnol 5(2):91–91
Díaz J, Estrada L, Acuña Y, González E (2019) The bio-nano offer and its impact on environment. Energy Agric Health Nano Sci Technol 5:14
Diaz-Angulo J, Lara-Ramos J, Mueses M, Hernández-Ramírez A, Puma GL, Machuca-Martínez F (2020) Enhancement of the oxidative removal of diclofenac and of the TiO_2 rate of photon absorption in dye-sensitized solar pilot scale CPC photocatalytic reactors. Chem Eng J 381:122520
Díez-Pascual AM (2019) Synthesis and applications of biopolymer composites. Int J Mol Sci 20:2321, MDPI
Dong J, Zhao H, Qiao F, Liu P, Wang X, Ai S (2013) Quantum dot immobilized acetylcholinesterase for the determination of organophosphate pesticides using graphene-chitosan nanocomposite modified electrode. Anal Methods 5(11):2866–2872
Dwivedi N, Dwivedi S (2022) Bionanotechnology towards sustainable management of environmental pollution. CRC Press
Eke J, Wagh P, Escobar IC (2018) Ozonation, biofiltration and the role of membrane surface charge and hydrophobicity in removal and destruction of algal toxins at basic pH values. Sep Purif Technol 194:56–63

Elemike EE, Uzoh IM, Onwudiwe DC, Babalola OO (2019) The role of nanotechnology in the fortification of plant nutrients and improvement of crop production. Appl Sci 9(3):499
Eleuterio L, Batista JR (2010) Biodegradation studies and sequencing of microcystin-LR degrading bacteria isolated from a drinking water biofilter and a fresh water lake. Toxicon 55(8):1434–1442
Elgarahy AM, Eloffy M, Guibal E, Alghamdi HM, Elwakeel KZ (2023) Use of biopolymers in wastewater treatment: A brief review of current trends and prospects. Chin J Chem Eng
Food and A. O. of the U. Nations (2017) The future of food and agriculture: trends and challenges. FAO
Federici G, Shaw BJ, Handy RD (2007) Toxicity of titanium dioxide nanoparticles to rainbow trout (Oncorhynchus mykiss): gill injury, oxidative stress, and other physiological effects. Aquat Toxicol 84(4):415–430
Fischhoff DA et al (1987) Insect tolerant transgenic tomato plants. Bio/technology 5(8):807–813
Gandara-Loe J, Pastor-Perez L, Bobadilla L, Odriozola J, Reina T (2021) Understanding the opportunities of metal–organic frameworks (MOFs) for CO_2 capture and gas-phase CO_2 conversion processes: a comprehensive overview. React Chem Eng 6(5):787–814
Gao Y et al (2020) Thermoresponsive polymer-encapsulated hollow mesoporous silica nanoparticles and their application in insecticide delivery. Chem Eng J 383:123169
Gholivand MB, Heydari H, Abdolmaleki A, Hosseini H (2015) Nanostructured CuO/PANI composite as supercapacitor electrode material. Mater Sci Semicond Process 30:157–161
Ghosh S, Webster TJ (2021) Nanobiotechnology: microbes and plant assisted synthesis of nanoparticles, mechanisms and applications. Elsevier
Giroto AS, Guimarães GG, Foschini M, Ribeiro C (2017) Role of slow-release nanocomposite fertilizers on nitrogen and phosphate availability in soil. Sci Rep 7(1):46032
Girotti S, Ghini S, Ferri E, Bolelli L, Colombo R, Serra G et al (2020) Bioindicators and biomonitoring: honeybees and hive products as pollution impact assessment tools for the Mediterranean area. Euro-Med J Environ Integr 5:1–16
Gong J, Liu J, Wen X, Jiang Z, Chen X, Mijowska E (2014) Upcycling waste polypropylene into graphene flakes on organically modified montmorillonite. Ind Eng Chem Res 53(11):4173–4181
Goodsell DS (2004) Bionanotechnology: lessons from nature. Wiley
Goswami A, Roy I, Sengupta S, Debnath N (2010) Novel applications of solid and liquid formulations of nanoparticles against insect pests and pathogens. Thin Solid Films 519(3):1252–1257
Goudarzi M, Mir N, Mousavi-Kamazani M, Bagheri S, Salavati-Niasari M (2016) Biosynthesis and characterization of silver nanoparticles prepared from two novel natural precursors by facile thermal decomposition methods. Sci Rep 6(1):32539
Grasso G, Zane D, Dragone RJN (2019) Microbial nanotechnology: challenges and prospects for green biocatalytic synthesis of nanoscale materials for sensoristic and biomedical applications. Nanomaterials 10(1):11
Gu P, Xing J, Wen T, Zhang R, Wang J, Zhao G et al (2018) Experimental and theoretical calculation investigation on efficient Pb(II) adsorption on etched Ti_3AlC_2 nanofibers and nanosheets. Environ Sci Nano 5(4):946–955
Guchhait R, Sarkar B (2023) Increasing growth of renewable energy: a state of art. Energies 16(6):2665
Habibpanah AA, Pourhashem S, Sarpoolaky H (2011) Preparation and characterization of photocatalytic titania–alumina composite membranes by sol–gel methods. J Eur Ceram Soc 31(15):2867–2875
Hagger JA, Jones MB, Leonard DP, Owen R, Galloway TS (2006) Biomarkers and integrated environmental risk assessment: are there more questions than answers? Integr Environ Assess Manag Int J 2(4):312–329
Hassan MN, Islam MR, Faisal F, Semantha FH, Siddique AH, Hasan M (2020) An IoT based environment monitoring system. Paper presented at the 2020 3rd international conference on intelligent sustainable systems (ICISS)

Haw C, Chiu W, Rahman SA, Khiew P, Radiman S, Shukor RA et al (2016) The design of new magnetic-photocatalyst nanocomposites ($CoFe_2O_4$–TiO_2) as smart nanomaterials for recyclable-photocatalysis applications. New J Chem 40(2):1124–1136

Hooda H, Singh P, Bajpai S (2020) Effect of quercitin impregnated silver nanoparticle on growth of some clinical pathogens. Mater Today Proc 31:625–630

Hu D, Yu S, Yu D, Liu N, Tang Y, Fan Y et al (2019) Biogenic Trichoderma harzianum-derived selenium nanoparticles with control functionalities originating from diverse recognition metabolites against phytopathogens and mycotoxins. Food Control 106:106748

Ibrahim ID, Sadiku ER, Jamiru T, Hamam Y, Alayli Y, Eze AA (2020) Prospects of nanostructured composite materials for energy harvesting and storage. J King Saud Univ Sci 32(1):758–764

Im J, Löffler FE (2016) Fate of bisphenol A in terrestrial and aquatic environments. Environ Sci Technol 50(16):8403–8416

Jacobstein N (2006) Foresight guidelines for responsible nanotechnology development. Inst Mol Manuf. [Online]. http://69.164.201.73/guidelines/ForesightGuidelinesV6.pdf. Accessed 26 Apr 2024

Jain N, Kejariwal M (2022) Case studies on applications of green nanotechnology. In: Innovations in green nanoscience and nanotechnology. CRC Press, pp 257–271

Jamei E, Vrcelj Z (2021) Biomimicry and the built environment, learning from nature's solutions. Appl Sci 11(16):7514

Jha AK, Prasad K (2012) Synthesis of nanomaterials using expired medicines: an eco-friendly option. Nanotechnol Dev 2(1):e7–e7

Ji J, Xiong H, Zhu Z, Li L, Huang Y, Yu X (2018) Fabrication of polypyrrole/chitosan nanocomposite aerogel monolith for removal of Cr (VI). J Polym Environ 26:1979–1985

Johnson JL, Behnam A, Pearton SJ, Ural A (2010) Hydrogen sensing using Pd-functionalized multi-layer graphene nanoribbon networks. Adv Mater 22(43):4877–4880. https://doi.org/10.1002/adma.201001798

Jones HD (2015) Regulatory uncertainty over genome editing. Nat Plants 1(1):1–3

Jung C, Phal N, Oh J, Chu KH, Jang M, Yoon Y (2015) Removal of humic and tannic acids by adsorption–coagulation combined systems with activated biochar. J Hazard Mater 300:808–814

Kadam A, Patil S, Patil S, Tumkur A (2016) Pharmaceutical waste management an overview. J Pharm Pract 9(1)

Kah M, Hofmann T (2014) Nanopesticide research: current trends and future priorities. Environ Int 63:224–235

Kah M, Tufenkji N, White JC (2019) Nano-enabled strategies to enhance crop nutrition and protection. Nat Nanotechnol 14(6):532–540

Kalimuthu K, Cha BS, Kim S, Park KS (2020) Eco-friendly synthesis and biomedical applications of gold nanoparticles: a review. Microchem J 152:104296

Kapiel T (2023) Nanobiotechnology: an introduction

Karami M, Bahabadi MA, Delfani S, Ghozatloo A (2014) A new application of carbon nanotubes nanofluid as working fluid of low-temperature direct absorption solar collector. Sol Energy Mater Sol Cells 121:114–118

Kerry RG, Rout JR, Das G, Fraceto LF, Paramithiotis S, Patra JK (2019) Applications of nanotechnology in food and agriculture. In: Food molecular microbiology. CRC Press, pp 223–249

Khademian E, Salehi E, Sanaeepur H, Galiano F, Figoli A (2020) A systematic review on carbohydrate biopolymers for adsorptive remediation of copper ions from aqueous environments-part A: classification and modification strategies. Sci Total Environ 738:139829

Khanna P (1996) Environmental biotechnology in India: prospects and case studies. World J Microbiol Biotechnol 12:457–461

Kokila T, Ramesh P, Geetha DJE, Safety E (2016) Biosynthesis of AgNPs using Carica Papaya peel extract and evaluation of its antioxidant and antimicrobial activities. Ecotoxicol Environ Saf 134:467–473

Kottegoda N, Munaweera I, Madusanka N, Karunaratne V (2011) A green slow-release fertilizer composition based on urea-modified hydroxyapatite nanoparticles encapsulated wood. Curr Sci 73–78

Kumar A, Dhawan A (2013) Genotoxic and carcinogenic potential of engineered nanoparticles: an update. Arch Toxicol 87(11):1883–1900. https://doi.org/10.1007/s00204-013-1128-z

Kumar V, Guleria P (2020) Application of DNA-nanosensor for environmental monitoring: recent advances and perspectives. Curr Pollut Rep 1–21

Kumar GD, Natarajan N, Nakkeeran S (2016) Antifungal activity of nanofungicide Trifloxystrobin 25%+ Tebuconazole 50% against Macrophomina phaseolina. Afr J Microbiol Res 10(4):100–105

Kumar P, Kim K-H, Bansal V, Kumar P (2017) Nanostructured materials: a progressive assessment and future direction for energy device applications. Coord Chem Rev 353:113–141

Kumar V, Sachdev D, Pasricha R, Maheshwari PH, Taneja NK (2018) Zinc-supported multiwalled carbon nanotube nanocomposite: a synergism to micronutrient release and a smart distributor to promote the growth of onion seeds in arid conditions. ACS Appl Mater Interfaces 10(43):36733–36745. https://doi.org/10.1021/acsami.8b13464

Li P et al (2021) Eco-friendly biomolecule-nanomaterial hybrids as next-generation agrochemicals for topical delivery. EcoMat 3(5):e12132. https://doi.org/10.1002/eom2.12132

Liu K, Catchmark JM (2019) Enhanced mechanical properties of bacterial cellulose nanocomposites produced by co-culturing Gluconacetobacter hansenii and Escherichia coli under static conditions. Carbohydr Polym 219:12–20

Liu R, Lal R (2014) Synthetic apatite nanoparticles as a phosphorus fertilizer for soybean (Glycine max). Sci Rep 4(1):5686

Liu S, Yu B, Wang S, Shen Y, Cong H (2020) Preparation, surface functionalization and application of Fe_3O_4 magnetic nanoparticles. Adv Colloid Interface Sci 281:102165

Madkhali N, Prasad C, Malkappa K, Choi HY, Govinda V, Bahadur I, Abumousa R (2023) Recent update on photocatalytic degradation of pollutants in waste water using TiO_2-based heterostructured materials. Results Eng 17:100920

Malhotra A, Sharma N, Kumar N, Dolma K, Sharma D, Nandanwar HS, Choudhury AR (2014) Multi-analytical approach to understand biomineralization of gold using rice bran: a novel and economical route. RSC Adv 4(74):39484–39490

Malik S, Dhasmana A, Preetam S, Mishra YK, Chaudhary V, Bera SP et al (2022) Exploring microbial-based green nanobiotechnology for wastewater remediation: a sustainable strategy. Nanomaterials 12(23):4187

Mallakpour S, Nezamzadeh Ezhieh A (2018) Citric acid and vitamin C as coupling agents for the surface coating of ZrO_2 nanoparticles and their behavior on the optical, mechanical, and thermal properties of poly (vinyl alcohol) nanocomposite films. J Polym Environ 26:2813–2824

Milinkovitch T, Geffard O, Geffard A, Mouneyrac C, Chaumot A, Xuereb B et al (2019) Biomarkers as tools for monitoring within the water framework directive context: concept, opinions and advancement of expertise. Environ Sci Pollut Res 26:32759–32763

Miller RJ, Bennett S, Keller AA, Pease S, Lenihan HS (2012) TiO_2 nanoparticles are phototoxic to marine phytoplankton. PLoS One 7(1):e30321

Mirkin CA, Niemeyer CM (2007) Nanobiotechnology II: more concepts and applications. Wiley

Mishra AK, Arockiadoss T, Ramaprabhu S (2010) Study of removal of azo dye by functionalized multi walled carbon nanotubes. Chem Eng J 162(3):1026–1034. https://doi.org/10.1016/j.cej.2010.07.014

Moitra P, Alafeef M, Dighe K, Frieman MB, Pan D (2020) Selective naked-eye detection of SARS-CoV-2 mediated by N gene targeted antisense oligonucleotide capped plasmonic nanoparticles. ACS Nano 14(6):7617–7627

Moreira AF, Rodrigues CF, Reis CA, Costa EC, Correia IJ (2018) Gold-core silica shell nanoparticles application in imaging and therapy: a review. Micropor Mesopor Mater 270:168–179

Mughal B, Zaidi SZJ, Zhang X, Hassan SU (2021) Biogenic nanoparticles: synthesis, characterisation and applications. Appl Sci 11(6):2598

Naimi-Shamel N, Pourali P, Dolatabadi S (2019) Green synthesis of gold nanoparticles using Fusarium oxysporum and antibacterial activity of its tetracycline conjugant. Journal de mycologie medicale 29(1):7–13

Nasrollahzadeh M, Sajjadi M, Iravani S, Varma RS (2021) Carbon-based sustainable nanomaterials for water treatment: state-of-art and future perspectives. Chemosphere 263:128005

Ni M, Leung MKH, Leung DYC, Sumathy K (2007) A review and recent developments in photocatalytic water-splitting using TiO_2 for hydrogen production. Renew Sustain Energy Rev 11(3):401–425. https://doi.org/10.1016/j.rser.2005.01.009

Niemeyer CM, Mirkin CA (2004) Nanobiotechnology: concepts, applications and perspectives, vol 1. Wiley

Oberdörster E, Zhu S, Blickley TM, McClellan-Green P, Haasch ML (2006) Ecotoxicology of carbon-based engineered nanoparticles: effects of fullerene (C60) on aquatic organisms. Carbon 44(6):1112–1120

Omran BA (2020) Fundamentals of nanotechnology and nanobiotechnology. In: Omran BA (ed) Nanobiotechnology: a multidisciplinary field of science. Springer International Publishing, Cham, pp 1–36

Otles S, Yalcin B (2010) Nano-biosensors as new tool for detection of food quality and safety. LogForum 6 4, vol 7. [Online]. http://www.logforum.net/vol6/issue4/no7/6_4_7_10.pdf. Accessed 26 Apr 2024

Ouchi K, Henzie J (2017) Hydrogen generation sailing ship: conceptual design and feasibility study. Paper presented at the OCEANS 2017-Aberdeen

Pamula AS, Ravilla A, Madiraju SVH (2022) Applications of the Internet of Things (IoT) in real-time monitoring of contaminants in the air water, and soil. Eng Proc 27(1):26

Pan G, Kurumada K-I (2008) Hybrid gel reinforced with coating layer for removal of phenol from aqueous solution. Chem Eng J 138(1–3):194–199

Pandey SS (2020) Nano-terrorism: shaping Indian response to a foreseeable threat. Int J Nanomater Nanostruct 6(1):1–7

Panduman YYF, Funabiki N, Fajrianti ED, Fang S, Sukaridhoto S (2024) A survey of AI techniques in IoT applications with use case investigations in the smart environmental monitoring and analytics in real-time IoT platform. Information 15(3):153

Pang H, Wu Y, Huang S, Ding C, Li S, Wang X et al (2018) Macroscopic and microscopic investigation of uranium elimination by Ca–Mg–Al-layered double hydroxide supported nanoscale zero valent iron. Inorg Chem Front 5(10):2657–2665

Pankaj, Shakil NA, Kumar J, Singh MK, Singh K (2012) Bioefficacy evaluation of controlled release formulations based on amphiphilic nano-polymer of carbofuran against Meloidogyne incognita infecting tomato. J Environ Sci Health Part B 47(6):520–528. https://doi.org/10.1080/03601234.2012.665667

Parmar N et al (2017) Genetic engineering strategies for biotic and abiotic stress tolerance and quality enhancement in horticultural crops: a comprehensive review. 3 Biotech 7(4):239. https://doi.org/10.1007/s13205-017-0870-y

Patra P, Mitra S, Debnath N, Goswami A (2012) Biochemical-, biophysical-, and microarray-based antifungal evaluation of the buffer-mediated synthesized nano zinc oxide: an in vivo and in vitro toxicity study. Langmuir 28(49):16966–16978. https://doi.org/10.1021/la304120k

Patra JK, Das G, Baek K-H (2016) Phyto-mediated biosynthesis of silver nanoparticles using the rind extract of watermelon (Citrullus lanatus) under photo-catalyzed condition and investigation of its antibacterial, anticandidal and antioxidant efficacy. J Photochem Photobiol B Biol 161:200–210

Paz-Trejo C, Flores-Márquez AR, Gómez-Arroyo S (2023) Nanotechnology in agriculture: a review of genotoxic studies of nanopesticides in animal cells. Environ Sci Pollut Res 30(25):66473–66485. https://doi.org/10.1007/s11356-023-26848-y

Peigneux A, Valverde-Tercedor C, López-Moreno R, Pérez-González T, Fernández-Vivas M, Jiménez-López C (2016) Learning from magnetotactic bacteria: a review on the synthesis of biomimetic nanoparticles mediated by magnetosome-associated proteins. J Struct Biol 196(2):75–84

Pérez-de-Luque A, Rubiales D (2009) Nanotechnology for parasitic plant control. Pest Manag Sci 65(5):540–545. https://doi.org/10.1002/ps.1732

Peteu SF, Oancea F, Sicuia OA, Constantinescu F, Dinu S (2010) Responsive polymers for crop protection. Polymers 2(3):229–251

Pham N-D, Duong M-M, Le M-V, Hoang HA (2019) Preparation and characterization of antifungal colloidal copper nanoparticles and their antifungal activity against Fusarium oxysporum and Phytophthora capsici. Comptes Rendus Chimie 22(11–12):786–793

Poreddy R, Engelbrekt C, Riisager A (2015) Copper oxide as efficient catalyst for oxidative dehydrogenation of alcohols with air. Catal Sci Technol 5(4):2467–2477

Pradhan S et al (2013a) Entomotoxicity and biosafety assessment of PEGylated acephate nanoparticles: a biologically safe alternative to neurotoxic pesticides. J Environ Sci Health Part B 48(7):559–569. https://doi.org/10.1080/03601234.2013.774891

Pradhan S et al (2013b) Photochemical modulation of biosafe manganese nanoparticles on Vigna radiata: a detailed molecular, biochemical, and biophysical study. Environ Sci Technol 47(22):13122–13131. https://doi.org/10.1021/es402659t

Ragaei M, Sabry AH (2014) Nanotechnology for insect pest control. Int J Sci Environ Technol 3(2):528–545

Rai M, Ingle A (2012) Role of nanotechnology in agriculture with special reference to management of insect pests. Appl Microbiol Biotechnol 94(2):287–293. https://doi.org/10.1007/s00253-012-3969-4

Rai V, Acharya S, Dey N (2012) Implications of nanobiosensors in agriculture. [Online]. https://www.scirp.org/html/10-3200181_18992.htm. Accessed 26 Apr 2024

Rajarao R, Ferreira R, Sadi SHF, Khanna R, Sahajwalla VJ (2014) Synthesis of silicon carbide nanoparticles by using electronic waste as a carbon source. Mater Lett 120:65–68

Rana A, Yadav K, Jagadevan S (2020) A comprehensive review on green synthesis of nature-inspired metal nanoparticles: mechanism, application and toxicity. J Clean Prod 272:122880

Rao M, Jha B, Jha AK, Prasad K (2017) Fungal nanotechnology: a pandora to agricultural science and engineering. In: Fungal nanotechnology: applications in agriculture, industry, and medicine, pp 1–33

Rathjen NA, Shahbodaghi SD (2021) Bioterrorism. Am Fam Phys 104(4):376–385

Ravelli D, Dondi D, Fagnoni M, Albini A (2009) Photocatalysis. A multi-faceted concept for green chemistry. Chem Soc Rev 38(7):1999–2011

Reddy DA, Choi J, Lee S, Ma R, Kim TK (2015) Self-assembled macro porous ZnS–graphene aerogels for photocatalytic degradation of contaminants in water. RSC Adv 5(24):18342–18351

Regassa Y, Dabasa T, Amare G, Lemu HG (2023) Bio-inspired design trends for sustainable energy structures. Paper presented at the IOP conference series: materials science and engineering

Resh VH, Cardé RT (2009) Encyclopedia of insects. Academic Press. Accessed 25 Apr 2024

Rodotà S, Capurro R (2005) Ethical aspects of ICT implants in the human body. Opin Eur Group Ethics Sci New Technol Eur Comm N 20 Adopt 16032005

Rop K, Karuku GN, Mbui D, Njomo N, Michira I (2019) Evaluating the effects of formulated nano-NPK slow release fertilizer composite on the performance and yield of maize, kale and capsicum. Ann Agric Sci 64(1):9–19

Rouse I, Power D, Brandt EG, Schneemilch M, Kotsis K, Quirke N et al (2021) First principles characterisation of bio–nano interface. Phys Chem Chem Phys 23(24):13473–13482

Roy Choudhury S et al (2011) Surface-modified sulfur nanoparticles: an effective antifungal agent against Aspergillus niger and Fusarium oxysporum. Appl Microbiol Biotechnol 90(2):733–743. https://doi.org/10.1007/s00253-011-3142-5

Sagadevan S (2013) Recent trends on nanostructures based solar energy applications: a review. Rev Adv Mater Sci 34(1):44–61

Sahoo S, Kalita H, Mohanty S, Nayak SK (2018) Degradation study of biobased polyester–polyurethane and its nanocomposite under natural soil burial, UV radiation and hydrolytic-salt water circumstances. J Polym Environ 26(4):1528–1539

Serrano E, Rus G, García-Martínez J (2009) Nanotechnology for sustainable energy. Renew Sustain Energy Rev 13(9):2373–2384. https://doi.org/10.1016/j.rser.2009.06.003

Shahcheraghi N, Golchin H, Sadri Z, Tabari Y, Borhanifar F, Makani S (2022) Nano-biotechnology, an applicable approach for sustainable future. Biotech 12(3):65

Shankar S, Jaiswal L, Aparna R, Prasad R (2014) Synthesis, characterization, in vitro biocompatibility, and antimicrobial activity of gold, silver and gold silver alloy nanoparticles prepared from Lansium domesticum fruit peel extract. Mater Lett 137:75–78

Shankar PD, Shobana S, Karuppusamy I, Pugazhendhi A, Ramkumar VS, Arvindnarayan S et al (2016) A review on the biosynthesis of metallic nanoparticles (gold and silver) using bio-components of microalgae: formation mechanism and applications. Enzyme Microb Technol 95:28–44

Sharma S, Kalita G, Hirano R, Shinde SM, Papon R, Ohtani H, Tanemura M (2014) Synthesis of graphene crystals from solid waste plastic by chemical vapor deposition. Carbon 72:66–73

Sharma G, Kumar A, Sharma S, Naushad M, Dwivedi RP, ALOthman ZA, Mola GT (2019) Novel development of nanoparticles to bimetallic nanoparticles and their composites: a review. J King Saud Univ Sci 31(2):257–269

Shoseyov O, Levy I (2008) Nanobiotechnology overview. In: Shoseyov O, Levy I (eds) NanoBioTechnology: bioinspired devices and materials of the future. Humana Press, Totowa, NJ, pp 3–15

Shrivastava P (2018) Environmental technologies and competitive advantage. In: Business ethics and strategy, vol I and II. Routledge, pp 317–334

Siddiqi KS, Husen A (2017) Recent advances in plant-mediated engineered gold nanoparticles and their application in biological system. J Trace Elements Med Biol 40:10–23

Stadler T, Buteler M, Weaver DK (2010) Novel use of nanostructured alumina as an insecticide. Pest Manag Sci 66(6):577–579. https://doi.org/10.1002/ps.1915

Stark WJ (2011) Nanoparticles in biological systems. Angew Chem Int Ed 50(6):1242–1258. https://doi.org/10.1002/anie.200906684

Su F, Lu C, Hu S (2010) Adsorption of benzene, toluene, ethylbenzene and p-xylene by NaOCl-oxidized carbon nanotubes. Colloids Surf A 353(1):83–91. https://doi.org/10.1016/j.colsurfa.2009.10.025

Suresh Kumar RS et al (2013) Distinctive effects of nano-sized permethrin in the environment. Environ Sci Pollut Res 20(4):2593–2602. https://doi.org/10.1007/s11356-012-1161-0

Suriyaprabha R, Karunakaran G, Kavitha K, Yuvakkumar R, Rajendran V, Kannan N (2014) Application of silica nanoparticles in maize to enhance fungal resistance. IET Nanobiotechnol 8(3):133–137. https://doi.org/10.1049/iet-nbt.2013.0004

Suryawanshi A, Biswal M, Mhamane D, Gokhale R, Patil S, Guin D, Ogale S (2014) Large scale synthesis of graphene quantum dots (GQDs) from waste biomass and their use as an efficient and selective photoluminescence on–off–on probe for Ag+ ions. Nanoscale 6(20):11664–11670

Syed A, Ahmad AJC (2012) Extracellular biosynthesis of platinum nanoparticles using the fungus fusarium oxysporum. Colloids Surf B Biointerfaces 97:27–31

Tao S, Pang R, Chen C, Ren X, Hu S (2012) Synthesis, characterization and slow release properties of O-naphthylacetyl chitosan. Carbohydr Polym 88(4):1189–1194

Tiwari SK, Sahoo S, Wang N, Huczko A (2020) Graphene research and their outputs: status and prospect. J Sci Adv Mater Dev 5(1):10–29

Tsekhmistrenko S, Bityutskyy V, Tsekhmistrenko O, Horalskyi L, Tymoshok N, Spivak MJBD (2020) Bacterial synthesis of nanoparticles: a green approach. Biosyst Divers 28(1):9–17

Tyczewska A, Twardowski T, Woźniak-Gientka E (2023) Agricultural biotechnology for sustainable food security. Trends Biotechnol 41(3):331–341

Valenkov A, Gofman I, Nosov K, Shapovalov V, Yudin V (2011) Polymeric composite systems modified with allotropic forms of carbon. Russ J Appl Chem 84:735–750

Venkatesan J, Manivasagan P, Kim S-K, Kirthi AV, Marimuthu S, Rahuman AA (2014) Marine algae-mediated synthesis of gold nanoparticles using a novel Ecklonia cava. Bioprocess Biosyst Eng 37:1591–1597

Vikesland PJ (2018) Nanosensors for water quality monitoring. Nat Nanotechnol 13(8):651–660

Wang Y, Zhang X, Yue H (2024) Two-dimensional nanomaterials induced nano-bio interfacial effects and biomedical applications in cancer treatment. J Nanobiotechnol 22(1):67

Williams CM (2013) Poultry waste management in developing countries. Role Poultry Hum Nutr 46

Willner MR, Vikesland PJ (2018) Nanomaterial enabled sensors for environmental contaminants. J Nanobiotechnol 16:1–16

Wu Q, Zhao J, Qin G, Wang C, Tong X, Xue S (2013) Photocatalytic reduction of Cr (VI) with TiO_2 film under visible light. Appl Catal B Environ 142:142–148

Yan L, Da H, Zhang S, López VM, Wang W (2017) Bacterial magnetosome and its potential application. Microbiol Res 203:19–28

Yang N, Li W-H (2013) Mango peel extract mediated novel route for synthesis of silver nanoparticles and antibacterial application of silver nanoparticles loaded onto non-woven fabrics. Ind Crops Prod 48:81–88

Yang L, Li Z, Jiang H, Jiang W, Su R, Luo S, Luo Y (2016) Photoelectrocatalytic oxidation of bisphenol A over mesh of TiO_2/graphene/Cu_2O. Appl Catal B Environ 183:75–85

Zähr M, Friedrich D, Kloth TY, Goldmann G, Tributsch H (2010) Bionic photovoltaic panels bio-inspired by green leaves. J Bionic Eng 7(3):284–293. https://doi.org/10.1016/S1672-6529(10)60252-1

Zeltner WA, Tompkins DT (2005) Shedding light on photocatalysis. Ashrae Trans 111:523

Zhang D, Wei A, Zhang J, Qiu R (2015) The photocatalytic interaction of Cr (VI) ions and phenol on polymer-modified TiO_2 under visible light irradiation. Kinet Catal 56:569–573

Zhang G, Zhang X, Meng Y, Pan G, Ni Z, Xia S (2020) Layered double hydroxides-based photocatalysts and visible-light driven photodegradation of organic pollutants: a review. Chem Eng J 392:123684

Zhao X, Zhu Y (2006) Synergetic degradation of rhodamine B at a porous ZnWO4 film electrode by combined electro-oxidation and photocatalysis. Environ Sci Technol 40(10):3367–3372. https://doi.org/10.1021/es052029e

Zhao M, Zhou H, Chen L, Hao L, Chen H, Zhou X (2020) Carboxymethyl chitosan grafted trisiloxane surfactant nanoparticles with pH sensitivity for sustained release of pesticide. Carbohydr Polym 243:116433

Chapter 3
Wealth Creation by Solubilization of Heavy Metals from E-Waste Using Indigenous Microorganisms—A New Approach

Kavya Rebecca Mathew, A. Prashanth Rajan, K. Harshan, Michael Pillay, S. Jose, and Anand Prem Rajan

Abstract Electronic waste or e-waste is one of the rapidly growing environmental problems of the world. Recent statistics showed that there will be an increased need for various renewal substances for manufacturing machinery and automobiles, medicines, items for the construction industry, and many more industries. This calls for more efficient solutions to manage e-waste. E-waste has both toxic and valuable materials. Printed circuit boards (PCBs) are the most complex, hazardous, and valuable components of e-waste. Through the utilization of indigenous microorganisms, e-waste can be recycled and leading to avenues such as for wealth generation. Biological methods like bioleaching are used commonly because it is cost-efficient and environmentally friendly. Bioleaching also offers diverse ways to extract metals from the printed circuit board. This chapter presents a new paradigm for e-waste management, emphasizing the importance of indigenous microorganisms in wealth creation and environmental sustainability.

M. Kavya Rebecca · A. Prashanth Rajan · K. Harshan · A. P. Rajan (✉)
School of BioSciences and Technology, Vellore Institute of Technology, Vellore 632014, Tamil Nadu, India
e-mail: aprdbt@gmail.com

M. Kavya Rebecca
e-mail: kavyarebecca.mathew2021@vitstudent.ac.in

A. Prashanth Rajan
e-mail: prashanth07052001@gmail.com

M. Pillay
Department of Life and Consumer Sciences, College of Agriculture and Environmental Sciences, University of South Africa, Florida Campus, Johannesburg, South Africa
e-mail: michaelpillay1950@gmail.com

S. Jose
School of Mechnical Engineering, Vellore Institute of Technology, Vellore 632014, Tamil Nadu, India

© The Author(s), under exclusive license to Springer Nature Switzerland AG 2025
A. Jeyaseelan and K. Murugasen (eds.), *Sustainable Environmental Remediation: Avenues in Nano and Biotechnology*, Environmental Science and Engineering,
https://doi.org/10.1007/978-3-031-78483-5_3

Keywords Electronic waste · Printed circuit board · Heavy metals · Bioleaching · Zinc

3.1 Introduction

3.1.1 E-Waste

E-waste is short for electronic waste and is defined as any electrical or electronic equipment that has been discarded either because it has reached the end of its life or it could not be sold as a second-hand item. It could be in a working condition or broken. In today's technologically advanced world, waste electrical and electronic equipment (WEEE) and electronic waste (E-waste) are growing concerns that require immediate action. The lack of appropriate procedures for tracking this type of garbage makes it difficult to determine the precise amount of e-waste created globally. For example, although India is one of the world's leading producers of e-waste, precise and up-to-date data on the amount of e-waste generated, collected, and processed each year is lacking. According to the United Nations University, 44.7 million tons of e-waste was produced in 2016 (Forti et al. 2020). The lifespan of most electronics is reduced over time and there is a general tendency for the population to purchase newer versions or models of their current electronics. This results in a tremendous increase in e-waste. Enhanced overall user experience through better aesthetic of newer models, accelerated operating speeds, advanced graphics and capabilities like AI computing and Internet of Things connectivity influence consumers to purchase the latest models. The technological advancement of the world in every field of life has led to a situation where a device is used for a short term and then discarded when a newer model arrives or it is used till the end of its life which is no longer than 4.5 years on average (Ghimire and Ariya 2020; Lebbie et al. 2021).

E-waste can include the following:

Household appliances

Refrigerators, microwaves, ovens, dishwashers, fans, lights, washing machines, heaters etc.

Communication, IT, and entertainment devices

Smartphones, laptops, televisions, computers, hard drives, DVDs, video games, stereos, fax machines, printers, copiers, smart watches, remotes, exercise equipment, etc.

Medical and diagnostic devices

Imaging equipment, diabetes testing equipment, ECG machines, etc.

3.1.2 *Composition of E-Waste*

E-waste can be quite broad and depends on a variety of factors. The source of e-waste plays a significant role in its composition. E-waste can come from smartphones, televisions, remotes, refrigerators, game consoles, etc. In addition, the models, manufacturers, date of manufacture, and age determine the toxicity of the product once it is discarded. The various parts present in e-waste can pose different threats (Liu et al. 2023). E-waste in general cannot be classified as hazardous or non-hazardous but the several components present in it can be toxic to the environment and health. In general, e-waste from IT sources and telecommunication systems contains a higher number of precious metals than the waste from household appliances. These components of e-waste can be retrieved more effectively from e-waste than from an ore. For example, a cell phone comprises more than forty elements, including precious metals like silver (Ag), gold (Au), and palladium (Pd), special metals like lithium (Li), cobalt (Co), indium (In), and antimony (Sb), and base metals like copper (Cu) and tin (Sn). Recycling e-waste is an important topic of discussion (Ismail and Hanafiah 2020).

Overall, printed circuit boards, plastics, glass, wood, rubber, concrete, ceramics, and other materials are some of the main components of E-waste (Fig. 3.1). Ferrous and non-ferrous metals, polymers, flame retardants, heavy metals, polychlorinated biphenyls, and other solid components are all included in the homogeneous E&E products. E-waste was also found to have toxic flame retardants in many of its plastic components. In addition, the circuit boards of these electronics were found to contain toxic elements such as Arsenic (As), Chromium (Cr), Cadmium (Cd), Lead (Pb), Mercury (Hg), and other toxic chemicals. Modern-day refrigerators, freezers, and air coolers do not contain CFCs (chlorofluorocarbons) as they were banned in 2000 by the Montreal Protocol but they were a major concern for the depletion of the ozone layer while still in use (Hamsavathi et al. 2020). Cathode ray tubes (CRTs) are another source of toxic e-waste. The average lead content in CRT monitors (TVs, Computers, Video Games, etc.) is 1.6–3.2 kg. CRTs in landfills are a huge problem as leaching of lead into the soil can cause major health issues for those living in and around the area. While CRTs are no longer used on such a large scale, they were a major component of televisions and computers before LCDs and LED screens were developed and made up a sizeable portion of e-waste. Many developed countries banned the disposal of CRTs in landfills because of their toxicity. The complex composition of e-waste includes a range of elements, including wood and ceramics, plastics (21%), copper (7%), glass (5%), and steel (47% by weight). Nasier (2021) called E-waste an urban mine since it contains dangerous metals and a variety of precious metals (Au, Ag, Pt, Pd, and Nd) (Nasier 2021).

Components of e-waste that can be reused

Metals, plastic, and glass are some of the things that can be reused from e-waste.

- Metals like gold, silver, copper, aluminium, and palladium recovery and reuse do not require extensive processing and can be obtained from circuit boards. These

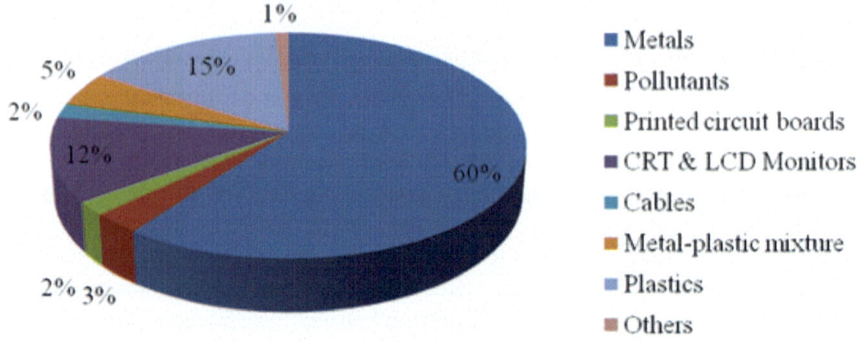

Fig. 3.1 Depicts the percentage of components in E-waste generated globally

metals, once extracted from e-waste, can be recycled into new electronics or other products.

- Plastic in e-waste can be easily recycled, especially the plastic in casings and covers. Recovered plastic can then be recycled into new electronic devices.
- Screens and monitors of many electrical devices are made of glass which can be reused or recycled into new e-displays or even be used to make showpieces.

Components that require processing/recycling

Many components of e-waste require proper processing before they can be recycled or reused. For instance,

- Printed Circuit Boards (PCBs) contain gold, silver, copper, and palladium. Processing methods are required to extract these metals. PCBs are usually ground, shredded and then subjected to chemical procedures for the extraction of these precious metals.
- Batteries contain toxic elements like lead, cadmium, and lithium. The disposal of batteries requires caution as these elements could lead to dangerous health defects—which is also why the recycling of batteries needs to be done carefully. The disassembly and sorting of the components of batteries allows us to reuse what is in them to make new batteries or other goods.
- As mentioned earlier, the newer flat-screen TVs and computers no longer have cathode ray tubes and instead have LCD or LED displays. These contain glass, liquid crystals, and semiconductors. These need to be dismantled properly and categorized for further recovery and use in new displays.
- Cables and wiring make up a substantial portion of e-waste and contain metals like copper, aluminium, and gold. Copper and aluminium are used in several ways in the electrical industry once it has been extracted and the gold can either be used in electrical engineering or be reused by jewellers.

- Most electronics like laptops have a cooling system (fans, heat sinks, or thermal compounds). These cooling systems prevent the device from overheating while in use. Employing shredding, smelting, and refining, copper and aluminium can be extracted and reused to make other electrical devices.

3.1.3 *Hazardous Effects of E-Waste*

E-waste has several hazardous chemicals that can cause many health issues. E-waste from developed countries like the USA and UK is usually exported illegally to developing countries like Kenya, India, Indonesia, Vietnam, Egypt, Ghana, China, etc., where it gets informally recycled. Only 17.4% of e-waste is formally collected and recycled (Rautela et al. 2021).

E-waste is usually collected by men women and children who hope to make some money out of the waste. Usable materials are sold to electrical and electronic stores where they are used to fix existing electronics; the rest undergo burning or acid washing which exposes the handlers to high concentrations of toxic chemicals (Li and Achal 2020). The elements present in the e-waste and its primary source are shown in Fig. 3.2.

The elements and some of their effects are mentioned below:

1. Lithium: Can cause nausea, vomiting, diarrhoea, fatigue, and muscle weakness.
2. Lead: Can cause kidney failure, central and peripheral nervous system (CNS & PNS) damage, blood and reproductive impairment, anaemia, and chronic neurotoxicity.
3. Cadmium: Long-term exposure is poisonous; causes bone, reproductive and kidney damage, and lung emphysema.
4. Mercury: Chronic brain, kidney, liver injury, foetal development issues such as underdeveloped CNS and PNS, anaemia and neurological disorders.
5. Chromium 6: Carcinogenic element that can cause lung cancer and DNA damage. It also damages the endocrine and reproductive systems.

3.2 Dealing with E-Waste

E-waste, in continents like Africa, is often recycled. Ghana and Egypt are known for importing e-waste from European countries and the USA so that parts from the e-waste can be reused to fix existing devices. Kenya and Nigeria sell redeemable e-waste as second-hand devices to students and internet cafes. Recycling of e-waste is done informally in these countries (Ahirwar and Tripathi 2021; Maphosa and Maphosa 2022). While this is one way to deal with e-waste, it is not the best since parts that cannot be salvaged—which often make up a significant percentage of e-waste—end up in landfills polluting the environment and putting those who live around the area at a health risk. Electronics are known to contain heavy metals

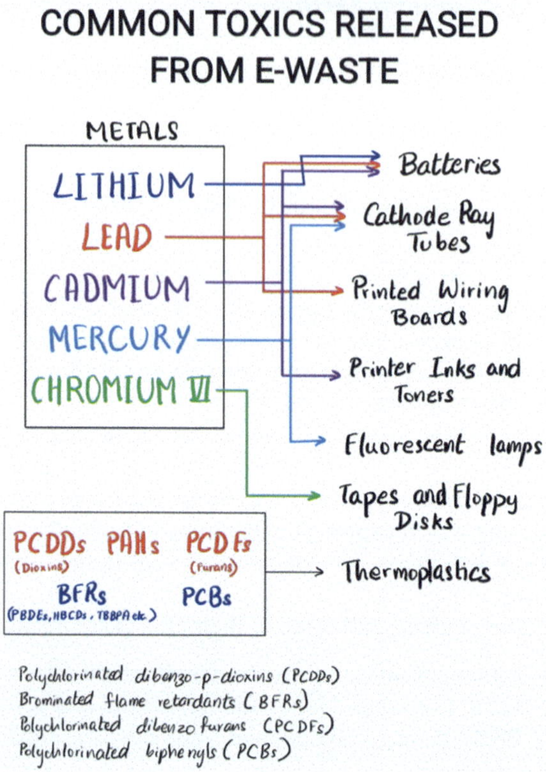

Fig. 3.2 Shows different elements released and their primary sources

like mercury that could seriously affect those who live around the area—especially children. Additionally, as the handlers come into contact with the e-waste, they put themselves at risk of heavy metal toxicity and other potentially hazardous chemicals. Due to the lack of knowledge on the health impacts and the disposal procedures of these materials they not only put themselves but also those around them at risk as dump sites are found all over Africa. Here women and children scavenge for e-waste to extract high-value metals such as gold, silver, copper, and aluminum using methods such as open burning and acid leaching (Andeobu et al. 2023).

In Chandigarh, India, 1.2 kg of gold, 6.2 kg of silver, and 0.3 kg of palladium are recovered every year. These precious metals are extracted primarily from ICT and consumer electronics (make up 96% of the source of extraction).

There are multiple ways that materials can be recovered from e-waste. Discussed ahead are the physical, chemical, and biological methods used to separate the components of e-waste.

3.2.1 Physical Methods

Physical separation methods allow for the e-waste to be categorized more efficiently. Preprocessing of WEEEs is necessary for optimal recycling and wealth creation from E-waste. This is also required for the removal of hazardous materials for proper disposal. As a first step of the recycling process chain, preprocessing determines to which recovery processes the materials are fed (Chancerel et al. 2009). E-waste is sometimes manually dismantled and sorted before it undergoes further fragmentation for easier processing.

Grinding and shredding

To separate e-waste as efficiently as possible, the materials must be separated from each other as much as possible. This is possible by mechanically grinding them. Grinding or shredding increases the surface area of the different components of WPCBs and other e-waste. Post grinding the mixed powders need to be separated efficiently. For example, waste-printed circuit boards undergo grinding before processes like electrostatic separation, gravity separation, etc. (Hou et al. 2010).

Electrostatic separation

Electrostatic separators run on the principle that dissimilar materials react differently when exposed to an electrical field. Their conductivity properties cause them to be attracted or repelled by an electric field allowing them to be separated based on their electric conductivity (Fig. 3.3). In the case of PCBs, metallic components like copper and aluminium gain a positive charge and get attracted to the negative electrode while non-metallic components like plastics gain a net negative charge and are attracted to the positive electrode (Islam et al. 2021; Li et al. 2023; Wu et al. 2008).

Pyrometallurgical processing

Smelting of WPBCs results in the enrichment or refinement of metals like copper in the metal phase. Grinding and shredding increases the surface area which allows for better heat distribution. This material is then placed in a furnace where organic materials, plastics, and resins are burned off while metal, metal oxides, and inorganic substances remain. The metal oxides are reduced using coke or carbon monoxide after which they undergo further refining by processes like gravity separation, floatation, or magnetic separation. This method has proven effective for the removal of copper from transistors and waste printed circuit boards (WPCBs). While pyrolysis is a good method, if done improperly, it can cause more damage than required. The gas phase should be disposed of properly. Pyrolysis can be good for retrieving energy and valuable metals. This method can help by turning plastic parts of e-waste into gas and liquid which makes the separation of metals easier (Sahle-Demessie et al. 2021).

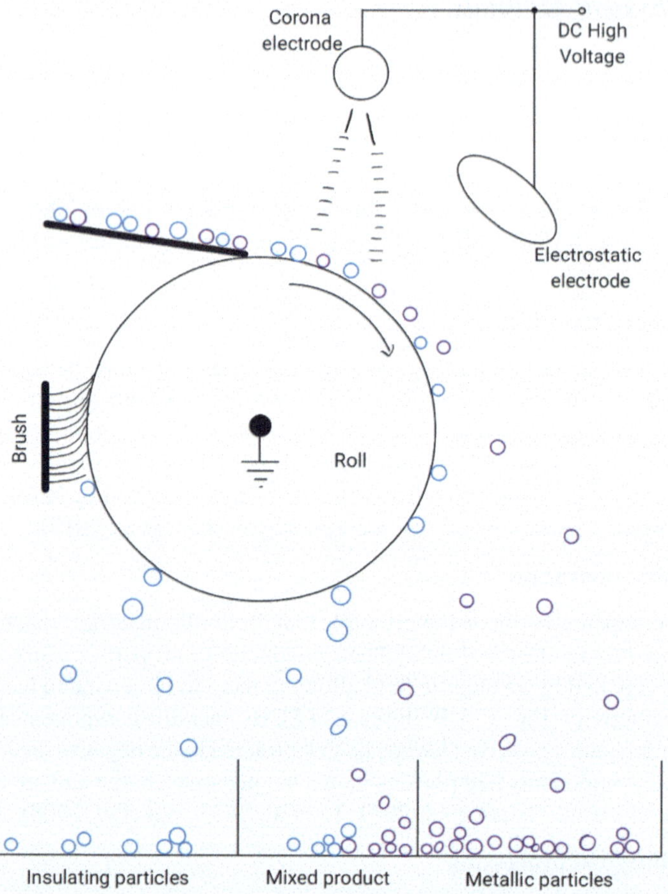

Fig. 3.3 Schematic of the role-type electrostatic separator (Amar et al. 2009)

Gravity separation

Gravity separation works on the principle that there are differences in density that allow for the separation of materials with different compositions. Gravitational separation uses gravity to separate heavier materials like metals from lighter materials like plastics. This method is useful during the early phases but to separate metals from other metals (a metal alloy) a technique called supergravity separation is used. This method uses centrifugal forces to increase the gravitational force experienced by the components of the metal. The usage of elevated temperatures to separate the metals was found to be especially effective. There was a linear increase in the recovery percentage of lead and tin from WPCBs between the temperatures of 700–900 °C after which the recovery percentage was stable (Table 3.1). The recovery percentage

Table 3.1 Shows the gravity separation and the relationship between temperature and recovery percentage

Temperature (°C)	Cu (wt%)	Zn (wt%)	Pb (wt%)	Sn (wt%)
700	36.83	1.05	19.11	23.86
800	47.01	1.71	14.56	20.77
900	59.96	3.40	9.44	11.55
1000	62.55	3.96	5.09	6.06
1100	65.83	3.71	4.60	5.44
1200	72.30	3.89	4.16	4.90
1300	78.74	4.07	3.01	3.89

of copper on the other hand was found to be the highest at 1300 °C (Meng et al. 2018).

Two crucibles are used for gravitational or supergravitational separation. The upper crucible contains the materials that are to be subjected to heating and centrifugation. It has small pores through which the liquid can seep down into the lower crucible. The lower crucible then collects the molten metal.

Magnetic separation

This method relies on the magnetic properties of materials for separation. Ferrous and non-ferrous materials are separated. Magnets attract iron and steel and leave behind metals like aluminium, lead, copper, zinc, etc. (Zhu et al. 2020).

3.2.2 Chemical Methods

In many cases of informal recycling, burning, and acid washing of Waste Printed Circuit Boards (WPCBs) occurs (Fig. 3.4). These methods are highly hazardous and not environmentally friendly. Chemical methods that will harm the life of the handlers and the environment must be adopted for the proper recovery of materials of waste electrical and electronic equipment (WEEE). Chemical methods like leaching are really useful as they are not as harmful when compared to the physical methods that are used informally (Roy et al. 2022).

Leaching

Acid leaching: Hydrochloric acid can be used as a leaching agent. It is often used for leaching out tin (metallic tin–alloy). HCl is not particularly good with leaching out copper because copper is significantly leached out only in an oxidative environment and HCl is a non-oxidative acid so there is not a significant reaction between copper and HCl.

$$Cu + 2HCl_{(aq)} \rightarrow CuCl_2 + H_{2(g)}\text{------}\Delta G^0_{80} = 92.577\ \text{kJ/mol}$$

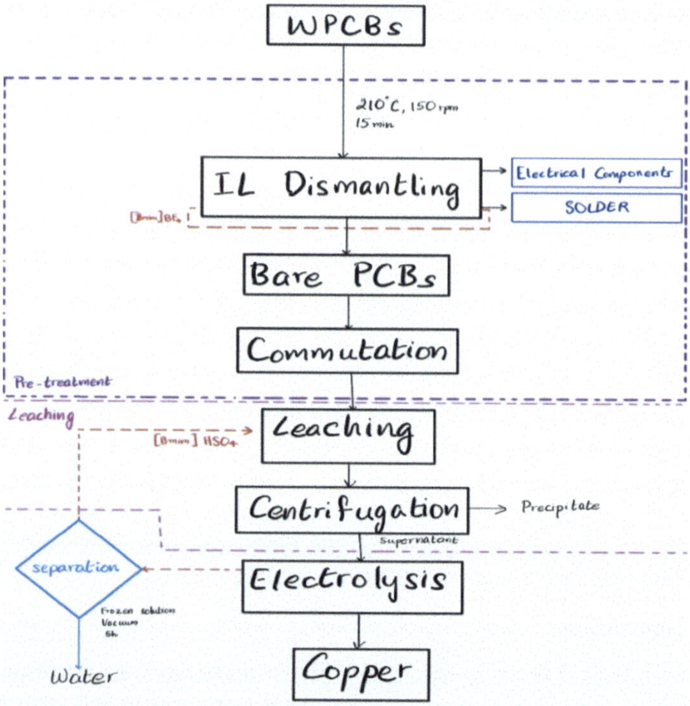

Fig. 3.4 shows the leaching of copper from Waste Printed Circuit Boards (WPCBs)

$$2Cu + 2HCl_{(aq)} \rightarrow 2CuCl_2 + H_{2(g)}\text{------}\Delta G^0_{80} = 1.68\ \text{kJ/mol}$$

During the leaching, the oxygen from the surrounding atmosphere acts as an oxidizing agent.

$$4Cu + 4HCl_{(aq)} + O_{2(aq)} \rightarrow 4CuCl_{(aq)} + 2H_2O\ \text{------}\Delta G^0_{80} = -63.479\ \text{kJ/mol}$$

However, it was noted that it was easier to leach out copper using inorganic acids post-pyrolysis. One reason for this could be that copper gets released from the composite during pyrolysis making it more accessible during the leaching process (Yaashikaa et al. 2022).

Copper is oxidized to copper oxides during combustion (not pyrolysis) which are well leached in HCl compared to metallic copper (Havlik et al. 2010)

$$CuO + 2HCl_{(aq)} + e^{+} \rightarrow CuCl + H_2O + Cl^{-}\ \text{------}\ \Delta G^0_{80} = -108.985\ \text{kJ/mol}$$

$$CuO + 2HCl_{(aq)} \rightarrow CuCl_2 + H_2O\ \text{------}\ \Delta G^0_{80} = -33.730\ \text{kJ/mol}$$

Leaching of tin in HCl:

$$Sn + 2HCl_{(aq)} \rightarrow SnCl_2 + H_{2(g)} \text{------} \Delta G^0_{80} = -56.091\ \text{kJ/mol}$$

Or in the presence of oxygen

$$2Sn + 4HCl_{(aq)} + O_{2(aq)} \rightarrow 2SnCl_{2(aq)} + 2H_2O \text{------} \Delta G^0_{80} = -294.376\ \text{kJ/mol}$$

HNO_3 is another popular inorganic acid that is used for leaching.

While acid leaching is popular it is not sustainable as it is toxic to handle and expensive. Alkaline leaching methods are being researched for retrieving copper from WPCBs and spent lithium-ion batteries (Gopikrishnan et al. 2020).

Alkaline leaching: When ammonia is used to extract copper from a solution, a particular complex of copper called $Cu(NH_3)_4{}^{2+}$ acts as an oxidizing agent to facilitate the dissolution of metallic copper. This process works best when the pH of the solution is between 9 and 11. The $Cu(NH_3)_4{}^{2+}$ complex reacts with the dissolved oxygen in the solution to produce $Cu(NH_3)^{2+}$. The newly formed $Cu(NH_3)_4{}^{2+}$ diffuses throughout the solution as this reaction quickly approaches equilibrium, making copper extraction easier (Jadhao et al. 2021; Feng et al. 2022).

DES leaching: In the case of $LiCoO_2$ batteries, the choline chloride (ChCl)/ethylene glycol (EG)-based DES is a classical system for leaching of valuable metals. It turns out that effective dissolution occurs when the radical β-hydrogen in the ChCl component lowers the Co^{3+} from LCO to soluble Co^{2+}. By encouraging the creation of oxidizing Cl^{3-} species, the dissolving capacity of ChCl/EG DES is enhanced, leading to 80% Li and >90% Co/Mn/Ni leaching efficiency from the ternary NCM battery (Alotaibi et al. 2021). Nevertheless, the extremely high reaction temperature (>180 °C) and comparatively low solid-to-liquid ratio (S/L, 20 g/L) restrict this method. The goal of developing several acidic DESs has been to achieve effective metal leaching (Guo et al. 2024).

Citrate leaching: The use of organic solvents instead of inorganic ones is preferred as organic solvents do not form harmful side products. It is a much safer method to use. Citrate is less corrosive and works between a pH of 3–8 as an effective complexing agent. Given below are the oxidation reactions of three metals using Sodium citrate solution. Hydrogen peroxide was used as an oxidant as it is relatively clean and leaves behind no products other than water. The extraction rates are higher at temperatures greater than 20 up to 30 °C. Above 30 °C the leaching rates decrease (Torres et al. 2018)

$$2Cu^\circ + 2H_2O_2 + 2HCit^{2-} + H^+ \rightarrow Cu_2(Cit)_2OH^{3-} + 3H_2O$$

$$Fe^\circ + H_2O_2 + HCit^{2-} + 2H^+ \rightarrow FeHCit + 2H_2O$$

$$Pb^{\circ} + H_2O_2 + HCit^{2-} + 2H^+ \rightarrow PbHCit + 2H_2O$$

$$\text{Note : Cit}^- \text{ is the citrate ion}$$

Spent NiMH battery: A recycling company in Finland examined the composition of the battery powder in NiMH batteries. Rare earth elements like La (10 wt%), Ce (6 wt%), and Nd (3 wt%) were found in addition to carbon, hydrogen and nitrogen. Three organic acids–gluconic acid sodium salt, 2-keto-D-gluconic acid hemicalcium salt, and 5-keto-D-gluconic acid potassium salt—that are produced by biological organisms was evaluated to see which ones would leach out the rare earth elements best. Gluconate was found to be the best for dissolving the rare earth elements while 5-ketogluconate was best for dissolving the base metals (Rasoulnia et al. 2021).

Electrochemical Separation

The use of electricity to separate metals from the composite. Metals are extracted by conducting electricity through a solution that then results in the accumulation of the element (copper) in the cathode. On the anode, regeneration of the leaching agent occurs (Fogarasi et al. 2014).

Cathode:

$$Cu^{2+} + 2e^- \leftrightarrow Cu \text{ ------ } E^0 = 0.337V$$

$$Cu^+ + e^- \leftrightarrow Cu \text{ ------ } E^0 = 0.506V$$

Secondary reactions:

$$Cu^{2+} + e^- \leftrightarrow Cu^+ \text{------ } E^0 = 0.167V$$

$$Sn^{2+} + 2e^- \rightarrow Sn \text{ ------ } E^0 = -0.137V$$

$$Sn^{4+} + 2e^- \rightarrow Sn^{2+} \text{ ------ } E^0 = 0.154V$$

$$Fe^{3+} + e^- \rightarrow Fe^{3+} \text{ ------ } E^0 = 0.77V$$

Anode:

$$Sn^{2+} \rightarrow Sn^{4+} + 2e^-$$

$$2H_2O \rightarrow 4H^+ + 4e^- + O_2 \text{ ------ } E^0 = 1.24V$$

$$2Cl^- \rightarrow Cl_2 + 2e^- \text{ ------ } E^0 = 1.36V$$

Using this method, scientists were able to retrieve copper and gold at high purity rates without using much energy. The best conditions for this process were found to be using 4 mA/cm^2 of electricity and a solution containing acid and iron chloride (0.37 M). Pure copper of about 99.04% and gold was extracted successfully. The amount of energy used was around 1.75 kW per kilogram of copper.

3.2.3 Biological Methods (Microbes)

Bioleaching

A study by Singh et al. (2016), researched the ability of native *Lactobacillus* strains to bio-adsorb lead using standardized bio-adsorption assays. A total of forty-one strains of lactobacilli were involved in this in vitro experiment and the amount of lead ions adsorbed was measured. The results of this experiment were promising as nearly ten strains of lactobacilli proved to have a lead retention rate exceeding 80%. After further investigation, *Lactobacillus plantarum* HD51 was found to be the most effective strain to adsorb lead amongst the forty-one that were examined (Han et al. 2023). Scanning electron microscopy (SEM) was used to observe the binding of lead ions to the cell wall of this bacterium. This allows us the possibility of using these microbes to leach lead out of e-waste (Table 3.2). Lead is a toxic element that has dangerous side effects for humans. Usually, e-waste is managed in developing countries by men, women, and children who lack the awareness of the risk that goes into dismantling and sorting e-waste. Removing hazardous chemicals before giving them to handlers would significantly improve their quality of life.

Waste to wealth creation, using the E-waste

Post-processing, the materials extracted from e-waste can be used for various purposes. From the above-mentioned methods, many metals, plastics, and other elements are extracted from e-waste. These can now be used to make many different products (Kaur et al. 2022; Leong et al. 2016).

Rare earth metals are on the verge of depletion. Many elements like lithium will not be available as easily over the next few decades. Extracting these elements and reusing them makes it possible for us to continue advancing technologically while also being conscious of the environment. Using products that have been retrieved from e-waste prevents the depletion of natural sources of elements while also preventing the accumulation of e-waste in landfills resulting in health issues and informal recycling (Arya and Kumar 2020).

Manufacturing of electrical and electronic products

Components such as circuit boards, wiring, and other small parts that have not been extensively damaged can be reused in newer electrical devices or they could be used

Table 3.2 Shows the bioleaching abilities of various microorganisms

S. No	Organism	Researcher	Year	Result
1	*Cupriavidus necator*	Morlino et al.	2023	*Cupriavidus necator* is a multipurpose bacterium that is well-known for its resistance to exposure to heavy metals, especially copper. The main mechanisms responsible for its resistance to copper stress include chelation, efflux pumps, and maybe enzymatic processes. While chelating drugs bind to copper ions and lessen their toxicity, efflux pumps assist in removing copper from the cell. Detoxification may potentially be aided by enzymatic activities, although more investigation is required. A new strain of *C. necator* (B9) was found near a medieval foundry in a recent study, underscoring the need to examine its genome to gain an understanding of its processes of copper resistance
		Casida et al.	1988	
		Xie et al.	2023	
2	*Bacillus* spp.	Wróbel et al.	2023	The *Bacillus* species have shown marvellous heavy metal tolerance in addition to which researchers have found that this species is capable of bioaccumulation—meaning that they can accumulate high concentrations of heavy metals without any ill effects themselves. While less explored, some *Bacillus* might even have enzymatic detoxification. These "detox enzymes" modify metals into less harmful forms, further reducing their threat. This could be a way to detoxify the environment or even become a source of essential nutrients for other organisms. By understanding these mechanisms, we can harness the power of Bacillus for bioremediation. These metal-tolerant microbes can be used to clean up contaminated environments, paving the way for a cleaner future
		Syed et al.	2015	
		Karmakar et al.	2024	
		Alotaibi et al.	2021	
3	*Lactobacillus plantarum* HD51	Singh et al.	2022	The results of this experiment were promising as nearly ten strains of lactobacilli proved to have a lead retention rate exceeding 80%
4	*Lactobacillus plantarum* HIF33			
5	Halomonas BVR1	Manasi et al.	2017	This study proved that the bacteria along with a biopolymer were capable of adsorbing more lead and cadmium than the polymer alone. The combined system was found to hold 24.15 mg of lead and 23.88 mg of cadmium for every gram of material used

to fix existing ones. The metals that are extracted through physical and chemical processes can be recycled to provide a circular economy. Batteries recovered from e-waste should be checked for damages before being dismantled. If damages are minimal, the batteries may be refurbished and reused. In other cases, the battery must be dismantled and whatever is salvaged can be used to make new batteries (Rajesh and Rajesh 2015).

Construction material

Recycled plastic from electronic garbage can be shaped into bricks, flooring tiles, roofing tiles, and insulation panels. These materials offer strength and insulating properties, reducing the requirement for virgin polymers. Glass recovered from e-waste can be crushed and used as aggregates in concrete, or it can be made into architectural glass for walls, doors, and windows. For example, clay roof tiles coated with waste glass were produced in a single fire process using soda lime silica waste glass (WG1) from recovered hollow glass and lead silica waste glass (WG2) from television sets (REF). Waste glass was cleaned, ground, sieved, and dried throughout processing (Ravindra and Mor 2019). The clay substrate was prepared using a local clay paste that possessed a high degree of environmental resistance. An opalescent solution called Cerograf was layered between the substrate and the waste glass coating to improve adhesion. The test results showed that the tiles met the industry standards for both flat and curved flexural strength, with curved tiles having greater strength. The utilization of waste glass improved flexural strength and solar reflectance (WG2 outperformed WG1 in this aspect as well as for solar reflectance). The tiles complied with the relevant requirements for fire and frost resistance, and high impermeability. A leaching test confirmed that employing waste glass did not pose a risk to public health or the environment because there was little lead leaked from CRT waste glass. This study proved that waste glass from CRTs can be used in the making of tiles for roofs (Mourou et al. 2023).

Automotive parts

Steel, aluminum, and copper that are recovered from e-waste can be used to make recycled metals for vehicle components like engine parts, chassis, body panels, and interior trim. Plastic waste gathered from electronic equipment can be used to mould automotive parts such as dashboards, bumpers, interior trim, and insulating panels (Van Yken et al. 2021).

Jewellery

Precious metals extracted from e-waste like gold, silver, and platinum can be processed and used to make jewellery and accessories. Recycled precious metals are not lacking in quality and using them instead of virgin metals makes no difference in how they appear on the outside. Recycled precious metals are, however, much more environmentally friendly than virgin metals (Rene et al. 2021).

Raw materials

Recycled metals extracted from e-waste can be used in a variety of production processes, including casting, forging, machining, and fabrication. These metals can be used to make a wide range of industrial items, such as tools, machinery, equipment, and components. Plastic recovered from e-waste can be recycled and made into new products for a range of applications, including consumer goods, packaging, vehicle parts, and home appliances. Repurposed components from e-waste, such as glass, ceramics, and composites, can be used to make composite panels, tiles, ceramics, and insulating materials.

Conclusion

This chapter presents a promising and innovative strategy for addressing the environmental and economic challenges associated with electronic waste (e-waste). By harnessing the power of indigenous microorganisms, this approach offers a sustainable solution for the extraction and recovery of valuable metals from e-waste while simultaneously mitigating the environmental hazards posed by heavy metal contamination. Furthermore, the utilization of indigenous microorganisms underscores the importance of bio-based solutions in addressing complex environmental challenges. By leveraging the natural capabilities of microorganisms, this approach offers a cost-effective and environmentally friendly alternative for heavy metal recovery. However, while the potential of this approach is evident, further research is required to optimize the process parameters, enhance metal solubilization efficiency, and scale up the technology for industrial applications. Additionally, the economic feasibility and environmental impact of this approach need to be evaluated to ensure its viability on a larger scale. In summary, the chapter provides valuable insights into a new approach for wealth creation through the solubilization of heavy metals from e-waste using indigenous microorganisms. By combining environmental sustainability with economic viability, this approach has the potential to revolutionize e-waste management practices, paving the way for a more sustainable and resource-efficient future.

Acknowledgment The authors would like to acknowledge Sponsored Research and Industrial Consultancy (SpoRIC), Vellore Institute of Technology, Vellore for providing financial support through SEED grant (SG20240102).

References

Ahirwar R, Tripathi AK (2021) E-waste management: a review of recycling process, environmental and occupational health hazards, and potential solutions. Environ Nanotechnol Monit Manag 15:100409

Alotaibi BS, Khan M, Shamim S (2021) Unraveling the underlying heavy metal detoxification mechanisms of Bacillus species. Microorganisms 9(8):1628

Amar T, Salah-Eddine B, Muhammed Y, Dascalescu L (2009) Experimental analysis and optimisation of a free-fall triboelectric separator of granular plastic particles. Int J Sustain Eng 2(3):184–191

Andeobu L, Wibowo S, Grandhi S (2023) Informal E-waste recycling practices and environmental pollution in Africa: what is the way forward?. Int J Hyg. Environ. Health 252:114192

Arya S, Kumar S (2020) Bioleaching: urban mining option to curb the menace of E-waste challenge. Bioengineered 11(1):640–660

Casida LE (1988) Response in soil of Cupriavidus necator and other copper-resistant bacterial predators of bacteria to addition of water, soluble nutrients, various bacterial species, or *Bacillus thuringiensis* spores and crystals. Appl Environ Microbiol 54. https://doi.org/10.1128/aem.54.9.2161-2166.1988

Chancerel P, Meskers CE, Hagelüken C, Rotter VS (2009) Assessment of precious metal flows during preprocessing of waste electrical and electronic equipment. J Ind Ecol 13(5):791–810

Feng P, Wang Z, Meng L, Guo Z (2022) Supergravity-enhanced liquation crystallization for metal recovery from waste printed circuit boards. Chem Eng Process Process Intensif 173:108813

Fogarasi S, Imre-Lucaci F, Imre-Lucaci Á, Ilea P (2014) Copper recovery and gold enrichment from waste printed circuit boards by mediated electrochemical oxidation. J Hazard Mater 273:215–221

Forti V, Baldé CP, Kuehr R, Bel G (2020) The global e-waste monitor 2020. United Nations University (UNU), International Telecommunication Union (ITU) & International Solid Waste Association (ISWA), Bonn/Geneva/Rotterdam, p 120

Ghimire H, Ariya PA (2020) E-wastes: bridging the knowledge gaps in global production budgets, composition, recycling and sustainability implications. Sustain Chem 1(2):154–182

Gopikrishnan V, Vignesh A, Radhakrishnan M, Joseph J, Shanmugasundaram T, Doble M, Balagurunathan R (2020) Microbial leaching of heavy metals from e-waste: opportunities and challenges. In: Biovalorisation of wastes to renewable chemicals and biofuels, pp 189–216

Guo M, Deng R, Gao M, Xu C, Zhang Q (2024) Sustainable recovery of metals from E-Waste using deep eutectic solvents: advances, challenges, and perspectives. Curr Opin Green Sustain Chem 100913

Hamsavathi K, Prakash KS, Kavimani V (2020) Green high strength concrete containing recycled Cathode Ray Tube Panel Plastics (E-waste) as coarse aggregate in concrete beams for structural applications. J Build Eng 30:101192

Han Z, Levett A, Edraki M, Jones MW, Howard D, Southam G (2023) Accelerating bioleaching of tungsten mining wastes using indigenous acidophilic bacteria. J Hazard Mater 454:131490

Havlik T, Orac D, Petranikova M, Miskufova A, Kukurugya F, Takacova Z (2010) Leaching of copper and tin from used printed circuit boards after thermal treatment. J Hazard Mater 183(1–3):866–873

Hou S, Wu J, Qin Y, Xu Z (2010) Electrostatic separation for recycling waste printed circuit board: a study on external factor and a robust design for optimization. Environ Sci Technol 44(13):5177–5181

Hu M, Wang J, Xu Z (2017) Pyrolysis-based technology for recovering copper from transistors on waste printed circuit boards. ACS Sustain Chem Eng 5(12):11354–11361

Islam A, Swaraz AM, Teo SH, Taufiq-Yap YH, Vo DVN, Ibrahim ML, Awual MR (2021) Advances in physiochemical and biotechnological approaches for sustainable metal recovery from e-waste: a critical review. J Clean Prod 323:129015

Ismail H, Hanafiah MM (2020) A review of sustainable e-waste generation and management: present and future perspectives. J Environ Manag 264:110495

Jadhao PR, Pandey A, Pant KK, Nigam KDP (2021) Efficient recovery of Cu and Ni from WPCB via alkali leaching approach. J Environ Manage 296:113154

Karmakar D, Magotra S, Negi R, Kumar S, Rustagi S, Singh S et al (2024) Bacillus species for sustainable management of heavy metals in soil: current research and future challenges. J Appl Biol Biotechnol 12(2):22–35

Kaur P, Sharma S, Albarakaty FM, Kalia A, Hassan MM, Abd-Elsalam KA (2022) Biosorption and bioleaching of heavy metals from electronic waste varied with microbial genera. Sustainability 14(2):935

Lebbie TS, Moyebi OD, Asante KA, Fobil J, Brune-Drisse MN, Suk WA, Carpenter DO (2021) E-waste in Africa: a serious threat to the health of children. Int J Environ Res Public Health 18(16):8488

Leong SS, Yeap SP, Lim J (2016) Working principle and application of magnetic separation for biomedical diagnostic at high-and low-field gradients. Interface Focus 6(6):20160048

Li W, Achal V (2020) Environmental and health impacts due to e-waste disposal in China—a review. Sci Total Environ 737:139745

Li J, Yan S, Li Y, Wang Z, Tan Y, Li J et al (2023) Recycling Si in waste crystalline silicon photovoltaic panels after mechanical crushing by electrostatic separation. J Clean Prod 415:137908

Liu K, Tan Q, Yu J, Wang M (2023) A global perspective on e-waste recycling. Circ Econ 2(1):100028

Maphosa M, Maphosa V (2022) A bibliometric analysis of the effects of electronic waste on the environment. Glob J Environ Sci Manag 8(4):589–606

Manasi, Mohapatra S, Rajesh N, Rajesh V (2017) Impact of heavy metal lead stress on polyamine levels in Halomonas BVR 1 isolated from an industry effluent. Sci Rep 7(1):13447

Meng L, Zhong Y, Guo L, Wang Z, Chen K, Guo Z (2018) Recovery of Cu and Zn from waste printed circuit boards using super-gravity separation. Waste Manage 78:559–565

Morlino MS, García RS, Savio F, Zampieri G, Morosinotto T, Treu L, Campanaro S (2023) Cupriavidus necator as a platform for PHA production: An overview of strains, metabolism, and modeling approaches. Biotechnol Adv 108264

Mourou C, Zamorano M, Ruiz DP, Martín-Morales M (2023) Characterization of ceramic tiles coated with recycled waste glass particles to be used for cool roof applications. Constr Build Mater 398:132489

Nasier S (2021) Utilization of recycled form of concrete, E-wastes, glass, quarry rock dust and waste marble powder as reliable construction materials. Mater Today Proc 45:3231–3234

Rajesh V, Rajesh N (2015) An indigenous Halomonas BVR1 strain immobilized in crosslinked chitosan for adsorption of lead and cadmium. Int J Biol Macromol 79:300–308

Rasoulnia P, Barthen R, Puhakka JA, Lakaniemi AM (2021) Leaching of rare earth elements and base metals from spent NiMH batteries using gluconate and its potential bio-oxidation products. J Hazard Mater 414:125564

Rautela R, Arya S, Vishwakarma S, Lee J, Kim KH, Kumar S (2021) E-waste management and its effects on the environment and human health. Sci Total Environ 773:145623

Ravindra K, Mor S (2019) E-waste generation and management practices in Chandigarh, India and economic evaluation for sustainable recycling. J Clean Prod 221:286–294

Rene ER, Sethurajan M, Ponnusamy VK, Kumar G, Dung TNB, Brindhadevi K, Pugazhendhi A (2021) Electronic waste generation, recycling and resource recovery: technological perspectives and trends. J Hazard Mater 416:125664

Roy JJ, Rarotra S, Krikstolaityte V, Zhuoran KW, Cindy YDI, Tan XY et al (2022) Green recycling methods to treat lithium-ion batteries E-waste: a circular approach to sustainability. Adv Mater 34(25):2103346

Sahle-Demessie E, Mezgebe B, Dietrich J, Shan Y, Harmon S, Lee CC (2021) Material recovery from electronic waste using pyrolysis: emissions measurements and risk assessment. J Environ Chem Eng 9(1):104943

Singh N, Li J, Zeng X (2016) Global responses for recycling waste CRTs in e-waste. Waste Manage 57:187–197

Singh A, Pal DB, Mohammad A, Alhazmi A, Haque S, Yoon T, Gupta VK et al (2022) Biological remediation technologies for dyes and heavy metals in wastewater treatment: new insight. Bioresour Technol 343:126154

Syed S, Chinthala P (2015) Heavy metal detoxification by different Bacillus species isolated from solar salterns. Scientifica (2015)

Torres R, Segura-Bailón B, Lapidus GT (2018) Effect of temperature on copper, iron and lead leaching from e-waste using citrate solutions. Waste Manage 71:420–425

Van Yken J, Boxall NJ, Cheng KY, Nikoloski AN, Moheimani NR, Kaksonen AH (2021) E-waste recycling and resource recovery: a review on technologies, barriers, and enablers with a focus on Oceania. Metals 11(8):1313

Wróbel M, Śliwakowski W, Kowalczyk P, Kramkowski K, Dobrzyński J (2023) Bioremediation of heavy metals by the genus Bacillus. Int J Environ Res Public Health 20(6):4964

Wu J, Li JIA, Xu Z (2008) Electrostatic separation for recovering metals and nonmetals from waste printed circuit board: problems and improvements. Environ Sci Technol 42(14):5272–5276

Xie Z, Wang D, Ben Fekih I, Yu Y, Li Y, Alwathnani H, Rensing C (2023) Whole genome sequence analysis of Cupriavidus necator C39, a multiple heavy metal (loid) and antibiotic resistant bacterium isolated from a gold/copper mine. Microorganisms 11(6):1518

Yaashikaa PR, Priyanka B, Kumar PS, Karishma S, Jeevanantham S, Indraganti S (2022) A review on recent advancements in recovery of valuable and toxic metals from e-waste using bioleaching approach. Chemosphere 287:132230

Zhu XN, Nie CC, Wang SS, Xie Y, Zhang H, Lyu XJ et al (2020) Cleaner approach to the recycling of metals in waste printed circuit boards by magnetic and gravity separation. J Clean Prod 248:119235

Chapter 4
Nanobioremediation: A Path to Sustainable Pesticide Remediation

Reshma Raviuday Pednekar, Arumugam Vijaya Anand, Chandramohan, and Anand Prem Rajan

Abstract The widespread use of pesticides in industry and agriculture persists in environmental bodies posing significant risks to human health and the environment. Conventional bioremediation has limitations in terms of precision and efficiency but nano bioremediation is a unique approach that combines the properties of nanomaterials with the augment of pesticide bioremediation efforts. It is an innovative strategy integrating nanobiotechnology and bioremediation techniques called nano bioremediation which offers a viable approach to deal with the urgent problem of pesticide pollution in the environment with enhanced efficiency for treating these pollutants. The nanostructures act as carriers with immobilized or encapsulated remediating microbial enzymes which can facilitate the bioremediation of such hazardous materials with improved stability and delivery to the target sites. The elimination of harmful pesticide contaminants using the synergistic effect of nanoparticles and microbial technologies along with the application is the primary aim of this chapter. Also, the importance of unique nano bioremediation techniques as a sustainable and cleaner technology for pollutant remediation will be discussed.

Keywords Nanobiotechnology · Nano bioremediation · Microbes · Pesticide · Bioremediation

---

R. R. Pednekar · A. P. Rajan (✉)
School of Biosciences and Technology, Vellore Institute of Technology, Vellore 632014, Tamil Nadu, India
e-mail: aprdbt@gmail.com

A. V. Anand
Department of Human Genetics and Molecular Biology, Bharathiar University, Coimbatore, 641046, Tamil Nadu, India

Chandramohan
Department of Community Health Sciences, College of Applied Medical Sciences, King Saudi University, 10219, Riyadh, Saudi Arabia

© The Author(s), under exclusive license to Springer Nature Switzerland AG 2025
A. Jeyaseelan and K. Murugasen (eds.), *Sustainable Environmental Remediation: Avenues in Nano and Biotechnology*, Environmental Science and Engineering,
https://doi.org/10.1007/978-3-031-78483-5_4

4.1 Introduction

With the growing population, the demand for food is also increasing and to meet the requirements and protect the plants farmers use different strategies. One such strategy is the application of pesticides to protect the crop from pests. The agricultural sector holds the most demand for pesticides accounting for 85% of total output worldwide, while the remainder is utilized for other purposes such as controlling seasonal vector-borne diseases and pest elimination in home and industrial settings. Pesticides are chemical substances used to kill, control or suppress plants and organisms that cause loss to human health and the environment (Mishra et al. 2020). It can be classified according to their target pest species, chemical and physical states (solid, gas or liquid) and source of origin (natural or synthetic). Based on the target organism they can be classified as insecticides (for insect pests), herbicides (unnecessary vegetation) fungicides (pathogenic fungi) and others. Organochlorines, organophosphorus, pyrethroid, etc. are also pesticides that are classified based on their chemical nature.

However, more accurate criteria for pesticide classifications may be based on the class of chemical compounds and their functional groups, toxicity and different modes of action (Table 4.1). The World Health Organization (WHO) divided pesticides into four main classes based on their type and level of toxicity: Ia (extremely hazardous), Ib (highly hazardous), II (moderately hazardous), III (slightly hazardous) and IV (product unlikely to present acute toxicity/hazard in normal use). However, the use of pesticides comes with several problems such as agricultural pollution which is a global environmental hazard, mostly caused by the use of many farming inputs (such as pesticides and fertilizers) and multiple practices such as excessive soil tillage and runoff. These agro-pollutants are highly toxic and linked to significant negative impacts on water, soil and air (Pednekar and Rajan 2024). Additionally, these chemicals also contribute to various health issues upon exposure such as eye, skin and nervous-related problems with severity leading to the formation of cancer.

4.2 Impact of Pesticides on Human Health and Environment

People who live near farmlands have an increased risk of inhaling pesticide spray fumes than the general population. Furthermore, nursing moms and pregnant women from such populations are at higher risk of transferring the pesticides via the placenta to the growing fetus in the womb and through breast milk to the nursing newborns leading to spontaneous abortions, delayed neurodevelopment during childhood and permanent change in the child's behaviour (Saxena et al. 1981; Allsop et al. 2015). Deaths induced by organophosphorus (OPs) are called cholinergic syndrome where in the person suffers from headache, slurred speech, coma, blockage of the respiratory center, blurred vision and delayed neuropathy. They increase the risk of neurodegenerative diseases such as Alzheimer's and Parkinson's disease (Eskenazi et al. 2006).

Table 4.1 Type of pesticide, examples and its mode of action against the target organism

Pesticide category	Pesticide class	Use	Mode of action	References
Herbicides	Chlorphenoxy herbicides Phosphates Carbamates	Kill unwanted plants and weeds	Acetylcholinesterase (AChE) inhibitors Lipid biosynthesis inhibitor Cell membrane disruptors	Sherwani et al. (2015)
Insecticides	Organochlorines Organophosphates Carbamates Pyrethroids	Kill or repel insects	GABA-gated chloride channel blockers Sodium channel modulators	Sparks et al. (2020)
Fungicides	Thiocarbamates Triazoles Pyrimidinamines	Kills fungi and moulds	Disrupts lipids, sterol and other membrane components NADH oxidoreductase (complex I) inhibitor	Yang et al. (2011)
Rodenticides	Endrin	Kill rodents	Chloride blocking channels Inhibitors epoxy reductase and Vitamin K reductase	Berny (2011)
Algaecides	Sodium carbonate peroxyhydrate	Kill or prevent the growth of algae	Disruption of genetic material	Berthold et al. (2020)
Avicides	Endrin	Kill birds	Chloride blocking agent	Rahman et al. (2019)
Others	*B. thuringiensis*	Includes seed disinfectants (protect from fungus, insects, birds) Biopesticides Plant growth-promoting chemicals for increasing growth and yield	Microbial disruptors of midgut membranes e.g. *B. thuringiensis*	Rajput et al. (2020)

Impair the endocrine function, enzymatic function and signaling mechanisms at the cellular level. It also affects gene expression which may transfer to generation through epigenetic inheritance (Nicolella and Assis 2022).

Pesticides may not kill only the target species but also the surrounding species which are an essential part of the food chain as 98% of sprayed pesticides directly or indirectly affect them. Amphibians the most threatened and rapidly decreasing species on Earth are sensitive to pesticide toxicity (Özkara et al. 2016). Soil respiration is decreased by 35% and almost 90% of water sources in agro lands are polluted with pesticides due to the microbial threat. IN European nations 70% decline in insect biomass and a 50% decline in farmland birds were reported. Besides 30% honey population is reported in American and European countries (Ali et al. 2021). In recent years green and sustainable remediation strategies have been in trend prioritizing the decrease of pollutant concentrations, limiting toxicity levels and eliminating environmental consequences. As a result, significant efforts and research into the development of technology for the rehabilitation of polluted sites with faster schedules and lower decontamination costs are progressively supported.

4.3 Nanotechnology in Pesticide Remediation

Nanotechnology has attracted a lot of attention in recent years. It encompasses materials sized between 1–100 nm (nm) (Yap et al. 2020). Although a compound's size is what defines it as a nanomaterial, its shape and geometry also have a big impact on its properties (Albanese et al. 2012). With the use of nanotechnology, materials intended for use can be transformed by nanoparticles leading to notable advancements in mechanical, thermal and barrier properties. A multitude of applications are made possible by the meticulous development of different nanoparticle morphologies such as spheres, rods, quantum dots and particles which allow for variation in the application and can result in endless opportunities for technological growth (Chen et al. 2024).

4.3.1 Tools for Nano Remediation of Pesticide Pollution

Nanomaterials use a wide range of materials for effective pesticide remediation. For that, it follows a sequential mechanism leading to pollutant pollutant-free environment. Thus, the mechanism for pesticide removal varies depending on the material used for the remediation of pesticides as explained in Fig. 4.1.

4.3.1.1 Photocatalysts

In photocatalysis light and semiconductor are used wherein the light of a particular wavelength falls on a semiconductor creating positive holes via the transfer of electrons to the conduction band (Sengupta and Parijat 2019). As there is a sequential movement of electron–hole pairs towards the surface of semiconductors, organic pollutants are transferred into less harmful compounds (Guo et al. 2023). The most widely used photocatalyst materials for pesticide remediation include nanomaterials such as silver, titanium dioxide and zinc oxide (ZnO).

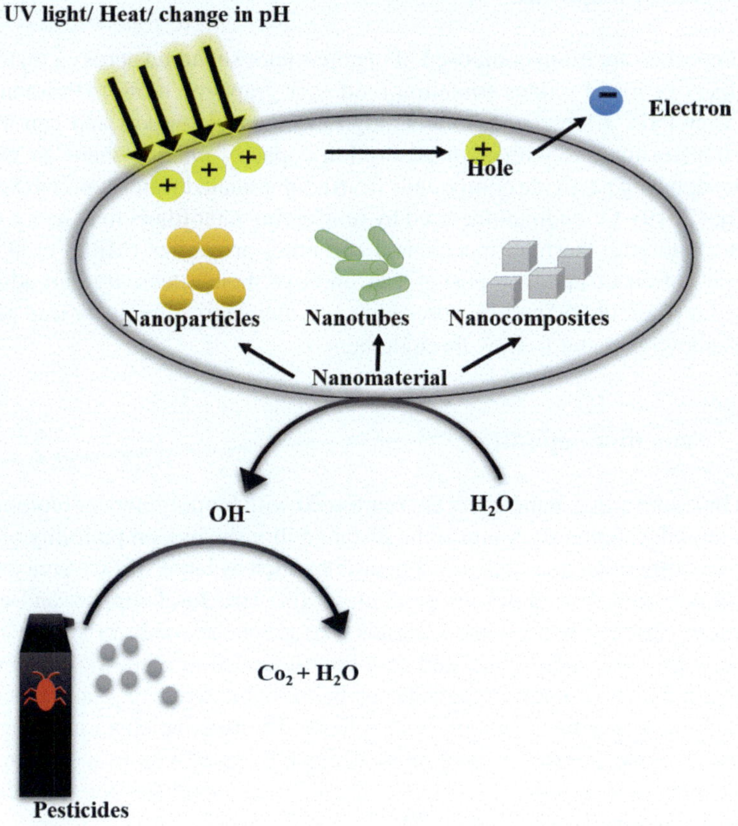

Fig. 4.1 Mechanism of pesticide remediation by nanomaterials such as nanoparticles, nanocomposite, and nanotubes upon trigger factor such as UV light, heat or change in pH

4.3.1.2 Nanofilters

Nanomaterials with large surface areas and adsorption capacities such as metal oxides, graphene oxide and carbon nanotubes are used as nanofilters (Chadha et al. 2022). Through physical and chemical interactions, pesticide molecules are successfully trapped within the nanomaterial matrix obstructing their passage through the filter. As a result, pesticides are retained in nanofilters and water is filtered through the nano holes of the filter (Moreira et al. 2022). Furthermore, these nanomaterials can be functionalized with certain ligands and coatings to improve pesticide adsorption and selectivity.

4.3.1.3 Nano Composites

Nanocomposites are often composed of nanoparticles scattered across a matrix such as polymers or metal oxides (titanium oxides or graphene oxide) (Prasanna et al. 2019). They have a large surface area, high mechanical support and can manifest radical transformations with high adsorption capacity making them an excellent option is capturing pesticide compounds from contaminated water sources (Sengupta and Parijat 2019). Often manufactured by reinforcing nanofillers to enhance optical, magnetic, catalytic, thermal, mechanical and other properties (Mittal et al. 2024). Functionalization of nanocomposites can improve their selectivity and adsorption capacity for pesticides. The tiny pores with less diameters effectively trap pesticide particles allowing water to flow through them.

4.3.1.4 Nano Biocomposites

In nano biocomposites, nanofillers are reinforced with biopolymers or biomolecules. The cutting-edge materials combine the sustainability and biocompatibility of natural polymers with special qualities of nanoparticles such as large surface area and reactivity. Biopolymers like chitosan or cellulose give structural support and environmental compatibility while nanomaterials like graphene oxide or iron nanoparticles enable effective adsorption and catalytic degradation of pesticides. Pesticide degrading enzymes or microorganisms can be immobilized using nano biocomposites, which increases their stability and activity for effective clean up. Alginate or starch are the most common biopolymers which are mixed with iron oxide nanoparticles for improved stability, reactivity and catalytic capabilities for the breakdown of pesticides (Nasrollahzadeh et al. 2021).

4.4 Bioremediation of Pesticides

The term ‘Bioremediation’ refers to the process of stimulating microbial growth for a quick breakdown of hazardous contaminants to an environmentally safe level in the soil, air and water (Niti et al. 2013). It provides an economical and environmentally beneficial substitute for conventional remediation techniques by using the metabolic powers of living things to cleanse contaminated areas. There are two ways by which bioremediation can be carried out namely Ex situ where in the contaminated materials are removed and treated off-site and In-situ where the contaminants are treated directly at the site of contamination (Singh 2008). Bioremediation techniques have become an attractive alternative in environmental cleanup when compared with physio-chemical methods due to their relatively low capital costs, least disruptive methods and minimal secondary pollution of by-products (Ahmad and Ahmad 2014).

4.4.1 Mechanism of Pesticide Degradation

Bioremediation of pesticides can be achieved by microorganisms through enzymatic reactions and metabolic pathways. Because of their metabolic capabilities bacteria, algae and fungi can either use pesticides as carbon and energy sources or break them into less hazardous compounds as seen in Fig. 4.2. In the initial step, the microbe uptakes the pesticide through its cells by passive diffusion or active transport mechanism. Inside the cell, microorganisms produce specific enzymes for the target pesticide and transform the chemical structure of the pollutant to less toxic compound with the production of carbon dioxide, inorganic ions and water (Stoytcheva et al. 2011). Enzymes such as hydrolases (hydrolyses ester bonds), oxidoreductases (electron transfer process) and transferases (transfer of functional groups between molecules) are most commonly identified in this process (Ortiz-Hernández et al. 2013).

Alcaligenes, Flavobacterium, Pseudomonas, Bacillus and *Rhodococcus* are some bacterial species known to be effective against pesticide degradation (Varjani et al. 2018). Besides white rot fungi such as *Phanerochaete, Chrysosporium and Trametes* are known to be effective against lindane, atrazine, diuron, terbuthylazine, metalaxyl, DDT, gamma–gamma-hexachlorocyclohexane, aldrin, heptachlor, etc. (Bhadouria et al. 2020). Furthermore, algae such as *Spirulina, Chlorella* and *Scenedesmus* are known for their pesticide removal capacities via bioaccumulation, adsorption and photosynthetic degradation process (Castellanos-Estupiñan et al. 2022; Abdel-Razek

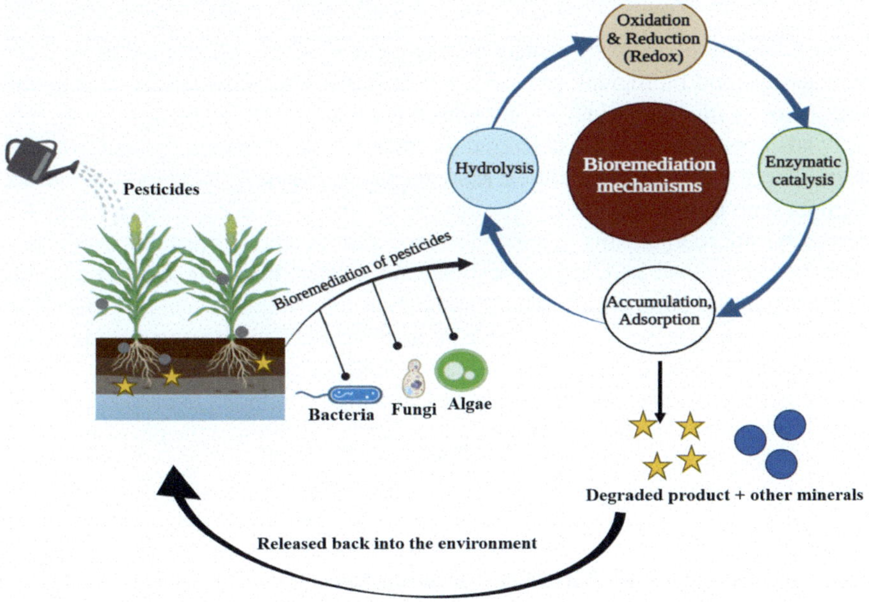

Fig. 4.2 Mechanism involved in bioremediation of pesticides by various microorganisms and the release of degraded product into the environment

et al. 2019). However, the most effective remediation is known to happen when the application of microbial consortia is done as organisms work synergistically exhibiting enhanced pesticide degrading capacity due to their metabolic diversity and cooperative interactions (Rodríguez-Castillo et al. 2019).

4.5 Basics of Nano-bioremediation of Pesticides

Nano bioremediation has come into the limelight in the field of research in recent years due to its unique combination of nanomaterials and bioremediation. It is a nanotechnology based approach to eliminating environmental pollutants from contaminated areas using nanoparticles generated from prokaryotes (Gram-negative, actinobacteria) and eukaryotes (fungi, algae and plants) (Rajput et al. 2021). *Burkholderia xenovorans* bacteria and zinc nanoparticles have been found to dechlorinate and biodegrade biphenyls (Hidangmayum et al. 2023). Additionally, plant-derived nanomaterials such as Noaea mucronate, have been used in the bioremediation of heavy metal contamination in groundwater and rivers (Choudhary et al. 2022).

4.5.1 Mechanism

The mechanisms of nano bioremediation involve synergistic interaction between nanomaterials and biological systems for transformations, degradation and immobilization of pesticides (Sunanda and Ghosh Sachan 2023). Pesticides can be efficiently adsorbed, catalysed or delivered to microbial or enzymatic breakdown sites due to its large surface area to volume ratio, increased reactivity and modified surface structures such as that of nanocomposites or nanoparticles as explained in Fig. 4.3. Conversely, microorganisms or enzymes have the metabolic capacity to identify, recognize, metabolise and uptake pesticide molecules, breaking them down into simpler less toxic compounds. The combination of nanomaterials with microbial consortia or enzymes increases pesticide breakdown rates, improves substrate accessibility and provides protection against environmental stress or inhibitory factors. Thus, physical, chemical and biological processes are combined in nano bioremediation mechanisms in a way that minimizes pesticide pollution in the environment by working synergistically.

4.5.2 Advantages of Nano Bioremediation

Nanobioremediation is a combination technique of bioremediation and nanotechnology thus, holds several advantages. As they are made by fusing the metabolic

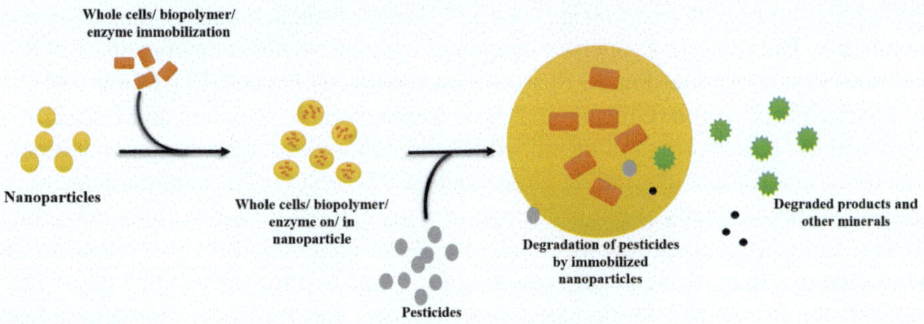

Fig. 4.3 Schematic presentation of nano bioremediation of pesticides by biological system immobilized Nanoparticles and the formation of degraded products

capacities of living organisms with the catalytic qualities of nanomaterials, they speed up the reaction rate of pesticide breakdown. The synergetic effects help in pesticide breakdown and the efficiency can be improved by using nanomaterials as transporters or immobilization matrices for microbial cells or enzymes. These materials offer protection, stability and increased activity. Additionally, the surface area to volume ratio of nanomaterials is high enabling the contact time between pesticide and nanomaterials to increase for effective pesticide degradation. The pesticide remediation capacity is improved in nano bioremediation as it targets selective pesticides without impacting the non-target substances thus lowering the possibility of collateral damage to beneficial biota of the ecosystem. Furthermore, it lowers the possibility of pesticide bioaccumulation in the future by dissolving it into harmless chemicals, thus leaving fewer residues for further contamination.

Nanomaterials offer flexibility in tackling various pesticide contaminants in various environmental contexts by enabling incorporation into remediation strategies suited to particular contamination scenarios. Besides, the use of natural components obtained by natural processes lowers energy consumption directly lowering environmental impact and resource utilization. By application of in-situ nano bioremediation, the pesticide-contaminated areas can be treated without the need to evacuate or transport materials thus, minimizing the disturbances to ecosystems and lowering expense incurred by ex-situ. Overall, the advantages associated with nanobioremediations make it a feasible and economical approach for efficient pesticide remediation.

4.6 Application of Nano Bioremediation in Pesticide Treatment

Researchers have developed many nanobio advanced materials once such is the development of a biocatalyst for degrading ethyl-paraoxon, a common pesticide by functionalizing the magnetic membrane with a phosphorhydrolase. They fused the

Opd gene from *Flavobacterium* sp. Atcc 27551 into a protein mamC in magnetosome membrane. The catalytic properties compared with the purified enzymes showed Km and Kcal values of 58 μM and 178 s^{-1} for immobilized Opd and 43 μM and 314 s^{-1} for the purified enzyme (Ginet et al. 2011). Another study by Salam and Das (2015) evaluated the effectiveness of an embedded bio-nanohybrid system using nanoscale zinc oxide and lindane-degrading yeast candida VITJzNO4. The nanoparticles were embedded in yeast cells and translocated into the cell cytoplasm without any lethal effects. Lindane degradation was studied and the nanobiohybrid was found to be more effective than native yeasts, resulting in complete removal within 3 days. This suggests the use of nZNO-mediated dichlorination and the novel bio-nano hybrid system for treating lindane-contaminated wastewater. Reports revealed remediation of polychlorinated biphenyl (PCB) Arcoclor 1248 by an integrated system using chemical catalysis and biodegradation. The process dechlorinated 99, 92, 84 and 28% of PCB congeners and the resulting biphenyl was further degraded by *B. xenovorans* LB400 with 90% efficiency (Le et al. 2015).

Recent studies developed an ecofriendly method for removing 4-chlorophenols from wastewater using horseradish peroxidase (HRP) immobilized onGo-Fe_3O_4/Au@CA nanoparticles. On investigating the reaction, it was revealed that the immobilized HRP retained 95% of the initial activity for first 3 cycles due to the presence of gold (Sarno and Iuliano 2020). Similarly, a phosphotriesterase (PTE) was hybridized with copper ions forming a CU-PTE hybrid nanoflower with high catalytic activity and efficiency towards biodegradation of organophosphorus pesticides. Compared with the free PTE, the hybrid showed Kcat/Vmax of 1.76 folds which is 2.2 fold more than free PTE with 72.3% relative activity after 10 consecutive reaction runs (Chen et al. 2021). Another, cost effective and eco friendly method for the degradation of chlorophenols in water and wastewater is by using amino functionalized superparamagnetic FE_3O_4@SiO_2-NH_2 nanoparticles to immobilize laccase from *Trametes versicolor*. The degradation rate for 2,4 dichlorophenol (2,4-DCP) and 2,4,6-trichlorophenol (2,4,6-TCP) were found to be 5439 and 68.7% respectively (Chen et al. 2022). A low cost synthesis of nanobiocatalysts is discovered using aryloxyphenoxypropionate herbicide-hydrolyzing enzyme QpeH embedded into zeolitic imidazole frameworks (ZIFs). The composite showed improved stability and reusability after 10 reuse cycles with QpeH@ZIF-10 showing 88% degradation efficiency (Yu et al. 2021).

A stable enzyme biocatalyst is developed by covalently immobilizing multi enzymes onto the nanosilica using glutaraldehyde. Lignolytic enzymes were produced from *P. ostreatus* and its immobilization revealed stability and activity across various temperature and pH ranges with effective degradation of DDt within 12 h (Salem et al. 2023). Another recent study revealed the synthesis of metal oxide nanocatalysts with 4 set manganese and cerium ratios on zeolite nanoparticles for the degradation of gaseous toluene using *B. subtilus* biofilm. The results showed the highest degradation of 99% with 83% CO_2 release with ZMnCe-20% biofilm (EPS) nano biocatalyst (Mohsin et al. 2024). A novel nanocomposite is been synthesized for bioremediation of multi-pesticide pollution by encapsulating hydrolase enzymes for aryloxyphenoxypropionate herbicide (AOPPs) and organophosphorus

(Ops) insecticide in a magnetic zeolitic imidazolate frameworks-8 (mZIF-8). The study provided efficient degradation results for chlorpyrifos and Quizalofop-P-ethyl degradation with high efficiency and economic benefits (Li et al. 2023).

4.7 Environmental Implications and Challenges

Nano bioremediation covers numerous advantages over conventional methods yet several complications need to be taken care of. It has been observed that the presence of NPs in the soil changes soil pH, one of the key factors influencing soil nutrient availability, microbial dynamics, general soil health and plant growth and development (Ali et al. 2022). Furthermore, to biosynthesize nanoparticles on a wide scale and increase the scope and efficacy of their applications, new techniques must yet be developed (Dikshit et al. 2021).

The previously available research data indicates that these technologies can be successfully applied to clean up soils and streams that have been contaminated by pesticides and herbicides. Still, numerous studies are simultaneously worked on a laboratory scale under carefully regulated simulations (Hidangmayum et al. 2023). However, Research indicates that the quality and duration of bioremediation will be improved as nanotechnology develops and adapts in the future. This significant correlation can ensure increased treatment efficacy and decreased failure rate (Maravi and Nistala 2021).

Many researchers are finding new ways to increase the applicability of nano bioremediation from theoretical to practical knowledge. In the future, the researchers can focus more on improving the applicability of enzyme-based nanobioremediation as it increases target-based remediation of pesticides leaving the surrounding processes unaltered. For the enzyme to be stable inert material nanoparticles need to be identified which will aid the process speed and will remain stable till the remediation is complicated without itself posing an environmental risk. Although it presents a great challenge the interdisciplinary nature of nanobioremediation provides a wide opportunity for application in real-time.

4.8 Conclusion

Pesticide pollution is been a major environmental concern for years now. Not only does it pose human health risks but also are harmful to other environmental biota. Numerous research are conducted to find an effective technique for pesticide remediation. Nanotechnology is explored widely for green synthesis of remediating agents for pesticides. Simultaneously bioremediation techniques are also applied for pesticide remediation owing to their ecofriendly nature. However, nano bioremediation is viewed to be a better alternative in comparison to the two as it is a technology developed by considering the advantages of nanotechnology and bioremediation techniques. However, research needs to be carried out for its application in field studies to avoid secondary pollution.

Acknowledgment The authors would like to acknowledge Sponsored Research and Industrial Consultancy (SpoRIC), Vellore Institute of Technology, Vellore for providing financial support through SEED grant (SG20240105).

References

Abdel-Razek MA, Abozeid AM, Eltholth MM, Abouelenien FA, El-Midany SA, Moustafa NY, Mohamed RA (2019) Bioremediation of a pesticide and selected heavy metals in wastewater from various sources using a consortium of microalgae and cyanobacteria

Ahmad M, Ahmad I (2014) Bioremediation of pesticides. In: Biodegradation and bioremediation. Studium Press LLC, Boca Raton, pp 125–165

Albanese A, Tang PS, Chan WC (2012) The effect of nanoparticle size, shape, and surface chemistry on biological systems. Annu Rev Biomed Eng 14:1–16

Ali S, Ullah MI, Sajjad A, Shakeel Q, Hussain A (2021) Environmental and health effects of pesticide residues. In: Sustainable agriculture reviews 48: pesticide occurrence, analysis and remediation, vol 2 Analysis, pp 311–336

Ali A, Ahmed Z, Maqbool R, Shahzad K, Shah ZH, Ali MZ et al (2022) Nanobioremediation of insecticides and herbicides. In: Biodegradation and biodeterioration at the nanoscale. Elsevier, pp 655–674

Allsop M, Huxdorff C, Johnston P, Santillo D, Thompson K (2015) Pesticides and our health. A growing concern. Greenpeace Research Laboratories

Berny P (2011) Challenges of anticoagulant rodenticides: resistance and ecotoxicology. In: Pesticides in the modern world–pests control and pesticides exposure and toxicity assessment. Tech Europe, Rijeka, pp 441–468

Berthold DE, Elazar A, Lefler F, Marble C, Laughinghouse HD IV (2020) Control of algal growth on greenhouse surfaces using commercial algaecides. Sci Agric 78:e20180292

Bhadouria R, Das S, Kumar A, Singh R, Singh VK (2020) Mycoremediation of agrochemicals. In: Agrochemicals detection, treatment and remediation. Butterworth-Heinemann, pp 593–620

Castellanos-Estupiñan MA, Carrillo-Botello AM, Rozo-Granados LS, Becerra-Moreno D, García-Martínez JB, Urbina-Suarez NA et al (2022) Removal of nutrients and pesticides from agricultural runoff using microalgae and cyanobacteria. Water 14(4):558

Chadha U, Selvaraj SK, Thanu SV, Cholapadath V, Abraham AM, Manoharan M, Paramsivam V (2022) A review of the function of using carbon nanomaterials in membrane filtration for contaminant removal from wastewater. Mater Res Express 9(1):012003

Chen J, Guo Z, Xin Y, Shi Y, Li Y, Gu Z et al (2021) Preparation of efficient, stable, and reusable copper-phosphotriesterase hybrid nanoflowers for biodegradation of organophosphorus pesticides. Enzyme Microb Technol 146:109766

Chen Z, Yao J, Ma B, Liu B, Kim J, Li H et al (2022) A robust biocatalyst based on laccase immobilized superparamagnetic Fe_3O_4@ SiO_2–NH_2 nanoparticles and its application for degradation of chlorophenols. Chemosphere 291:132727

Chen L, Zhang S, Duan Y, Song X, Chang M, Feng W, Chen Y (2024) Silicon-containing nanomedicine and biomaterials: materials chemistry, multi-dimensional design, and biomedical application. Chem Soc Rev

Choudhary R, Hiti-Bandaralage JCA, Ahlawat J, Gaur N, Diwan B (2022) Nanobioremediation: an introduction. In: Nano-bioremediation: fundamentals and applications. Elsevier, pp 263–279

Dikshit PK, Kumar J, Das AK, Sadhu S, Sharma S, Singh S et al (2021) Green synthesis of metallic nanoparticles: applications and limitations. Catalysts 11(8):902

Eskenazi B, Marks AR, Bradman A, Fenster L, Johnson C, Barr DB, Jewell NP (2006) In utero exposure to dichlorodiphenyltrichloroethane (DDT) and dichlorodiphenyldichloroethylene (DDE) and neurodevelopment among young Mexican American children. Pediatrics 118(1):233–241

Ginet N, Pardoux R, Adryanczyk G, Garcia D, Brutesco C, Pignol D (2011) Single-step production of a recyclable nanobiocatalyst for organophosphate pesticides biodegradation using functionalized bacterial magnetosomes. PLoS One 6(6):e21442

Guo Y, Zhou Q, Zhu B, Tang CY, Zhu Y (2023) Advances in organic semiconductors for photocatalytic hydrogen evolution reaction. EES Catal 1(4):333–352

Hidangmayum A, Debnath A, Guru A, Singh BN, Upadhyay SK, Dwivedi P (2023) Mechanistic and recent updates in nano-bioremediation for developing green technology to alleviate agricultural contaminants. Int J Environ Sci Technol 20(10):11693–11718

Le TT, Nguyen KH, Jeon JR, Francis AJ, Chang YS (2015) Nano/bio treatment of polychlorinated biphenyls with evaluation of comparative toxicity. J Hazard Mater 287:335–341

Li M, Li L, Sun Y, Ma H, Zhang H, Li F (2023) Facile synthesis of dual-hydrolase encapsulated magnetic ZIF-8 composite for efficient removal of multi-pesticides induced pollution in water. Chemosphere 314:137673

Maravi P, Nistala S (2021) Nano-bioremediation and its promising role in pesticide remediation. NewBioWorld 3(2):22–25

Mishra A, Saini RK, Bajpai AK (2020) Polymer formulations for pesticide release. In: Controlled release of pesticides for sustainable agriculture, pp 185–206

Mittal M, Tripathi S, Shin DK (2024) Biopolymeric nanocomposites for wastewater remediation: an overview on recent progress and challenges. Polymers 16(2):294

Mohsin MZ, Mohsin A, Zaman WQ, Zhu X, Zhao X, Abbas Z et al (2024) Efficient toluene degradation using Bacillus subtilis biofilm supported Mn-Ce/zeolite catalysts. Green Synth Catal

Moreira JB, Santos TD, Zaparoli M, de Almeida ACA, Costa JAV, de Morais MG (2022) An overview of nanofiltration and nanoadsorption technologies to emerging pollutants treatment. Appl Sci 12(16):8352

Nasrollahzadeh M, Sajjadi M, Iravani S, Varma RS (2021) Starch, cellulose, pectin, gum, alginate, chitin and chitosan derived (nano) materials for sustainable water treatment: a review. Carbohyd Polym 251:116986

Nicolella HD, de Assis S (2022) Epigenetic inheritance: intergenerational effects of pesticides and other endocrine disruptors on cancer development. Int J Mol Sci 23(9):4671

Niti C, Sunita S, Kamlesh K, Rakesh K (2013) Bioremediation: an emerging technology for remediation of pesticides. Res J Chem Environ 17:4

Ortiz-Hernández ML, Sánchez-Salinas E, Dantán-González E, Castrejón-Godínez ML (2013) Pesticide biodegradation: mechanisms, genetics and strategies to enhance the process. Biodegrad-Life Sci 10:251–287

Özkara A, Akyıl D, Konuk M (2016) Pesticides, environmental pollution, and health. In: Environmental health risk-hazardous factors to living species. IntechOpen

Pednekar RR, Rajan AP (2024) Unraveling the contemporary use of microbial fuel cell in pesticide degradation and simultaneous electricity generation: a review. Environ Sci Pollut Res 31(1):144–166

Prasanna SS, Balaji K, Pandey S, Rana S (2019) Metal oxide based nanomaterials and their polymer nanocomposites. In: Nanomaterials and polymer nanocomposites. Elsevier, pp 123–144

Rahman MR, Islam MS, Rahman MM, Prodhan MDH, Uddin MK (2019) Organochlorine pesticide residue status in dry fish of Bangladesh and their risk assessment: a review. Int J Sci Res Publ 9(10):619–638

Rajput VS, Jhala J, Acharya VS (2020) Biopesticides and their mode of action against insect pests: a review. Int J Chem Stud 8:2856–2862

Rajput VD, Singh A, Minkina T, Rawat S, Mandzhieva S, Sushkova S et al (2021) Nano-enabled products: challenges and opportunities for sustainable agriculture. Plants 10(12):2727

Rodríguez-Castillo G, Molina-Rodríguez M, Cambronero-Heinrichs JC, Quirós-Fournier JP, Lizano-Fallas V, Jiménez-Rojas C et al (2019) Simultaneous removal of neonicotinoid insecticides by a microbial degrading consortium: detoxification at reactor scale. Chemosphere 235:1097–1106

Salam JA, Das N (2015) Degradation of lindane by a novel embedded bio-nano hybrid system in aqueous environment. Appl Microbiol Biotechnol 99:2351–2360

Salem TM, Fetyan NA, Taha AA, Malhat FM, Ramadan AI (2023) Covalent immobilization of multi-enzyme biocatalysts on nano-silica for efficient p, p'-DDT degradation

Sarno M, Iuliano M (2020) New nano-biocatalyst for 4-chlorophenols removal from wastewater. Mater Today Proc 20:74–81

Saxena MC, Siddiqui MKJ, Seth TD, Murti CK, Bhargava AK, Kutty D (1981) Organochlorine pesticides in specimens from women undergoing spontaneous abortion, premature or full-term delivery. J Anal Toxicol 5(1):6–9

Sengupta S, Bhattacharya P (2019) Nano-remediation: carving solution to pesticide pollution. XVIII:134–135

Sherwani SI, Arif IA, Khan HA (2015) Modes of action of different classes of herbicides. In: Herbicides, physiology of action, and safety, vol 10, p 61779

Singh DK (2008) Biodegradation and bioremediation of pesticide in soil: concept, method and recent developments. Indian J Microbiol 48:35–40

Sparks TC, Crossthwaite AJ, Nauen R, Banba S, Cordova D, Earley F, Wessels FJ (2020) Insecticides, biologics and nematicides: updates to IRAC's mode of action classification-a tool for resistance management. Pestic Biochem Physiol 167:104587

Stoytcheva M (ed) (2011) Pesticides in the modern world: pesticides use and management. BoD–Books on Demand

Sunanda, Ghosh Sachan S (2023) Nanobioremediation of pesticides by immobilization technique: a review. Int J Environ Sci Technol 1–12

Varjani SJ, Agarwal AK, Gnansounou E, Gurunathan B (eds) (2018) Bioremediation: applications for environmental protection and management. Springer, New York, NY

Yang C, Hamel C, Vujanovic V, Gan Y (2011) Fungicide: modes of action and possible impact on nontarget microorganisms. International Scholarly Research Notices

Yap YH, Azmi AA, Mohd NK, Yong FSJ, Kan SY, Thirmizir MZA, Chia PW (2020) Green synthesis of silver nanoparticle using water extract of onion peel and application in the acetylation reaction. Arab J Sci Eng 45:4797–4807

Yu T, Ma H, Zhang H, Xiong M, Liu Y, Li F (2021) Fabrication and characterization of purified esterase-embedded zeolitic imidazolate frameworks for the removal and remediation of herbicide pollution from soil. J Environ Manage 288:112450

Chapter 5
Red Mud Cement Colloidal Nano Silica Composites for the Heavy Metal Ion Removal from the Wastewater

K. Athira and T. Shanmuga Priya

Abstract Industrial wastewater is a significant threat to aquatic and environmental habitats. The harsh chemicals and heavy metal ions present in industrial wastewater heavily impact the equilibrium of nature. This study uses red mud (RM), an industrial waste byproduct, as a metal adsorbent over thermal processing. A recent study examined the effects of calcined red mud and colloidal nano-silica (CNS) composite on metal extraction efficiency. This study measured the concentration of Cd, Pb, As, and Cu remaining in the wastewater solution after exposure to the ternary blended red mud CNS cement samples over time using atomic absorption spectroscopy. The Red mud cement CNS mortar samples are kept in the wastewater over a period of time, and the heavy metal ion concentration in the solution before and after the exposure is measured. The results showed that a combination of ternary blended mortar with cement replacement by 1.5% CNS and 15% red mud showed the highest adsorption. Calcination significantly enhanced the metal extraction efficiency of red mud. The metal extraction efficiency order for the proposed composites is in the order of Cr > Cd > Pb > Cu. The X-ray diffraction patterns of heavy metal ions loaded samples revealed changes in silicate crystallinity and an alumina phase transition, which shows the affinity of the proposed ternary blend mortar to the heavy metal ions. Therefore, the red mud CNS nano composites can alter the mineral phases, impacting its functionality as a metal adsorbent. Hence, red mud CNS cement composites can be used as a metal extraction agent for wastewater purification.

Keywords Heavy metal ion extraction · Red mud · Colloidal nano silica · Waste water treatment · Ternary blended mortar blocks

K. Athira · T. Shanmuga Priya (✉)
School of Civil Engineering, Vellore Institute of Technology, Vellore, Tamil Nadu, India
e-mail: shanmugapriya.t@vit.ac.in

© The Author(s), under exclusive license to Springer Nature Switzerland AG 2025
A. Jeyaseelan and K. Murugasen (eds.), *Sustainable Environmental Remediation: Avenues in Nano and Biotechnology*, Environmental Science and Engineering,
https://doi.org/10.1007/978-3-031-78483-5_5

5.1 Introduction

The chemical, textile, and pharmaceutical industries are crucial economic drivers for many countries. However, it generates a significant amount of wastewater that contains heavy metals, dyes, and chemicals (Ayachi et al. 2019; Sahu et al. 2015; Shahid et al. 2013). The large volume of waste water is produced due to the discharge of reactive dyes wasted during the dying process (Blackburn 2004). It is threat to the aquatic ecosystem due to the hinderance of sunlight penetration caused by the colour of the dye and the dissolved oxygen reduction in the water body (Ingrassia et al. 2023). Also, these dyes cause a considerable health issues such as skin irritation, allergic reactions, respiratory disorders and cancer (Rai et al. 2005; Malik et al. 2007). The International Dye Industry Wastewater Discharge Quality Standards have set a 1 mg/g limit for the discharge of dye wastewater into aquatic ecosystems (Katheresan et al. 2018). Therefore, it is crucial to remove reactive dyes from industrial wastewater effectively.

Another threat to the aquatic life is heavy metal discharges from the industries. Among the heavy metals, major cause of the water pollution around the world is Cadmium (Cd). The global pollution problem created by Cadmium (Cd)-containing wastewater is mainly due to the waste discharges from electroplating, printing, and dyeing industries. By water pollution with Cd the human health system also gets affected severely. Apart, from the water pollution Cd causes soil pollution when it enters in the soil (Peng et al. 2021; Xie et al. 2018; Yang et al. 2020). Biological, physiochemical and chemical processes are adopted to treat Cd affected waste water. Physicochemical methods include adsorption (Chigondo et al. 2019; Truong et al. 2017), solvent extraction (Bilińska et al. 2019), membrane separation (Asuquo et al. 2017; Weng et al. 2013), electrodialysis, chemical precipitation (Elzinga and Kretzschmar 2013), advanced oxidation, ion exchange (Yan et al. 2019), chemical precipitation (Elzinga and Kretzschmar 2013), photocatalysis (Liu et al. 2018). The Cd(II) is removed by converting it into $Cd(OH)_2$ by the chemical method. However, the chemical process ends up in secondary pollution due to the production of sludge as a Cd(II) enriched byproduct. These methods limits by the technical difficulty and low efficiency (Kong et al. 2022). Therefore, adsorption is one of the most commonly used methods (Athira and Shanmugapriya 2023). It is currently efficient and economical due to its high selectivity, environmentally friendly, efficiency, and cost-effectiveness. During the adsorption process, certain parameters must be controlled. Industrial, agricultural, and biological waste materials along with activated carbon are recommended as adsorbents to remove heavy metals (Athira and Shanmugapriya 2023).

One such industrial waste is red mud, which can adsorb heavy metal ions from the wastewater. Bauxite residue, or red mud (RM), is a solid waste produced large amounts during the Bayer process. A significant problem with highly alkaline industrial waste is appropriate disposal since it creates storage problems. The leaching of red mud causes the percolation of toxic substances into the groundwater. Therefore, researchers are going on to find a proper disposal technique for red mud (Chang et al. 2020; Liu and Poon 2016; Liu et al. 2020; Mukiza et al. 2019; Patel and Barros 2015; Tang et al. 2019). Recycling a large amount of red mud is impractical since recovering valuable metals such as aluminum, titanium, and iron are essential (Evans 2016; Khairul et al. 2019). Waste management strategies are needed to ensure environmentally sustainable and eco-friendly solid waste disposal globally, thereby reducing risks to human health and environmental damage (Aydın et al. 2021; Shabnam et al. 2019; Teng et al. 2021; Zhang et al. 2021). Replacing cement, fine or coarse aggregate, with red mud, as in the construction industry, is a cleaner production approach (Shi et al. 2020; Wang et al. 2019). However, the red mud's highly porous nature negatively affects the structure's durability (Cheng et al. 2019). Therefore, its raw use as a cement replacement for structural elements is not advisable from a sustainable point of view. Incorporating the CNS with red mud to form cement composites enhances the strength and durability properties of the structure (Athira and Shanmugapriya 2022).

Researchers are exploring alternative options to using red mud with filter paper as an adsorbent for water treatment, as it is impractical. One approach is the creation of red mud blended geopolymer and polymer composites for the water treatment process (Hertel and Pontikes 2020; Vigneshwaran et al. 2020). For instance, Chen et al. (2019) developed a red mud blended geopolymer with granulated blast furnace slag for water treatment, improving the heavy metal removal ratio. They used a 1:1 ratio by weight of red mud and granulated blast furnace slag, which resulted in an increase of heavy metal removal. Additionally, Li et al. (2020) prepared immobilized red mud beads that removed 57.25 mg (Mn)/g (red mud) after running for 51 days in a lab-scale experiment. However, using red mud cementitious composites is limited, and more practical solutions are needed to implement the water treatment process using red mud on an industrial scale (dos Reis et al. 2020; Ho et al. 2021; Li et al. 2019). Holistic planning for environmentally sustainable cities can consider infrastructure, green roads, flood control, and rainwater management. Rainwater meeting with soil and cement-based adsorptive cubes can be a good candidate for this purpose (Dlamini et al. 2020; Gao et al. 2020).

Removing toxic metal ions from wastewater before discharging it into the environment is crucial. One practical and cost-efficient method is adsorption using materials like agricultural or industrial waste and activated carbon. This research discusses the utilization of a ternary blended cement-red mud-colloidal nano-silica as an adsorbent for heavy metal ion extraction. This method helps clean up metal ion-contaminated wastewater and provides a sustainable solution for managing this problematic industrial byproduct.

5.2 Materials and Microstructure

The materials used to create the binder in this investigation were cement, raw red mud, and colloidal nano-silica (CNS). The red mud was collected from HINDALCO, Belgaum, oven-dried for 24 h at 105 °C before being crushed until it passed through

Table 5.1 Chemical composition of cement and red mud

Compounds	Cement (%)	Red mud (%)
SiO_2	19.70	10.31
Al_2O_3	4.83	23.20
Fe_2O_3	4.71	43.10
CaO	65.70	6.35
MgO	0.97	0.55
Na_2O	0.22	8.92
K_2O	0.45	0.13

a 150 μm sieve. The powdered red mud was then calcined in an electrical tilting furnace at 700 °C for three hours. Cement of grade 53, with a specific gravity of 3.15, and river sand, with a specific gravity of 2.67, were also used. Table 5.1 shows the percentage by mass of the compounds present in cement and red mud. Compared to red mud, cement contains less alumina and iron oxide but a higher percentage of lime and silica.

According to Table 5.1, red mud has high Al_2O_3 and Fe_2O_3 but low levels of SiO_2 and CaO compared to cement. The CaO percentage is compensated by adding cement and SiO_2 with CNS. Levasil CB8 is a form of colloidal nano-silica that contains 50% solids by weight and has a specific gravity of 1.03. The average particle size of the nano-silica is 20 nm, with a specific surface area of 86 m^2/g. The specimens underwent several characteristic studies to analyze their crystallinity, pore structure, and hydration products. Figure 5.1 TGA analysis of red mud. Shows the FESEM image of red mud.

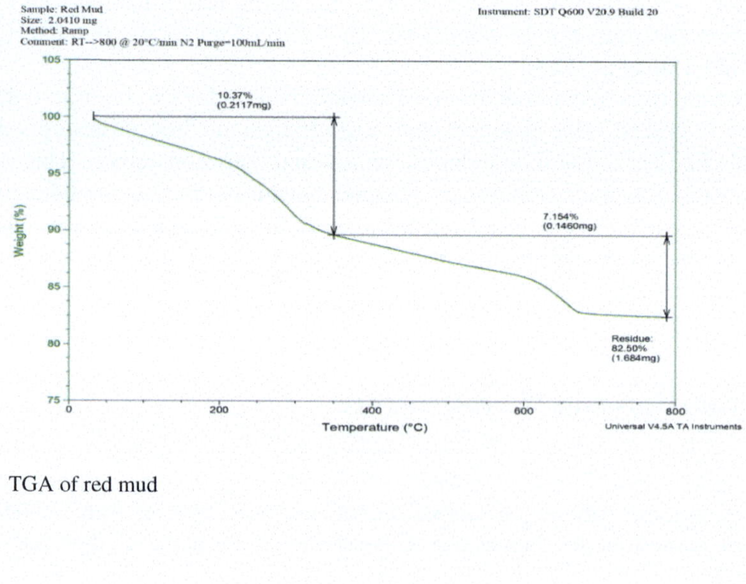

Fig. 5.1 TGA of red mud

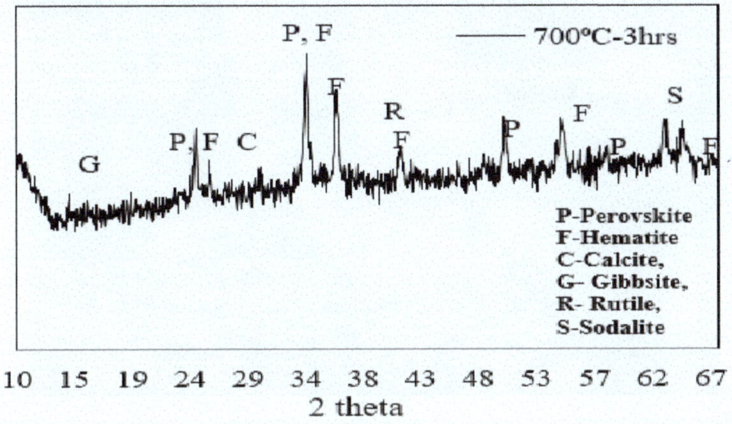

Fig. 5.2 XRD of the calcined red mud

At temperatures above 600 °C, the red mud helps form silica and alumina compounds, which is supported by the research conducted by Pera et al. (1997). However, if the calcination process goes beyond 700 °C, it will decarbonate calcite (calcium carbonate), resulting in quick lime formation and negatively impacting reactivity. The reaction between water and quick lime forms calcium hydroxide, leading to a faster hydration process and better binding properties in cement composites. As per the TGA results, the reactive compounds with the heavy metal ions are formed after reaching 700 °C. Therefore, the red mud samples are calcined at 700 °C for 3 h to achieve maximum metal extraction properties by red mud. Figure 5.2 shows the X-ray diffraction (XRD) result of calcined red mud.

Sharp peaks of X-Ray diffraction patterns (XRD) are mainly due to the presence of perovskite ($CaTiO_3$), hematite (Fe_2O_3), gibbsite ($Al(OH)_3$), calcite ($CaCO_3$), rutile (TiO_2), and sodalite ($Na_2O \cdot Al_2O_3 \cdot SiO_2$). Calcination decomposes compounds and turns most crystalline structures to the amorphous phase, which readily reacts with water and Portland cement to form C–S–H gel (Castaldi et al. 2008). Figure 5.3 shows the FESEM micrograph of red mud.

Red mud has various compounds, including goethite, magnetite, and γ-FeOOH. The FESEM micrograph shows the presence of Gibbsite and Goethite in the calcined red mud, which is marked in the image and has a high affinity towards the heavy metal ions.

The A highly reactive intermediate compound $[Fe\ (OH)_2(OH_2)_4]^0$, formed due to the hydroxylation of ferrous ions in calcined red mud, is the prime center of the heavy metal ion reaction. When the pH exceeds 7, and the $Fe(OH)_2$ precipitation is supported in calcined red mud, the resulting product is sensitive to oxidation. Also, compounds such as goethite, magnetite, and γ-FeOOH precipitation can occur, and the resulting product remains sensitive to oxidation. The volatile intermediate compound, ferrihydrite, has poor structural organization, ultimately transforming

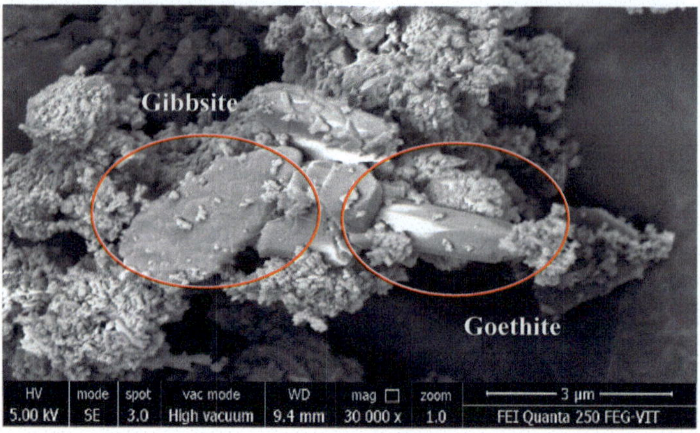

Fig. 5.3 FESEM image of red mud

into hematite particles. These unstable intermediate compounds have a high affinity towards heavy metal ions and form complex compounds.

5.3 Mix Design and Experimental Methods

5.3.1 Mix Design

Table 5.2, shows 5 red mud-cement-CNS mortar mixes that were used for the experiment. The mixes were made with and without CNS at a w/b ratio of 0.45. The base mix used for the experiment involved replacing 15 and 20% of cement with calcined red mud. To achieve the best strength and durability of the mix, CNS was added at 1.5 and 1% to this base mix. The nanoparticles were mixed with water for 3 min to ensure they were evenly dispersed. The w/b ratio was carefully chosen to ensure the mix had the desired workability and consistency. The cement and fine aggregate were mixed in a 1:3 ratio and added to the mortar mixer. The constituents were mixed for 4–5 min until a homogenous mix was achieved. The mix was then poured into moulds of standard dimensions and left for 24 h before normal water curing was carried out to reach the various testing periods.

Table 5.2 Mix identification

Mix designation	Cement	Red mud	Nano silica
R0	100	–	–
R15	85	15	–
R20	80	20	–
R15NS1.5	83.5	15	1.5
R20NS1	79	20	1

5.3.2 Experimental Methods

5.3.2.1 Compressive Strength

A compressive strength test was conducted on cubes that were cured for 28 days, in accordance with IS 516-2018 (25) guidelines. For the test, cement paste cubes measuring 50 mm were prepared. The cubes were then tested using a compression testing machine with an accuracy of ±1% and a capacity of 3000 kN.

5.3.2.2 Water Absorption and Water Permeable Porosity

The water absorption test for the mortar mixtures is conducted as per ASTMC642-13 (1997). The 50 mm^3 specimens were casted and cured for 28-days. The cured samples are weighed and kept in oven at 110 ± 5 °C for not less than 24 h and the weight is checked in the interval of 24 h. The reason for measuring the weight consecutively is to conform that, the difference in weight of the specimen before and after placing in the oven is not exceeding 5% of the total weight of the specimen. Once the specimen weight became constant it is marked as ‘A’ and is kept in water of 21 °C for another 48 h and the weight has taken and marked as B. The weighted samples are kept in water bath for 5 h at 90 °C and the suspended weight of the same sample is measured and are noted as C and D respectively. The suspended weight of the same sample is marked as D. Rate of water absorption, bulk density, apparent density, and volume of permeable pore space is obtained from the water absorption test. The equations for each parameter are as follows (5.1 and 5.2).

$$\text{Water absorption} = \frac{A - B}{B} \times 100 \tag{5.1}$$

$$\text{Volume of permeable pore space} = \frac{g2 - g1}{g2} \times 100 \tag{5.2}$$

where

$$g1 = \frac{A}{C - D}$$

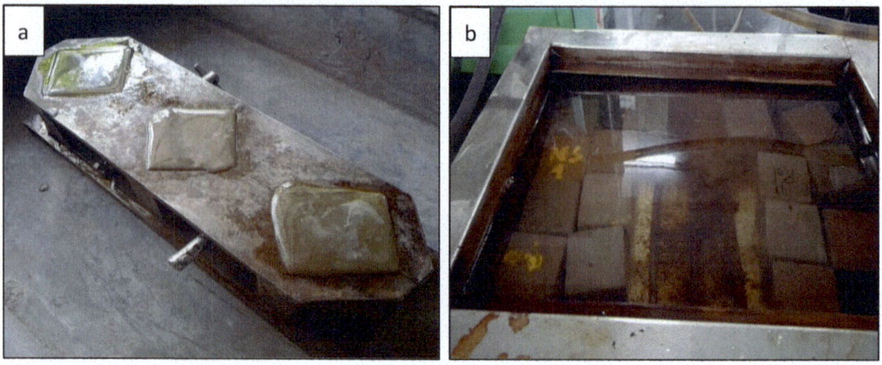

Fig. 5.4 **a** 50 mm Mortar cubes. **b** Mortar cubes immersed in waste water

$$g2 = \frac{A}{A - D}$$

5.3.2.3 Atomic Absorption Spectroscopy (AAS)

The waste water collected from nearby industrial area is tested for quantifying the heavy metal ions presented. The mortar cubes are immersed in the waste water for 10 days (Fig. 5.4b). The water is then again subjected to AAS to determine the heavy metal ion presence in mg/L (ppm). The obtained results are expressed in equivalence of the standard solution of corresponding metal ion. The metal ions considered are Copper (Cu), Chromium (Cr), Cadmium (Cd), and Lead (Pb).

The heavy metal ion concentration before and after is measured by the AAS method, and the results are expressed in ppm. The contact time with the red mud samples with heavy metal ions also influences the adsorption spectra significantly. This study considers a uniform contact time in all the samples with each heavy metal ion.

5.4 Results and Discussions

5.4.1 Compressive Strength

After 28 days of normal water curing, the compressive strength of red mud-cement mortar with a 0.45 w/b ratio was tested. The compressive strength decreased as the amount of red mud increased. All the values obtained were lower than the reference paste. This decrease in strength was due to the increased porosity and loose structure

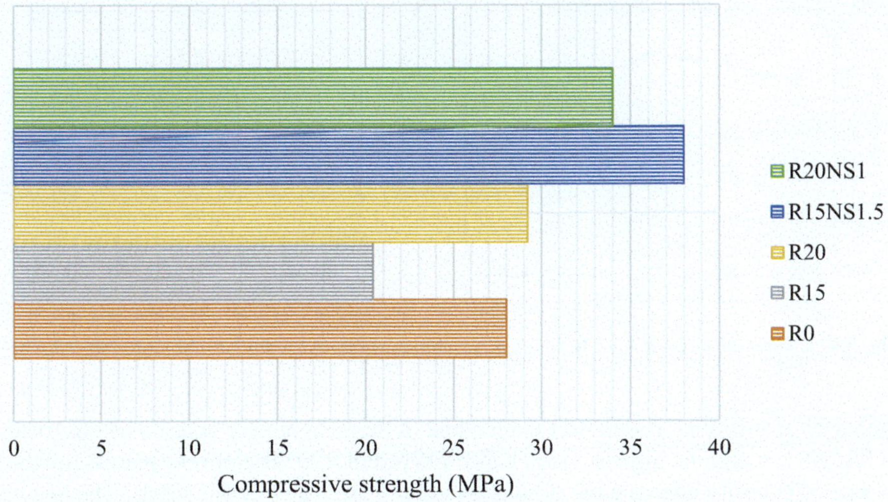

Fig. 5.5 28-days compressive strength of the mortar blocks

caused by the addition of red mud. The mix R15NS1.5 had the slightest reduction in compressive strength for 28-day curing compared to the reference mix, with a reduction of only 35% at 0.45 w/b ratio (Fig. 5.5). Therefore, R15NS1.5 was chosen as the optimal mix. In the 28-day cement paste, the addition of CNS showed an apparent increase in compressive strength compared to the red mud cement paste without CNS (R15 and R20). The binary mix with 1.5 and 1% addition of CNS recorded a maximum strength increase of 35 and 17%, respectively (Fig. 5.5). The mixes with CNS achieved higher compressive strength than the reference mix (R0) due to the refined pore network and dense microstructure. However, adding more than 1.5% CNS to the red mud cement paste decreased the strength due to excessive fluid ingression from the colloidal suspension of nano-silica to the pore network of the red mud cement paste structure, forming gelatinous C–S–H (21).

5.4.2 Water Absorption and Water Permeable Porosity

The water absorption test determined the water absorption and water permeable porosity of the 28-day cured red mud cement paste with and without CNS. The results are exhibited in Fig. 5.5. The percentage of water absorption was reduced from R15 to R15NS1.5 and is less than the reference mix, similar to R20 to R20NS1. Water absorption of cement composites should decrease when CNS is added to the mix. However, the red mud introduced to the cement matrix increased the permeable pore space of the mixture (Fig. 5.6b). Minimum water absorption was found in R15NS1.5, which has a 10.07% lesser absorption than R15. R20 reported the maximum water absorption and is even greater than R0. Higher water permeable pore volume connects

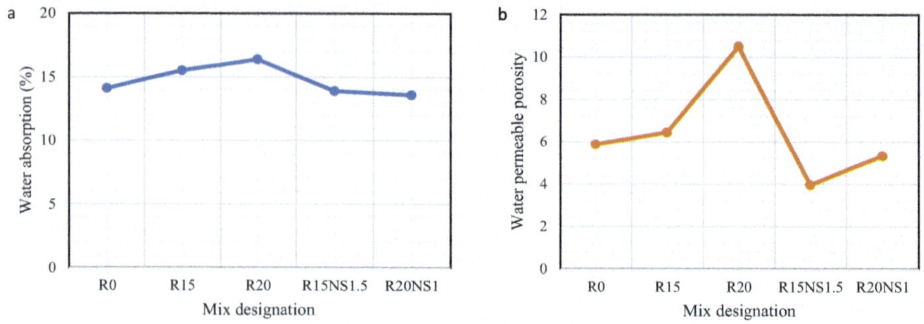

Fig. 5.6 **a** Water absorption (%). **b** Water permeable porosity of 28 days cured mortar specimens

the water absorption. The lowest percentage of permeable pore space was observed in R15NS1.5, and the highest value was observed in R20. The porous nature and the increased porosity enhance the flow of external wastewater to the inside. The heavy metal ions in the wastewater adhere to the compounds present in the mortar.

5.4.3 *Metal Extraction Property from AAS*

Figure 5.7 presents the metal extraction efficiency of red mud cement mortar, with and without CNS, obtained from atomic absorption spectroscopy (AAS). The overall metal extraction capacity of red mud samples is R15NS1.5 ≥ R20NS1 ≥ R15 ≥ R20.

The $CaCO_3$ content in red mud is improved during the calcination process. Copper is known to have more affinity to $CaCO_3$ as per sources (Ma et al. 2012). Therefore, raw red mud has the minimum capacity for Cu extraction. Conversely, calcined red mud has a better percentage of Cu extraction, as seen in the atomic absorption spectra (Eq. 5.3). In summary, Cu retention is tighter in calcined red mud ($CaCO_3$-dominated) compared to raw red mud (boehmite-dominated).

$$CuSO_4 + CaCO_3 + 2H_2O \rightarrow Cu_2CO_3(OH)_2 + CaSO_4 + CO_2 \quad (5.3)$$

Figure 5.7 demonstrates that red mud is more effective in extracting Cr(III) than other heavy metal ions. This process involves the precipitation of $Cr(OH)_3$ with red mud during the extraction of Cr(III). At higher temperatures, Fe(II) ions are oxidized in red mud, while Fe(III) ions react efficiently with Cr(III) to precipitate $Cr(OH)_3$ and co-precipitate aluminum hydroxide. However, at a pH above 5, Fe(II) is highly reactive to the precipitation of Cr(III). Raw red mud, with a pH above 5, is rich in Fe(II), making it highly suitable for the extraction of Cr(III). Additionally, red mud cement CNS mortar has Fe(III) ions with a lower pH than raw red mud, further enhancing the extraction process. Consequently, the mixes exhibit a greater affinity towards Cr than other heavy metal ions.

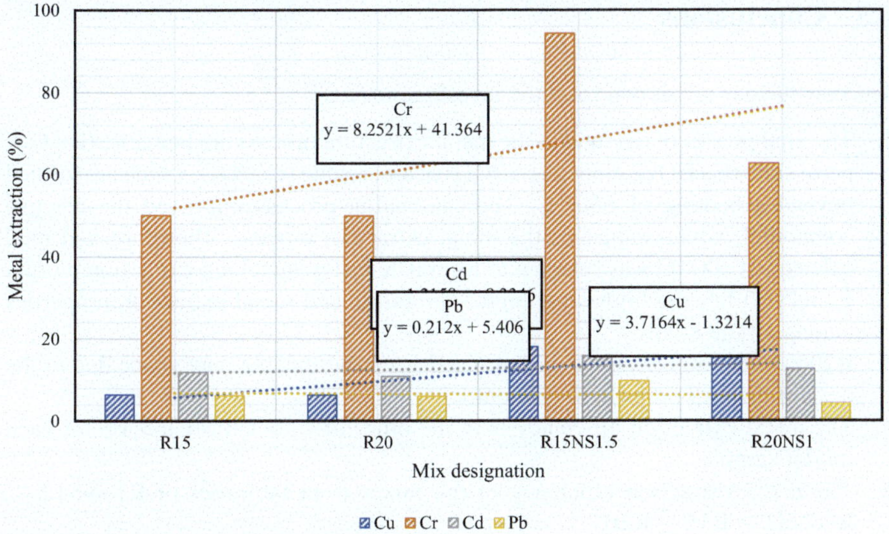

Fig. 5.7 Metal extraction capacity of red mud

$$CrO_4^{2-} + 3Fe_2 + 8H_2O \rightarrow Cr(OH)_3 + 3Fe(OH)_3 + 4H^+ \quad (5.4)$$

Over time, the concentration of Cd in the standard solution was reduced by different percentages by weight. R0, R15, R20, R15NS1.5, and R20NS1 achieved a reduction of 11.904, 10.89, 15.842, 12.692, and 15.842%, respectively. The maximum metal extraction property was observed in R15NS1.5. According to Eq. (5.3), the formation of goethite calcined red mud improved the Cd(II) adsorption rate (Rao et al. 2011). The Cd(II) adsorbed spectrum indicated a change in the crystallinity of silicate from 110 to 011. After adsorption, the alumina phase in the calcined red mud shifted from Al_2O_3 to σ-Al_2O_3 (Rao et al. 2011).

$$FeOH^{0.5+} + Cd^{2+} + H_2O \rightarrow Fe\,(OH)_2CdOH + H^+ \quad (5.5)$$

The mechanism of Pb(II) adsorption indicates that Pb(II) ions have a high affinity towards $CaCO_3$. The adsorption of Pb was contributed by a combination of ion exchange and precipitation. When $CaCO_3$ interacted with Pb(II) in the solution, it resulted in the formation of Pb $(CO_3)_2(OH)_2$ and $PbCO_3$ (Lyu et al. 2021).

$$Pb^{2+} + CO_3^{2-} \rightarrow PbCO_3 \text{ (Lyu et al.2021)} \quad (5.6)$$

$$3Pb^{2+} + 2HCO^{3-} + 4OH^- \rightarrow Pb\,(CO_3)_2(OH)_2 + H_2O \text{ (Lyu et al.2021)} \quad (5.7)$$

5.5 Conclusions

The conclusions are drawn from the investigations.

1. The vital phases in red mud in the metal extraction process are hematite (Fe_2O_3), perovskite ($CaTiO_3$), gibbsite ($Al(OH)_3$), and calcite ($CaCO_3$), which are formed beyond a heating of 650 °C. The calcination process improved the reactive compound, which enhances the metal extraction process. Therefore, red mud calcined at 700 °C was selected to formulate heavy metal ion extraction mortar.
2. R15NS1.5 has the highest strength, durability, and metal extraction properties among all the mixes.
3. R20 exhibits high water absorption and water-permeable pore space due to the development of pores in the structure.
4. The metal extraction efficiency is in the order of Cr > Cd > Cu > Pb for each mortar sample.
5. The metal extraction efficiency of the mixes is in the order of R15NS1.5 $\geq$ R20NS1 $\geq$ R15 $\geq$ R20.

References

ASTM C642-21 (1997) Standard test method for density, absorption, and voids in hardened concrete C642-97. ASTM Int 1–3. https://doi.org/10.1520/C0642-13.5

Asuquo E, Martin A, Nzerem P, Siperstein F, Fan X (2017) Adsorption of Cd(II) and Pb(II) ions from aqueous solutions using mesoporous activated carbon adsorbent: equilibrium, kinetics and characterisation studies. J Environ Chem Eng 5:679–698. https://doi.org/10.1016/j.jece.2016.12.043

Athira K, Shanmugapriya T (2022) Investigation on effect of colloidal nano-silica on the strength and durability characteristics of red mud blended Portland cement paste through tortuosity. Mater Constr 72:1–14. https://doi.org/10.3989/mc.2022.01922

Athira K, Shanmugapriya T (2023) Investigation on properties and heavy metal ion extraction of thermally activated red mud incorporated cement mortar. Polish J Environ Stud 32:3959–3970. https://doi.org/10.15244/pjoes/162550

Ayachi F, Lima EC, Sakly A, Mejri H, Ben Lamine A (2019) Modeling of adsorption isotherms of reactive red RR-120 on spirulina platensis by statistical physics formalism involving interaction effect between adsorbate molecules. Prog Biophys Mol Biol 141:47–59. https://doi.org/10.1016/j.pbiomolbio.2018.07.004

Aydın S, Bedük F, Ulvi A, Aydın ME (2021) Simple and effective removal of psychiatric pharmaceuticals from wastewater treatment plant effluents by magnetite red mud nanoparticles. Sci Total Environ 784:147174. https://doi.org/10.1016/j.scitotenv.2021.147174

Bilińska L, Blus K, Gmurek M, Ledakowicz S (2019) Coupling of electrocoagulation and ozone treatment for textile wastewater reuse. Chem Eng J 358:992–1001. https://doi.org/10.1016/j.cej.2018.10.093

Blackburn RS (2004) Natural polysaccharides and their interactions with dye molecules: applications in effluent treatment. Environ Sci Technol 38:4905–4909. https://doi.org/10.1021/es049972n

Castaldi P, Silvetti M, Santona L, Enzo S, Melis P (2008) XRD, FTIR, and thermal analysis of bauxite ore-processing waste (red mud) exchanged with heavy metals. Clays Clay Miner 56:461–469. https://doi.org/10.1346/CCMN.2008.0560407

Chang G, Shi P, Guo Y, Wang L, Wang C, Guo Q (2020) Enhanced pyrolysis of palm kernel shell wastes to bio-based chemicals and syngas using red mud as an additive. J Clean Prod 272:122847. https://doi.org/10.1016/j.jclepro.2020.122847

Chen X, Guo Y, Ding S, Zhang H, Xia F, Wang J, Zhou M (2019) Utilization of red mud in geopolymer-based pervious concrete with function of adsorption of heavy metal ions. J Clean Prod 207:789–800. https://doi.org/10.1016/j.jclepro.2018.09.263

Cheng X, Long D, Zhang C, Gao X, Yu Y, Mei K, Zhang C, Guo X, Chen Z (2019) Utilization of red mud, slag and waste drilling fluid for the synthesis of slag-red mud cementitious material. J Clean Prod 238:117902. https://doi.org/10.1016/j.jclepro.2019.117902

Chigondo M, Paumo HK, Bhaumik M, Pillay K, Maity A (2019) Magnetic arginine-functionalized polypyrrole with improved and selective chromium(VI) ions removal from water. J Mol Liq 275:778–791. https://doi.org/10.1016/j.molliq.2018.11.032

Dlamini DS, Tesha JM, Vilakati GD, Mamba BB, Mishra AK, Thwala JM, Li J (2020) A critical review of selected membrane- and powder-based adsorbents for water treatment: sustainability and effectiveness. J Clean Prod 277:123497. https://doi.org/10.1016/j.jclepro.2020.123497

dos Reis G, Thue PS, Cazacliu B, Lima E, Sampaio C, Quattrone M, Ovsyannikova E, Kruse A, Dotto G (2020) Effect of concrete carbonation on phosphate removal through adsorption process and its potential application as fertilizer. J Clean Prod 256:120416. https://doi.org/10.1016/j.jclepro.2020.120416

Elzinga EJ, Kretzschmar R (2013) In situ ATR-FTIR spectroscopic analysis of the co-adsorption of orthophosphate and Cd(II) onto hematite. Geochim Cosmochim Acta 117:53–64. https://doi.org/10.1016/j.gca.2013.04.003

Evans K (2016) The history, challenges, and new developments in the management and use of bauxite residue. J Sustain Metall 2:316–331. https://doi.org/10.1007/s40831-016-0060-x

Gao J, Zhao J, Zhang J, Li Q, Gao J, Cai M, Zhang J (2020) Preparation of a new low-cost substrate prepared from drinking water treatment sludge (DWTS)/bentonite/zeolite/fly ash for rapid phosphorus removal in constructed wetlands. J Clean Prod 261:121110. https://doi.org/10.1016/j.jclepro.2020.121110

Hertel T, Pontikes Y (2020) Geopolymers, inorganic polymers, alkali-activated materials and hybrid binders from bauxite residue (red mud)—putting things in perspective. J Clean Prod 258:120610. https://doi.org/10.1016/j.jclepro.2020.120610

Ho H-J, Iizuka A, Shibata E (2021) Chemical recycling and use of various types of concrete waste: a review. J Clean Prod 284:124785. https://doi.org/10.1016/j.jclepro.2020.124785

Ingrassia EB, Lemos ES, Escudero LB (2023) Treatment of textile wastewater using carbon-based nanomaterials as adsorbents: a review. Environ Sci Pollut Res Int 30:91649–91675. https://doi.org/10.1007/s11356-023-28908-9

Katheresan V, Kansedo J, Lau SY (2018) Efficiency of various recent wastewater dye removal methods: a review. J Environ Chem Eng 6:4676–4697. https://doi.org/10.1016/j.jece.2018.06.060

Khairul MA, Zanganeh J, Moghtaderi B (2019) The composition, recycling and utilisation of Bayer red mud. Resour Conserv Recycl 141:483–498. https://doi.org/10.1016/j.resconrec.2018.11.006

Kong Y, Li M, Zhou Y, Pan R, Han Z, Ma J, Chen Z, Shen J (2022) Carbothermal synthesis of nano-iron-carbon composites for arsenate removal from high-arsenic acid wastewater. J Environ Chem Eng 10:107140. https://doi.org/10.1016/j.jece.2022.107140

Li Z, Zhang J, Li S, Gao Y, Liu C, Yanhai Q (2019) Effect of different gypsums on the workability and mechanical properties of red mud-slag based grouting materials. J Clean Prod 245:118759. https://doi.org/10.1016/j.jclepro.2019.118759

Li Y, Huang H, Xu Z, Ma H, Guo Y (2020) Mechanism study on manganese(II) removal from acid mine wastewater using red mud and its application to a lab-scale column. J Clean Prod 253:119955. https://doi.org/10.1016/j.jclepro.2020.119955

Liu R, Poon C (2016) Effects of red mud on properties of self-compacting mortar. J Clean Prod 135:1170–1178. https://doi.org/10.1016/j.jclepro.2016.07.052

Liu J, Zhu R, Liang X, Ma L, Lin X, Zhu J, He H, Parker SC, Molinari M (2018) Synergistic adsorption of Cd(II) with sulfate/phosphate on ferrihydrite: an in situ ATR-FTIR/2D-COS study. Chem Geol 477:12–21. https://doi.org/10.1016/j.chemgeo.2017.12.004

Liu Y, Qin Z, Chen B (2020) Experimental research on magnesium phosphate cements modified by red mud. Constr Build Mater 231:117131. https://doi.org/10.1016/j.conbuildmat.2019.117131

Lyu F, Niu S, Wang L, Liu R, Sun W, He D (2021) Efficient removal of Pb (II) ions from aqueous solution by modified red mud. J Hazard Mater 406:124678. https://doi.org/10.1016/j.jhazmat.2020.124678

Ma Y, Si C, Lin C (2012) Comparison of copper scavenging capacity between two different red mud types. Materials (Basel) 5:1708–1721. https://doi.org/10.3390/ma5091708

Malik R, Ramteke DS, Wate SR (2007) Adsorption of malachite green on groundnut shell waste based powdered activated carbon. Waste Manag 27:1129–1138. https://doi.org/10.1016/j.wasman.2006.06.009

Mukiza E, Zhang LL, Liu X, Zhang N (2019) Utilization of red mud in road base and subgrade materials: a review. Resour Conserv Recycl 141:187–199. https://doi.org/10.1016/j.resconrec.2018.10.031

Patel SA, Barros JA (2015) Introduction, advances in salivary diagnostics. Springer Berlin Heidelberg. https://doi.org/10.1007/978-3-662-45399-5_1

Peng Z, Lin X, Zhang Y, Hu Z, Yang X, Chen C, Chen H, Li Y, Wang J (2021) Removal of cadmium from wastewater by magnetic zeolite synthesized from natural, low-grade molybdenum. Sci Total Environ 772:145355. https://doi.org/10.1016/j.scitotenv.2021.145355

Pera J, Boumaza R, Ambrois J (1997) Development of a pozzolanic pigment from red mud. Cem Concr Res 27:1513–1522

Rai HS, Bhattacharyya MS, Singh J, Bansal TK, Vats P, Banerjee UC (2005) Removal of dyes from the effluent of textile and dyestuff manufacturing industry: a review of emerging techniques with reference to biological treatment. Crit Rev Environ Sci Technol 35:219–238. https://doi.org/10.1080/10643380590917932

Rao K, Mohapatra M, Anand S, Venkateswarlu P (2011) Review on cadmium removal from aqueous solutions. Int J Eng Sci Technol 2:81–103. https://doi.org/10.4314/ijest.v2i7.63747

Sahu M, Sahu U, Patel R (2015) Adsorption of safranin-O dye on CO_2 neutralized activated red mud waste: Process modelling, analysis and optimization using statistical design. RSC Adv 5. https://doi.org/10.1039/C5RA03777H

Shabnam N, Ahn Y, Maksachev A, Lee JH, Huang C-P, Kim H (2019) Application of red-mud based ceramic media for phosphate uptake from water and evaluation of their effects on growth of Iris latifolia seedling. Sci Total Environ 688:724–731. https://doi.org/10.1016/j.scitotenv.2019.06.267

Shahid M, Shahid-ul-Islam, Mohammad F (2013) Recent advancements in natural dye applications: a review. J Clean Prod 53:310–331. https://doi.org/10.1016/j.jclepro.2013.03.031

Shi W, Ren H, Huang X, Li M, Tang Y, Guo F (2020) Low cost red mud modified graphitic carbon nitride for the removal of organic pollutants in wastewater by the synergistic effect of adsorption and photocatalysis. Sep Purif Technol 237:116477. https://doi.org/10.1016/j.seppur.2019.116477

Tang WC, Wang Z, Donne SW, Forghani M, Liu Y (2019) Influence of red mud on mechanical and durability performance of self-compacting concrete. J Hazard Mater 379:120802. https://doi.org/10.1016/j.jhazmat.2019.120802

Teng X, Li J, Wang J, Liu J, Ge X, Gu T (2021) Effective degradation of atrazine in wastewater by three-dimensional electrochemical system using fly ash-red mud particle electrode: mechanism and pathway. Sep Purif Technol 267:118661. https://doi.org/10.1016/j.seppur.2021.118661

Truong HT, Nguyen TH, Lee MS (2017) Separation of molybdenum(VI), rhenium(VII), tungsten(VI), and vanadium(V) by solvent extraction. Hydrometallurgy 171:298–305. https://doi.org/10.1016/j.hydromet.2017.06.006

Vigneshwaran S, Uthayakumar M, Arumugaprabu V (2020) Potential use of industrial waste-red mud in developing hybrid composites: a waste management approach. J Clean Prod 276:124278. https://doi.org/10.1016/j.jclepro.2020.124278

Wang H, Zhang Y, Wang C (2019) Surface modification and selective flotation of waste plastics for effective recycling—a review. Sep Purif Technol 226:75–94. https://doi.org/10.1016/j.seppur.2019.05.052

Weng X, Lin S, Zhong Y, Chen Z (2013) Chitosan stabilized bimetallic Fe/Ni nanoparticles used to remove mixed contaminants-amoxicillin and Cd (II) from aqueous solutions. Chem Eng J 229:27–34. https://doi.org/10.1016/j.cej.2013.05.096

Xie W-M, Zhou F-P, Bi X-L, Chen D-D, Li J, Sun S-Y, Liu J-Y, Chen X-Q (2018) Accelerated crystallization of magnetic 4A-zeolite synthesized from red mud for application in removal of mixed heavy metal ions. J Hazard Mater 358:441–449. https://doi.org/10.1016/j.jhazmat.2018.07.007

Yan S, Cai Y, Li H, Song S, Xia L (2019) Enhancement of cadmium adsorption by EPS-montmorillonite composites. Environ Pollut 252:1509–1518. https://doi.org/10.1016/j.envpol.2019.06.071

Yang T, Wang Y, Sheng L, He C, Sun W, He Q (2020) Enhancing Cd(II) sorption by red mud with heat treatment: performance and mechanisms of sorption. J Environ Manage 255:109866. https://doi.org/10.1016/j.jenvman.2019.109866

Zhang Y, Qian W, Zhou P, Liu Y, Lei X, Li B, Ning P (2021) Research on red mud-limestone modified desulfurization mechanism and engineering application. Sep Purif Technol 272:118867. https://doi.org/10.1016/j.seppur.2021.118867

Chapter 6
Extraction of Lead and Copper Metals Using Microbial Leaching Process from Electronic Waste

N. Mahesh, T. Maheshwari, and S. Balakumar

Abstract Electronic wastes, sometimes referred to as garbage electrical and electronic equipments (WEEE), are a significant kind of urban garbage that contributes to widespread contamination on a global scale. E-waste is a significant threat to the environment and can cause many health hazards. This includes the release of harmful substances through incineration, pollution of food and water, and long-term exposure to recycling facilities leading to organ imbalances. These risks affect all living organisms. Typically, e-waste is produced as a result of the manufacturing of new electronic or electrical products. Furthermore, electronic trash contains a variety of useful and recyclable components, particularly metals. Precious metals such as silver, copper, gold, lead, mercury, and cadmium, which are classified as heavy metals, are obtained through the process of recycling. There are three distinct methods for recycling e-waste: the physical approach, the chemical method, and the biological method. The recycling process, using both physical and chemical methods, can lead to the generation of smoke and dust, which pose various hazardous risks. Biological leaching is the most likely and natural method for cleaning the environment and extracting various metals, as well as removing hazardous substances like Persistent Organic Pollutants (POPs). Thus, the present study focuses on utilizing cyanogenic organisms to biologically extract copper and lead metals from electronic trash through a process known as leaching. *Chromobacterium violaceum* MTCC-2656, *Pseudomonas aeruginosa*, and *Pseudomonas fluorescens*. The microbes MTCC-103 and *Aspergillus niger* MXPE6 are employed for metal extraction by generating secondary metabolites, namely Hydrogen cyanide (HCN) and citric acid, which function as leaching agents. The results obtained from Flame Atomic Absorption Spectroscopy (FAAS) indicated that the cyanogenic microbe, *C. violaceum* MTCC-2656, was able to extract up to 50% of copper and 60% of lead metals from 1 g of electronic waste. Additionally, *P. aeruginosa* was shown to leach 50% of lead metal per 0.5 g of e-waste.

Keywords Electronic waste · Metal ions · Microbial leaching · Microorganisms

N. Mahesh (✉) · T. Maheshwari · S. Balakumar
Department of Chemistry and Biosciences, SASTRA Deemed University, Srinivasa Ramanujan Centre, Kumbakonam, Tamil Nadu, India
e-mail: magi.mibt@gmail.com

© The Author(s), under exclusive license to Springer Nature Switzerland AG 2025
A. Jeyaseelan and K. Murugasen (eds.), *Sustainable Environmental Remediation: Avenues in Nano and Biotechnology*, Environmental Science and Engineering,
https://doi.org/10.1007/978-3-031-78483-5_6

6.1 Introduction

The development of electronic garbage is increasing rapidly in today's growing globe. Of all the types of garbage, electronic waste poses a significant environmental challenge. Our day is incomplete without the usage of electronic gadgets, no matter how little they may be. However, when these devices become obsolete, they pose a significant challenge. Based on the literature review, the generation of electronic garbage (E-waste) is increasing rapidly on a daily basis. E-waste, also known as waste electrical and electronic equipment (WEEE), refers to electrical and electronic gadgets that are no longer in use. It is a threat to the environment and poses significant risks to human health. The electronic waste consists of six types: monitors (10%), televisions (10%), computers, telephones, and their components such as printed circuit boards (PCB), resistors, LED lamps, transistors, capacitors, batteries (15%), DVD/VCR players, CD players, radios and hi-fi sets (15%), refrigerators (20%), washing machines, air conditioners, and vacuum cleaners (30%). E-waste also contains a significant quantity of metals and heavy metals. Heavy metals such as lead, mercury, arsenic, cadmium, selenium, and chromium (cr^{6+}) are present in e-waste at levels higher than usual. These metals pose a considerable risk to the environment (Pant et al. 2012). Only a small proportion of electronic trash can be effectively recycled. Furthermore, the detrimental impacts are exacerbated by the inadequate management of the e-waste transportation infrastructure. This issue arises in both governmental and non-governmental management systems.

Incinerating waste can result in contamination of the soil and atmosphere, while landfilling can lead to groundwater pollution. These activities eventually contribute to biomagnification, impacting humans, animals, and plants through cyclic reactions (Awasthi et al. 2016). The electronic trash, which includes hazardous substances including polybrominated biphenyls and polybrominated diphenyl ethers known as flame retardants (Chaughan and Upadhyay 2015). The version is updated daily, causing the newer release to replace the previous one to facilitate development. Individuals innovate and develop sophisticated goods based on existing ones, resulting in a significant accumulation of electronic garbage. The e-waste includes metals with concentrations that are ten times greater than those found in commercial ores.

E-waste often originates from several aspects of our daily lives worldwide. Between 1994 and 2003, almost 500 million personal computers (PCs) were abandoned worldwide. In 2016, the global e-waste production reached a staggering 44.7 million tons. The United States, China, Japan, and Germany are the countries that have gained popularity for manufacturing millions of tons of electronic garbage. The United States produces 7072 metric kilo tonnes of e-waste per year, whereas China produces 6033 metric kilo tonnes and Japan produces 2200 metric kilo tonnes (Anshu and Subrata 2016). Within the European Union, each individual in 15 nations produces an average of 3.3–4.3 kg of electronic garbage. Europe produces more than 8.7 million tons of electronic garbage. Out of these, only 25% are officially recycled, while the rest are disposed of in an unknown manner (Perkins et al. 2014). According to the United Nations Environmental Programme (UNEP), the annual generation of electronic garbage (e-waste) amounts to 50 million tons. Additionally, it is estimated that annually, 20–50 million metric tonnes of waste electrical and electronic equipment (WEEE) are created. However, only 10% of this material gets recycled. The amount of electronic garbage produced is projected to reach 120 million tons by the year 2050. According to literature, India produced 41.8 million tons of electronic garbage (e-waste) in 2014, which increased to 49.8 million tonnes in 2018. The states of Maharashtra, Tamil Nadu, Andhra Pradesh, Uttar Pradesh, West Bengal, Delhi, Karnataka, Gujarat, Madhya Pradesh, and Punjab are the leading contributors to E-waste in India, accounting for up to 70% of the total. Norway is the top purchaser of e-waste from the USA for recycling purposes. They provide $40,000 (USD) for e-waste weighing over 30 kg. Every hour, around 4000 tons of electronic garbage (e-waste) is dumped worldwide, with nearly 80% of this material being illegally trafficked to Asian nations through unlawful cross-border trafficking. Similarly, Switzerland would generate 110,000 tons while Germany would create 1.5 million tonnes. Consequently, the production of garbage is steadily increasing.

Guidelines for global world according to WEEE:

Currently, the handling of electronic trash in India is regulated by the Environmental Protection Act of 1986 and the Hazardous Material Management and Transporting Facility of 2008, in accordance with legislative norms and regulations. The WEEE Directive (2002/96/EC), commonly referred to as the RoHS guidelines, is a European Council directive that mandates nations and states to transform WEEE (Waste Electrical and Electronic Equipment) into valuable elements through metal recovery in order to mitigate environmental risks. This will also decrease the financial strategy of the electrical and electronic products maker by reusing specific components in a certain product. This will reduce the duration required for product design and manufacture. However, this will only apply to designers who create their own brand and not to dealers who act on behalf of patenting rights. The RoHS standards prohibit the use of electric and electronic equipment that contains any of the six hazardous compounds, namely Pb, Cr IV, Cd, Poly brominated biphenyls, mercury, and Poly brominated diphenyl ethers, over their permissible levels. Table 6.1 outlines the acceptable thresholds for six hazardous substances that may be present in electronic equipment. The RoHS directive, originally known as 2011/65/EU, was updated in 2011 and is now referred to as "RoHS 2" or "RoHS recast." This update was made to maintain compliance with regulations for medical and other monitoring devices.

Furthermore, industrialized nations have devised many strategies for managing e-waste, including MFA (material flow analysis), LCA (life cycle assessment), EPR (extended producer responsibility), and ARF (advanced recycling fee). These techniques are designed for the installation of membrane flow analysis devices by waste management industries. The purpose is to measure the amount of recycled materials or estimate the presence of important materials in the trash by passing them through a membrane and recording the data digitally. Similarly, the life cycle assessment (LCA) determines the duration for which a product can be used in electronic devices based on its energy generation. It also specifies the type of material used and the method of disposal. When the product is no longer functional, extended producer

Table 6.1 Concentration of HCN produced by microbes estimated through silver nitrate titration method

S. no	Microorganisms	Titrate value (ml)	Mass of bacterial sample (ml)	Conc. of HCN (ppm)
1	*C. violaceum* MTCC-2656	A1: 0.6	1	8.1
		A2: 0.6	1	8.1
2	*P. fluorescens* MTCC-103	B1: 0.4	1	5.4
		B2: 0.4	1	5.4
3	*P. aeruginosa*	C1: 0.4	1	5.4
		C2: 0.5	1	6.75

responsibility (EPR) and advanced recycling fees (ARF) benefit both consumers and manufacturers or resellers by allowing them to create new products from the old ones and paying certain taxes.

6.2 Materials and Methods

6.2.1 Collection of Electronic Waste

E-wastes are gathered from waste marts in the vicinity of Kumbakonam, a town situated in the Thanjavur district of Tamil Nadu. It comprises a combination of components such as resistors, capacitors, transistors, batteries, and PCBs. The sorting process was conducted according to the categorization of components into metallic and non-metallic, as well as their variance in size. The physical approach is employed to extract plastic and other non-metallic compounds from electronic trash. Next, the sorted electronic debris can be fragmented into smaller bits using a hammer. The electronic garbage can be pulverized by the waste mart steel firm situated in Kumbakonam, located in the Thanjavur district of Tamil Nadu. To achieve particles smaller than 1 mm, the granulated electronic trash can be sifted using a sieve with a pore size of 0.1 mm.

6.2.2 Procurement of Microbial Cultures

The Meenatchi lab in Thanjavur district, Tamil Nadu, provided samples of *Pseudomonas aeruginosa* and *Klebsiella pneumoniae*. Additionally, *Chromobacterium violaceum* MTCC-2656, *Pseudomonas fluorescens* MTCC-103, and fungi like *Aspergillus niger* MXPE6 are obtained from IMTech Chandigarh. Next, bacteria are transferred to nutritional agar slant and fungi to PDA agar slant for further growth. These cultures are then maintained in a refrigerator at a temperature of 10 °C. The

LB medium is composed of 10 g of tryptone, 5 g of yeast extract, and 10 g of sodium chloride per 1000 ml, with the pH set to 7.0. Initially, each bacterial culture is introduced into a sterile 10 ml Luria Bertani Broth from the culture that has been cultivated overnight. Subsequently, the sample is placed in an incubator set at a temperature of 37 °C for a duration of 24 h. Glycine is made as a 10% solution by dissolving 10 g of glycine in 100 ml of distilled water. This solution is then used as a supplement for bacteria to produce Hydrogen cyanide (HCN).

6.2.3 Media Preparation for HCN Production

Each of the 4 small sterile tubes is filled with 1.5 ml of a solution containing 10% glycine. Then, 1.5 ml of a bacterial culture broth that has been incubated for 24 h (specifically, *Chromobacterium violaceum* MTCC-2656, *Pseudomonas fluorescens* MTCC-103, *Pseudomonas aeruginosa*, and *Klebsiella pneumoniae*) is added to each tube, resulting in a total volume of 3 ml. These tubes are labeled as T1, T2, T3, and T4, and are stored at a temperature of −2 °C for a duration of 12 h. Finally, eight sterile 250 ml conical flasks (with two flasks per bacterial culture) are filled with 50 ml of LB medium and sterilized at 121 °C for 15 min under 15 pounds per square inch (LBS) pressure. These flasks are then labeled as follows: (A1, A2) for *Chromobacterium violaceum* MTCC-2656, (B1, B2) for Pseudomonas fluorescens MTCC-103, (C1, C2) for Pseudomonas aeruginosa, and (D1, D2) for Klebsiella pneumoniae. As a result, the culture that has been cultured for 12 h is transferred into a sterile 50 ml LB broth. The mixture was agitated in a shaker at a speed of 35 revolutions per minute at a temperature of 30 °C for a duration of 2 days.

6.2.4 HCN Estimation by Titrimetric Method

Initially, the cultures that have been cultured for 2 days may be assessed for their optical growth density using a spectrophotometer. Next, it undergoes HCN analysis using the titrimetric approach as described by Iliya and Madumelu (2019). A sterile conical flask is filled with 10 ml of a solution containing 0.1% KCN. A little amount of a solution containing 5% KI is added to the flask. This mixture is then referred to as the Positive control (P1). To conduct the test, a 1 ml sample of bacterial culture broth (A1, A2, B1, B2, C1, C2, D1, D2) that has been cultured for 2 days is collected. This sample is then diluted with 100 ml of distilled water. 10 ml of each diluted sample is transferred into individual conical flasks, and a little amount of 5% KI is added to each diluted test sample. A burette solution containing a concentration of 0.02 M $AgNO_3$ solution is used, whereas the pipette solution consists of diluted bacterial culture samples. The positive control is titrated against a 0.02 M solution of $AgNO_3$, resulting in a light-yellow color with a muddy appearance as the endpoint of the analysis. Consequently, the test samples (A1, A2, B1, B2, C1, C2 and D1, D2) are likewise subjected to titration using 0.02 M AgNO3. This results in the formation of a faint yellow color with turbidity, which serves as the endpoint for analysis. Repeated the process to obtain the consistent values. Thus, this test verifies the existence of the HCN. The *Aspergillus niger* MXPE6 is initially introduced into a sterile potato dextrose agar medium and incubated at a temperature of 30 °C for a duration of 48 h.

6.2.5 Media Preparation for Citric Acid Production

Two sterile 250 ml conical flasks were obtained and 50 ml of potato dextrose broth medium (PDB) was poured to each flask. The flasks were then autoclaved at a temperature of 120 °C for a duration of 15 min at a pressure of 15 lbs. The flasks were labeled as F1 and F2. A little amount of culture that has been cultivated overnight (*Aspergillus niger* MXPE6) is transferred and introduced into a sterile PDB medium (F1 and F2). Subsequently, the sample is maintained at a temperature of 30 °C for a duration of 7 days while being subjected to a rotational speed of 30 revolutions per minute in a rotatory shaker. Afterwards, it is retained for the purpose of estimating citric acid using the titrimetric technique.

6.2.6 Citric Acid Estimation by Titrimetric Method

To establish a positive control (P2), 10 ml of a pure solution of 0.1% Citric acid anhydrous is placed in 250 ml conical flasks. A little amount of phenolphthalein indicator is then added. Subsequently, it is subjected to titration using a 0.1N NaOH solution contained in a burette. The pale pink color is produced as the result of endpoint analysis. Repeated this process to obtain consistent values. For the test, a 1 ml sample of a 7-day fermented broth is transferred into a 250 ml conical flask from F1 and F2. The sample is then diluted with 10 ml of distilled water after verifying the pH value. A little amount of Phenolphthalein indicator was added and then titrated against the 0.1N NaOH solution. The pale pink color gradually diminishes, resembling the appearance of a positive control. Repeated the process to obtain consistent values. Therefore, this test has verified the existence of citric acid in this broth. To implement abiotic control, obtain four 250 ml sterile conical flasks and label them as X1, X2, X3, and X4. Subsequently, 50 ml of LB medium is introduced into X1 and X2. Then, 50 ml of Potato dextrose medium (PDB broth) is added to X3 and X4. Next, the item is subjected to sterilization at a temperature of 121 °C for a duration of 15 min, while maintaining a pressure of 15 lbs. Finally, the item is cooled down. For biotic control, five sterile conical flasks with a volume of 250 ml each are used. These flasks are labeled as Y1, Y2, Y3, Y4, and Y5. Next, 50 ml of LB medium is poured into each of the four 250 ml conical flasks labeled Y1, Y2, Y3, and Y4. Then, 50 ml of Potato Dextrose medium (PDB broth) is added to Y5. Next, the item is subjected to sterilization at a temperature of 121 °C for a duration of 15 min, while maintaining a pressure of 15 lbs. Finally, it is cooled down.

6.2.7 E-waste Inoculation

Prior to introducing e-waste into the broth medium, it is necessary to disinfect the substrate using ethanol (acetone). A little amount of acetone is introduced to the e-waste and allowed to evaporate for a short period of time. Subsequently, it is incorporated into the broth medium.

6.2.8 Inoculation Based on Pulp Density

0.5 g of sterilized electronic trash is introduced into the bacterial samples (A1, B1, C1, and D1) and the abiotic control (X1 and X3) conical flasks. Similarly, a quantity of 1.0 gm of electronic trash is introduced into the bacterial samples being tested in wells A2, B2, C2, and D2, as well as in the abiotic control wells X2 and X4. For control, 1.5 ml of bacterial broth cultures, namely *Chromobacterium violaceum* MTCC-2656, *Pseudomonas fluorescens* MTCC-103, *Pseudomonas aeruginosa*, and *Klebsiella pneumoniae*, are added to each of the four sterilized LB medium (Y1, Y2, Y3, and Y4). The mixture is then cultured for a period of 10 days. An inoculum of *Aspergillus niger* MXPE6 is added to a sterile PDB medium, also known as Y5-biotic control. Afterwards, 0.5 g and 1.0 g of electronic trash (e-waste) are introduced into the fungal cultures for testing (F1 and F2). Subsequently, the specimens are placed in a rotary shaker for a duration of 20 days to undergo incubation.

6.2.9 Leaching by Chemical Method

This approach facilitates the comparative analysis of the leaching process between microorganisms and chemical reagents, namely acids. A beaker holding 20 ml of concentrated HNO_3 is used to hold 0.5 g of e-waste scrapings, which is referred to as V1. Urea is then added to deactivate the active nitric acid. The addition continues until the reaction ceases. Next, the mixture is filtered and rinsed with a small quantity of distilled water. It is then left overnight to allow any remaining particles to settle to the bottom of the beaker. The remains were dried using a hot air oven and then weighed. It was conducting both qualitative and quantitative analysis. Similarly, the aforementioned technique is applied on 1.0 g of electronic waste and recorded as V2.

6.2.10 Qualitative Analysis of Metals from Leached Cultures

Qualitative analysis is mostly employed to ascertain the presence of certain metals in the solution. Therefore, by employing certain chemical reagents that detect the existence of metals in a solution through a change in color. After a 10-day incubation period, the fermented bacterial culture is collected and filtered. The filtered culture broth is then separated into different substrates, each containing *Chromobacterium violaceum* MTCC-2656 (A1 and A2), *Pseudomonas fluorescens* MTCC-103 (B1 and B2), *Pseudomonas aeruginosa* (C1 and C2), and *Klebsiella pneumoniae* (D1 and D2). Subsequently, each filtrate is individually gathered and subjected to centrifugation at a speed of 3000 revolutions per minute for a duration of 15 min. The pellet is retrieved and the liquid above it is discarded. The bacterial sample is analyzed qualitatively using this pellet. Similarly, the fungal culture (F1 and F2) undergoes filtration, and the resulting filtrate is used for subsequent processing. Likewise, the chemically extracted material has likewise experienced this process.

6.2.11 Estimation of Lead and Copper

In a standard solution, a little quantity of lead nitrate ($PbNO_4$) and a few drops of acetic acid are combined, and then potassium chromate (K_2CrO_4) is added. The same technique is followed for all tests (A1, A2, B1, B2, C1, C2, D1, D2, F1, F2, V1, and V2), where the pellets are replaced with residues in standard solutions. To prepare a standard solution, a little quantity of copper sulphate ($CuSO_4$) and a few drops of acetic acid are added, followed by the addition of ammonium hydroxide (NH_4OH). The same process is followed for all tests (A1, A2, B1, B2, C1, C2, D1, D2, F1, F2, V1, and V2), wherein pellets and residues are replaced in standard solutions.

6.2.12 Quantitative Analysis for Metals from Leached Cultures

The metals in the analytic solution are concentrated using the process of atomic absorption spectroscopy (AAS) analysis. Meanwhile, the samples underwent acid digestion by being mixed with 10 ml of 0.1N Nitric acid. Hence, the chemicals extracted from both chemical and biological process and control samples are examined using flame atomic absorption spectroscopy (FAAS) to compare their leaching efficiency.

6.3 Results and Discussion

Electronic garbage, sometimes known as e-waste, is a significant kind of urban waste that is rapidly increasing on a global scale. Contamination of the environment may occur through the soil, air, and water, which in turn can have adverse effects on biotic life. This contamination can lead to biomagnification, resulting in various health problems. Consequently, the majority of discarded electrical goods may be reused by fixing broken components or recycled to extract various metals and polymers. This can be accomplished by many recycling methods, including physical leaching, chemical leaching, and bioleaching. Despite the significant pollution caused by the physical and chemical leaching process, most research has focused on the bioleaching method for metal extraction due to its environmentally benign nature and cost-effectiveness. Bioleaching is a likely sluggish procedure that exhibits excellent efficiency and reduces both labor and economic costs.

Both heavy metals and precious metals should be mined because of their economic worth. The majority of the studies utilized Printed Circuit Boards (PCBs) for the purpose of metal extraction. The transistors, capacitors, and resistors comprised various metals, such as copper (Cu), silver (Ag), lead (Pb), tin (Sn), zinc (Zn), aluminum (Al), gold (Au), and numerous rare metals. The computer PCB, after digestion, often includes concentrated amounts of metals such as aluminum (Al), copper (Cu), lead (Pb), nickel (Ni), tin (Sn), and zinc (Zn). The concentrations of these elements in a 1 kg sample were evaluated using ICP-AES-MS (inductively coupled plasma-atomic emission mass spectroscopy) analysis, with values of 237 g, 80 g, 20 g, 15 g, 23 g, and 26 g respectively (Brandl et al. 2001). The concentration of copper metal in mobile PCB was determined to be 34.49% using ICP-OES (inductively coupled plasma-optical emission spectroscopy), while the concentration of lead metals was calculated to be 1.87%. In computer PCB, the gold metal content was found to be 0.13%. Willner and Fornalczyk (2013) conducted an analysis using atomic absorption spectroscopy (AAS) to examine the mobile PCB. They found that the PCB included 0.68% lead metal and 35.9% copper metals from a 1-g sample of e-scrappings with a particle size of less than 0.5. Based on the studies, ICP-AES (inductively coupled atomic emission spectroscopy) was used to analyze 35% of Cu metal in a camera. Consequently, all electrical equipment have been composed of both lead and copper metals.

6.3.1 Qualitative Estimation of Copper and Lead from Transistors, Capacitors and Resistors

Following the disassembly of the transistors, the capacitors and resistors were pulverized and subjected to qualitative testing to determine the specific metals included inside the electronic waste. Moreover, the extraction of metals such as lead and copper is facilitated by the straightforward and uncomplicated method of bioleaching.

Eliminating or extracting lead metal would reduce the toxicity of heavy metals and enhance the recycling potential for producing lead-coated objects. The extraction of copper facilitates the reutilization and recycling of products that are covered with copper. Our recent research indicates that the microbial leaching method is used to recover copper and lead metals. This process involves the creation of leaching agents, specifically hydrogen cyanide and citric acid. Currently, cyanogenic microorganisms such as *C. violaceum*, *P. aeruginosa*, and *P. fluorescens* are extensively employed for metal extraction.

6.3.2 HCN Determination by Silver Nitrate Titrimetric Method

In their 1986 experiment, Knowles and Bunch observed that Pseudomonas species may create cyanide at levels of up to 300 μM by the oxidative decarboxylation of glycine. The silver nitrate titrimetric approach produces a pale-yellow tint as a result of the presence of HCN, indicating a favorable outcome for this method. This occurs as a result of the reaction between silver ions from silver nitrate and cyanide ions from hydrogen cyanide, leading to the formation of silver cyanide under alkaline conditions. Iliya and Madumelu (2019) indicated that the quantities of the titrant used in the determination of HCN in apple seeds by the silver nitrate titration technique were 0.64, 0.58, and 0.62 ml. The synthesis of hydrogen cyanide by some microorganisms is significantly influenced by nutrition and other parameters. In their study, Zhihui et al. (2018) investigated the production of cyanide by *P. fluorescens*. They found that maintaining the bacteria in a nutrient-rich medium with a suitable pH led to an increase in cyanide production. Specifically, when the bacteria were cultured in a Luria Bertani (LB) medium containing 10 gm/l of NaCl, a cyanide production of 6.25 mg/l was observed after 16–21 h of incubation at pH 8.5. The cyanide production was determined using the Calorimetric method. The current study demonstrated that the estimation of Hydrogen cyanide (HCN) using the silver nitrate titration method resulted in a pale-yellow color. The titration values were found to be up to 0.6 ml for *C. violaceum* MTCC-2656 (A1 and A2), 0.4 ml and 0.5 ml for *P. aeruginosa* (C1 and C2), 0.4 ml for *P. fluorescens* MTCC-103 (B1 and B2), and *Klebsiella pneumoniae* (D1 and D2) consumed 1.0 ml of the burette solution. Consequently, the formation of a green color indicated a negative result for this method. The findings are displayed in Table 6.1.

Table 6.2 Estimation of citric acid by titrimetric method produced by *A. niger* MXPE6

S. no	Sample description	Pipette solution volume (ml)	Burette solution volume (ml)	Concordant value (ml)	Indicator	Final color
1	Positive control: citric acid	P2: 10	Initial: 0, final: 5.6	5.6	Phenolphthalein	Pale pink
2	Test: *A. niger* MXPE6	F1: 10	Initial: 0, final: 6.9	6.9	Phenolphthalein	Pale pink
		F2: 10	Initial: 0, final: 7.0	7.0	Phenolphthalein	Pale pink

6.3.3 Calculation for Cyanide Concentration

Micorbial sample titrated against 0.2 ml of $AgNO_3$ burette solution. Therefore,

$$\text{Concentration of HCN} = \frac{\text{Titrate value} \times 13.5(\text{constant value})}{\text{Mass of sample used(bacterial sample)}}$$

Hence, the dosage of HCN creation by the microbes was verified before the e-waste inclusion perceived by means of the calculation using the above formula and it is presented in Table 6.1.

6.3.4 Results for Citric Acid Estimation from Aspergillus niger *MXPE6 Culture Medium by Titrimetric Method*

Fungi generate several enzymes and organic acids as byproducts of their metabolism, which are studied in the field of literature. The titrimetric approach used for citric acid quantification results in the formation of a pale pink color. Table 6.2 displays the positive results obtained by *A. niger* MXPE6.

6.3.5 Results for Qualitative Estimation of Copper and Lead Metals After the Bioleaching Process

Following the leaching procedure, the leachate is gathered to qualitatively determine the presence of estimated metals in the solution using straightforward chemical methods. The presence of copper is shown by the production of a brown-colored ring with brown precipitation.

Table 6.3 Qualitative estimation for copper (Cu) and lead (Pb) metal after bioleaching

S. no	Microorganism sample	E-waste concentration in culture medium	Presence of copper	Presence of lead
1	*C. violaceum* MTCC-2656	A1: 1 g	Yes	Yes
		A2: 0.5 g	No	No
2	*P. fluorescens* MTCC-103	B1: 1 g	Trace	Yes
		B2: 0.5 g	No	Trace
3	*P. aeruginosa*	C1: 1 g	No	Yes
		C2: 0.5 g	No	Yes
4	*Klebsiella pneumonia*	D1: 1 g	No	No
		D2: 0.5 g	No	No
5	*Aspergillus niger* MXPE6	F1: 1 g	Yes	No
		F2: 0.5 g	Yes	No
6	By chemical leaching method	V1: 0.5 g	Yes	Yes
		V2: 1.0 g	Yes	Yes

Note Present (+); Negative (−)

This color formation is then compared to a reference copper sample. Yellow precipitation is created when lead is present and it is compared to the standard lead. The findings obtained from the microorganisms and chemical processes are recorded and organized in Table 6.3, based on the visual observation of color.

6.3.6 Results Predicted for Quantitative Assay After Bioleaching

Jatindra and Sudhir (2014) found that *Pseudomonas fluorescens* leached up to 19.366 mg/l (13%) of copper from 1 g of e-waste, while *Pseudomonas aeruginosa* leached 74.65 mg/l (52%) and *Chromobacterium violaceum* leached 110.79 mg/l (79.3%). These estimates were obtained through ANOVA analysis of triplicate analyses conducted over a 30-day incubation period in a two-step bioleaching process. When the pH level rose, the bacteria were more efficient at leaching, indicating the existence of a leaching agent (Pham and Ting 2009). Similarly, Brandl et al. (2001) reported that Aspergillus niger was able to leach 65% of copper metal and 40% of lead metal from computer PCBs (printed circuit boards). This was achieved by the production of several organic acids such as citrate, gluconate, and oxalate, which acted as leaching agents. The leaching process occurred when a 1% w/v solution of e-scraps was used.

The concentration of the leached solution is determined quantitatively using flame atomic absorption spectroscopy (FAAS). This investigation aims to compare the

Table 6.4 Quantitative estimation for copper (Cu) metal by AAS analysis after bioleaching of e-waste by microbes:

Sample ID	Signal absorbance	RSD %	Concentration (mg/l)	Corrected concentration (mg/l)
Copper Blank	0.0000	>99	0.0	
Copper Standard 1	0.0893	0.4	1.0	
Copper Standard 2	0.1769	0.3	2.0	
Copper Standard 3	0.3721	0.2	5.0	
Copper Standard 4	0.7667	0.3	10.0	
Copper Sample ID A Low	0.0887	0.3	1.0486	1.0486
Copper Sample ID A High	0.0544	0.4	0.5934	0.5934

leaching process between *C. violaceum* MTCC-2656 using 1 g of pulp density and *P. aeruginosa* using 0.5 g of pulp density. The leached concentration of metals was evaluated at a concentration of 1 ml of microbial culture pellet. At a low dilution rate, the signal absorption passed above 0.0887 (Cu Sample Id A low), indicating a correlation with a concentration of 1.0486 mg/l. Therefore, with a greater dilution rate, the absorption signal of 0.0544 (Cu Sample Id A HI) surpassed the concentration of 0.5934 mg/l during the quantification of copper by *C. violaceum* MTCC-2656 (A2). In lead determination, the absorption signal 0.1863 (Pb Sample A REP) indicated a value of 7.0649 mg/l, whereas the absorption signal 0.0186 (Pb Sample Id A REP) indicated a concentration of 0.6034 mg/l by *C. violaceum* MTCC-2656, at different dilution rates. *Pseudomonas aeruginosa* surpassed the absorption signal of 0.017 (Pb Sample B) and achieved a lead metal concentration of up to 0.5413 mg/l at a lower dilution rate. At a greater dilution rate, the concentration of lead metal increased to 0.5298 mg/l, resulting in an absorption signal of 0.0167 (Pb Sample Id B REP) by *P. aeruginosa.* The concentrations obtained from the flame atomic absorption spectroscopy (FAAS) examination are organized in Tables 6.4 and 6.5.

Chromobacterium violaceum extracted a much greater quantity of lead metal compared to *P. aeruginosa. Pseudomonas aeruginosa* (C1) extracted 0.5298 mg/l of lead metal from a pulp density of 0.5 g, even with low concentration of HCN generation. On the other hand, *C. violaceum* MTCC-2656 (A2) extracted 0.6034 mg/l of lead metal from a pulp density of 1 g. From here on, *P. aeruginosa* has a greater leaching efficiency than *Chromobacterium* MTCC-2656, unless *C. violaceum* MTCC-2656 (A2) is able to remove 0.5934 mg/l of copper metal from 1 g of pulp density. Reducing heavy metal contamination is a significant achievement in improving leaching efficiency. Hence, *C. violaceum* MTCC-2656 achieved a leaching efficiency of 60% for copper and lead extraction, whereas *P. aeruginosa* achieved a 50% extraction efficiency for lead metal from 0.5 g of e-waste. Due to its significant synthesis of HCN, C. violaceum has a relatively modest leaching ability compared to *P. aeruginosa.* The comparison is presented in Table 6.6 and Fig. 6.1.

Table 6.5 Quantitative estimation for lead (Pb) metal by AAS analysis after bioleaching of e-waste by microbes

Sample identifier	Absorbance signal	RSD (%)	Lead concentration (mg/l)	Adjusted concentration (mg/l)
Lead Blank	−0.0008	39.5	0.0	
Lead Standard 1	0.0274	0.8	1.0	
Lead Standard 2	0.0563	0.4	2.0	
Lead Standard 3	0.1401	0.1	5.0	
Lead Standard 4	0.2585	0.2	10.0	
Lead Sample B	0.0170	1.2	0.5413	0.5413
Lead Sample A REP	0.1863	0.1	7.0649	7.0649
Lead Sample ID A REP	0.0186	0.7	0.6034	0.6034
Lead Sample ID B REP	0.0167	0.4	0.5298	0.5298

Table 6.6 Comparison of leaching efficiency between *P. aeruginosa* (C1) and *C. violaceum* MTCC-2656 (A2)

Serial no	Microorganism sample	HCN concentration (mg/ml)	Copper content (mg/l)	Lead content (mg/l)
1	*C. violaceum* MTCC-2656 A2	8.1	0.5934	0.6034
2	*P. aeruginosa* C1	5.4	Not detected	0.5298

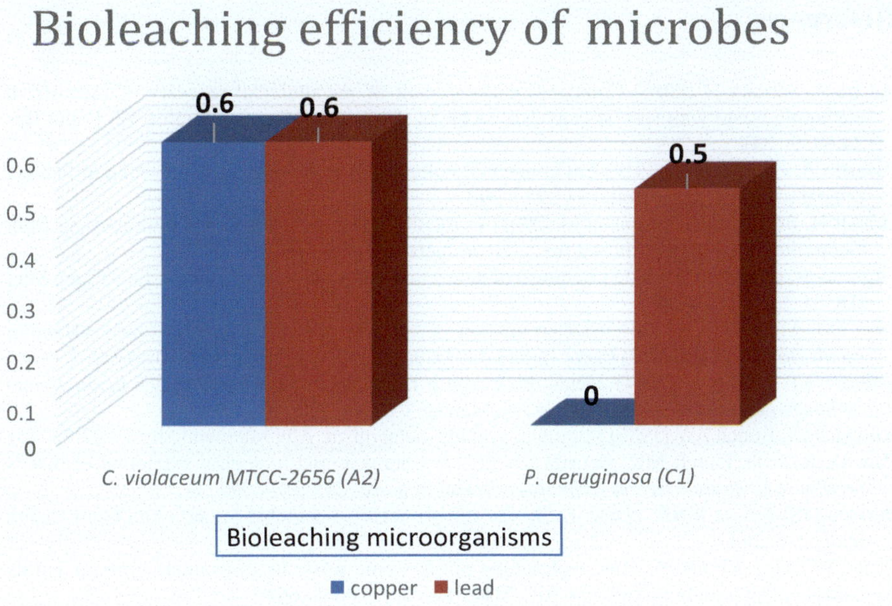

Fig. 6.1 Comparision of bioleaching efficiency in e-waste between *P. aeruginosa* (C1) and *C. violaceum* MTCC-2656 (1 g) A2

6.4 Conclusion

The current investigation identified the existence of lead and copper metals in electronic trash, such as transistors, capacitors, and resistors. This study investigated the extraction of metals using certain microbes, including *C. violaceum* MTCC-2656, *P. aeruginosa*, *P. fluorescens* MTCC-103, and *A. niger* MXPE6. These bacteria were found to create hydrogen cyanide (HCN) and citric acid, which acted as leaching agents. The presence of these agents was validated using the titrimetric technique. The presence of hydrogen cyanide (HCN) in the incubated media was verified using the silver nitrate titrimetric technique. In addition, citric acid may also be determined using the titrimetric technique. Following the introduction of e-waste, the incubated cultures underwent centrifugation to determine the presence of lead and copper metal using qualitative estimate. The concentration of metals collected was accurately anticipated using AAS Spectroscopy. The extraction of lead and copper metals by chemical leaching is also assessed using qualitative analysis. Therefore, bioleaching is currently the predominant method utilized in mining sectors for extracting metals. The use of physical and chemical methods to enhance efficiency can lead to environmental pollution, but it is a slow process that ultimately yields long-term results. Thus, employing a biological system will exhibit the verdant nature of the earth.

References

Anshu P, Subrata H (2016) Comparative assessment of metallurgical recovery of metalsfrom electronic waste with special emphasis on bioleaching-review article. Environ Sci Pollut Res 24:6989–7008

Awasthi A, Zeng X, Li J (2016) Environmental pollution of electronic waste recycling in India: a critical review. Environ Pollut 211:259–270

Brandl H, Bosshard R, Wegmann M (2001) Computer-munching microbes: metal leaching from electronic scrap by bacteria and fungi. Hydrometallurgy 59:319–326

Chauhan R, Upadhyay K (2015) Removal of heavy metal from E-waste: a review. Int J Chem Stud (IJCS). 3(3):15–21

Iliya JM, Madumelu M (2019) Determination of cyanide content in cassava tubers (*Manihot esculenta*) and apple seed (*Pyrus malus*). Int J Curr Res Chem Pharm Sci 6(2):14–20

Jatindra KP, Sudhir K (2014) Metals bioleaching from electronic waste by *Chromobacterium violaceum* and *Pseudomonads* sp. Waste Manage Res 30(11):1151–1159

Knowles CJ, Bunch AW (1986) Microbial cyanide metabolism. Adv Microbial Physiol 27:73–111

Pant D, Joshi D, Upreti MK, Kotnala RK (2012) Chemical and biological extraction of metals present in E waste: a hybrid technology. Waste Manage 32(5):979–990

Perkins DN, Drisse MNB, Nxele T, Sly PD (2014) WEEE—a global hazard. Ann Glob Health 80(4):286–295

Pham VA, Ting YP (2009) Gold bioleaching of electronic waste by cyanogenic bacteria and its enhancement with bio-oxidation. Adv Mater Res 71(73):661–664

Willner J, Fornalczyk A (2013) Extraction of metals from electronic waste by bacterial leaching. Environ Protect Eng 39(1):197–208

Zhihui Y, Jujun R, Yaying L, Rongliang Q (2018) A new model for simulating microbial cyanide production and optimizing the medium parameters for recovering precious metals from waste printed circuit boards. J Hazard Mater 353:135–141

Chapter 7
A Comparative Analysis on the Decomposition and Mitigation of Azo Dyes in Industrial Discharge Using Microbes, Nanoparticles and Nanozymes

R. Rajeswari, S. Balakumar, and N. Mahesh

Abstract Release of industrial effluent into waterbodies is the major reason for the pollution of aquatic ecosystem. The major contaminant in this effluent is the azo dyes. These azo dyes are employed to impart the characteristic colour to the textiles, leather, cosmetics, paint, paper, plastics and all. Azo dyes are the man-made synthetic compounds that have the specific azo bond (N = N). They are also called as xenobiotics, the foreign compounds that are often refractory to degradation. Release of used dyes into the water resources is highly hazardous. These staining dyes reduce the transparency of water and prevent the light to penetrate into it. Further it reduces the dissolved oxygen content in the waterbodies. Without light and oxygen, all the organisms in the aquatic ecosystem are greatly affected. Further, human consumption of that water or the animals from that waterbody is highly toxic as these azo dyes can induce mutation and cause cancer. Several mechanical and chemical techniques are being used to eliminate the dye from the effluent and to reduce its toxicity. But these physiochemical techniques didn't remove the hazardous compounds completely and produce toxic by-products. Biodegradation is the method of breaking down of carbon based compounds using microbes like bacteria, algae and fungi. As the microbes produce several enzymes, they have the metabolic ability to degrade them. Thus, microbes can be used to reduce these azo dyes. This method is cheap and eco-friendly. Owing to the emerging advancements in nanotechnology, nanoparticles (Nps) serve to be a promising means of degradation of azo dyes. Nps with their increased surface area, get oxidized and reduce the dye molecules. Further, nanozymes are the nano-sized particles with enzyme-like activity. It is a nano-metal that act like an enzyme and degrade the toxic dye compounds. It is believed to be the product of biotechnological and nano technological innovations. This review is aimed to give a critical analysis on the toxic nature of industrial effluent and to make

R. Rajeswari · S. Balakumar · N. Mahesh (✉)
Department of Chemistry and Biosciences, SASTRA Deemed University, Kumbakonam, Tamil Nadu 612001, India
e-mail: magi.mibt@gmail.com

© The Author(s), under exclusive license to Springer Nature Switzerland AG 2025
A. Jeyaseelan and K. Murugasen (eds.), *Sustainable Environmental Remediation: Avenues in Nano and Biotechnology*, Environmental Science and Engineering,
https://doi.org/10.1007/978-3-031-78483-5_7

a comparative study on the efficiency of using microbeds, Nps and Nz to degrade the toxic compounds.

Keywords Azo dyes · Nanozyme · Nano particles · Biodegradation

7.1 Introduction

The environment has been severely affected through the expanding industrialization and increasing discharge of industrial effluent into aquatic environment. Water bodies become polluted by the improper handling and subsequent discharge of industrial waste. Polluted water containing hazardous compounds is unsuitable for human consumption and other uses. Human life and animal life are at risk if it is consumed as such. Intake of water containing toxic compounds cause allergy, physiological disorders, oxidative stress, organ damage, reflex loss and even cancer. It impairs the nervous system and affect the growth & development (Ahmed et al. 2021). One such major toxic compound in industrial effluent is azo dyes. They are the coloring agents containing chromophore groups used to impart colour to the desired materials. Around 50 percent of the azo dyes are discharged in free natural state and affect the p^H, BOD, COD of the waterbody (Dwivedi et al. 2021). This molecule is defined by the presence of one or more azo (–N = N–) groups linked to phenyl and naphthyl radicals. They are the derivative of diazene (HN = NH). Typically, these aromatic radicals contain a variety of substituent groups, such as sodium salts ($-SO_3Na$), nitro ($-NO_2$), methyl ($-CH_3$), hydroxyl (–OH), and sulfonic acid ($-SO_3H$) (Alderete et al. 2021). It is necessary for a color imparting dye to have both a chromophore and an auxochrome group. Chromophores give the dye its characteristic color by absorbing light in the visible spectrum, whereas auxochromes deepen the color of a colored molecule (Bafana et al. 2011). As per the data published by the National institute of occupational safety and health, these synthetic azo dyes and their metabolic intermediate compounds are linked with cancer. They react to DNA, RNA, protein and induce mutations. They also affect the soil fertility, plant growth, seed germination, chlorophyll pigment production. Bioaccumulation of these toxic substances in the food chain is a serious threat to mankind (Puvaneswari et al. 2006). Several disadvantages are associated with conventional methods of elimination of azo dyes. This review paper gives the basic idea about azo dyes and their toxic effects. It examines the drawbacks associated with conventional techniques used to degrade azo dyes, and also provide a comprehensive description of the degradation of dyes using microbes, nanoparticles and nanozymes.

7.1.1 *Classification of Azo Dyes*

A class of synthetic dyes known as "azo dyes" is distinguished by the looking for one or more azo groups (–N = N–) in its molecular structure. These dyes are widely used in food, medical, cosmetics, and textile sectors due to their vibrant colors and simplicity of synthesis. Azo dyes can be grouped based on several characteristics, such as their properties, uses, and chemical composition. The following are some typical azo dye classifications:

7.1.1.1 Based on the Chemical Structure

Mono-azo dyes: The most basic type of azo dyes that have just one azo group (–N = N–).

Di-azo dyes: when compared to monoazo dyes, these have a greater color range and are complicated due to their two azo groups (–N = N–).

Tri-azo dyes: These are considerably more complex than diazo dyes since they have three azo groups (–N = N–).

Poly-azo dyes: These dyes are extremely complex since their structure includes several azo groups.

7.1.1.2 Based on the Application

Textile dyes: Azo dyes are commonly used to color textiles, including silk, cotton, wool, and synthetic materials, in the textile industry.

Food dyes: A few azo dyes are permitted to be added to certain foods as food additives to give them color.

Printing inks: Azo dyes are utilized in the formulation of printing inks for a variety of printing techniques.

Cosmetic dyes: A few azo dyes are used in makeup and personal hygiene items such nail paints, lipsticks, and hair dyes.

Medical dyes: Azo dyes are used for histological staining and medical diagnostics.

7.1.1.3 Based on Properties

Acid dyes: These dyes are used to color protein fibers in acidic environments, such wool and silk.

Basic dyes: These dyes are used to color paper and synthetic fabrics under simple circumstances.

Direct dyes: Without the use of a mordant, these azo dyes can be used to directly color cellulosic fibers like cotton.

Reactive dyes: During the dyeing process, these azo dyes create covalent bonds with the fibers, giving the fabric exceptional wash fastness.

Disperse dyes: Hydrophobic fabrics like polyester and acetate can be dyed with these azo dyes.

7.1.1.4 Based on Color

A vast array of hues, such as reds, oranges, yellows, blues, greens, purples, and browns, are available in azo dyes. The dye molecule's chemical structure and the substituents bonded to the azo groups determine the particular color (Benkhaya et al. 2020; Gregory 1990) (Fig. 7.1).

7.1.2 Carcinogenicity of Azo Dyes and Their Metabolites

For many azo dyes to cause cancer, azo reduction is a crucial metabolic activity. Metabolism of azo dyes produces little aromatic amines.

Sources of azo dyes → Names of aromatic amines

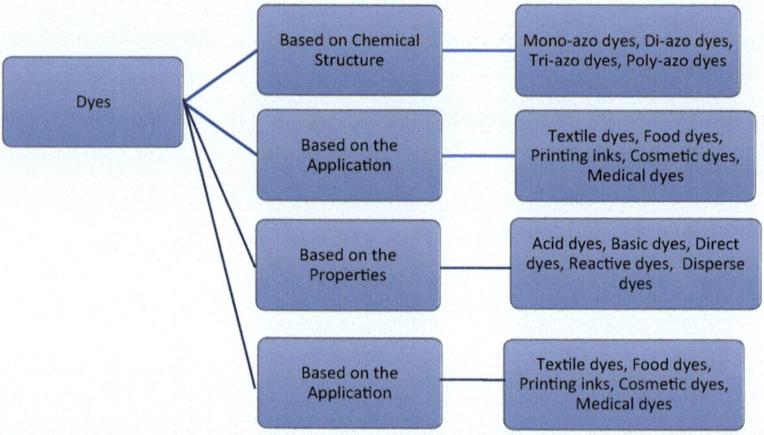

Fig. 7.1 Classification of azo dyes

- Ponceau BS, Orange II, Methyl Orange, Ponceau S → 4-Aminobenzenesulfonic acid
- Acid Red 88, Orange II, Para Red, Ponceau BS, Lithol Red, 1-Phenylazonaphthol → 1-Amino-2-Naphthol
- Orange G, Methyl Yellow, 1-Phenylazo-2-naphthol 4-Aminoazobenzene → Aniline
- Congo Red, Direct Black 38, Direct Blue 6, Direct Brown 95 → Benzidine
- Ponceau BS, Ponceau S → 2,5-Diaminobenzenesulfonic acid
- 1-[2, 4-(Dimethylphenyl)azo]-2-naphthalenol → 2,4-Dimethylaniline
- Methyl Orange, Methyl Yellow, Methyl Red → *N*,*N*-Dimethyl-p-phenelenediamine (p-PDA)
- Para Red → p-Nitroaniline
- 4-Hydroxy-4-aminoazobenzene, Sudan IV, 4-Aminoazobenzene 4-Hydroxy-4-monomethylaminoazbenzene, → p-Phenylenediamine (p-PDA)
- Sunset Yellow, Tartrazine → Sulphanilinic
- 1-[2-Methyl-4-[(2-methylphenyl)azo]phenylazo-2- naphthaleneol → Toluidine
- Ponceau 3 R → 2,5-Trimethyaniline

Some azo dyes can be carcinogenic directly without being cleaved into aromatic amines. They are 4-Aminoazobenzene, o-Aminoazotoluene, 4-Chloro-4-dimethylaminoazobenzene, 3′-Methyl-4-monomethylaminoazobenzene, Methyl Yellow, 4-Ethylmethylaminoazobenezene, Sudan I, Sudan II, Sudan III, Sudan IV, Para Red, Ponceau 3R Orasol Navy Blue 2RB, 6-(p-Dimethylaminophenyl)azobenzothiazole3′-Nitro-4-dimethylaminoazobenzene, 3′-Methyl-4-dimethylaminoazobenzene, 2′-Nitro-4-dimethylaminoazobenzene, 3-Chloro-4-dimethylaminoazobenzene. Thiodiphenyl-4, 4-diazobissalicyclic Acid.

Azo Dyes listed as Carcinogens are Solvent Yellow 1, also called p-(phenylazo)aniline; paminoazobenzene; Solvent Yellow 2, also called 4-(dimethylamino)azobenzol; Solvent Yellow 3; Pigment Orange 5; Solvent Orange 2; Pigment Red 3; Solvent Red 80, commonly called Citrus Red 2; Pigment Red 53 also called D&C Red No. 8, Pigment 53:1, barium salt; Acid Red 26, also called Ponceau 26; xylidine; Ponceau 2R; Acid Dye, also called Ponceau 3R; Direct Red 28, also called Congo Red; Direct Blue 6; Acid Red 114; Direct Blue 14; Direct Blue 53; Direct Blue 15; Direct Blue 218; Direct Brown 95; Direct Black 38; Basic Red 9; Basic Red 9 hydrochloride; Disperse Blue 1; Pigment Yellow 34; and Pigment Red 104 (Chung 2016; Chequer et al. 2011).

7.1.3 Various Methods of Treatment

Physical treatment—Coagulation–flocculation, Electrocoagulation, Adsorption, Ion exchange, Irradiation, Ultrasound, Membrane filtration.

Chemical treatment—Ozonation, Sodium hypochlorite, Redox-active metals, Photochemical treatment, Electrochemical destruction, photolysis using a H_2O_2 and O_3,

TiO_2 photolysis, radiolysis, Corona process, wet oxidation and the use of electronic beams.

Combined methods—O_3/UV, H_2O_2/UV, Ozone/H_2O_2, Ozone/UV/H_2O_2.

Biological treatment—Using microbes completely degrade the azodyes

Anaerobic reduction of azo links gives colorless aromatic amines, later they are easily broken down aerobically. Consequently, the mineralization of azo compounds is suggested to occur under a mix of anaerobic and aerobic circumstances (Bafana et al. 2011; Puvaneswari et al. 2006; Benkhaya et al. 2020; Katheresan et al. 2018; Mishra et al. 2019; Selvaraj et al. 2021; Stolz 2001).

7.1.4 *Demerits Associated with Conventional Methods*

Incomplete degradation—The dye molecule may not be fully degraded by these methods leading to persistance of coloured compound in the environment. High energy consumption—These methods need high energy input and thus they require more investments. Generation of harmful byproducts—The intermediate compounds formed during the elimination of azo dyes by these methods are toxic raising the concern about environmental effects. Limited applicability—Certain methods are suitable only for specific type of dye and affectsits application on broad spectrum. Slow degradation rates—These methods take long time to complete the process. Requirement for catlyst, pH and temperature sensitivity—Specific catalyst and maintainance of optimum pH, temperature are required for efficient elimination of azo dyes and this add to the complexity of the method.

Formation of resistant dyes, Limited scalability—Scaling up conventional methods from laboratory scale to industrial scale is difficult because of various factors like reactor design, mass transfer limitations and process control.

Dependency on harsh chemicals—Certain methods depends on the use of hazardous chemicals which put the employers at risk.

Generation of sludge—The sludge formed during the process need to be disposed properly and this complicates the overall treatment (Katheresan et al. 2018; Mishra et al. 2019; Selvaraj et al. 2021).

7.2 Degradation of Azo Dyes Using Microbes

As microorganisms reduce the azo dyes by synthesizing enzymes like laccase, peroxidase, azoreductase and hydrogenase, they can be used to degrade dyes. Bacteria, actinomycetes, fungi and yeast have enzyme systems that, in specific conditions decolorize and mineralize azo dyes (Table 7.1).

Under anaerobic conditions, the azo bond cleaves to produce aromatic amines and in aerobic conditions, non-specific enzymes cleave the azo bond to mineralize

Table 7.1 List of recognized dye degrading microorganisms

Bacteria	Fungi
Bacilus subtilis	*Ligninolytic fungal sp*
Pseudomonas sp	*Phanerochaete chrysosporium*
Escherichia coli	*Coriolus versicolor*
Rhabdobacter sp	*Trametes versicolar*
Enterococcus sp	*Fungalia trogii*
Staphylococcus sp	*Penicillium geastrivous*
Xenophilus sp	*Rhizopus oryzar*
Corneybaterium sp	*Pleurotus ostreatus*
Clostridium sp	*Rigidoporus lignosus*
Micrococcus dermacoccus sp	*Pycnoporus sanguineus*
Acinetobacter sp	*Aspergillus flavus*
Geobacillus	*Aspergillus niger*
Lactobacillus	*Fungalia trogii*
Rhizobium	
Proteus sp	
Alcaligenes sp	
Aeromonas sp	
Morganella sp	
Klebsiellla sp	

it through ring cleavage. As a result, an effective technique for degrading azo dyes is to combine anaerobic and aerobic treatments. A mixed culture can degrade more quickly than a single strain because of synergistic action (Stolz 2001; Sudha et al. 2014).

During this reduction process, 4 electrons are being donated to the azo dye in a 2 step process i.e., 2 electrons transferred in each step. This make the dye to decolorize and form colorless solutions. The resultant metabolic intermediate compounds, such as aromatic amines, are subsequently subjected to additional anaerobic or aerobic degradation (Saratale et al. 2011) (Table 7.2).

Enzymes synthesised by the microbes play a vital role in microbial decomposition of azo dyes. For example, laccases, peroxidases, phenolic oxidases, hydroxylases and azo-dye reductases enzymes released by many microorganisms mediate the microbial breakdown of azo dyes (Sudha et al. 2014).

7.2.1 Laccase

(Benzenediol: oxygen oxidoreductase; EC 1.10.3.2) is an oxidoreductase enzyme of multicopper oxidase family. Through a redox reaction driven by copper ions, it

Table 7.2 Efficiency of various microbes in degrading dyes (Ngo and Tischler 2022)

Dye	Microbe used	Enzyme involved	Decolouration (%)
Crystal violet	*Marasmius scorodonius*	Laccases	69
	Cerrena unicolor BBP6	Manganese peroxidase	81
Malachite Green	*Marasmius scorodonius*	Laccases	82
Reactive Orange 16	*Marasmius scorodonius*	Laccases	48
Congo red	*Marasmius scorodonius*	Laccases	90
	Phanerochaete chrysosporium	Lignin peroxidases	54
	Micrococcus luteus 24 M		99

oxidizes substrates like azo dyes (Ghobadi Nejad et al. 2019; Joshni and Subramaniam 2011). Its redox potential is around 780 mV andso it can oxidize azo dyes without the use of hydrogen peroxide or additional secondary metabolites by converting molecular oxygen to water directly. The incorporation of redox mediators such ABTS, HBT, TEMPO, polyoxometalates, osmium-based redox polymers, or substances released by lignolytic fungus might further increase the relatively broad substrate specificity of laccases (Almansa et al. 2004). It has a total of 4 copper ions organised into three different copper-binding sites-T1, T2, and T3. T1 site is the substrate binding site. T2/T3 site activate O_2. Upon binding of substrate to T1 site, the enzyme undergoes conformational shift and catalyses the oxidation of the bound substrate. T2/T3 site reduces molecular oxygen to reactive intermdiates which eventually interact with the substrate resulting in formation of radical species. These radicals are highly reactive that further undergo degradation (Ghobadi Nejad et al. 2019). Laccase displays good stability in a variety of environmental conditions. Further it can be immobilized on a variety of supports, improving its operational stability and making it easier to reuse in continuous operations.

7.2.2 *Peroxidase*

(Donor: hydrogen-peroxide oxidoreductase; EC 1.11.1.7) catalyses variety of substrates using H_2O_2 as a co-substrate. It oxidizes another substrate while reducing H_2O_2. The highly reactive oxygen intermediates generated during the process attack and further degrade the dye (Bilal et al. 2018; Joshni and Subramaniam 2011). To improve the efficiency of the enzyme, redox mediators can be used by conjugating it with the enzyme. Redox mediators help to oxidize the dye molecule by shifting electrons between the enzyme and the dye (Jamal et al. 2011). Peroxidases and laccases may work together to degrade dye. Certain substrates can be oxidized by laccases to create compounds that peroxidases can use as substrates, which leads to the production of reactive oxygen species (ROS) and further dye molecule destruction.

7.2.3 Azo Reductase

(EC 1.7.1.6) an oxidoreductase enzyme that catalyses the reduction reaction of azo bond present in azo dyes and produce hydrazines or primary aromatic amines based on the azo compound's structure. These metabolites are further degraded (Sandhya 2010; Joshni and Subramaniam 2011). Depending on how dependent they are on flavin cofactors, azo reductases can be classified as either flavin-free or flavin-dependent. While flavin-free azo reductases do not require flavin cofactors and instead use electron transfer pathways through redox-active amino acid residues, metal ions, or organic cofactors, flavin-dependent azo reductases use flavin cofactors (FAD, FMN) as electron carriers to reduce the azo bond in azo dyes (Chen 2006).

FDAR—Azo dyes bind to active site of enzyme. The flavin cofactor in the enzyme goes through a reduction process by the transfer of electrons from a reducing agent (like NADH or NADPH) to the flavin cofactor. Thus the oxidized form of the flavin (FAD or FMN) is reduced to flavin hydroquinone intermediate. This intermediate transfers electron to azo bond of azo dyes resulting in breakage of azo bond and forming hydrazines or primary aromatic amines later they are further degraded. Later flavin cofactor is regenerated back.

FFAR—Azo dyes bind to active site of enzyme. This enzyme begin the electron transfer process by using metal ions or amino acid residues in the active site, in addition to other cofactors or coenzymes. Just like FDAR, electron is initially donated by NADH or NADPH. Later these electrons are directed towards the azo bond present in azo dyes leading to breakage of azo bond and forming hydrazines or primary aromatic amines and later they are further degraded (Krithika et al. 2021; Russ et al. 2000).

Aerobic azoreductases—When oxygen is present during the reduction process, it works by employing it as an electron acceptor.

Anaerobic azoreductases—It doesn't need oxygen and can function without it during the reduction process (Chen 2006).

7.2.4 Factors Influencing Microbial Degradation of Azo Dyes

Microbial strain, substrate specificity, dye structure and concentration, temperature, p^H, oxygen availability, nutrient concentration and presence of inhibitors. For effective dye degradation, all these factors must be optimized (Ajaz et al. 2020).

This process makes it easier for azo bonds to break, which causes the color to fade, but it also causes harmful and cancer-causing by products to not fully mineralize. Nevertheless, it turns out that these traditional biological treatment methods don't work well for synthetic dyes that are recalcitrant (Ajmal et al. 2014).

7.3 Degradation of Azo Dyes Using Nanoparticles

Nanoparticles possessing a large surface area exhibit greater reactivity towards chemical compounds, making them efficient instruments for treating effluent quickly (Saravanan et al. 2017). Its mode of action is based on the idea of reductive degradation, which is an inexpensive, quick, and simple technique for industrial-scale dye removal (Sha et al. 2016). By combining H_2O_2 and nanoparticles, the degradation efficiency of organic dyes could be increased (Jeon et al. 2021).

Types of Nps

Metal Nps—Au Nps, Ag Nps, Fe Nps, Cu Nps, Co Nps, Pt Nps

Metal oxide Nps—TiO_2 Nps, ZnO Nps, SiO_2 Nps, Al_2O_3 Nps

Carbon based Nps—Graphene Nps, Fullerenes Nps

Polymeric Nps—Dendrimers (Hasan 2015; Ealia and Saravanakumar 2017)

There exist two distinct methods for synthesizing nanoparticles: "top-down" and "bottom-up." Top-down methods entail breaking down bigger structures into nanoparticles, whereas bottom-up methods assemble nanoparticles from smaller building pieces. Numerous techniques available for synthesizing nanoparticles, each having pros and cons of its own.

7.3.1 Methods of Synthesis of Nps

7.3.1.1 Chemical Reduction Methods

Solvent reduction—This process involves dissolving metal salts in a solvent and then adding a reducing agent to the mixture. As a result, metal ions are reduced and turn into nanoparticles.

Polyol Process—As reducing agents and solvents, polyols like glycerol and ethylene glycol are employed. After dissolving metal salts in polyol, heating starts the reduction reaction, which forms nanoparticles.

Microemulsion Technique—A microemulsion system comprising of water, oil, surfactant, and co-surfactant is employed in this case to regulate the reaction environment. Nanoparticles production is caused by the reduction of metal salts within the microemulsion droplets.

1. Physical Vapor Deposition (PVD)

   Evaporation—Techniques such as thermal evaporation and laser ablation are used to evaporate metal. The vapor forms nanoparticles when it condenses on a substrate.

   Sputtering—When powerful ions assault a solid target material, metal atoms are expelled from it. Once on a substrate, these atoms condense to create nanoparticles.
2. Mechanical Milling—This method produces nanoparticles by mechanically deforming and breaking down bulk materials in an attritor or ball mill.
3. Chemical Vapor Deposition (CVD)—On the surface of a heated substrate, precursor gasses break down to produce nanoparticles.
4. Template-Assisted Synthesis—A template structure, such as a self-assembling molecular structure or a porous substance, is where nanoparticles are produced. The template is removed once the nanoparticles have formed, leaving the nanoparticles behind.
5. Green Synthesis—In order to aid in the production of nanoparticles, biological entities like plant extracts, microbes, or enzymes are used in this process. It is said to be eco-friendly.
6. Electrochemical Synthesis—Metal ions from a solution are electrochemically reduced onto an electrode surface to generate nanoparticles.
7. Hydrothermal/Solvothermal Synthesis—Using this technique, nanoparticles are produced in a solvent or aqueous solution under high pressure and temperature (Ealia and Saravanakumar 2017; Rane et al. 2018; Dhand et al. 2015; Rajput 2015).

Each synthesis method has unique features. The required size, suitable surface characteristics, and the kind of material all play a significant role in the choice of synthesis technique to produce the necessary nanoparticles. Based upon the needs, one can choose the best method.

7.3.2 Mechanism of Action of Nps

The basic mechanism behind the degradation of azo dyes using Nps typically involves the generation of ROS (reactive oxygen species) in response to certain stimuli.

Following their interaction with the azo dye molecules, these ROS either mineralize the dye molecules into safe byproducts like carbon dioxide and water or break down their chemical structure into smaller, less hazardous chemicals. Certainly, dye degradation using nanoparticles employs various mechanisms, each designed to take advantage of unique characteristics of the nanoparticles and the makeup of the dye molecules.

Photocatalysis—Photocatalytic degradation frequently makes use of nanoparticles with semiconductivity such as zinc oxide (ZnO), tungsten trioxide (WO_3), and titanium dioxide (TiO_2). These nanoparticles produce electron–hole pairs when exposed to light, usually UV light for TiO_2 and both UV and visible light for ZnO and WO_3. These charge carriers have the ability to take part in redox processes and generate reactive oxygen species (ROS). As these ROS are so reactive, they attack the dye molecules that have been deposited on the surface of the Nps, causing the dye molecules to degrade.

Sonochemical Degradation—Cavitation bubbles can be produced by applying ultrasonic vibrations to a solution containing dye molecules and nanoparticles. Because of the localized high pressure and temperature created by the collapse of these bubbles, extremely reactive substances such hydroxyl radicals (·OH) are produced. As sonocatalysts, nanoparticles can speed up the oxidation and other chemical events that lead to dye molecule breakdown, hence improving the sonochemical process.

Chemical Degradation—Fenton-like reactions can be catalyzed by iron-based nanoparticles, for example iron oxide (Fe_2O_3) or zero-valent iron (FeO). In these processes hydroxyl radicals (·OH) are formed from hydrogen peroxide (H_2O_2) in the presence of nanoparticles. The oxidation of dye molecules results in their breakdown into less toxic, smaller chemicals.

Enzymatic Degradation—Enzymes that naturally degrade dyes, like laccases or peroxidases, can be used to functionalize nanoparticles. For efficient dye degradation, these enzymes are immobilized onto the surfaces of nanoparticles, which also increases their stability and reusability.

Adsorption-Coupled Degradation—Since Nps have high surface area and surface energy, dye molecules can physically adsorb onto their surfaces. After being adsorbed, the dye molecules are in close proximity to catalytic sites present on the surface of the Nps, which makes it easier for processes like oxidation or reduction to occur later on.

Combination Strategies—The dye degradation efficiency can be enhanced by combining many nanoparticles or coupling nanoparticles with other unique materials (such as carbon-based organic materials like graphene or carbon nanotubes) to produce synergistic results. For instance, through plasmonic effects and improved charge separation, the combination of semiconductor nanoparticles and noble metal nanoparticles (such as silver or gold) can boost photocatalytic performance (Ahsan et al. 2022; Khan et al. 2017; Jagadale et al. 2012; Chauhan et al. 2020; Truskewycz

Table 7.3 Efficiency of various Nps in degrading dyes (Joseph and Mathew 2015; Rasool et al. 2023; Thomas et al. 2016)

Nps	Dye	Degradation (%)	Incubation
AgNPs	Methyl Orange	95	4 h
ZnO-NPs	Acid Red 18	85	6 h
TiO_2-NPs	Congo Red	75	8 h under UV irradiation

et al. 2016; Alsaiari et al. 2021; Darwesh et al. 2019; Mohajershojaei et al. 2015; Nasrollahzadeh et al. 2021; Zada et al. 2020) (Table 7.3).

7.3.3 Factors Affecting the Efficiency of Nps

When using nanoparticles, a variety of factors might affect how effectively and efficiently the color degradation process proceeds. Here are a few crucial elements.

7.3.3.1 Properties of Nps

Kind of Nps—Different kinds of Nps have different characteristics that might affect the effectiveness and methods of dye degradation. Examples of these include semiconductor Nps (TiO_2, ZnO), metal nanoparticles (Fe, Ag, Au), and carbon-based organic Nps (graphene, carbon nanotubes).

Size and Morphology—The shape and size of Nps influence their surface area, surface energy, and reactivity, which in turn affects their ability to adsorb dyes and their catalytic activity.

Surface Chemistry—Depending on the degradation process used, surface functionalization or modification of Nps with particular functional groups might improve stability, stimulate catalytic activity, or enhance dye adsorption.

7.3.3.2 Dye Characteristics

Chemical Structure—The molecular weight, functional groups, aromaticity, and chromophoric moieties of the dye molecules can all affect how susceptible they are to being broken down by Nps.

Concentration and Solubility—Dye molecule availability for adsorption onto Nps surfaces and subsequent degradation events is influenced by their initial concentration and solubility in the solution.

7.3.3.3 Conditions of the Environment

p^H—The pH of the solution can affect the surface charge of dye compounds and Nps, which can consequently affect chemical reactivity, adsorption capacity, and electrostatic interactions.

Temperature—By altering molecular kinetic energy and the activation energy needed for chemical reactions, temperature affects the pace at which dye degradation reactions occur. On the other hand, aggregation or destruction of Nps may also result from overly high temperatures.

Presence of Oxygen—oxygen is necessary for a number of degradation processes, including photocatalysis and Fenton-like reactions. It act as oxidizing agent Thus, the availability of oxygen can affect how well dye degradation processes work.

7.3.3.4 Light and Energy Source

Wavelength and Intensity—In photocatalytic degradation, the wavelength and intensity of the light radiation (such as UV, visible, or solar irradiation) affects the excitation of nanoparticles and the creation of electron–hole pairs, which in turn affects the rate of dye degradation.

Ultrasound Irradiation—Cavitation effects, sonochemical reactions, and the production of reactive species can all be influenced by the frequency and strength of ultrasonic waves delivered to a solution containing nanoparticles and dyes.

7.3.3.5 Catalyst Loading and Dosage

Nanoparticle Concentration—Based on the concentration of nanoparticles in the solution, surface area accessible for catalytic activity and dye adsorption is determined. On the other hand, too high a concentration of nanoparticles could cause agglomeration or less stable dispersion.

Catalyst Dosage—The effectiveness of dye degradation and the total amount of catalyst consumed can be affected by the quantity of catalyst (nanoparticles) applied in relation to the concentration of dye molecules.

7.3.3.6 Reaction Time and Kinetics

Reaction Time—The amount of dye degradation and the rate at which the degradation process is finished are both influenced by the duration of the degradation reaction. While longer reaction periods could promote dye degradation, they also run the risk of producing undesirable byproducts or deactivating the catalyst.

Reaction Kinetics—Numerous variables, including the qualities of the dye, the environment, and the properties of the nanoparticles, affect the kinetics of the degradation process, including reaction rate constants and processes (Ahmed et al. 2020; Akbari et al. 2020; Mahmoodi et al. 2017; Devadi et al. 2014; Mas.soudinejad et al. 2019).

For efficient and sustainable treatment of dyes using Nps, these should be considered to optimize the design and operation of dye degradation process.

Concern has grown over the presence of nanoparticles in wastewater after degradation. It is necessary for nanoparticles to be recoverable and reusable (Nandhini et al. 2019).

7.4 Degradation of Azo Dyes Using Nanozymes

Nanozymes are the nanomaterials that have enzyme-like catalytic activities like those of like peroxidase, oxidase, or catalase, but without the intricate structure. Because of their potential uses in environmental cleanup, particularly the breakdown of azo dyes, nanozymes have attracted a lot of attention recently. When compared to natural enzymes, nanozymes have far better benefits, including great stability, cost-effectiveness, and a wide range of potential applications (Feng et al. 2023).

Types

Metal based Nz—Au Nz, Ag Nz, Cu Nz, Pt Nz, Pd Nz

Metal oxide Nz—Fe_3O_4 Nz, CeO_2 Nz, MnO Nz

Carbon based Nz—Graphene oxide Nz, Carbon nanotube Nz

Hybrid Nz—Metal organic framework Nz, Polymer Nz hybrids (Wang et al. 2020).

7.4.1 Methods of Synthesis of Nz

Here are some common methods used to synthesize nanozymes—Chemical Reduction, Hydrothermal/Solvothermal Synthesis, Template-Assisted Synthesis, Self-Assembly, Biomimetic Mineralization, Atomic Layer Deposition (ALD), Biological Synthesis and Electrochemical Synthesis (Wang et al. 2020).

With the use of these techniques, the characteristics of nanozymes, including their size, composition, form, and surface chemistry can be easily modified, in order to produce the appropriate catalytic activity.

7.4.2 Mechanism of Action of Nz

When dye compounds are treated with nanozymes, the azo linkages present in them are often broken down through catalytic reactions. The brilliant colors of azo dyes, a family of synthetic dyes, are caused by the existence of one or more azo (–N = N–) groups. The following describes the general mechanism by which nanozymes degrade azo dyes:

Adsorption—Initially, physical interactions like electrostatic attraction, hydrogen bonding, or π–π stacking may cause azo dyes to adhere to the surface of the nanozyme particles.

Turning on Peroxide—When it comes to catalyzing the breakdown of hydrogen peroxide (H_2O_2) into reactive oxygen species (ROS) like hydroxyl radicals (·OH) and superoxide radicals ($\cdot O_2^-$), many nanozymes, especially those based on metals, have peroxidase-like activity. The catalytic cycle of the nanozyme is frequently triggered by the presence of H2O2.

Catalytic Degradation—The dye molecules' azo bonds (–N = N–) are attacked by the produced ROS, which have the ability to behave as strong oxidants. The azo group's electron-rich double bond is attacked by the ROS species, causing it to split into smaller, lighter-colored pieces. The azo dye is eventually degraded by this process, which breaks the dye molecules down into simpler, frequently colorless or less strongly colored chemicals.

Oxidation of Byproducts—The ROS produced by the nanozyme may oxidize any intermediates or byproducts created during the degradation process, causing them to fully mineralize into harmless substances like carbon dioxide, water, and nitrogen gas.

Desorption and Release—After the dye degradation products are broken down, they may desorb from the nanozyme particles' surface, enabling the nanozymes to catalyze more degradation events.

Regeneration—Based on its stability and activity, the nanozyme may regenerate, allowing it to take part in several cycles of dye degradation without experiencing a discernible loss in activity (Zandieh and Liu 2024; Huang et al. 2019; Liang and Yan 2019).

7.4.3 Unique Features of Nz

Because of their special qualities and catalytic activity that resembles that of an enzyme, nanozymes are superior to regular nanoparticles in a number of ways when it comes to degrading dyes, especially azo dyes. The following features of the nanozymes make them to be more efficient at eliminating dyes:

Table 7.4 Efficiency of various Nz in degrading dyes (Feng et al. 2023; Jangi 2023; Geng et al. 2021)

Nz	Dye	Degradation (%)	Incubation
Au Nz	Acid Orange 7	90	2 h
Fe_3O_4 Nz	Reactive Red 120	80	4 h
CuO Nz	Direct Blue 15	75	6 h
Cu Nz	Methyl orange	93	10 min
Mn-SAzyme	Methylene blue	90	5 h

Biocompatibility and Environmental Friendliness—Certain nanozymes exhibit biocompatibility and environmental friendliness, rendering them appropriate for utilization in biomedical and environmental settings. When compared to some typical nanoparticles, biocompatible or green chemistry-synthesised nanozymes have less toxicity and less of an adverse effect on the environment.

Synergistic Effects—The catalytic activity and efficiency of hybrid nanozymes, which are made up of several nanomaterials or nanoparticles, can be further increased by these synergistic effects.

When combining several nanoparticles forms or adding organic elements, one can produce nanozymes that function better than when using individual nanoparticles (Huang et al. 2019; Liang and Yan 2019).

In general, nanozymes are more advantageous than traditional nanoparticles in breaking down dyes, which makes them attractive options for a range of environmental remediation tasks, such as pollution management and wastewater treatment. Their exceptional efficiency, regenerability, flexibility, and catalytic activity that resembles that of an enzyme is what gives them their potency in dye degradation processes (Table 7.4).

7.4.4 Factors Affecting the Efficiency of Nz

The following factors could have an impact on how well nanozymes break down dyes:

Type of nanozyme—The catalytic potential of various nanozymes varies. The choice of nanozyme depends on the specific dye molecule to be broken down and the conditions of the reaction.

Catalytic activity—An important factor influencing how well dyes degrade is the nanozyme's intrinsic catalytic activity. Several parameters can influence the catalytic activity of a nanozyme, including its surface area, shape, content, and crystallinity.

p^H—The ionization state of the dye molecules and the surface charge of the nanozyme are both influenced by the pH of the reaction media, which in turn influences how the two interact. The easiest way to determine the optimal pH range for nanozyme activity is to carry out experiments. Temperature—The temperature affects both the kinetics of the dye degradation reaction and the stability of the nanozyme. Higher temperatures generally accelerate the reaction's kinetics but can potentially destabilize some nanozymes.

Nanozyme concentration—The amount of nanozyme in the reaction solution affects the rate at which the dye breaks down. There might be a threshold for nanozyme concentration, though, beyond which any increases have minimal effect on the rate of disintegration.

Dye concentration—The quantity of dye molecules in the combination affects how quickly the dye degrades. Greater dye concentrations may require more nanozyme or longer reaction times for efficient breakdown.

Co-reactant involvement—Certain dye degradation processes may require the presence of co-reactants or oxidants in order to increase the catalytic activity of nanozymes. Examples include molecular oxygen (O_2), hydrogen peroxide (H_2O_2), and ozone (O_3).

Surface functionalization—Nanozymes can be made more effective by modifying their catalytic activity and selectivity towards specific dye molecules by adding specific ligands or functional groups to their surface.

Light irradiation—Light irradiation can enhance the catalytic activity of some nanozymes, especially those based on semiconductor materials like titanium dioxide (TiO_2), through processes like photocatalysis.

Existence of inhibitors—Some substances in the reaction media may have the ability to block processes, preventing nanozymes from catalyzing them. Understanding the effects of inhibitors and being dedicated to minimizing them are necessary for optimizing dye degradation efficiency.

By carefully assessing and optimizing these factors, we may increase the effectiveness of nanozymes in dye degradation situations, which will further the development of more durable and effective wastewater treatment techniques (Liu et al. 2022; Yang et al. 2023).

7.5 Conclusion

In a nutshell, nanozymes with excellent enzyme-like activity show good degradation efficiency and may have applications in environmental protection. They are adaptable and may be made to target particular dye molecules or classes of dyes, making them useful for a variety of industrial effluent streams that contain various dye kinds. They are often easily retrieved and reused multiple times without considerably diminishing their catalytic activity. All things considered, nanozymes present a fresh and intriguing approach to the degradation of dyes in wastewater, potentially resolving the drawbacks of traditional chemical or enzyme-based methods. Their excellent stability, cost-effectiveness, strong catalytic activity, and tunability make them desirable options for industrial applications that try to reduce dye pollution. By using these benefits, nanozymes present encouraging prospects for the creation of cutting-edge and environmentally friendly dye wastewater treatment systems, tackling the problems related to dye pollution in a range of industrial sectors.

References

Ahmed A, Usman M, Yu B, Ding X, Peng Q, Shen Y, Cong H (2020) Efficient photocatalytic degradation of toxic Alizarin yellow R dye from industrial wastewater using biosynthesized Fe nanoparticle and study of factors affecting the degradation rate. J Photochem Photobiol B 202:111682

Ahmed J, Thakur A, Goyal A (2021) Industrial wastewater and its toxic effects

Ahsan H, Shahid M, Imran M, Mahmood F, Siddique MH, Ali HM, Niazi MB, Hussain S, Shahbaz M, Ayyub M, Shahzad T (2022) Photocatalysis and adsorption kinetics of azo dyes by nanoparticles of nickel oxide and copper oxide and their nanocomposite in an aqueous medium. PeerJ 10:e14358

Ajaz M, Shakeel S, Rehman A (2020) Microbial use for azo dye degradation—a strategy for dye bioremediation. Int Microbiol 23:149–159

Ajmal A, Majeed I, Malik RN, Idriss H, Nadeem MA (2014) Principles and mechanisms of photocatalytic dye degradation on TiO_2 based photocatalysts: a comparative overview. RSC Adv 4(70):37003–37026

Akbari A, Sabouri Z, Hosseini HA, Hashemzadeh A, Khatami M, Darroudi M (2020) Effect of nickel oxide nanoparticles as a photocatalyst in dyes degradation and evaluation of effective parameters in their removal from aqueous environments. Inorg Chem Commun 115:107867

Alderete BL, da Silva J, Godoi R, da Silva FR, Taffarel SR, da Silva LP, Garcia ALH, Júnior HM, de Amorim HLN, Picada JN (2021) Evaluation of toxicity and mutagenicity of a synthetic effluent containing azo dye after advanced oxidation process treatment. Chemosphere 263:128291

Almansa E, Kandelbauer A, Pereira L, Cavaco-Paulo A, Guebitz GM (2004) Influence of structure on dye degradation with laccase mediator systems. Biocatal Biotransform 22(5–6):315–324

Alsaiari NS, Amari A, Katubi KM, Alzahrani FM, Harharah HN, Rebah FB, Tahoon MA (2021) The biocatalytic degradation of organic dyes using laccase immobilized magnetic nanoparticles. Appl Sci 11(17):8216

Bafana A, Devi SS, Chakrabarti T (2011) Azo dyes: past, present and the future. Environ Rev 19(NA):350–371

Benkhaya S, M'rabet S, El Harfi A (2020) Classifications, properties, recent synthesis and applications of azo dyes. Heliyon 6(1)

Bilal M, Rasheed T, Iqbal HM, Yan Y (2018) Peroxidases-assisted removal of environmentally-related hazardous pollutants with reference to the reaction mechanisms of industrial dyes. Sci Total Environ 644:1–13

Chauhan M, Kaur N, Bansal P, Kumar R, Srinivasan S, Chaudhary GR (2020) Proficient photocatalytic and sonocatalytic degradation of organic pollutants using CuO nanoparticles. J Nanomater 2020:1–15

Chen H (2006) Recent advances in azo dye degrading enzyme research. Curr Protein Pept Sci 7(2):101–111

Chequer FMD, Dorta DJ, de Oliveira DP (2011) Azo dyes and their metabolites: does the discharge of the azo dye into water bodies represent human and ecological risks. Adv Treating Textile Effluent 48:28–48

Chung KT (2016) Azo dyes and human health: a review. J Environ Sci Health C 34(4):233–261

Darwesh OM, Matter IA, Eida MF (2019) Development of peroxidase enzyme immobilized magnetic nanoparticles for bioremediation of textile wastewater dye. J Environ Chem Eng 7(1):102805

Devadi MAH, Krishna M, Murthy HN, Sathyanarayana BS (2014) Statistical optimization for photocatalytic degradation of methylene blue by Ag-TiO_2 nanoparticles. Procedia Mater Sci 5:612–621

Dhand C, Dwivedi N, Loh XJ, Ying ANJ, Verma NK, Beuerman RW, Lakshminarayanan R, Ramakrishna S (2015) Methods and strategies for the synthesis of diverse nanoparticles and their applications: a comprehensive overview. RSC Adv 5(127):105003–105037

Dwivedi N, Dwivedi S, Adetunji CO (2021) Efficacy of microorganisms in the removal of toxic materials from industrial effluents. Microbial Rejuven Pollut Environ 3:325–358

Ealia SAM, Saravanakumar MP (2017) A review on the classification, characterisation, synthesis of nanoparticles and their application. In: IOP conference series: materials science and engineering, vol 263, no 3. IOP Publishing, p 032019

Feng Q, Wang G, Xue L, Wang Y, Liu M, Liu J, Zhang S, Hu W (2023) Single-atom nanozyme based on Mn-center with enhanced peroxidase-like activity for organic dye degradation. ACS Appl Nano Mater 6(6):4844–4853

Geng X, Xie X, Liang Y, Li Z, Yang K, Tao J, Zhang H, Wang Z (2021) Facile fabrication of a novel copper nanozyme for efficient dye degradation. ACS Omega 6(9):6284–6291

Ghobadi Nejad Z, Borghei SM, Yaghmaei S (2019) Biodegradation of synthetic dye using partially purified and characterized laccase and its proposed mechanism. Int J Environ Sci Technol 16:7805–7816

Gregory P (1990) Classification of dyes by chemical structure. Chem Appl Dyes 17–47

Hasan S (2015) A review on nanoparticles: their synthesis and types. Res J Recent Sci 2277:2502

Huang Y, Ren J, Qu X (2019) Nanozymes: classification, catalytic mechanisms, activity regulation, and applications. Chem Rev 119(6):4357–4412

Jagadale T, Kulkarni M, Pravarthana D, Ramadan W, Thakur P (2012) Photocatalytic degradation of azo dyes using Au: TiO_2, γ-Fe_2O_3: TiO_2 functional nanosystems. J Nanosci Nanotechnol 12(2):928–936

Jamal F, Qidwai T, Pandey PK, Singh R, Singh S (2011) Azo and anthraquinone dye decolorization in relation to its molecular structure using soluble *Trichosanthes dioica* peroxidase supplemented with redox mediator. Catal Commun 12(13):1218–1223

Jangi SRH (2023) Introducing a high throughput nanozymatic method for eco-friendly nanozyme-mediated degradation of methylene blue in real water media. Sustain Chem Eng 90–99

Jeon S, Ko JW, Ko WB (2021) Synthesis of Gd_2O_3 nanoparticles and their photocatalytic activity for degradation of azo dyes. Catalysts 11(6):742

Joseph S, Mathew B (2015) Facile synthesis of silver nanoparticles and their application in dye degradation. Mater Sci Eng, B 195:90–97

Joshni TC, Subramaniam K (2011) Enzymatic degradation of azo dyes-a review. Int J Environ Sci 1(6):1250–1260

Katheresan V, Kansedo J, Lau SY (2018) Efficiency of various recent wastewater dye removal methods: a review. J Environ Chem Eng 6(4):4676–4697

Khan ZUH, Khan A, Chen Y, Ullah Khan A, Shah NS, Muhammad N, Murtaza B, Tahir K, Khan FU, Wan P (2017) Photo catalytic applications of gold nanoparticles synthesized by green route and electrochemical degradation of phenolic Azo dyes using AuNPs/GC as modified paste electrode. J Alloys Compd 725:869–876

Krithika A, Gayathri KV, Kumar DT, Doss CGP (2021) Mixed azo dyes degradation by an intracellular azoreductase enzyme from alkaliphilic *Bacillus subtilis*: a molecular docking study. Arch Microbiol 203:3033–3044

Liang M, Yan X (2019) Nanozymes: from new concepts, mechanisms, and standards to applications. Acc Chem Res 52(8):2190–2200

Liu Q, Zhu X, Zhong L, Zhang S, Luo X, Liu Q, Tang L, Lu Y (2022) Recent advances in the applications of nanozymes for the efficient detection/removal of organic pollutants: a review. Environ Sci Nano 9(4):1212–1235

Mahmoodi NM, Keshavarzi S, Ghezelbash M (2017) Synthesis of nanoparticle and modelling of its photocatalytic dye degradation ability from colored wastewater. J Environ Chem Eng 5(4):3684–3689

Massoudinejad M, Sadani M, Gholami Z, Rahmati Z, Javaheri M, Keramati H, Sarafraz M, Avazpour M, Shiri S (2019) Optimization and modeling of photocatalytic degradation of Direct Blue 71 from contaminated water by TiO_2 nanoparticles: response surface methodology approach (RSM). Iran J Catal 9(2):121–132

Mishra S, Chowdhary P, Bharagava RN (2019) Conventional methods for the removal of industrial pollutants, their merits and demerits. Emerg Eco-Friendly Approaches Waste Manage 1–31

Mohajershojaei K, Mahmoodi NM, Khosravi A (2015) Immobilization of laccase enzyme onto titania nanoparticle and decolorization of dyes from single and binary systems. Biotechnol Bioprocess Eng 20:109–116

Nandhini NT, Rajeshkumar S, Mythili S (2019) The possible mechanism of eco-friendly synthesized nanoparticles on hazardous dyes degradation. Biocatal Agric Biotechnol 19:101138

Nasrollahzadeh M, Sajjadi M, Iravani S, Varma RS (2021) Carbon-based sustainable nanomaterials for water treatment: state-of-art and future perspectives. Chemosphere 263:128005

Ngo ACR, Tischler D (2022) Microbial degradation of azo dyes: approaches and prospects for a hazard-free conversion by microorganisms. Int J Environ Res Public Health 19(8):4740. https://doi.org/10.3390/ijerph19084740.PMID:35457607;PMCID:PMC9026373

Puvaneswari N, Muthukrishnan J, Gunasekaran P (2006) Toxicity assessment and microbial degradation of azo dyes

Rajput N (2015) Methods of preparation of nanoparticles-a review. Int J Adv Eng Technol 7(6):1806

Rane AV, Kanny K, Abitha VK, Thomas S (2018) Methods for synthesis of nanoparticles and fabrication of nanocomposites. In: Synthesis of inorganic nanomaterials. Woodhead publishing, pp 121–139

Rasool A, Kiran S, Gulzar T, Abrar S, Ghaffar A, Shahid M, Nosheen S, Naz S (2023) Biogenic synthesis and characterization of ZnO nanoparticles for degradation of synthetic dyes: a sustainable environmental cleaner approach. J Clean Prod 398:136616

Russ R, Rau J, Stolz A (2000) The function of cytoplasmic flavin reductases in the reduction of azo dyes by bacteria. Appl Environ Microbiol 66(4):1429–1434

Sandhya S (2010) Biodegradation of azo dyes under anaerobic condition: role of azoreductase. Biodegrad Azo Dyes 39–57

Saratale RG, Saratale GD, Chang JS, Govindwar SP (2011) Bacterial decolorization and degradation of azo dyes: a review. J Taiwan Inst Chem Eng 42(1):138–157

Saravanan C, Rajesh R, Kaviarasan T, Muthukumar K, Kavitake D, Shetty PH (2017) Synthesis of silver nanoparticles using bacterial exopolysaccharide and its application for degradation of azo-dyes. Biotechnol Rep 15:33–40

Selvaraj V, Karthika TS, Mansiya C, Alagar M (2021) An over review on recently developed techniques, mechanisms and intermediate involved in the advanced azo dye degradation for industrial applications. J Mol Struct 1224:129195

Sha Y, Mathew I, Cui Q, Clay M, Gao F, Zhang XJ, Gu Z (2016) Rapid degradation of azo dye methyl orange using hollow cobalt nanoparticles. Chemosphere 144:1530–1535

Stolz A (2001) Basic and applied aspects in the microbial degradation of azo dyes. Appl Microbiol Biotechnol 56:69–80

Sudha M, Saranya A, Selvakumar G, Sivakumar N (2014) Microbial degradation of azo dyes: a review. Int J Curr Microbiol App Sci 3(2):670–690

Thomas M, Naikoo GA, Sheikh MUD, Bano M, Khan F (2016) Effective photocatalytic degradation of Congo red dye using alginate/carboxymethyl cellulose/TiO_2 nanocomposite hydrogel under direct sunlight irradiation. J Photochem Photobiol, A 327:33–43

Truskewycz A, Shukla R, Ball AS (2016) Iron nanoparticles synthesized using green tea extracts for the fenton-like degradation of concentrated dye mixtures at elevated temperatures. J Environ Chem Eng 4(4):4409–4417

Wang Y, Liang M, Wei T (2020) Types of nanozymes: materials and activities. Nanozymol: Connecting Biol Nanotechnol 41–77

Yang T, Liu X, Zeng Z, Wang X, Zhang P, Feng B, Tian K, Qing T (2023) Efficient and recyclable degradation of organic dye pollutants by CeO_2@ ZIF-8 nanozyme-based non-photocatalytic system. Environ Pollut 316:120643

Zada A, Muhammad P, Ahmad W, Hussain Z, Ali S, Khan M, Khan Q, Maqbool M (2020) Surface plasmonic-assisted photocatalysis and optoelectronic devices with noble metal nanocrystals: design, synthesis, and applications. Adv Func Mater 30(7):1906744

Zandieh M, Liu J (2024) Nanozymes: definition, activity, and mechanisms. Adv Mater 36(10):2211041

Chapter 8
Comparative Analysis and Performance Evaluation of Sonolysis in Demineralizing Hazardous Polyquaterium Compound Under Optimum Conditions Against Sodium Laureth Sulfate

Poojari Prasanna and Palanisamy Suresh Babu

Abstract Aim: To assess the sonolytic breakdown of polyquaterium in artificial aqueous media under controlled conditions. Materials and methods: Distilled water and polyquaterium were combined to create a chemical solution that ranges in pH from 4.4 to 7. Results were recorded for each of five different time intervals, and pH values between 4.5 and 8.5 were listed in a table for each interval of ten minutes. Group 1 and group 2 of each five samples were taken for each pH solution, and a one-way Anova test was performed using IBM SPSS Version 26 to compare the samples using a numerical type of analysis with the confidence interval 95%, G-power 80%, and enrollment ratio all set to one ($N = 10$). Results: Freshly made polyquaterium stock solution was examined for ideal temperature and pH range. The results were examined and recorded. The initial concentration of the sample solution was set at 100 mg/L, and a pH range of 4.5 to 8.5 was taken into account. The solution's absorbance at 540 nm wavelength was measured using a UV–Vis spectrophotometer. For statistical observation, the mean–variance was compared between Group 1 and Group 2 using a one way ANOVA test that reveals the significance value $p = 0.001$ ($p < 0.05$). Conclusion: The effectiveness of removal and the frequency of response during the sonolysis process may both be impacted by temperature. A pH and temperature were indeed two of the factors that affect how effectively polyquaterium can be taken down by sonolysis. For the breakdown of polyquaterium sonolysis operates efficiently.

P. Prasanna · P. S. Babu (✉)
Department of Biotechnology, Saveetha School of Engineering, Saveetha Institute of Medical and Technical Sciences, Chennai, Tamil Nadu 602105, India
e-mail: sureshbabup.sse@saveetha.com

P. Prasanna
e-mail: prassannap19@saveetha.com

© The Author(s), under exclusive license to Springer Nature Switzerland AG 2025
A. Jeyaseelan and K. Murugasen (eds.), *Sustainable Environmental Remediation: Avenues in Nano and Biotechnology*, Environmental Science and Engineering,
https://doi.org/10.1007/978-3-031-78483-5_8

Keywords Personal care products · Wastewater treatment · Advanced oxidation process · Sonolysis · Polyquaterium

8.1 Introduction

One of the most difficult problems facing the environment is wastewater pollution. Due to its excessive use, contaminants in wastewater, such as personal care products, have been rising recently. Personal care items have the potential to be developing pollutants that have negative effects on both humans and aquatic as well as terrestrial species (Liu et al. 2006). Personal care items and medications were challenging to treat and minimize because of their extreme toxicity. One of the most promising methods for destroying the dangerous organic debris was sonolysis, one of numerous sophisticated oxidation processes (Prasad et al. 2019). A class of synthetic polymers known as polyquaternium was frequently used in personal care and home goods like shampoos, conditioners, and cleaning supplies. These polymers were often created from a mixture of monomers, such as vinylpyrrolidone, acrylamide, and diallyldimethylammonium chloride (Maldonado et al. 2023).

In the science direct database, it had about 280 research articles, 70 review articles were found to carry out the experiments. A literature review on sonolysis of polyquaterium would involve a comprehensive analysis of published research articles, studies and reviews on this topic its applications and its methods. In google scholar, nearly 130 review articles were listed and about 730 research articles were presented for the study of the sonolysis and their features.

The positively charged quaternary ammonium groups in the structure of polyquaternium polymers make them highly cationic (positively charged), enabling them to bind and attract to negatively charged surfaces like hair and skin. Due to its capacity to moisturize and condition the skin, polyquaternium polymers were useful in hair and skin care products. They can also aid to improve the texture, conditioning, and manageability of hair (Kamali 2010). Many personal care items improve our quality of life and, in some situations, offer cures that can save our lives. Yet, they have an impact on the ecosystem. Several of these chemicals have been detected in our streams and finished drinking water at subtherapeutic levels, raising concerns about the possible environmental and public health effects of very low, chronic exposure. Personal care goods may contain fragrances, sunscreen UV filters, deodorants, and disinfectants. Because the chemicals in personal care products were complex in nature and have numerous physicochemical qualities, it was challenging to remove them from the environment (Prasad et al. 2019).

In addition to the difficulty of their degradation, the presence of micropollutants in water sources had become a crucial concern for the safety of drinking water due to their detrimental effects on health, which include neurotoxicity, developmental and reproductive toxicity, and metabolic interference. Due to their toxicity and durability in cells, micropollutants, in particular pharmaceutical and personal care products (PPCPs), have recently attracted a lot of attention. Since the majority of the present WWTPs were not built to remove micropollutants, many micropollutants can pass through wastewater treatment systems. Moreover, the majority of WWTPs lack performance monitoring and well-established micropollutant safeguards (Lun et al. 2022). Chemical synthesis, wastewater treatment, and sample preparation for analysis were just a few of the industrial and laboratory uses for sonolysis. It has also been researched as a viable technique for removing toxins from polluted water and soil (Fayos-Solà and Cooper 2018). However, the frequency and strength of the sound waves, the characteristics of the solution, and the kind of molecule being destroyed all affect (Wang and Meng 2023). The present study was focused to evaluate the performance of sonolysis in demineralizing the hazardous polyquaterium under standard conditions.

8.2 Materials and Methods

The study was carried out in a bioengineering laboratory at Saveetha School of Engineering, Chennai, India. Two groups were taken totally; each group carried 5 samples and the total sample size was 10 and this had been calculated based on (Kane, Phar, and BCPS n.d.). The value of G power was determined to be 80% and the coincidence interval of 95% and the sample preparation was given below.

Distilled water was combined with polyquaterium. In the process, ultrasonic vibrations were applied to create chemical or physical changes in a solution. Polyquaterium can be broken down into smaller molecules. In sonochemistry, tiny bubbles in a solution were produced by high-frequency sound waves, which caused them to form and burst. These bubbles can trigger chemical reactions in the solution when they rupture because high temperatures and pressures were created. Cavitation was the name of this procedure.

8.2.1 Quantitative Analysis

We used a UV–Vis spectrophotometer to measure the solution's absorbance at 540 nm. A sample's ability to absorb or pass through a particular wavelength of UV or visible light was measured quantitatively using UV–Vis spectroscopy in comparison to a standard or control sample.

Table 8.1 Comparison of effect of varying pH and time in demineralizing polyquaternium compound

S. no	Time intervals (min)	pH 4.5 solution (%)	pH 8.5 solution (%)
1	25	0.21	0.19
2	35	0.35	0.36
3	45	0.47	0.41
4	55	0.53	0.57
5	65	0.63	0.65

8.2.2 Statistical Analysis

The IBM SPSS Software Version 26 application was used to estimate results such as mean, standard deviation and standard deviation error. Analyzing statistical significance $p = 0.001 (p < 0.05)$ involved using a ONE WAY ANOVA test, sonicated pH 4.5 and pH 8.5 solutions with time and temperature were the independent parameters.

8.3 Results

8.3.1 Effects of pH and Temperature

The specific temperature required to break down cetyl alcohol in an ultrasonic bath may depend on a number of factors. Chemical elimination typically increases with increase in temperature. The efficiency of polyquaterium elimination increases as a result. Certain temperatures between 30 and 70 °C were examined to see how well the polyquaterium was eliminated. Table 8.1. Comparison of effect of varying pH and temperature in demineralizing polyquaternium compound. Table 8.2. Comparison of effect of varying pH and time in demineralizing polyquaternium compound. Tables 8.3 and 8.4. Comparison of the one way ANOVA test for varying temperature and pH shows the significant value $p < 0.05$ using SPSS analysis. Figures 8.1, 8.2 shows a comparison of the typical breakdown rate over time and pH ranges. When pH was raised over time, the percentage of chemical compound removal gradually rised.

8.3.2 Effects of pH and Time

Polyquaterium and other types of organic components were distributed in the wastewater in various ways, depending on the effect of the pH value, which can be interpreted as having an impact on this. It had been observed that the pH levels of the

Table 8.2 Comparison of effect of varying pH and temperature in demineralizing polyquaternium compound

S. no	Temperature (°C)	pH 4.5 solution (%)	pH 8.5 solution (%)
1	30	0.19	0.23
2	40	0.27	0.36
3	50	0.34	0.49
4	60	0.46	0.55
5	70	0.59	0.68

Table 8.3 Comparison of the one way ANOVA test for varying temperature and pH shows the significant value $p < 0.05$ using SPSS analysis

ANOVA						
		Sum of squares	df	Mean square	F	Sig
pH4.5	Between groups	0.307	4	0.077	884.731	0.000
	Within groups	0.001	10	0.000		
	Total	0.308	14			
pH8.5	Between groups	0.361	4	0.090	902.100	0.000
	Within groups	0.001	10	0.000		
	Total	0.362	14			

sonicated polyquaterium were 4.5 and 8.5, respectively. It was common knowledge that a substance's pH can affect its physicochemical characteristics. polyquaterium can be removed more effectively when the pH rises (Elsayed 2015). Tables 8.5 and 8.6. Comparison of the one way ANOVA test for varying time and pH shows the significant value of $p < 0.05$ using SPSS analysis.

8.4 Discussion

In this present study, sonolysis was employed to treat wastewater from the personal care products company. Using sonolysis or ultrasonic waves for wastewater treatment was an uncomplicated technique that explored appealing applications in the search for more effective and beneficial to the environment methods (Elsayed 2015). Figure 8.1 shows the outcome of comparing the typical breakdown of polyquaterium removal percentage at various temperature and pH levels. It was evident that polyquaterium does not initially have an effective removal percentage, but that percentage steadily increases as temperature and pH rise. Figure 8.2 shows a comparison of the typical breakdown rate over time and pH ranges. When pH was raised over time, the percentage of chemical compound removal gradually rised (Edquist et al. 2015).

Table 8.4 Comparison of the one way ANOVA test for multiple comparisons varying temperature and pH shows the significant value of $p < 0.05$ using SPSS analysis

Multiple comparisons

Tukey HSD

Dependent variable	(I) Temperature (°C)	(J) Temperature (°C)	Mean Difference (I-J)	Std. Error	Sig	95% confidence interval	
						Lower bound	Upper bound
pH 4.5	30	40	−0.08667*	0.00760	0.000	−0.1117	−0.0617
		50	−0.15667*	0.00760	0.000	−0.1817	−0.1317
		60	−0.27667*	0.00760	0.000	−0.3017	−0.2517
		70	−0.40667*	0.00760	0.000	−0.4317	−0.3817
	40	30	0.08667*	0.00760	0.000	0.0617	0.1117
		50	−0.07000*	0.00760	0.000	−0.0950	−0.0450
		60	−0.19000*	0.00760	0.000	−0.2150	−0.1650
		70	−0.32000*	0.00760	0.000	−0.3450	−0.2950
	50	30	0.15667*	0.00760	0.000	0.1317	0.1817
		40	0.07000*	0.00760	0.000	0.0450	0.0950
		60	−0.12000*	0.00760	0.000	−0.1450	−0.0950
		70	−0.25000*	0.00760	0.000	−0.2750	−0.2250
	60	30	0.27667*	0.00760	0.000	0.2517	0.3017
		40	0.19000*	0.00760	0.000	0.1650	0.2150
		50	0.12000*	0.00760	0.000	0.0950	0.1450
		70	−0.13000*	0.00760	0.000	−0.1550	−0.1050
	70	30	0.40667*	0.00760	0.000	0.3817	0.4317
		40	0.32000*	0.00760	0.000	0.2950	0.3450
		50	0.25000*	0.00760	0.000	0.2250	0.2750
		60	0.13000*	0.00760	0.000	0.1050	0.1550
pH 8.5	30	40	−0.13000*	0.00816	0.000	−0.1569	−0.1031
		50	−0.26000*	0.00816	0.000	−0.2869	−0.2331
		60	−0.32000*	0.00816	0.000	−0.3469	−0.2931
		70	−0.45000*	0.00816	0.000	−0.4769	−0.4231
	40	30	0.13000*	0.00816	0.000	0.1031	0.1569
		50	−0.13000*	0.00816	0.000	−0.1569	−0.1031
		60	−0.19000*	0.00816	0.000	−0.2169	−0.1631
		70	−0.32000*	0.00816	0.000	−0.3469	−0.2931
	50	30	0.26000*	0.00816	0.000	0.2331	0.2869
		40	0.13000*	0.00816	0.000	0.1031	0.1569
		60	−0.06000*	0.00816	0.000	−0.0869	−0.0331
		70	−0.19000*	0.00816	0.000	−0.2169	−0.1631

(continued)

Table 8.4 (continued)

Multiple comparisons							
Tukey HSD							
Dependent variable	(I) Temperature (°C)	(J) Temperature (°C)	Mean Difference (I-J)	Std. Error	Sig	95% confidence interval	
						Lower bound	Upper bound
	60	30	0.32000*	0.00816	0.000	0.2931	0.3469
		40	0.19000*	0.00816	0.000	0.1631	0.2169
		50	0.06000*	0.00816	0.000	0.0331	0.0869
		70	−0.13000*	0.00816	0.000	−0.1569	−0.1031
	70	30	0.45000*	0.00816	0.000	0.4231	0.4769
		40	0.32000*	0.00816	0.000	0.2931	0.3469
		50	0.19000*	0.00816	0.000	0.1631	0.2169
		60	0.13000*	0.00816	0.001	0.1031	0.1569

*The mean difference is significant at the 0.05 level

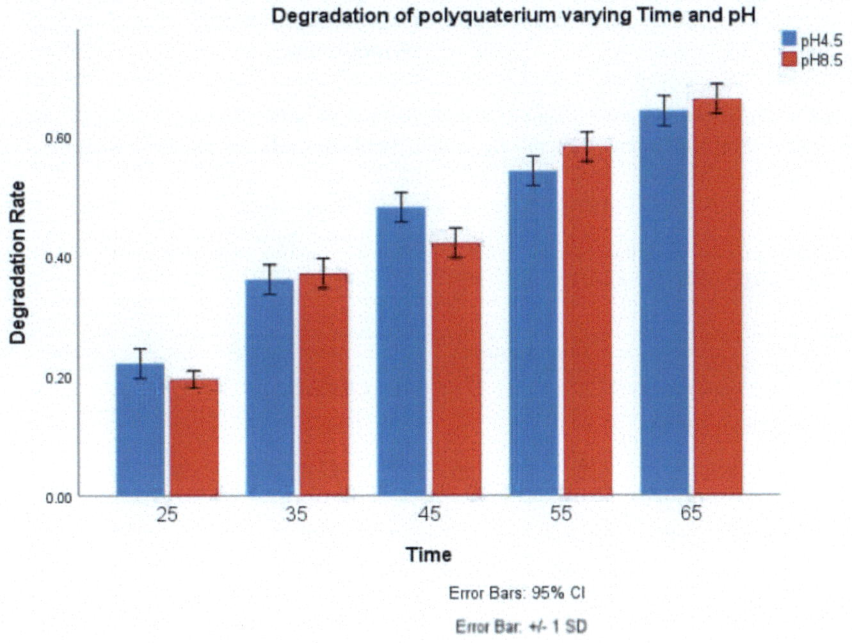

Fig. 8.1 Comparison of mean amount of degradation rate of polyquaterium solution at regular time intervals. X axis represents the time and Y axis represents the degradation rate Cl: 95%; SD ± 1

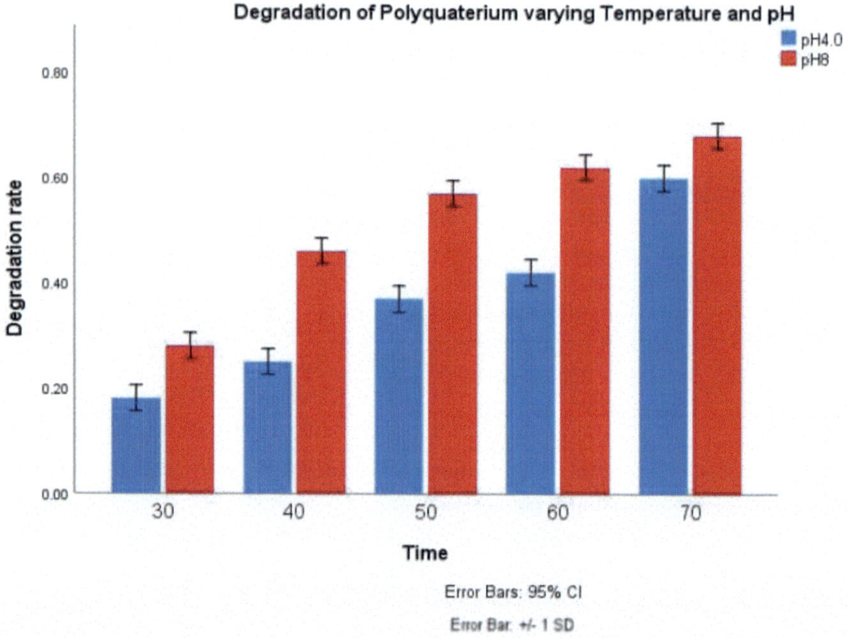

Fig. 8.2 Comparison of mean amount of degradation rate of polyquaterium solution at different temperatures. X axis represents the temperature and Y axis represents the degradation rate Cl: 95%; SD ± 1

Table 8.5 Comparison of the one way ANOVA test for varying time and pH shows the significant value of $p < 0.05$ using SPSS analysis

ANOVA						
		Sum of squares	df	Mean square	F	Sig
pH4.5	Between groups	0.318	4	0.080	795.600	0.000
	Within groups	0.001	10	0.000		
	Total	0.319	14			
pH8.5	Between groups	0.402	4	0.101	1159.923	0.000
	Within groups	0.001	10	0.000		
	Total	0.403	14			

A comparative analysis had been performed to determine when temperature disrupts solutions. In Fig. 8.3, cocamidopropyl betaine and sodium laureth sulfate were contrasted with a pH 8. The rate at which substances degrade often increases

Table 8.6 Comparison of the one way ANOVA test for multiple comparisons varying time and pH shows the significant value of $p < 0.05$ using SPSS analysis

Multiple Comparisons

Tukey HSD

Dependent variable	(I) Time	(J) Time	Mean difference (I-J)	Std. error	Sig	95% confidence interval	
						Lower bound	Upper bound
pH 4.5	25	35	−0.14000*	0.00816	0.000	−0.1669	−0.1131
		45	−0.26000*	0.00816	0.000	−0.2869	−0.2331
		55	−0.32000*	0.00816	0.000	−0.3469	−0.2931
		65	−0.42000*	0.00816	0.000	−0.4469	−0.3931
	35	25	0.14000*	0.00816	0.000	0.1131	0.1669
		45	−0.12000*	0.00816	0.000	−0.1469	−0.0931
		55	−0.18000*	0.00816	0.000	−0.2069	−0.1531
		65	−0.28000*	0.00816	0.000	−0.3069	−0.2531
	45	25	0.26000*	0.00816	0.000	0.2331	0.2869
		35	0.12000*	0.00816	0.000	0.0931	0.1469
		55	−0.06000*	0.00816	0.000	−0.0869	−0.0331
		65	−0.16000*	0.00816	0.000	−0.1869	−0.1331
	55	25	0.32000*	0.00816	0.000	0.2931	0.3469
		35	0.18000*	0.00816	0.000	0.1531	0.2069
		45	0.06000*	0.00816	0.000	0.0331	0.0869
		65	−0.10000*	0.00816	0.000	−0.1269	−0.0731
	65	25	0.42000*	0.00816	0.000	0.3931	0.4469
		35	0.28000*	0.00816	0.000	0.2531	0.3069
		45	0.16000*	0.00816	0.000	0.1331	0.1869
		55	0.10000*	0.00816	0.000	0.0731	0.1269
pH 8.5	25	35	−0.17667*	0.00760	0.000	−0.2017	−0.1517
		45	−0.22667*	0.00760	0.000	−0.2517	−0.2017
		55	−0.38667*	0.00760	0.000	−0.4117	−0.3617
		65	−0.46667*	0.00760	0.000	−0.4917	−0.4417
	35	25	0.17667*	0.00760	0.000	0.1517	0.2017
		45	−0.05000*	0.00760	0.000	−0.0750	−0.0250
		55	−0.21000*	0.00760	0.000	−0.2350	−0.1850
		65	−0.29000*	0.00760	0.000	−0.3150	−0.2650
	45	25	0.22667*	0.00760	0.000	0.2017	0.2517
		35	0.05000*	0.00760	0.000	0.0250	0.0750
		55	−0.16000*	0.00760	0.000	−0.1850	−0.1350
		65	−0.24000*	0.00760	0.000	−0.2650	−0.2150

(continued)

Table 8.6 (continued)

Multiple Comparisons							
Tukey HSD							
Dependent variable	(I) Time	(J) Time	Mean difference (I-J)	Std. error	Sig	95% confidence interval	
						Lower bound	Upper bound
	55	25	0.38667*	0.00760	0.000	0.3617	0.4117
		35	0.21000*	0.00760	0.000	0.1850	0.2350
		45	0.16000*	0.00760	0.000	0.1350	0.1850
		65	−0.08000*	0.00760	0.000	−0.1050	−0.0550
	65	25	0.46667*	0.00760	0.000	0.4417	0.4917
		35	0.29000*	0.00760	0.000	0.2650	0.3150
		45	0.24000*	0.00760	0.000	0.2150	0.2650
		55	0.08000*	0.00760	0.000	0.0550	0.1050

*The mean difference is significant at the 0.05 level

with warmth. When compared to cocamidopropyl betaine, the efficacy of the breakdown of sodium laureth sulfate increased since it can be broken down at a pH of 8 and a temperature of 65 °C.

Limitations: Polyquaternium polymers were synthetic chemicals and may not be readily biodegradable, which could raise environmental concerns when they were

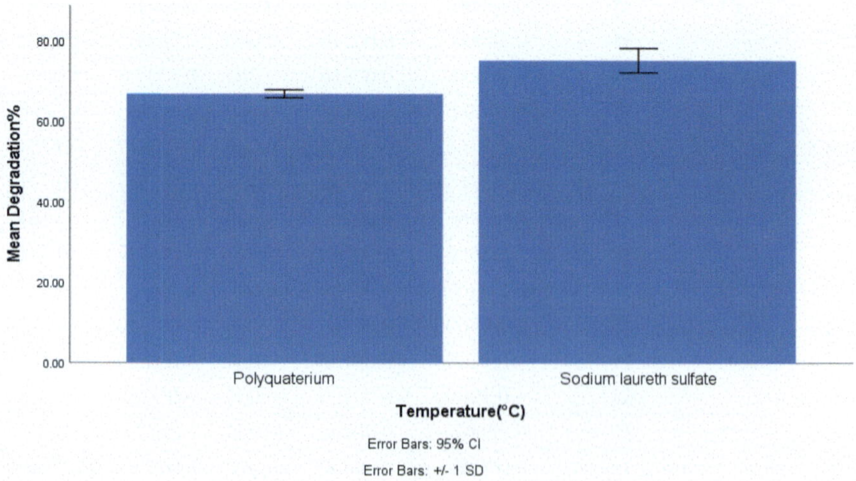

Fig. 8.3 Comparison of pH 7 solutions of polyquaterium and sodium laureth sulfate while X-axis with temperature and Y-axis with degradation rate Cl: 95%; SD ± 1

released into the environment, such as through wastewater. Their persistence in the environment could contribute to pollution and potentially harm aquatic life (Badami et al. 2023). **Future scope**: The breakdown of developing pollutants, like pharmaceuticals, personal care items, and microplastics, which were a rising concern in wastewater treatment, had been demonstrated to be successful by sonolysis. Sonolysis may be further researched and used as a workable method for emerging pollutants removal from wastewater as awareness and regulatory requirements for them continue to rise.

8.5 Conclusion

Ultrasonicator had the potential use for the treatment of wastewater. It was one of the advanced oxidation process methods eliminating or degrading the organic substances. Ultrasonicator and its technology decompose the organic compounds into simpler compounds present in the wastewater. As personal care products were plausible emerging contaminants containing toxicity which causes adverse effects to the human being. In this study sonolysis degradation of polyquaternium, a personal care product chemical with different parameters was conducted. This results in the breakdown of Polyquaternium in 8.5 and temperature 70 °C was effective. It's shown that using this sonolysis technology degradation of chemicals was suitable and the process was easy to operate. Analyzing statistical significance $p = 0.001 (p < 0.05)$ involved using a ONE WAY ANOVA test.

Acknowledgements The Saveetha Institute of Medical and Technological Sciences' SIMATS School of Engineering, for offering the equipment required to finish this work successfully, is acknowledged by the authors.

Authors Contribution Sample gathering, preparation, data analysis, and paper writing were all carried out by the author's PP. The concept was processed, the data was verified, and the article underwent a critical review by author PSB.

Funding The following organisations helped us financially so we were able to complete the research, and for that we are grateful. (1) Greenmeds Lab, Chennai. (2) Saveetha Institute of Medical and Technical Sciences. (3) Saveetha School of Engineering.

Declarations

Conflicts of Interest There are no conflicts of interest in this paper.

References

Badami V, Akarapu S, Kethineni H, Mittapalli SP, Bala KR, Fatima SF (2023) Efficacy of laser-activated irrigation versus ultrasonic-activated irrigation: a systematic review. Cureus 15(3):e36352

Edquist C, Vonortas NS, Zabala-Iturriagagoitia JM, Edler J (2015) Public procurement for innovation. Edward Elgar publishing

Elsayed MA (2015) Ultrasonic removal of pyridine from wastewater: optimization of the operating conditions. Appl Water Sci 5(3):221–227

Fayos-Solà E, Cooper C (2018) The future of tourism: Innovation and sustainability. Springer

Kamali MA (2010) Removal of heavy metal from industrial wastewater using ultrasonic assisted by tea waste as adsorbent

Kane SP, Phar, and BCPS. (n.d) Sample size calculator. Accessed https://clincalc.com/stats/sample size.aspx

Liu XM, Wei Gao E, Maziarz P, Salamone JC, Duex J, Xia E (2006) Detailed characterization of cationic hydroxyethylcellulose derivatives using aqueous size-exclusion chromatography with on-line triple detection. J Chromatogr A 1104(1–2):145–153

Lun YE, Abdullah SRS, Hasan HA, Othman AR, Kurniawan SB, Imron MF, Al Falahi OA, Said NSM, Sharuddin SSN, Ismail N (2022) Integrated emergent-floating planted reactor for textile effluent: removal potential, optimization of operational conditions and potential forthcoming waste management strategy. J Environ Manage 311:114832

Maldonado J, Porto E, Barbero JMR, Tello I, Rodas A, Vivas EX, Mattox DE, Solares CA, Garzon Muvdi T, Pradilla G (2023) Thermal analysis of an ultrasonic aspirator micro claw tool compared with standard high-speed drilling during internal auditory canal opening in a cadaveric model. Oper Neurosurg. https://doi.org/10.1227/ons.0000000000000721

Prasad MNV, Vithanage M, Kapley A (2019) Pharmaceuticals and personal care products: waste management and treatment technology: emerging contaminants and micro pollutants. Butterworth-Heinemann

Wang H, Meng Y (2023) Application value of peripheral blood IgG and IgM combined with ultrasonic echo parameters of substantia nigra in the diagnosis of Parkinson's disease. Biotechnol Genet Eng Rev 1–9

Chapter 9
Safety Evaluation of Hygienic Sanitary Pads Made Sustainably Using Novel *Pseudosasa japonica* Fibre Against Super Absorbent Polymer Containing Commercial Napkins

Peter Sharon and Palanisamy Suresh Babu

Abstract **Aim**: The main aim of the study was to develop a sustainable serviette using a novel *Pseudosasa japonica* fiber sample with superior absorption and compare it with the efficiency of a locally available sanitary napkin. **Materials and Methods**: Processed bamboo fibres, neem leaves, cotton fabric and citric acid binder and other essential chemicals were used. Brauner-Emmett-Teller isotherm was used to evaluate the porosity and surface adsorption efficiency. Fourier Transform Infrared Spectroscopy was used to analyse the presence of functional groups present in the novel bamboo fibers. The prepared napkin was evaluated for its absorbency, test fluid, rewet under load, retention % and G power was 80%, the coincidence interval of 95% including group 1 and group 2 of 2 samples each. **Results**: The present study reported the performance of sustainable and phthalate free napkins made up of novel bamboo fibres and other natural ingredients. The observed results were highly promising in terms of porosity, adsorption index (7.86), antimicrobial efficacy and biodegradability. SPSS statistical analysis results depicted that the value of p was 0.001 ($p < 0.05$) using sample T test and which confirmed the statistical significance. **Conclusion**: Thus, this study was helpful to develop a simple bio-friendly sanitary napkin which was free from toxic substances with exclusive adsorbing efficacy.

Keywords Novel bamboo fibre · BET analysis · Biodegradable · Bisphenol · Sanitation · Menstrual · Phthalate free

---

P. Sharon · P. Suresh Babu (✉)
Department of Biotechnology, Saveetha School of Engineering, Saveetha Institute of Medical and Technical Sciences, Chennai, Tamil Nadu 602105, India
e-mail: sureshbabup.sse@saveetha.com

P. Sharon
e-mail: sharonp19@saveetha.com

© The Author(s), under exclusive license to Springer Nature Switzerland AG 2025

A. Jeyaseelan and K. Murugasen (eds.), *Sustainable Environmental Remediation: Avenues in Nano and Biotechnology*, Environmental Science and Engineering,
https://doi.org/10.1007/978-3-031-78483-5_9

9.1 Introduction

According to various recent research reports, around 64% of women in India use disposable sanitary napkins, 50% use cloth, and 15% use locally produced napkins (Peterson and Warshaw 2023). In all, 78% of women in this age range defend against their periods in a hygienic way. Nearly, ~1.8 billion people menstruate every month in the world. Sanitation needs to be maintained. Numerous girls, women, transgenders, and non-binary people were unable to effectively control/maintain their menstrual cycle. Doing the napkin math, if a person bleeds every five to twenty-eight days, that was 130–260 pads or tampons annually, or 5,000–10,000 during the course of their reproductive years. These conditions were posing a serious threat leading to plausibilities like fungal infections, urinary infections, and even infertility paving in a need for biodegradable napkins. Menstrual cups, which last 10 years, can replace up to 2500 pads or tampons in a woman. If she utilised reusable phthalate free pads, each pad would replace 33–65 disposables (needing 4–5 pads to last a year). Whereas 800–1600 million pads and tampons that would be used daily if all women utilised disposable goods. Several absorbent fibres were readily available to make pads and tampons (Peterson and Warshaw 2023), which were natural or synthetic bisphenol, including cotton and rayon (Tierno and Hanna 1998; Er 2021). To absorb the menstrual fluid, pads that were adhered lay on the inside of your underpants (Scott et al. 2013). The primary components of disposable menstrual pads were often bleached rayon (cellulose generated from wood pulp) (Burge 2019), cotton, and polymers (Woeller and Hochwalt 2015). Despite the fact that manufacturers were typically unwilling to disclose the precise composition of their products creating a need for biodegradable napkins. It's important to sustain sanitation. Antibacterial and aroma ingredients were some of the added features in selected commercially available expensive napkins (Fischer et al. 2020).

In the last two years, sanitary napkins had been the subject of over 275 publications in the Pubmed database, as well as over 644 publications in the Science Direct databases. During the literature review, numerous publications from various journals in different databases were considered. Phthalate-containing commercial sanitary napkins (Shaikh et al. 2019) pose a major threat to the humans and their health owing to the presence of BPA and other toxic chemicals like bisphenol that, over time (Lum et al. 2017), can disrupt the reproductive system and can induce cancer (Dixit et al. 2013; Viguié et al. 2018). As an alternate solution, novel bamboo fibre was selected, owing to the unique tunable attributes like microgaps, porosity, soft nature like cotton, eco friendly, biodegradable, and exclusive absorbing capacity (Prakash 2020; Lo et al. 2008).

Cellulose enriched fibre derived from bamboo plants (Murugesh Babu and Chandrasekhara 2022; Muthu 2022). Such processed fibre was achieved by converting starch rich pulp through alkaline hydrolysis and multi-phase bleaching process (Jia and Zhou 2011; Mokeramin 2020). These novel bamboo fibres possess bacteriostatic, antifungal (Mokeramin 2020), antibacterial, hypoallergenic (Murugesh Babu and Chandrasekhara 2022; Bährle-Rapp 2007), hygroscopic, natural deodorizer and resistant against ultraviolet light. The aim of the present study was focused to create a sustainable napkin with exclusive absorbing efficiency and free from hazardous BPA (bisphenol A) and BPS (bisphenol S) chemicals and to evaluate the performance of sustainable and novel bamboo fibre based napkin for its antimicrobial, biodegradable, and absorption efficacy.

9.2 Materials and Methods

The research was carried out at the Saveetha School of Engineering's Toxicology Laboratory in the Department of Biotechnology. The enrollment ratio was set at 80% for continuous testing using clinical analysis. The size of each group was determined using an online sample calculator (clinical.com) based on prior focus on results, with G power set to 80%, edge set to 0.05, and certainty span set to 95%. The samples were divided into two groups of two, totaling four samples.

9.2.1 Preparation of Test Fluid

Six liters of distilled water were continuously mixed while four grams of methyl paraben were added. About 740 g of acacia gum was added and mixed until it dissolved. By adding boiled water, the liquid was increased to 8 L and then left undisturbed for at least 24 h. The liquid was then filtered using a glass wool filter. While the mixture was thoroughly agitated for sanitation, 840 mL of distilled water, 1.470 mL of glycerin, and 9 g of methylene blue were added again. It had been at least 24 h since the liquid had been disturbed.

9.2.2 Fabricating the Base

To begin, finished nonwoven textiles were treated with a neem extract solution made from freshly harvested neem leaves to form phthalate free pads. Fresh neem leaves were washed with distilled water, cut into small pieces, and boiled in 100 mL of water with Whatman No. 1 filter paper. Three different concentrated neem extracts were created for further research. The pad-dry-cure method was used to incorporate grafting of the neem leaf extract substrate for maintaining the sanitation quotient. In three different conical flasks, the nonwoven fabric (Mokeramin 2020) was soaked in solutions containing 10, 25, and 50% neem leaf extract. After 40 s, the citric acid binder (1%) was added to all three flasks. The excess solution from the test cloth was wiped away with a Laboratory Padding Mangle at room temperature and 2 kg/cm^2 pressure. All test materials were dried for 20 min at 80 °C after padding before being cured for 5 min at 120 °C.

9.2.3 Manufacturing Bamboo Fibre

Bamboo leaves and soft inner pith were extracted and crushed from the hard bamboo trunk. Crushed bamboo cellulose was soaked in a solution of 15–20% sodium hydroxide (NaOH) for one to three hours at a temperature of 20–25 °C to form alkali cellulose. The excess sodium hydroxide solution was then pressed out of the bamboo alkali cellulose. A grinder was used to crush the alkali cellulose, which was then dried for 24 h. To sulfurize the compound and cause it to jell, roughly one-third as much carbon disulfide (Dixit et al. 2013) was added to the bamboo alkali cellulose. Any remaining carbon disulfide or bisphenol was removed by evaporation as a result of decompression, yielding cellulose sodium xanthogenate. A phthalate free diluted solution of sodium hydroxide was added to the cellulose sodium xanthogenate to dissolve it, resulting in a viscous solution containing approximately 5% sodium hydroxide and 7–15% novel bamboo fibre cellulose.

9.2.4 Statistical Analysis

The IBM SPSS Version 26 Software was used to estimate results such as mean, standard deviation and standard deviation error. Analyzing statistical significance involved using Sample T test for equality of variance. The SPSS software was used to analyze the statistical comparison of group 1 water bulk density and group 2 porosity data using an Independent Samples Test. Bulk density was the dependent variable. Similarly, porosity was considered as the independent variable.

9.3 Results

9.3.1 Estimation of Specific Surface Area, Pore Volume and Density

In Table 9.1, the observed results of *Pseudosasa japonica* samples in the present study was compared with Guadua bamboo. The BET analysis clearly depicts the specific surface area and porosity of our test sample as 138 m^2/g and 0.031.

9.3.2 FTIR Spectra

The Fourier Transform Infrared (FTIR) spectra of the collected bamboo fibres as shown in Fig. 9.1. While the band at 2940 cm^{-1} was linked to the C–H symmetrical vibration of the methylene ($-CH_2$) group, the broad absorption band at 3427 cm^{-1}

Table 9.1 Estimation of specific surface area, pore volume and density of *Pseudosasa japonica compared with Guadua* bamboo fibre using BET analysis

Sample	Specific surface area A-BET (m^2/g)	Pore volume micropores (cm^3/g)	Density (g/cm^3)	References
Guadua Bamboo	144 + 7	0.050	1.764 + 0.010	Masis-Melendez et al. (2020)
Pseudosasa japonica	138 + 3	0.031	1.22 + 0.041	Present study

was –OH stretching vibration. The stretching vibrations of non-conjugated carbonyl and conjugate carbonate, or carboxylic acid and ketone group, respectively, were associated with the peaks at 1710 and 1654 cm^{-1}. The delignification in the novel bamboo fibre was shown by the vibration of aromatic rings, which were represented by the peaks at 1592, 1506, and 1420 cm^{-1}. According to the methoxyl C–H bending and C–C stretching in the aromatic skeleton, the intensity at 1461 cm^{-1} was caused. The p-hydroxy phenylpropane (H), syringyl (S), and guaiacyl (G) units, respectively, of typical HGS-lignin were represented by the absorption peaks at 1159, 1120, and 830 cm^{-1}. The bands at 1326 and 1265 cm^{-1} bands were attributed to condensed G-units at G5 and/or aromatic C–O stretching of S-units. The nonether-ified phenolic –OH produced as a result of the cleavage at the –O–4′ and –O–4′ linkage was attributed to the weak shoulder peak at 1362 cm^{-1}. The low intensity of this signal indicates that the cleavage of the connection between –O–4′ and –O–4′ in the isolated lignin macromolecules was not significantly caused by the extraction process used. The peak at 1030 cm^{-1} was attributed to the side group and glycosidic bond stretchings of C–OH and C–O–C, respectively.

9.3.3 *Absorption Capacity and Liquid Strike Through Rate and Rewet Under Load Conditions*

In the case of liquid strike-through, the napkins made of pure organic cotton take less time (0.75 s) to convey the liquid from their surface than the sanitary napkins manufactured of bamboo/organic cotton mixes in the proportions of 70/30 and 50/50. Due to the weak wicking ability of the fibre, pure bamboo napkins require 1.44 s longer to remove moisture from their surface. As a result, napkins made of bamboo/organic cotton blends in proportions of 70/30 and 50/50 rapidly absorb liquids and keep the skin dry, preventing dampness. The test result demonstrates that the bamboo/organic cotton (70/30), pure organic cotton, and bamboo/organic cotton (50/50) napkin take less time than the diapers made of pure bamboo to transport the recognised liquid from their surface. Additionally, compared to all other pads, the bamboo/organic cotton (70/30) napkin had a lower rewet value. Pure bamboo napkins had the lowest rewet value (5.8 g), while bamboo/organic cotton blend napkins (50/50) had the

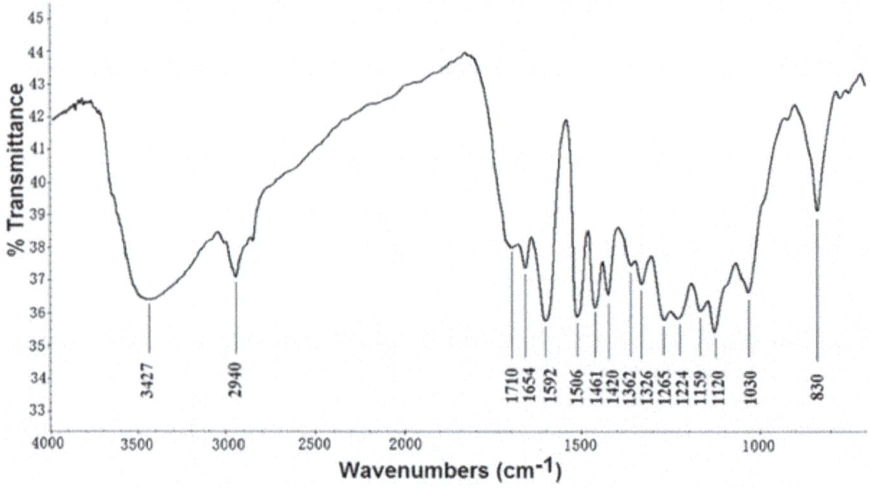

Fig. 9.1 FTIR spectra of *Pseudosasa japonica* bamboo fibre. X axis: wavenumber, Y axis: percentage transmittance

highest rewet value (6.3 g) (2.4 g). The performance of the diaper was improved by a low rewet value. If the absorbed liquid leaks onto the napkin, it would result in moistness and a skin infection (Fig. 9.2).

9.3.4 SPSS Analysis

According to the statistical analysis, the mean value in the group statistics for groups 1 and 2 was found to be 1.6412 and 3.9025 respectively. The standard deviation for group 1 was found to be 0.8012 for group 1 and 2.63113 for group 2. The standard error mean was found to be 0.3271 for group 1 and 1.31556 for group 2. The mean difference for sig 2-tailed test for the equal variances assumed was found to be 2.26133 for group 1 and 2 and when the equal variances not assumed the mean difference was found to be 2.26133 (Tables 9.2, 9.3 and 9.4).

9.4 Discussion

Some of the common issues experienced by women around the world during the menstrual cycle include the lack of affordable, safe napkins, unsanitary environments, and related cervical illnesses. Due to the inclusion of various hazardous compounds like bisphenol, commercially produced napkins were harmful. Many academics

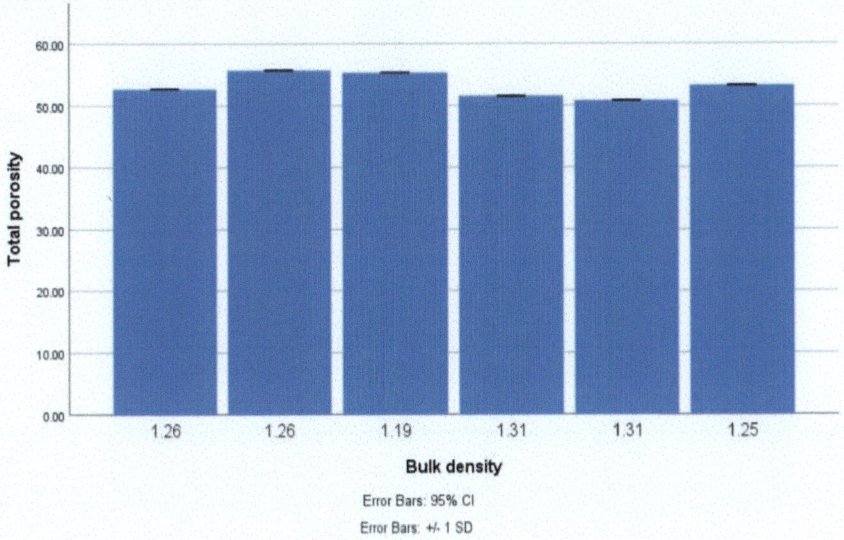

Fig. 9.2 Effect of bulk density and the total porosity of the *Pseudosasa japonica* sample. X-axis represents the absorption time taken between the *Pseudosasa japonica* fiber samples 1 and 2 while Y-axis represents the water absorption rate between the *Pseudosasa japonica* fiber samples 1 and 2. Cl 95% SD ± 1

Table 9.2 Highlights the absorption capacity, liquid strike through rate and rewet under load conditions onto novel *Pseudosasa japonica* bamboo fibre, pure organic cotton and in varying ratio

Sample	Pure bamboo	Pure organic cotton	Bamboo/organic cotton (70/30)	Bamboo/organic cotton (50/50)
Absorption capacity (g/g)	9.21	12.01	11.69	11.73
Liquid strike through (s)	1.23	0.72	1.20	0.89
Rewet under load (g)	4.74	5.56	1.61	1.98

Table 9.3 Comparison of the mean, standard deviation, and standard error for sustainably prepared sanitary napkins using SPSS analysis

Group statistics					
Parameter	Group	N	Mean	Std. deviation	Std. error mean
Water absorption rate	Equal variances assumed	6	1.6412	0.8012	0.3271
	Equal variances not assumed	4	3.9025	2.63113	1.31556

Table 9.4 Comparison of the Independent sample T test for the chemical constituent percentage shows the significant value of $p < 0.05$ using SPSS Analysis

Independent samples test										
Levene's test for equality of variance				t-Test for equality of means						
				t	df	Sig. (2-tailed)	Mean difference	Std. error difference	95% Confidence interval of the difference	
									Lower	Upper
Chemical constituents %	Equal variances assumed	68.760	0.001	2.173	8	0	2.26133	1.04085	4.661537	0.13886
	Equal variances not assumed			1.178	3.004	0.003	2.26133	1.31597	6.44641	1.92375

were interested in finding ways to create sanitary napkins that were both safe and biodegradable. Our current work focused on the performance assessment of bamboo fibres as an alternative to commercially available adsorbents. Sanitary napkins were made from bamboo, a resource that was renewable and wholly biodegradable, absorbs more CO_2 than trees and generates 30% more oxygen (Prakash 2020).

Novel bamboo fibre really offers better absorption (Jia and Zhou 2011) than cotton when compared to cotton in the competition between bamboo and cotton pads for flow protection (Lo et al. 2008) Cotton might still be a decent option for you even if your flow was lighter. Bamboo was regarded as a resource that was particularly sustainable because of its rapid growth and ability to recover after harvest. It was also a fantastic material and a hygienic, environmentally responsible choice because it was resistant (Shaikh et al. 2019) to pests and diseases.

Limitations: Species variation and seasonal factors can drastically influence the structural characteristics of bamboo. There was no standardized or comprehensive regulation, making it challenging to get permits and structural calculations. **Future scope**: Amalgamation of other natural fibres like flax, hemp, ramie, sunn etc. with bamboo may enhance the absorption capacity.

9.5 Conclusion

Within the limitation of study, the bamboo pulp, which was 300 times more absorbent than wood pulp, fills the inner layer. Compared to wood pulp, they were safer and more environmentally friendly. Because they were leak-proof and also stop seepage, bamboo sanitary pads were more affordable than conventional pads. Toxin free sustainable sanitary napkins can thus be made biodegradable using novel bamboo fibre. All of the materials used to make these pads were plant-based and biodegradable. In the contest between bamboo and cotton pads for flow protection, bamboo fibre actually provides up to 40% greater absorption than cotton. Even if your flow was lighter, cotton might still be a good choice for you. Due to its quick growth and capacity to rebound after harvest, novel bamboo fibre was seen as a resource that was very sustainable. Also, it was resistant to pests and diseases, making it a perfect material and sanitary, eco-friendly choice. The observed results were more impressive, still further developments were required to authenticate and to enhance the performance of the developed product. The independent sample test gives the mean, mean difference (2.26133) and standard deviation of parameters like absorption rate and chemical constituents percentage of the novel *Psuedosasa japonica* fibre used using sample T test for equality of variance. The p value was 0.001 ($p < 0.05$) which was statistically significant.

Acknowledgements The authors would like to express their sincere gratitude to Saveetha School of Engineering, Saveetha Institute of Medical and Technical Sciences for providing the necessary infrastructure to carry out this work successfully.

Author Contributions Author PS was involved in data collection, data analysis, manuscript writing. Dr PSB was involved in conceptualization, guidance and critical review of manuscript.

Funding We thank the following organizations for providing financial support that enabled us to complete the study. (1) Greenmeds Lab, Chennai. (2) Saveetha Institute of Medical and Technical Sciences. (3) Saveetha School of Engineering.

Declarations

Conflict of Interest The authors of this paper declare no conflict of interest.

References

Bährle-Rapp M (2007) Hypoallergenic. Springer Lexikon Kosmetik Und Körperpflege. https://doi.org/10.1007/978-3-540-71095-0_5134

Burge S (2019) Faculty opinions recommendation of occupational asthma associated with bleached chlorine-free cellulose dust in a sanitary pad production plant. Faculty opinions—Post-publication peer review of the biomedical literature. https://doi.org/10.3410/f.735015577.793556514

Dixit S, Fischer G, Wittekind C (2013) Recurrent menstrual toxic shock syndrome despite discontinuation of tampon use: is menstrual toxic shock syndrome really caused by tampons? Australas J Dermatol. https://doi.org/10.1111/j.1440-0960.2012.00938.x

Er A (2021) Tampon in-Situ. Radiopaedia.org. https://doi.org/10.53347/rid-92875

Fischer AM, Holmes P, Bahar YZ, Vacca S, Goldberg S, Gold MA (2020) Aroma acupoint therapy for symptom management with adolescent patients: early experiences from school-based health centers. Med Acupunct 32(5):287–292

Jia Y-D, Zhou X-F (2011) Mechanism of co-salen biomimetic catalysis bleaching of bamboo pulp. Adv Biomim. https://doi.org/10.5772/13544

Lo TY, Cui HZ, Tang PWC, Leung HC (2008) Strength analysis of bamboo by microscopic investigation of bamboo fibre. Constr Build Mater. https://doi.org/10.1016/j.conbuildmat.2007.03.031

Lum KJ, Sundaram R, Barr DB, Louis TA, Buck Louis GM (2017) Perfluoroalkyl chemicals, menstrual cycle length, and fecundity. Epidemiology. https://doi.org/10.1097/ede.0000000000000552

Mokeramin M (2020) Chemical effect on the mechanical properties of bamboo fiber for textile: a review. Int J Adv Trends Comput Sci Eng. https://doi.org/10.30534/ijatcse/2020/5191.42020

Murugesh Babu K, Chandrasekhara SM (2022) Bamboo fibres: processing, properties and applications. Woodhead Publishing

Muthu SS (2022) Sustainable approaches in textiles and fashion: fibres, raw materials and product development. Springer Nature

Peterson MY, Warshaw E (2023) Going green and adhesive-free: snap-in reusable menstrual pads for tampon and pad intolerance. Derm 34(1):60–61, Jan–Feb. https://doi.org/10.1089/DERM.0000000000000898

Prakash C (2020) Bamboo fibre. In: Handbook of natural fibres.https://doi.org/10.1016/b978-0-12-818398-4.00009-8

Scott L, Montgomery P, Steinfeld L, Dolan C, Dopson S (2013) Sanitary pad acceptability and sustainability study—long report. https://doi.org/10.35648/20.500.12413/11781/ii161.

Shaikh Z, Panda M, Gaur D (2019) Acceptability of low-priced, socially-marketed sanitary napkins amongst non-users of sanitary napkins in a slum population of Delhi. Int J Commun Med Public Health. https://doi.org/10.18203/2394-6040.ijcmph20191417

Tierno PM Jr, Hanna BA (1998) Viscose rayon versus cotton tampons. J Infect Dis

Viguié C, Mhaouty-Kodja S, Habert R, Chevrier C, Michel C, Pasquier E (2018) Evidence-based adverse outcome pathway approach for the identification of BPA as en endocrine disruptor in relation to its effect on the estrous cycle. Mol Cell Endocrinol 475(November):10–28

Woeller KE, Hochwalt AE (2015) Safety assessment of sanitary pads with a polymeric foam absorbent core. Regul Toxicol Pharmacol RTP 73(1):419–424

Chapter 10
Effect of Biopolymer and Polymeric Nanoparticles for Dye Degradation

Saranya Thayanithi, Kumar Janakiraman, Sridhar Alagesan, Abilesh Ramesh, Renugadevi Balu, Gracesuganthi Jayaraj, and Vaidevi Sethuraman

Abstract Environmental contamination has significantly increased when wastewater containing synthetic colors from the textile industry is discharged. These dyes pose a serious threat to aquatic ecosystems and human health. For resolving this problem, research has focused on developing sustainable and efficient methods for dye degradation. Biopolymers and polymeric nanoparticles have emerged as promising materials for dye degradation purpose. This abstract focuses on an overview of the research to investigate the effect of biopolymers and polymeric nanoparticles on dye degradation. Biopolymers, such as chitosan and alginate, have inherent biodegradability and low toxicity, making them eco-friendly options for dye removal. Additionally, polymeric nanoparticles, due to their high surface area and tunable properties, offer excellent adsorption and catalytic capabilities for dye degradation. To carry out this, various biopolymers and polymeric nanoparticles were synthesized and characterized, and their adsorption abilities for different synthetic dyes and their catalytic activities were evaluated. The use of biopolymers, when changed or combined with polymeric nanoparticles, will exhibit enhanced dye removal efficiency compared to traditional adsorbents. Moreover, incorporating catalysts into the polymeric nanoparticles improved the dyes' degradation kinetics. This synergistic approach provided an effective and environmentally friendly strategy for dye degradation in wastewater treatment. The research findings underscore the potential of biopolymers and polymeric nanoparticles as sustainable alternatives for dye removal and degradation, contributing to the reduction of environmental pollution. These innovative materials hold promise for future applications in water treatment processes, emphasizing the importance of further exploration and optimization to address the challenges of dye pollution in a sustainable and efficient manner.

Keywords Dye degradation · Biopolymers · Polymeric nanomaterials

S. Thayanithi · K. Janakiraman · S. Alagesan · A. Ramesh · R. Balu · G. Jayaraj · V. Sethuraman (✉)
Department of Biotechnology, Rathinam Technical Campus, Rathinam Techzone Campus, Eachanari, Coimbatore, Tamil Nadu 641021, India
e-mail: vaidevipavi@gmail.com

© The Author(s), under exclusive license to Springer Nature Switzerland AG 2025
A. Jeyaseelan and K. Murugasen (eds.), *Sustainable Environmental Remediation: Avenues in Nano and Biotechnology*, Environmental Science and Engineering,
https://doi.org/10.1007/978-3-031-78483-5_10

10.1 Introduction

Biopolymers are polymers made by living organisms. As a result, they are polymeric biomolecules, or lengthy chains of monomeric units that are covalently connected. Biopolymers are found in plants, algae, fungi, bacteria, and mammals (Dassanayake et al. 2021). Macromolecules include starch, alginate, and chitosan. Dextran and chitin Polysaccharides produced from plants and microbial biofilms, such as chitosan, starch, dextran, and cellulose, are commonly used in nanomaterials. Chitosan is commonly used to remove dyes and purify wastewater. Biopolymers have unique composition and physiological features. Attaching metals to biopolymers can result in the formation of nanomaterials. Biopolymers produce molecular capsules by intramolecular hydrogen bonding (Sarkar et al. 2017).

Biopolymers are sustainable and biodegradable, which sets them apart from fossil-fuel-derived polymers. Biodegradable biopolymers made from renewable resources offer alternatives to fossil-fuel-based polymers (Mohan et al. 2016). Some of the techniques such as adsorption, biodegradation, oxidative degradation, electrochemical destruction, ozonation, photochemical, and UV-assisted. These are the techniques used to assist biopolymer degradation (Dassanayake et al. 2021). Cellulose is one of the most well-known biopolymers. It is a common chemical found in plant cell walls, making it among the most abundant on Earth. This polysaccharide is composed of repeating units of glucose and can be found in natural fibers. It is insoluble in water and not digested by humans. Some animals, including termites and ruminants, may digest it with the help of microorganisms (Flórez et al. 2023).

Polymeric nanocomposites can be synthesized using several processes, including sol–gel, co-precipitation, chemical reduction, reverse micellar synthesis, microemulsion, hydrothermal, laser pyrolysis, and laser ablation. Biodegradable polymers including starch, cellulose, and PLA are blended into cosmetic products, reducing the need for disposal. Chemical modification of organically generated polymers can improve waste solubility and stability, resulting in extraordinary features (Matei et al. 2022).

10.2 Environmental Contamination—Causes and Impact on Human Health

Dyes discharged from industrial sources have deleterious impacts on the environment, even at minimal concentrations. Furthermore, the discharge of colored wastewater introduces highly toxic compounds, exacerbating environmental issues. It is widely recognized that the disposal of dyes into waterways diminishes sunlight penetration, elevates biochemical (BOD) and chemical (COD) oxygen demands, hinders photosynthesis, and stunts plant growth (Fig. 10.1). Synthetic dyes possess enduring properties, accumulate in organisms, and are hazardous, mutagenic, and carcinogenic (Ardila-Leal et al. 2021).

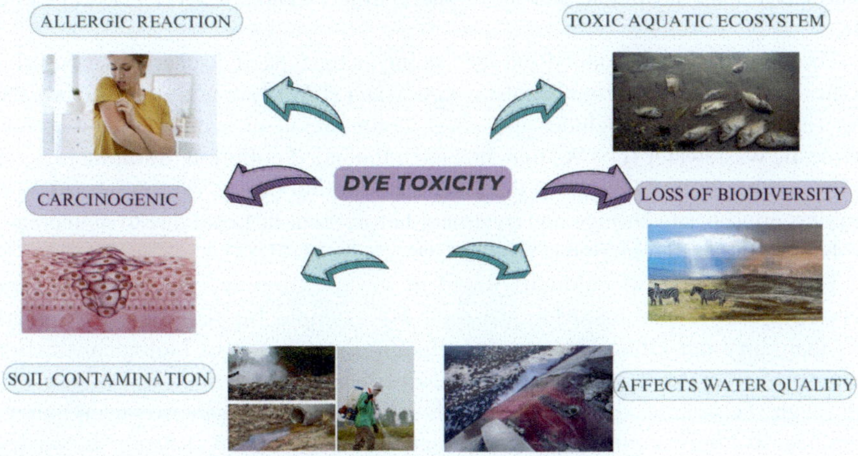

Fig. 10.1 Impact of dyes in environment and human health

Dyes, whether natural or synthetic, impart color to various fibers employed across industries such as textiles, pharmaceuticals, food, cosmetics, plastics, photography, and paper production. Colorants take the form of either pigments or dyes. Pigments, characterized by their near-insolubility, consist of particles ranging from 1 to 2 μm in size. In contrast, dyes readily dissolve in water, featuring particle sizes ranging from 0.025 to 1.0 μm (Braun 1983).

The utilization of artificial dyes in industry offers several advantages, including: (Paz et al. 2017).

(i) Chemical stability over time,
(ii) Resistance to physical, chemical, and biological degradation,
(iii) Consistent provision of color to the intended fiber through reproducible processes, ensuring color intensity is maintained, and
(iv) Cost-effectiveness.

The treatment of textile dye waste is concerned with both the bleaching, as well as the destruction and degradation of dye molecules. Decolourization can be achieved by either removing molecules from the solution itself or breaking the chromophore link. In both circumstances, the molecule itself (in the former) or the primary fragments (in the latter) remain unchanged. Consequently, the molecules' ability to absorb light changes from the visible range to either the ultraviolet or infrared areas of the electromagnetic spectrum.

The main dyes used on a large scale in industry include azo, anthraquinone, indigo, xanthene, and triarylmethane. Significantly, aromatic chemicals with complex structures such as anthraquinones have been remarkably prominent. Synthetic dyes are widely used in the textile and tanning industries to color a wide variety of things. In addition, a wide range of industries, including cosmetics, paper, food, pharmaceuticals, hospitals, and universities, utilize synthetic dyes (Tkaczyk et al. 2020).

The content of wastewater differs according on the type of discharge and the origin of the dye, whether it is from domestic wastewater (DWW) or non-domestic wastewater (nDWW), which includes industrial, commercial, or service-related activity. Domestic wastewater (DWW) may include unintended pollution, resulting in potentially lower pollution levels. On the other hand, non-domestic wastewater (nDWW) requires appropriate storage and treatment before final disposal due to its increased probability of contamination (Pedroza-Camacho et al. 2018).

The composition of colorants found in sewage from service providers varies depending on the staining techniques used. For example, biomedical research facilities that carry out clinical protocols including the staining of biological fractions dispose of dyes such as Haematoxylin, Eosin Y, Rose Bengal, or Auramine O. Although the volume of dye solutions is relatively modest, their concentration is significantly high, ranging from approximately 1 to 10 $g L^{-1}$. As a result, the wastewater produced has a high level of toxicity. Many other service providers generate large amounts of wastewater, which can be differentiated according on how it is discharged and the source of the dye. This wastewater is classified as either domestic wastewater (DWW) or non-domestic wastewater (nDWW). Domestic wastewater (DWW) can be unintentionally contaminated, leading to possibly reduced levels of pollution, while non-domestic wastewater (nDWW) comes from manufacturing, business, or service-related operations. Such wastewater requires storage in specialized containers for later collection by specialized businesses for treatment, which includes neutralization and final elimination (Ardila-Leal et al. 2021).

The textile industry primarily employs dyes that have a significant global environmental impact. Azo dyes, known for their affordability, vibrant colors, and colorfastness, dominate the market, accounting for approximately 60% of usage. Anthraquinone dyes rank second in popularity, valued for their dyeing efficiency, ready availability, and cost-effectiveness. The choice of dye is contingent upon various factors such as its compatibility and stability with the fabric, diffusion properties, reactivity, cost, and ability to fix colors. Meanwhile, in the paper industry, a substantial amount of wastewater originates from the dyeing process of handmade paper (Singh et al. 2016).

Within the cosmetics sector, various products like lipsticks, blushers, eye shadows, eyeliners, and nail polish incorporate one or more colorants (either dyes or pigments) at concentrations ranging from 1 to 25% to achieve the desired hues. However, the predominant activity within the industry revolves around the manufacturing of hair dyes, constituting approximately 80% of cosmetic production (Guerra et al. 2018).

Colorants used in the food business are employed to improve the visual attractiveness of food or to offset any changes in color that may occur during the processing of food. They provide pigmentation to a diverse range of products, such as carbonated beverages, fruit drinks, energy drinks, confectioneries, cereals, desserts, snacks, and other items. Currently, there are more than 60 synthetic dyes that have been authorized for use in food. Among the most prevalent are those belonging to the azo class, including Sunset Yellow (E-110), Tartrazine (E-102), and Ponceau 4R (E-124).

Moreover, dyes such as Brilliant Blue FCF (E-133) belonging to the triphenylmethane class and Quinoline Yellow (E-104) from the quinethazones group are extensively utilized. Regulations in each nation dictate the permissible type and concentration of colors in food. However, some investigations have shown cases of improper use of dyes, exceeding the maximum limit allowed. This poses potential dangers for customers (Tikhomirova et al. 2018).

Typically, food-grade colors approved according to the standards set by the Codex Alimentarius Commission (CAC) are also permissible for use in the pharmaceutical industry. As a result, it's commonplace to encounter dyes in pharmaceutical products, including Tartrazine (E-102), Sunset Yellow (E-110), Ponceau 4R (E-124), Azorubine/Carmoisine (E-122), Amaranth (E-123), Brilliant Blue (E-133), and Allura Red (E-129) (Ardila-Leal et al. 2021).

10.3 Impact of Synthetic Dyes on the Environment

The textile industry stands as the primary source of colored wastewater, estimated to generate the largest volume. Around 20% of the dyes utilized in textile fiber dyeing processes remain unfixed, ultimately being discharged into wastewater, leading to substantial pollution levels. However, the environmental impact isn't solely dictated by the volume of dye discharged; it also hinges on the mixture of dyes with other substances, all possessing toxic properties, comprising the effluents from industrial operations. Discharging wastewater that is colored and may contain dyes that can be seen without using any tools, with a concentration of less than 1 part per million (ppm), into bodies of water on the surface or underground, causes a decrease in the amount of oxygen dissolved in the water. This leads to an increase in various physical, chemical, and biological factors such as chemical oxygen demand (COD), biochemical oxygen demand (BOD), total dissolved solids (TDS), total nitrogen (TN), total phosphorus (TFP), and organic compounds that cannot be broken down by living organisms. Additionally, this wastewater displays significant variations in pH levels and may contain toxic heavy metals such as chromium (Cr), arsenic (As), and zinc (Zn) (Al-Tohamy et al. 2022).

Typically, synthetic dyes have low biodegradability due to their chemical composition and structure, leading to negative environmental impacts. The majority of artificial dyes exhibit resistance to degradation, hence presenting carcinogenic and toxic hazards to ecosystems. Furthermore, the adverse effects of dyes can be intensified, resulting in elevated levels of contamination in higher trophic levels. Nevertheless, it is necessary to assess the toxicity of each dye separately, as the extent of harm they cause is contingent upon their specific structure and the level of exposure. As a result, dyes can remain in the environment for long periods of time, typically lasting about 50 years or more. The durability of dyes is strongly associated with their chemical response, with unsaturated substances exhibiting reduced durability in comparison to saturated ones. In addition, the presence of aromatic compounds becomes more enduring as the quantity of chemical and halogen replacements grows. Thus, it is essential to evaluate the deterioration of dyestuffs both separately and in combination (Ferraz et al. 2011).

Dyes that are often employed are usually categorized under the azo, anthraquinone, or triarylmethane classes. Extensive research has been conducted on Azo class dyes to gain a comprehensive understanding of their uses as well as their potential negative impacts. Around 60–70% of azo dyes are recognized as having toxicity, carcinogenicity, and resistance to traditional physicochemical remedies. Azo dyes are harmful due to their chemical reduction, which results in the creation of aromatic amines such benzidine, dimethoxy-benzidine, and dimethyl-benzidine. The toxicity of these aromatic amines is mostly caused by metabolic oxidation, which produces electrophilic reductive intermediates known as diazonium ions. These intermediate substances possess the capability to form covalent bonds with DNA, leading to the occurrence of mutagenic consequences and illnesses, such as cancer. An alternative version of this process entails the chemical reduction of particular azo bonds found in specific dyes, resulting in the creation of corresponding harmful aromatic mono-azo amines (Varjani et al. 2020).

When azo ionic dyes are discharged into water, they can attach to floating organic matter by means of electrostatic interactions, causing them to stick to sediments or scum. This characteristic enhances their longevity in the environment. In addition, when colored water or contaminated scum comes into touch with aquatic creatures, it can lead to the transmission of harmful chemicals through the food chain to people. This transfer can result in a range of health issues, such as hypertension, cramping, nausea, bleeding, and ulceration of the skin or mucous membranes. Significant harm can be inflicted on important organs such as the kidney, reproductive system, liver, brain, and central nervous system (CNS) depending on the amount of dye exposure (Sarkar et al. 2017).

Anthraquinone dyes, although extensively utilized, are considered to be one of the most hazardous dyes. Nevertheless, the research on the biotransformation and toxicity of anthraquinones is relatively scarce when compared to that of azo dyes. Certain anthraquinoid dyes possess significant resistance to chemical oxidation because of the stability provided by the resonance of their aromatic structure. The durability of these colors allows them to stay in wastewater, leading to significant quantities of anthraquinoid dyes in textile effluent. Studies on different anthraquinones have shown varied toxicological findings (Shwe et al. 2014).

Triarylmethane dyes are generally recognized for their substantial effects on metabolism, their capacity to accumulate, and their ability to permeate the skin. When consumed or breathed in, they are recognized for their ability to cause irritation, trigger cancerous effects resulting in sarcomas, and lead to methemoglobinemia if exposed excessively. Out of the three well recognized triarylmethane dyes, Malachite Green (MG), Violet Crystal (CV), and Bright Green (BG), Malachite Green (MG) has received significant attention in research due to its designation as a hazardous substance affecting multiple organs. Methylglyoxal (MG) has been found to have negative effects on both the immunological and reproductive systems, demonstrating genotoxic and carcinogenic characteristics. The impacts of these consequences not only jeopardize the species that are directly exposed, but also constitute a threat to those species that are already at risk of extinction (Belpaire et al. 2015).

10.3.1 Role of Nanoparticles in Dye Degradation

Nanoparticles have special physical and chemical characteristics not seen in bulk materials, they have been employed in the reductive destruction of hazardous and poisonous colors. Versatile nanoparticles have a wide range of uses, such as bioremediation, medications, wastewater treatment, and energy production (Jyoti and Singh 2016). In reductive processes, nanoparticles function as powerful catalysts (Vanaamudan et al. 2016). Many nanoparticles that are used in many different industries, including Ag (Silver) (Kavitha et al. 2014), Au (Gold) (Paul et al. 2016), Zn (Zinc), and silica, are derived from biological sources.

Othman et al. utilized a straightforward hydrothermal treatment procedure to create Ti_3C_2Tx MXene adorned with silver and palladium nanoparticles. This was done with the aim of employing the resulting material for the photocatalytic degradation of methylene blue (MB) and rhodamine B (RhB). The photocatalytic degradation capabilities of the original MXene and the manufactured MXene composites were examined using ultraviolet and solar light irradiation. All oxidized MXene composites showed a notable enhancement in their photocatalytic activities compared to the pristine MXene. Among them, the AgNPs/TiO_2/Ti_3C_2Tx composite exhibited the highest degradation efficiency (Fig. 10.2). This study expands the potential uses of oxidized MXene composites, offering a viable material for the degradation of organic dyes and the treatment of wastewater.

Nanoparticles have a vast surface area, making them useful for dye removal from waste water. Its large surface area to volume ratio allows it to mix with more dye molecules to speed up the degradation process. Fe_3O_4 nanoparticles are magnetic, which makes it simple to remove them from water following cleanup (Ahmad et al. 2015). Fe_3O_4 nanoparticles were created using a variety of methods, but the chemical co-precipitation method is the most often used because it is simple to use and allows for easy size control by altering the salt content and reaction conditions (Dąbrowski 2001). Fe_3O_4 nanoparticles are produced in the chemical co-precipitation process by the reaction of ferrous and ferric ions with a base. Because of their inherent magnetic characteristics, bare magnetite nanoparticles tend to aggregate (Gupta et al. 2000), therefore it's important to coat them with a chemical substance to improve their physical stability and catalytic capabilities.

10.3.1.1 Gold Nanoparticle

Au nanoparticles are employed in the administration of drugs, gene delivery, and the treatment of some malignancies. Au nanoparticles efficiently contribute to dye degradation. Au nanoparticles produced by *Paderia foetida* Linn dramatically destroyed

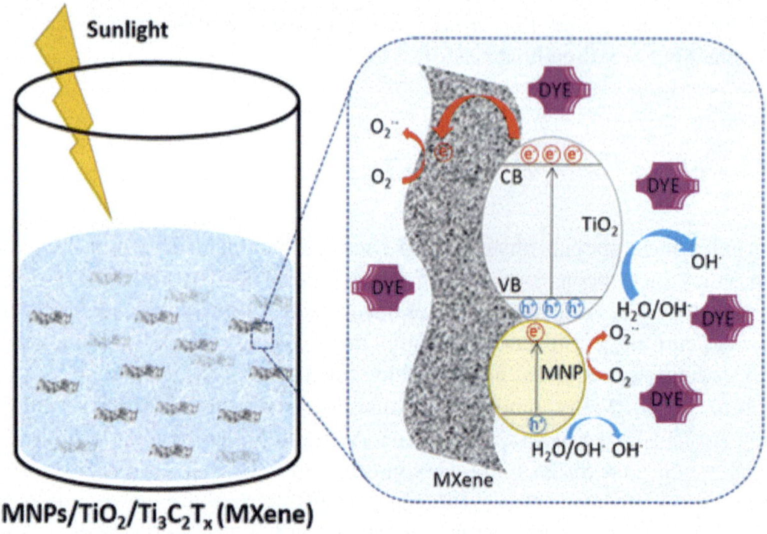

Fig. 10.2 Photocatalytic decomposition of organic dyes is achieved using composites of AgNPs/TiO_2/Ti_3C_2Tx MXene under both UV and solar light (Reprinted with permission from Othman et al.)

toxic colors such as methylene blue and rhodamine B in less than 12 min (Bhuyan et al. 2017). Organic dyes like methyl orange, methylene blue, and straight blue were destroyed by *Sterculia acuminata*-mediated Au nanoparticles; the reaction time was found to be 4 min (Kumari et al. 2016).

10.3.1.2 Silver Nanoparticles

Ag nanoparticles have outstanding anti-microbial and catalytic capabilities, which make them widely used. In addition to being utilized in a variety of industries, including biological labeling, catalysis, biosensors, optoelectronics, and photography, silver metal can be employed to create stable nanoparticles. Ag nanoparticles have a good capacity to degrade dyes. In the presence of a catalyst, Ag nanoparticles produced from *Actinidia deliciosa* fruit extract totally broke down the hazardous dye methylene blue in 33 min (Naraginti and Li 2017). Ag nanoparticles produced by the biomass of *Parkia roxburghii* leaf extract completely degraded methylene blue and rhodamine B in 12 min (Paul et al. 2016).

Kumar et al. utilized *Ulva lactuca*, a type of seaweed, to synthesize silver nanoparticles. These nanoparticles were then employed for the photocatalytic degradation of methyl orange dye. The silver nanoparticles were observed to have a spherical morphology, with an average size of around 48.59 nm, as validated by high-resolution scanning electron microscopy (HR-SEM). The nanoparticles exhibit a high level of stability in the colloidal solution, as indicated by the negative zeta potential value

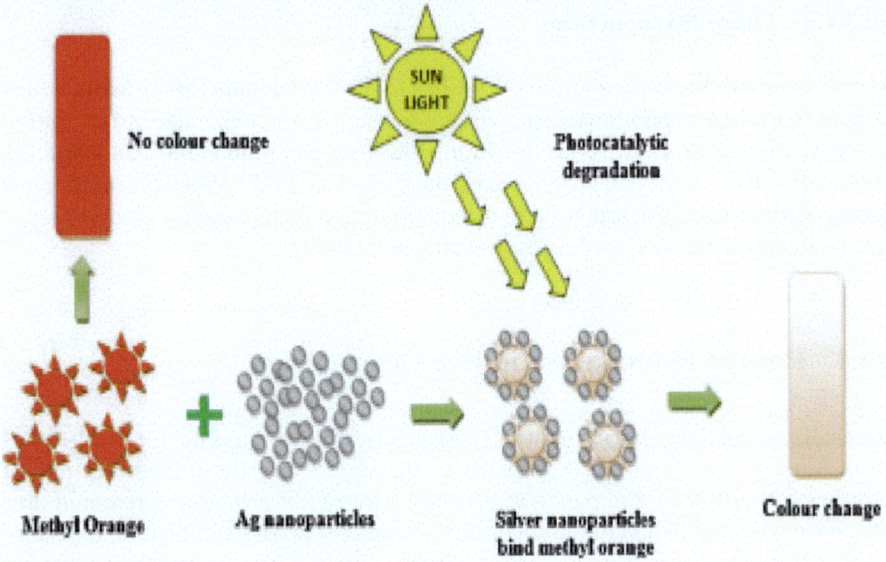

Fig. 10.3 Utilization of silver (Ag) nanoparticles derived from Ulva lactuca for the photocatalytic degradation of methyl orange dye (Kumari et al. 2016)

of -34 mV. Additionally, the spectrophotometric measurement of the photocatalytic degradation of methyl orange was conducted utilizing silver as a nanocatalyst under visible light exposure (Fig. 10.3). The findings demonstrated that silver nanoparticles produced with the use of *U. lactuca* were highly effective in decomposing methyl orange (Kumari et al. 2016).

10.3.1.3 Zinc Oxide Nanoparticles

The large surface area, non-toxicity, stable structure, photosensitivity, increased reactivity, and affordability of ZnO nanoparticles have led to their widespread application (Xie et al. 2011). Zinc oxide (ZnO) nanoparticles have been employed as strong photocatalysts in dye degradation (Peng et al. 2013). It has been discovered that *Artocarpus heterophyllus* leaf extract contains the necessary components to create ZnO nanoparticles, which completely broke down rose bengal dye in just one hour (Jain et al. 2014). The fungus *Aspergillus* sp. produces ZnO nanoparticles by biological means, which efficiently break down the methylene blue dye.

10.3.1.4 Other Nanoparticles

Metal nanoparticles such as nickel (Ni), silica (Si), molybdenum (Mb), titanium (Ti), copper (Cu), and others have emerged due to their distinct characteristics. *Catharanthus roseus* leaf extract was used for producing palladium (Pd) nanoparticles, which effectively degraded phenol red (Kalaiselvi et al. 2015). When *La gerstroemia* leaves were utilized for synthesis of zirconium oxide nanoparticles, methyl orange dye degradation reached 94.5% (Saraswathi et al. 2017).

10.4 Role of Biopolymers in Dye Degradation

Extensive research has been conducted to develop effective ways for removing colors from water-based solutions, in response to the significant rise in the utilization of industrial dyes during the past few decades. More than a thousand research articles have been published in the scientific community on dye elimination technologies, indicating a significant need for innovative procedures. Adsorption, advanced oxidation processes, chemical coagulation-flocculation, electrochemical treatment, membrane filtration, biological treatment, reverse micelle extraction, and other techniques represent the latest advancements in dye removal technologies (Ahmad et al. 2015).

Despite the development of several dye removal technologies, there are still many limits that need to be solved. The primary disadvantages of commonly employed color elimination methods are their exorbitant personnel expenses, elevated manufacturing expenses, and intricate nature. Because they depend on materials that cannot be renewed and do not break down naturally, several of these technologies raise concerns about their potential impact on the environment. A prime example of such a resource is petroleum. Moreover, due to the involvement of multiple sequential processes and the use of various perilous compounds, such as organic solvents, the chemically synthesized materials are virtually devoid of any practical value (Inyinbor et al. 2017). Recently, natural polymers, also known as biopolymers, such as cellulose, chitin, and chitosan, have gained attention for their abundance, affordability, and adjustable properties, including surface area, pore size, pore volume, ease of handling, and environmental friendliness. These polymers are being considered for their potential in dye removal technologies. In addition to their biodegradability, biopolymers provide an additional advantage over synthetic materials: they do not pose any harm to the environment or living organisms (Sharma and Kaur 2018). Biopolymers are viable alternatives to materials derived from petroleum or chemical processes due to their biocompatibility and renewable origin. The surface modification of biopolymers is relatively straightforward and does not compromise the biological and physicochemical properties of the surface due to the presence of functional groups such as hydroxyl (–OH), amine ($-NH_2$), carboxyl (–COO–), and amide ($-NHCOCH_3$). Surface-modified biopolymers exhibit enhanced compatibility, rigidity, pliability, heat and chemical resistance, as well as rapid reactivity (Kyzioł and Kyzioł

2018). This article provides an overview of the latest advancements in three primary methods for removing dyes using biopolymer-based technologies: adsorption and advanced oxidation processes (AOPs) (Strnad and Zemljič 2023).

10.4.1 Adsorption

Adsorption is a surface phenomenon when a substance, such as a molecule, ion, or atom, accumulates on a solid surface from its surrounding gaseous or liquid medium. The adsorbent is the solid surface where the process of adsorption occurs, whereas the adsorbate is the substance that accumulates on the surface of the solid (Ahmad et al. 2015). There are two types of adsorption processes: chemisorption, which involves the chemical adsorption of adsorbates onto the adsorbent surface, and physisorption, which involves the physical adsorption of adsorbates upon the adsorbent surface (Dąbrowski 2001).

Due to their comparatively simple design, safe handling, high treatment efficiency, affordability, and sludge-free cleaning operation, adsorption technologies have been widely used for the dye removal Process (Gupta et al. 2000). Research has investigated the elimination of dyes from water solutions utilizing different types of biological and polymeric substances that can adsorb them. These substances include activated carbon (AC), clays, zeolites, alumina, silica gel, composites, and biomasses (Malik 2004).

Reports indicate that cellulose, chitin, and chitosan, which are biopolymer-based adsorbents, have been used to remove various colors from aqueous solutions and industrial effluents. Extensive study has been conducted on the elimination of industrial dyes using different adsorbents based on biopolymers. However, the primary objective of this section is to offer a concise summary of the most thoroughly studied types of adsorbents. Biopolymer-based adsorbents such as AC, aerogels, hydrogels, microspheres, beads, sponges, metal-oxide and polymeric composites, and nanofibrils have been extensively employed for the purpose of removing dyes (Baysal et al. 2018).

10.4.2 Advanced Oxidation Process

In general, advanced oxidation processes (AOPs) are thought to be among the most enticing and successful dye removal strategies for treating surface water and groundwater soil remediation. William Glaze and associates first presented this AOP technology in 1987 (Dassanayake et al. 2018) (Wols and Hofman-Caris 2012). AOPs are predicated on the in-situ creation of potent oxidants, such as sulphate radicals, hydroxyl radicals, superoxide, and peroxides, for the chemical oxidation of stubborn organic molecules and the elimination of specific inorganic contaminants (Lofrano and Meric 2016).

Over the past few years, there has been a noticeable increase in interest in the use of biopolymers in AOPs. Biopolymers, such as chitin, cellulose, and chitosan, are frequently used as solid AOP substrates. Moreover, potent reducing and stabilizing agents, biopolymers have surface hydroxyl groups. Additionally, biopolymers can stop nanoparticles from self-aggregating while maintaining their morphology and structure. This section reviewed new applications of cellulose, chitin, and chitosan—three major biopolymer-based substrates—in semiconductor photocatalytic and catalytic oxidation processes for dye removal (Ogundare and van Zyl 2019).

MC and TiO_2+MC composites made from banana pseudostems were produced, and the photocatalytic degradation of cationic (methylene blue and methyl violet) and anionic (acid violet) dyes was studied (Lofrano and Meric 2016). Zhao et al. created a self-assembling gel of iron (Fe)-chitosan (CS)/montmorillonite nanosheets (FeCS/MMTNS) to study the degradation of methylene blue (MB) in the presence of H_2O_2 and visible light.

Biopolymers are not directly involved in the oxidative breakdown mechanisms during the dye removal process by AOPs. Biopolymers, however, are important because they act as a matrix that holds the oxidants in place. Increased surface characteristics of biopolymer matrices, such as large surface area and micropore volumes, offer more active sites for the adsorption of dyes. Effective dye oxidation requires the adsorption of dyes onto the catalytic surface (Lofrano and Meric 2016).

10.5 Role of Biopolymer in Dye Degradation

Biopolymers play a crucial role in dye degradation processes owing to their inherent biodegradability and diverse functional groups. These polymers, derived from natural sources such as cellulose, chitosan, alginate, and starch, possess unique chemical properties that make them effective agents in removing and degrading dyes from wastewater. Biopolymers can adsorb dye molecules onto their surfaces through electrostatic interactions, hydrogen bonding, and other mechanisms, effectively sequestering them from aqueous solutions. This adsorption process not only aids in the removal of dyes but also sets the stage for subsequent degradation mechanisms.

Furthermore, biopolymers serve as substrates for enzymatic degradation by facilitating the activity of dye-degrading enzymes produced by microorganisms. Enzymes such as laccase, peroxidase, and lignin peroxidase can break down dye molecules into smaller, less harmful compounds. Biopolymers like cellulose and lignin present in natural materials provide a supportive matrix for the growth and activity of these microorganisms, thereby enhancing the overall efficiency of dye degradation processes. Additionally, biopolymer-based materials can be functionalized with photocatalytic nanoparticles or incorporated into membrane filtration systems to synergistically enhance dye removal and degradation. In summary, biopolymers offer a sustainable and effective solution for mitigating the environmental impact of dye pollutants in wastewater streams.

Due to the extensive functionality of biopolymers, they find application in various aspects of wastewater treatment for several reasons: (Christina et al. 2024).

(i) Enhanced treatment efficacy: Biopolymers contribute to improving the efficiency and performance of wastewater treatment processes. They aid in the removal of suspended solids, heavy metals, and organic matter from wastewater, thereby mitigating the environmental repercussions of discharge.
(ii) Eco-friendly nature: Biopolymers are frequently biodegradable and sourced from natural origins, rendering them a more environmentally benign alternative to synthetic polymers or chemical agents.
(iii) Cost-effectiveness: Biopolymers often offer a more economical solution compared to synthetic polymers or chemicals utilized in wastewater treatment. They can be derived from waste materials like seaweed, thereby reducing production costs.
(iv) Sustainability: Integration of biopolymers into wastewater treatment practices fosters sustainable methodologies and diminishes dependence on non-renewable resources. Through biopolymer utilization, we can curtail environmental impact and advance a circular economy model wherein waste materials are repurposed or recycled.

Overall, employing biopolymers in wastewater treatment presents a viable, cost-efficient, and environmentally sound approach to addressing wastewater concerns and mitigating environmental discharge impacts.

Peramune et al., stated that the use of biopolymers and its derivatives in their nanoscale level has enhanced its properties due to its unique surface area and physiochemical stability. These nanobiopolymers were extensively used in food industry, energy sector, biomedical Industry and also in waste water treatment. Among these polysaccharide-based nanoparticles such as, nanochitin, nanocellulose and nanochitosan are widely used in dye degradation processes. These compounds were used along with Advanced Oxidation Process (AOP) specifically photocatalysis, were involved in dye removal applications. Photocatalysis involves use of photocatalyst which convert solar energy to chemical energy, thereby oxidizing organic pollutants. When combining Cellulose biopolymer with semiconductor photocatalyst metal oxide, it not only provides efficient contaminant adsorption, and also prevents leaching of catalytic particles into the treated water. Also, the presence of cellulose possess good thermal strength and stability to the composite material. The most commonly used metal oxides along with cellulose biopolymer for the purpose of dye degradation are TiO_2 and MnO_2 (Peramune et al. 2022).

A research work reported that cellulose nanofibers (CNFs) coated with MnO_2 causes effective decolorization of methylene blue MB dye within 2 min of application and the degree of degradation is more at lower pH ranges. Also, under dark conditions, MnO_2/CNF nanocomposites causes 95% degradation of indigo carmine dye within 60 min (Oliveira et al. 2021).

In addition to cellulose, chitin and chitosan biopolymers have also been demonstrated to possess exceptional color degradation characteristics. While chitin itself does not have a highly effective dye degradation process, a study has demonstrated that a hybrid material called Chitin-MnO_2 is capable of achieving 100% degradation of MB (methylene blue) from an aqueous solution in about 2.5 min. Chitosan, a very effective biopolymer for dye degradation, can be enhanced by combining it with a filler molecule like cellulose. This combination improves the material's tensile strength and helps to address its limitations, such as its susceptibility to mechanical forces and certain chemical conditions. The utilization of chitosan/metal-oxide based nanocomposites exhibits the greatest level of degradation for the anionic dye (reactive red 2) in a mixture of dyes, including the anionic dye (reactive red 2), cationic dye (methylene blue), and zwitterionic dye (Rhodamine B). Furthermore, the combination of Chitosan/cellulose with TiO_2 has been identified as a highly effective option for the elimination of dyes (Farzana and Meenakshi 2014).

Other example of biopolymers includes gelatin, silk fibroin, alginate, starch, zein, albumin, and lignin are shown to have some dye degradation properties. Gelatin when combined with Zirconium (Zr) phosphate nanocomposite can be shown to degrade fast green dye by means of photocatalytic degradation. Gelatin/Zr nanocomposite having a high adsorption removal capacity, can be shown to degrade nearly 89 and 44% of dye under solar light irradiation and dark conditions respectively. Silk fibroin, extracted from silk worms has been used for biomedical applications to water purification processes. Silk fibroin when conjugated with Ag and AgCl along with cellulose acetate can be shown to have good dye degradation capabilities. A study revealed that the use of Ag/AgCl—CA/SF (Cellulose acetate—Silk fibroin) has good dye removal capacity due to surface plasmon resonance (SPR) exhibited by silver nanoparticle. When compared to other combinations such Cellulose acetate/Silk fibroin film (CA/SF) and CA/SF, Ag/AgCl—CA/SF has shown to posses excellent dye degradation capacities (Peramune et al. 2022).

Alginate, a component extracted from seaweed has a strong affinity for charged dye molecules. Alginates when combined with photocatalyst as floating beads, helps in absorbing UV light. Alginates were combined with TiO_2 and bentonite clay shows 80% of methylene blue dye removal and this combination has shown to have good recyclability nature for upto7 cycles (Dallabona et al. 2021). Another polysaccharide such as starch, when capped with ZnO quantum dots helps in removal of Rhodamine B dyes (Vidhya et al. 2015).

Santhosh et al., says that the dye adsorption capacity of cellulose was primarily facilitated by physical adsorption and electrostatic attraction between positively charged azo dye compounds and negatively charged cellulose sheets. The cellulose sheets made from water hyacinth by alkali treatment and bleaching to remove hemicellulose and lignin and thereby decreasing cellulose fibre crystallinity. This cellulose sheets efficiently removed reactive red dye, having an adsorption capacity of 210 min and can be used for wastewater dye removal applications. Further, cellulose and its derivatives can be blend with other biopolymers and their efficacy in absorbing heavy metals (Santhosh et al. 2024).

Mennani et al. (2024) says that lignin, a bio-polyphenol when functionalized with cobalt catalyst (Lig-Co) can be used for the degradation of toxic organic dyes such as methylene blue and methyl orange dyes from the simple and binary systems of industrial effluents. The stability of Lig-Co catalyst is retained even after 6 successive catalytic runs of which nearly 80% of activity has been retained. Based on which, in a hybrid binary system, the conversion rate of Methylene blue and Methyl orange dyes under mild conditions were found to be 94.12% and 98.15% respectively (Mennani et al. 2024).

Zheng et al. (2023) states that use of adsorption enhanced catalyst (MnOx-PP) made from biopolymer pomelo peels along with manganese oxide (MnO_2) metal-oxide catalyst can be used for catalytic organic dyes oxidative-degradation. The biopolymer (here plant peel) can be used as an adsorbent for organic pollutant removal due to its porous structure and number if oxygen containing functional groups, also they posses negatively charged polar sites which aids in adsorption of organic dyes in wastewater. The MnOx-PP, when used under a dosage of 0.027 g/L shows better methylene blue (MB) and total carbon content (TOC) removal efficiency of 99.5% and 66.31% respectively, under 60 C. Based on the studies, it shows that the use of MnOx-PP (adsorption enhanced catalyst) shows good efficiency for organic dyes removal than the use of pure biopolymer (PP) or pure catalyst (MnOx) alone.

Christiana et al. (2024) explains that the natural polymers such as Xanthan gum, Guar gum can also be used for contaminants removal from wastewater. In general, Guar gum can be used to remove Acid Blue 193, Direct black 22 and Reactive red 238 dyes with the efficiency of 60%, 70.3% and 78.6% respectively. Also, modified guar gum aids in the removal of cationic carcinogen (Crystal Violet CV) from the environment. According to Mandal et al. (2023), the use of acrylic acid-grafted Guar gum along with silane modified nano clay can be used to absorb 89% of crystal violet dye during pseudo-second order rate reaction. Also, Guar cum can be linked with Glutathione (GSH) and iron oxide (Fe_3O_4) and can be used for the removal of crystal violet and phenol red dyes effectively. The use on Magnetic Nanoparticles MNPs along with guar gum, glutathione and Fe_3O_4 nanochains forms Guar gum glutathione@Fe_3O_4 MNPs. Of which, Glutathione activates functional groups such as carboxyl, amine, carbonyl and hydroxyl groups, thereby helping in the removal of brilliant green dye and crystal violet dyes. The magnetic nanocomposites can effectively remove 97% of crystal violet dye and 79% of phenol red dyes respectively. Xanthan gum, a natural biopolymer, on the other hand has number of hydroxyl groups, which when conjugated with polyaluminium chloride (PAC) [Primary coagulant], aids in the removal of Congo red dye from waste water.

Shan et al., studied a natural biopolymer present in the pods of *Moringa olifera* (MO) can be used as an effective water coagulant and an organic sorbent. The use of MO seed cake reduces the turbidity of waste water by 85–94%. Also, it helps in the removal of heavy metals such as Iron from waste water completely, while other heavy metals such as copper and cadmium were removed by 98% and lead was reduced to 78.1%. On the whole, heavy metals can be completely removed from water samples on using 1% MO seed cake (Shan et al. 2017).

Angelin et al., developed that X-ray photoelectron spectroscopy (XPS) analysis revealed that bio-adsorbent derived from carbonized wattle humus can be used to remove fluoride ions from waste water by chemisorption. Based on the regulation of factors such as pH, contact time, dosage and the inferring ions, the maximum

(98%) of fluoride adsorption was found at $pH > 6$ at five cycles of recycling, both in monolayer and multilayer forms (Angelin et al. 2021).

Grenda et al., defies Quebracho tannin obtained from *Acacia mearnsii* barks and Schinopsisbalansae woods were modified using Mannich amino methylation technique can be used for industrial scale waste water treatment. The effectiveness of such tannins was checked by putting it onto test jars of waste water to check any changes in colour and turbidity of waste water. Upon the treatment of effluent from both single and multiple systems, by using 5 ppm of cationic, 40% charged polyacrylamide, it shows effective reduction of turbidity and colour by 93% and 89% respectively from effluents in pilot plant scale (Grenda et al. 2020).

Wu et al., postulates that the use of Chitosan-Benzaldehyde/SiO_2 (CTA-BZA/SiO_2) helps in the adsorption of organic dyes from the industrial waste water, especially acidic dye (Acid red 88, AR88). In general, Acid red dye can be completely degraded, but the byproducts produced during the process have negative impact on environment. The use of hydrothermal-assisted synthesis of CTA-BZA/SiO_2 adsorbent helps in the decolorization of waste water containing Acidic dye up to 70.94% at pH 4.7 for the duration of 50.4 min. According to the physisorption of AR88 dye on to CTA-BZA/SiO_2 adsorbent, the maximum adsorption capacity for CTA-BZA/SiO_2 for AR88 dye was estimated to be 252.4 mg/g from industrial effluents (Wu et al. 2023).

10.6 Role of Polymeric Nanoparticles in Dye Degradation

Arif et al., 2023 demonstrated the synthesis of a microgel system employing free radical precipitation polymerization. The system consisted of Nisopropylmethacrylamide (NMA) as the monomer and methacrylic acid (MAa) as the comonomer. The P(NMA-MAa) microgels were modified by including Bimetallic (Ag/Ni) nanoparticles through the reduction of silver and nickel salts. Subsequently, $NaBH_4$, which acts as a reductant, is injected along with both Ag^+ and Ni^{2+} ions. In addition, the catalytic efficiency of Ag/Ni–P(NMA-MAa) was evaluated for the breakdown of methyl orange (MOr), congo red (CRe), eriochrome black T (EBlT), and methyl red (MRe) dyes in an aqueous solution. The rate constant (k0) values for MOr, CRe, EBlT, and MRe were 0.925 min^{-1}, 0.486 min^{-1}, 0.540 min^{-1}, and 0.525 min^{-1}, respectively (Arif 2023).

Gomes et al. (2023) produced Nb_2O_5 nanoparticles with the utilization of the Pechini method and precursor disintegration techniques. Subsequently, X-ray diffraction (XRD) analysis detected the existence of niobium pentoxide crystals in the TT, T, H, and non-stoichiometric $Nb_{12}O_{29}$ phases. Calcining samples at 500 °C results in a specific surface area (SSA) of 55–65 $m^2\ g^{-1}$, which enhances their effectiveness as adsorbents and photocatalysts for the degradation of crystal violet. The use of Nb_2O_5 as a catalyst resulted in the photodegradation of crystal violet dye. This process shown synergistic properties, where samples with higher SSA (specific surface area) and crystallization in the TT-phase were able to decolorize the solution in a time frame of

2 h and 30 min. Nevertheless, the combination of H-Nb_2O_5 and $Nb_{12}O_{29}$ exhibited favorable adsorption properties and successfully removed the CV pollutant within a time frame of 2 h and 30 min. This indicates that the effectiveness of adsorption and photodegradation processes heavily relies on the specific crystalline phase of niobium pentoxide (Gomes et al. 2023).

Aziz et al. (2023) nanoparticles can be produced by conventional techniques, including chemical and physical procedures. However, these approaches possess several drawbacks, such as the utilization of hazardous substances, the necessity for elevated temperature and pressure, and the substantial expenses involved. The utilization of *Grewia asiatica* extract was employed to synthesize zinc oxide nanoparticles at optimal conditions, including temperature, concentration, and pH, marking the first instance of such a process. In addition, the rate at which methylene blue degrades was investigated using a photocatalyst. Within a span of 70 min, the methylene blue saw a 92% reduction, with a rate constant of 0.0141 min^{-1}.

The study conducted by Iqbal et al., focuses on creating poly (*N*-isopropylmethacrylamide-co-methacrylic acid) [p(NMA)] microgels as microreactors. These microgels are used to manufacture stable Ag nanoparticles, which are then utilized for catalytic reductive degradation of dyes. The p(NMA) microgel and Ag–p(NMA) hybrid microgel systems were analyzed using Fourier transform infrared and dynamic light scattering techniques. The kapp value exhibited a progressive increase in direct correlation with the quantity of hybrid microgel employed as a catalyst (Fig. 10.4). Through a comparison of the kapp values of the above dyes, it was found that the Ag–p(NMA) hybrid microgel exhibited greater efficacy in degrading MO dye compared to the catalytic breakdown of CR and AY (Iqbal et al. 2020).

Preethi et al., suggested a method for producing an environmentally friendly chitosan/zinc oxide (CS/ZnO) nanocomposite using a bio-inspired methodology and *S. lycopersicum* leaf extract. Cotton fabrics treated with CS/ZnO nanocomposite exhibited antibacterial efficacy against *S. aureus*, *B. subtilis*, and *E. coli*. Moreover, the CS/ZnO nanocomposite acted as a photocatalyst for the degradation of Congo red in the presence of sunlight. Ultimately, the CS/ZnO nanocomposite that is produced can serve as an effective antibacterial agent in the textile sector, as well as a catalyst for the breakdown of colors by photocatalysis (Preethi et al. 2020).

Mousa et al., enhanced the efficacy of this method in terms of photodegradation and antibacterial characteristics in wastewater treatment. Hence, the resulting material primarily consists of titanium dioxide (TiO_2) and zinc oxide (ZnO), which are photocatalytic nanoparticles. A composite material called cellulose acetate CA@TiO_2 NPs (CTO) was created by employing electrospinning. This involved incorporating a core TiO_2 NP into a polymeric matrix. The resulting material was then placed in a suspension solution of ZnO NPs, causing ZnO NPs to be deposited on the surface of the nanofibers. This new material is referred to as CTZO. In addition, the composite nanoparticles exhibited significant photocatalytic degradation capabilities, with around 25% and 13% photodegradation rates for MB and MR dyes, respectively, under direct sunlight. The enhanced degradation of the dye is attributed to the formation of an S–S heterojunction structure with a lower bandgap energy

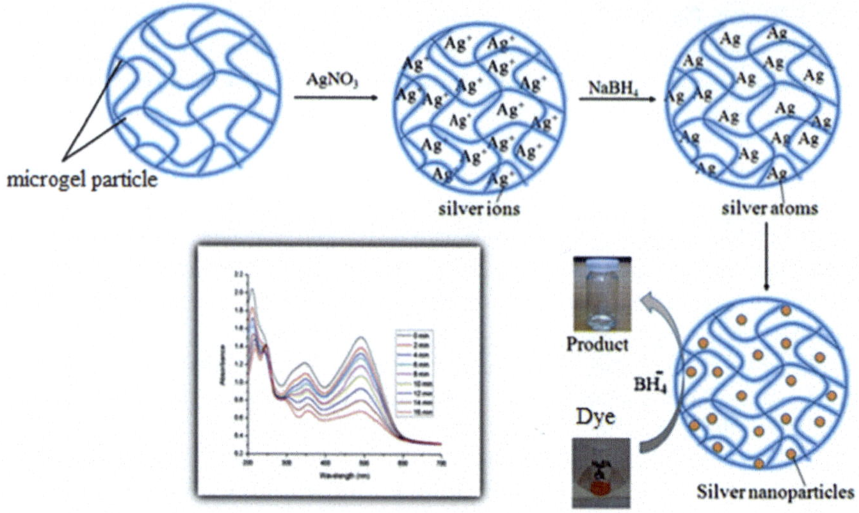

Fig. 10.4 Polymer hydrogels are used to stabilize silver nanoparticles for the purpose of catalytically degrading azo dyes

(1.17 eV) compared to the larger bandgap energy (3.52 eV) of TiO_2. This heterojunction structure facilitates electron transport, hence promoting photodegradation (Mousa et al. 2021).

A study conducted by Aziz et al., involved the production of chitosan-zinc sulfide nanoparticles (CS-ZnS-NPs) for use as a photocatalyst in the degradation of harmful pigments. Consequently, the degradation reactions exhibited pseudo-first-order kinetics. The rate constants (k) for Acid Brown 98 and Acid Black 234 are 0.01464 and 0.04096 min $^{-1}$, respectively. The R^2 values for Acid Brown 98 and Acid Black 234 are 0.98891 and 0.99406, correspondingly. Additionally, The CS-ZnS-NPs were readily retrieved and reused in four consecutive batches. These data indicate that CS-ZnS-NPs are an efficient, economical, and prospective photocatalyst for the degradation of contaminants over numerous cycles (Aziz et al. 2020).

10.7 Conclusion

The use of biopolymers and polymeric nanoparticles for dye degradation has attracted considerable interest because of their environmentally benign properties and exceptional effectiveness in wastewater treatment. Biopolymers, sourced from plants, animals, and microorganisms, have natural biodegradability and low toxicity, making them well-suited for environmental applications. Polymers, whether they occur naturally or are artificially created, have been modified into nanoparticles in order to increase their surface area, reactivity, and stability. This modification has resulted in improved performance when it comes to breaking down dyes. Biopolymers such as chitosan, alginate, and cellulose have high dye adsorption capabilities due to their plentiful functional groups and porous architectures. They can efficiently adsorb dyes from water-based solutions by utilizing mechanisms like as electrostatic interactions, hydrogen bonding, and surface complexation. Polymeric nanoparticles possess unique benefits in the degradation of dyes as a result of their adjustable characteristics and elevated surface-to-volume ratio. Biopolymer-based nanocomposites combine the benefits of both components, resulting in improved abilities to adsorb substances and break down materials. Ongoing research endeavors to enhance the efficiency, analysis, and utilization of nanomaterials derived from biopolymers have significant potential to tackle the issues related to dye contamination and promote the development of sustainable methods for treating wastewater.

References

Ahmad A, Mohd-Setapar SH, Chuong CS, Khatoon A, Wani WA, Kumar R et al (2015) Recent advances in new generation dye removal technologies: novel search for approaches to reprocess wastewater. RSC Adv 5(39):30801–30818

Al-Tohamy R, Ali SS, Li F, Okasha KM, Mahmoud YA-G, Elsamahy T et al (2022) A critical review on the treatment of dye-containing wastewater: ecotoxicological and health concerns of textile dyes and possible remediation approaches for environmental safety. Ecotoxicol Environ Saf 231:113160

Angelin A, Kalpana M, Govindan K, Kavitha S (2021) Characterizations and fluoride adsorption performance of wattle humus biosorbent. Environ Sci Pollut Res 1–14

Ardila-Leal LD, Poutou-Piñales RA, Pedroza-Rodríguez AM, Quevedo-Hidalgo BE (2021) A brief history of colour, the environmental impact of synthetic dyes and removal by using laccases. Molecules 26(13):3813

Arif M (2023) Catalytic degradation of azo dyes by bimetallic nanoparticles loaded in smart polymer microgels. RSC Adv 13(5):3008–3019

Aziz A, Ali N, Khan A, Bilal M, Malik S, Ali N et al (2020) Chitosan-zinc sulfide nanoparticles, characterization and their photocatalytic degradation efficiency for azo dyes. Int J Biol Macromol 153:502–512

Aziz A, Memon Z, Bhutto A (2023) Efficient photocatalytic degradation of industrial wastewater dye by *Grewia asiatica* mediated zinc oxide nanoparticles. Optik 272:170352

Baysal M, Bilge K, Yılmaz B, Papila M, Yürüm Y (2018) Preparation of high surface area activated carbon from waste-biomass of sunflower piths: kinetics and equilibrium studies on the dye removal. J Environ Chem Eng 6(2):1702–1713

Belpaire C, Reyns T, Geeraerts C, Van Loco J (2015) Toxic textile dyes accumulate in wild European eel *Anguilla anguilla*. Chemosphere 138:784–791

Bhuyan B, Paul A, Paul B, Dhar SS, Dutta P (2017) *Paederia foetida* Linn. promoted biogenic gold and silver nanoparticles: synthesis, characterization, photocatalytic and in vitro efficacy against clinically isolated pathogens. J Photochem Photobiol B: Biol 173:210–215

Braun H (1983) Particle size and solubility of disperse dyes. Rev Prog Color Relat Top 13(1):62–72

Christina K, Subbiah K, Arulraj P, Krishnan SK, Sathishkumar P (2023) A sustainable and eco-friendly approach for environmental and energy management using biopolymers chitosan, lignin and cellulose—a review. Int J Biol Macromol 128550

Dąbrowski A (2001) Adsorption—from theory to practice. Adv Coll Interface Sci 93(1–3):135–224

Dassanayake RS, Rajakaruna E, Abidi N (2018) Preparation of aerochitin-TiO_2 composite for efficient photocatalytic degradation of methylene blue. J Appl Polym Sci 135(8):45908

Dassanayake RS, Acharya S, Abidi N (2021) Recent advances in biopolymer-based dye removal technologies. Molecules 26(15):4697

Farzana MH, Meenakshi S (2014) Synergistic effect of chitosan and titanium dioxide on the removal of toxic dyes by the photodegradation technique. Ind Eng Chem Res 53(1):55–63

Ferraz ER, Grando MD, Oliveira DP (2011) The azo dye disperse orange 1 induces DNA damage and cytotoxic effects but does not cause ecotoxic effects in *Daphnia similis* and *Vibrio fischeri*. J Hazard Mater 192(2):628–633

Flórez M, Cazón P, Vázquez M (2023) Selected biopolymers' processing and their applications: a review. Polymers 15(3):641

Gomes GH, Olusegun SJ, Gabriel JB, Costa RC, Mohallem ND (2023) The role of crystalline Nb_2O_5 nanoparticles for enhanced dye adsorption and photodegradation. Ceram Int 49(4):6164–6176

Grenda K, Arnold J, Gamelas JA, Rasteiro MG (2020) Up-scaling of tannin-based coagulants for wastewater treatment: performance in a water treatment plant. Environ Sci Pollut Res 27:1202–1213

Guerra E, Llompart M, Garcia-Jares C (2018) Analysis of dyes in cosmetics: challenges and recent developments. Cosmetics 5(3):47

Gupta VK, Mohan D, Sharma S, Sharma M (2000) Removal of basic dyes (rhodamine B and methylene blue) from aqueous solutions using bagasse fly ash. Sep Sci Technol 35(13):2097–2113

Iqbal S, Zahoor C, Musaddiq S, Hussain M, Begum R, Irfan A et al (2020) Silver nanoparticles stabilized in polymer hydrogels for catalytic degradation of azo dyes. Ecotoxicol Environ Saf 202:110924

Jain N, Bhargava A, Panwar J (2014) Enhanced photocatalytic degradation of methylene blue using biologically synthesized "protein-capped" ZnO nanoparticles. Chem Eng J 243:549–555

Jyoti K, Singh A (2016) Green synthesis of nanostructured silver particles and their catalytic application in dye degradation. J Genetic Eng Biotechnol 14(2):311–317

Kalaiselvi A, Roopan SM, Madhumitha G, Ramalingam C, Elango G (2015) Synthesis and characterization of palladium nanoparticles using *Catharanthus roseus* leaf extract and its application in the photo-catalytic degradation. Spectrochim Acta Part A Mol Biomol Spectrosc 135:116–119

Kavitha S, Umadevi M, Janani S, Balakrishnan T, Ramanibai R (2014) Fluorescence quenching and photocatalytic degradation of textile dyeing waste water by silver nanoparticles. Spectrochim Acta Part A Mol Biomol Spectrosc 127:115–121

Kumari RM, Thapa N, Gupta N, Kumar A, Nimesh S (2016) Antibacterial and photocatalytic degradation efficacy of silver nanoparticles biosynthesized using *Cordia dichotoma* leaf extract. Adv Nat Sci: Nanosci Nanotechnol 7(4):045009

Kyzioł A, Kyzioł K (2018) Surface functionalization with biopolymers via plasma-assisted surface grafting and plasma-induced graft polymerization—materials for biomedical applications. Biopolym Grafting 115–151

Lofrano G, Meric S (2016) A comprehensive approach to winery wastewater treatment: a review of the state-of the-art. Desalin Water Treat 57(7):3011–3028

Malik P (2004) Dye removal from wastewater using activated carbon developed from sawdust: adsorption equilibrium and kinetics. J Hazard Mater 113(1–3):81–88

Mandal S, Hwang S, Shi SQ (2023) Guar gum, a low-cost sustainable biopolymer, for wastewater treatment: a review. Int J Biol Macromol 226:368–382

Matei E, Predescu AM, Râpă M, Țurcanu AA, Mateş I, Constantin N et al (2022) Natural polymers and their nanocomposites used for environmental applications. Nanomaterials 12(10):1707

Mennani M, Kasbaji M, Benhamou AA, Ablouh E-H, Grimi N, El Achaby M et al (2024) Lignin-functionalized cobalt for catalytic reductive degradation of organic dyes in simple and hybrid binary systems. Chemosphere 350:141098

Mohan S, Oluwafemi OS, Kalarikkal N, Thomas S, Songca SP (2016) Biopolymers–application in nanoscience and nanotechnology. Recent Adv Biopolym 1(1):47–66

Mousa HM, Alenezi JF, Mohamed IM, Yasin AS, Hashem A-FM, Abdal-Hay A (2021) Synthesis of TiO_2@ ZnO heterojunction for dye photodegradation and wastewater treatment. J Alloys Compd 886:161169

Naraginti S, Li Y (2017) Preliminary investigation of catalytic, antioxidant, anticancer and bactericidal activity of green synthesized silver and gold nanoparticles using *Actinidia deliciosa*. J Photochem Photobiol, B 170:225–234

Ogundare SA, van Zyl WE (2019) A review of cellulose-based substrates for SERS: fundamentals, design principles, applications. Cellulose 26:6489–6528

Oliveira LV, Limousy L, Bennici S, Josien L, Hajjar-Garreau S, Goddard M-L et al (2021) Facile elaboration of wet cellulose film as catalyst support of MnOx nanoparticles for the catalytic oxidation of dyes in absence of light. Clean Technol 3(2):288–298

Paul B, Bhuyan B, Purkayastha DD, Vadivel S, Dhar SS (2016) One-pot green synthesis of gold nanoparticles and studies of their anticoagulative and photocatalytic activities. Mater Lett 185:143–147

Paz A, Carballo J, Pérez MJ, Domínguez JM (2017) Biological treatment of model dyes and textile wastewaters. Chemosphere 181:168–177

Peng Y, Ji J, Zhao X, Wan H, Chen D (2013) Preparation of ZnO nanopowder by a novel ultrasound assisted non-hydrolytic sol–gel process and its application in photocatalytic degradation of CI Acid Red 249. Powder Technol 233:325–330

Peramune D, Manatunga DC, Dassanayake RS, Premalal V, Liyanage RN, Gunathilake C et al (2022) Recent advances in biopolymer-based advanced oxidation processes for dye removal applications: a review. Environ Res 215:114242

Santhosh AS, Umesh M, Kariyadan S, Suresh S, Salmen SH, Alharb SA et al (2024) Fabrication of biopolymeric sheets using cellulose extracted from water hyacinth and its application studies for reactive red dye removal. Environ Res 240:117466

Saraswathi VS, Kamarudheen N, BhaskaraRao K, Santhakumar K (2017) Phytoremediation of dyes using *Lagerstroemia speciosa* mediated silver nanoparticles and its biofilm activity against clinical strains *Pseudomonas aeruginosa*. J Photochem Photobiol, B 168:107–116

Sarkar S, Banerjee A, Halder U, Biswas R, Bandopadhyay R (2017) Degradation of synthetic azo dyes of textile industry: a sustainable approach using microbial enzymes. Water Conserv Sci Eng 2:121–131

Shan TC, Matar MA, Makky EA, Ali EN (2017) The use of *Moringa oleifera* seed as a natural coagulant for wastewater treatment and heavy metals removal. Appl Water Sci 7:1369–1376

Sharma S, Kaur A (2018) Various methods for removal of dyes from industrial effluents-a review. Indian J Sci Technol

Shwe MT, Dimaculangan MM, De Luna MDG (2014) Decolourization of simulated dye wastewater containing reactive blue 19 (RB19) by the electro-Fenton reaction in the presence of metal oxide-coated electrodes. Adv Mater Res 858:40–45

Strnad S, Zemljič LF (2023) Cellulose–chitosan functional biocomposites. Polymers 15(2):425

Tikhomirova TI, Ramazanova GR, Apyari VV (2018) Effect of nature and structure of synthetic anionic food dyes on their sorption onto different sorbents: peculiarities and prospects. Microchem J 143:305–311

Tkaczyk A, Mitrowska K, Posyniak A (2020) Synthetic organic dyes as contaminants of the aquatic environment and their implications for ecosystems: a review. Sci Total Environ 717:137222

Vanaamudan A, Soni H, Sudhakar PP (2016) Palm shell extract capped silver nanoparticles—as efficient catalysts for degradation of dyes and as SERS substrates. J Mol Liquids 215:787–794

Varjani S, Rakholiya P, Ng HY, You S, Teixeira JA (2020) Microbial degradation of dyes: an overview. Biores Technol 314:123728

Vidhya K, Saravanan M, Bhoopathi G, Devarajan V, Subanya S (2015) Structural and optical characterization of pure and starch-capped ZnO quantum dots and their photocatalytic activity. Appl Nanosci 5:235–243

Wols B, Hofman-Caris C (2012) Review of photochemical reaction constants of organic micropollutants required for UV advanced oxidation processes in water. Water Res 46(9):2815–2827

Wu R, Abdulhameed AS, Yong SK, Li H, ALOthman ZA, Wilson LD et al (2023) Functionalization of chitosan biopolymer with SiO_2 nanoparticles and benzaldehyde via hydrothermal process for acid red 88 dye adsorption: box-Behnken design optimization. Int J Biol Macromol 247:125806
Xie J, Li Y, Zhao W, Bian L, Wei Y (2011) Simple fabrication and photocatalytic activity of ZnO particles with different morphologies. Powder Technol 207(1–3):140–144
Zheng JY, He J, Han CB, Huang G, Sun BC, Zhao WK et al (2023) Adsorption-enhanced catalytic oxidation for long-lasting dynamic degradation of organic dyes by porous manganese-based biopolymeric catalyst. Int J Biol Macromol 237:124152

Chapter 11
Optimization of Chitin Extraction Using Fish Scale and Removal of Drimarene Yellow and Methyl Orange from Aqueous Solution

P. Muthukumaran, K. Niranjan, B. Mohith, T. Vishal, S. Ramasamy, and Aravind Jeyaseelan

Abstract Water is essential for resource, and it is used for day-to-day life. Due to increased industrial process, both fresh water and marine water ecosystem contaminated. In industrial effluents, huge variety and classes of both organic and inorganic dyes, bleaching components used and it is directly affects natural water bodies by polluting the water quality. Literature survey showed that, various types of biopolymers which include cellulose, chitosan, chitin, alginate, xanthan gum, dextrin, carrageenan, lignin, protein are as adsorbents to remove dye. In this experimentation, initially, chitin was extracted from fish scale by optimizing demineralization, deproteination, and contact time. To screen optimum condition, Normality of HCL, Molarity of NaOH and contact time (Hours) studied to extract chitin from fish scale. Based on yield, fish scale and chitin were used as natural bio adsorbent to remove methylene blue and acid green from aqueous solution.

Keywords Fish scale · Extraction · Chitin · Dye removal · Optimization

P. Muthukumaran (✉) · K. Niranjan · B. Mohith · T. Vishal · S. Ramasamy
Department of Biotechnology, Kumaraguru College of Technology, Coimbatore, Tamil Nadu 641049, India
e-mail: m.kumaran005@gmail.com; muthukumaran.p.bt@kct.ac.in

A. Jeyaseelan
Department of Biotechnology, Saveetha School of Engineering, Saveetha Institute of Medical and Technical Sciences (SIMATS), Thandalam, Chennai, Tamil Nadu 602 105, India

© The Author(s), under exclusive license to Springer Nature Switzerland AG 2025
A. Jeyaseelan and K. Murugasen (eds.), *Sustainable Environmental Remediation: Avenues in Nano and Biotechnology*, Environmental Science and Engineering,
https://doi.org/10.1007/978-3-031-78483-5_11

11.1 Introduction

Dye is one of the major threats for any ecosystem, especially manmade dyes and its highly challengeable to remove and destroy dye and its impact on any ecosystem. Each day, still several industries discharge wastewater with huge volume and concentration of dyes. Dyes always showcase complexity and diversity in chemical structure. Mostly, dye consists of chromophores and auxochromes components and which is the main reason to give intensity and color to dye (Bharathi and Ramesh 2013). Moreover, presence of both synthetic and aromatic structure complexity in dye, it shows inhibitory effect against biodegradation (Şayan 2006). The main objective of this experiment is to extract the chitin\chitosan from the fish waste as much as possible and to optimize the conditions for the extraction process. Extracting chitin from fish waste is a practical and sustainable way to utilize the byproducts of the fishing industry and reduce waste. Chitin is a natural biopolymer found in the exoskeletons of crustaceans and the cell walls of fungi. The extracted chitin finds applications in various industries. Chitosan, derived from chitin, possesses unique properties used in wound dressings, drug delivery systems, and water treatment. As proposed by Takka and Gürel (2010), either chitosan or alginate beads can be used as alternative for dye removal studies. Where in case of chemically synthesized and characterized chitosan/sodium alginate/carboxymethyl cellulose beads most effective adsorbent widely used to remove textile dye (Devi et al. 2014). Its is necessity to remove certain anionic dyes like azo dyes from wastewater and it can be recycled by alginate-chitosan beads as alternative adsorbent (Semeraro et al. 2017).

To remove maximum dye removal, few common adsorbents like activated carbon were used, but it possesses high running cost and shows challenges when activated carbon regenerated (Weng et al. 2009). The main objective this experimental study is to find out optimal condition to remove maximum minerals and protein from fish scale and ensure maximal release of chitin from fish scale. Based on the adsorption capacity of chitin, extracted chitin converted into chitosan beads and same was compared with activated charcoal, chitin flakes, chitin powder and chitosan beads to remove two anionic dye Drimarene Yellow and Methyl orange.

11.2 Materials and Method

11.2.1 Sample Collection

Fish scale collected from local fish market at Ukkadum, Coimbatore, Tamil Nadu to extract chitin from fish scale and dye removal studies.

11.2.2 Cleaning and Preparation of Fish Scale for Chitin Extraction

Initially fish scale collected from fish market and cleaned for extraction of chitin. To ensure uniformity of scale, sample collected at only one shop for the entire experimental studies. Collected and washed fish scale shade dried and packed in airtight container and stored at room temperature for further study.

11.2.3 Deprotination and Demineralization Study

For demineralization studies, 10 ml of HCl used to demineralize 1 g of dry fish scale at room temperature for 36 h. Similarly, for deproteination, Strong alkali NaOH used at 5 M concentration. For this study, 10 ml of 5 M NaOH used to deproteinize 1 g of fish scale under room temperature for 36 h as method described by Kumari et al (2015).

11.2.4 Optimization of Chitin Extraction from Fish Scale

To optimize chitin extraction from fish scale, demineralization, deproteination, and contact time varied by changing normality of HCL (0.5, 1, 1.5, 2, 2.5 N), Molarity of NaOH (0.2, 0.4, 0.6, 0.8, 1.0) and contact time (20, 40, 60, 80, 100, 120 min) varied and all experimental studies were triplicated. This all-experimental studies performed by using packed bed column reactor as described by Nainamalai et al (2018).

11.2.5 Preliminary Studies on Drimarene Yellow and Methyl Orange Adsorption

To check dye adsorption capacity of chitin from fish scale, two anionic dye namely Drimarene Yellow and Methyl orange were used. To compare dye removal efficiency (%), few more adsorbents like activated charcoal, commercially available Chitin powder, Chitin Flakes and Chitosan Beads used.

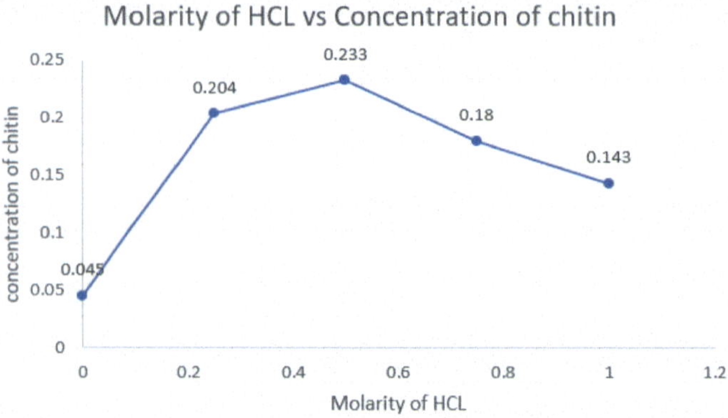

Fig. 11.1 Effect of molarity on extraction of chitin from fish scale waste

11.3 Results and Discussion

11.3.1 Optimization of Molarity of Hcl for Chitin Extraction

Optimizing chitin production by varying the concentration of hydrochloric acid (HCl) as a factor involves adjusting the acid treatment conditions to achieve desired outcomes. The choice of HCl molarity affects chitin yield, quality, and environmental impact. Lower molarities (e.g., 0–1 M) are gentler but may require longer extraction times, potentially leading to decreased yield. Higher molarities (e.g., 6–8 M) are more aggressive and yield chitin more rapidly but may cause degradation or loss of chitin's structural integrity.

Figure 11.1 effect of molarity on extraction of chitin from fish scale waste. 0.5 M of HCl showed maximal chitin release from fish scale. In 2006, Kumari et al., described various extraction and characterization method for chitin and chitosan from fishery waste by chemical method

11.3.2 Optimization with Normality of Naoh for Chitin Extraction

Optimizing chitin production by varying the concentration of Sodium hydroxide (NaOH) as a factor involves adjusting the acid treatment conditions to achieve desired outcomes. The choice of NaOH normality affects chitin yield, quality, and environmental impact. Lower molarities (e.g., 0–1 M) require longer extraction times, potentially leading to decreased yield. Higher molarities (e.g., 6–8 M) are more extensively and yield chitin more rapidly but may cause degradation or loss of chitin's structural

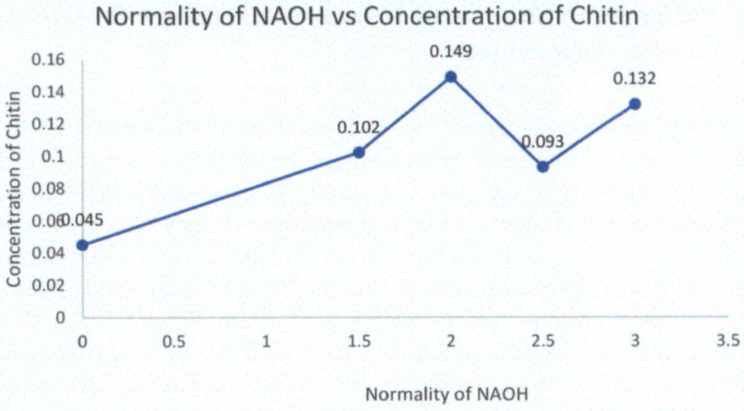

Fig. 11.2 Effect of NaOH on extraction of chitin from fish scale waste

integrity (Kumari et al. 2015). Figure 11.2 denote, effect of NaOH on extraction of chitin from fish scale waste and optimum condition was 2 N NaOH.

11.3.3 Optimizing Cntact-Time of Naoh for Chitin Extraction

Optimizing chitin production by varying the contact time of Sodium hydroxide (NaOH) as a factor involves adjusting the base treatment conditions to achieve desired outcomes. Here we have achieved the desired result in 24 h of contact time of NaOH (Fig. 11.3).

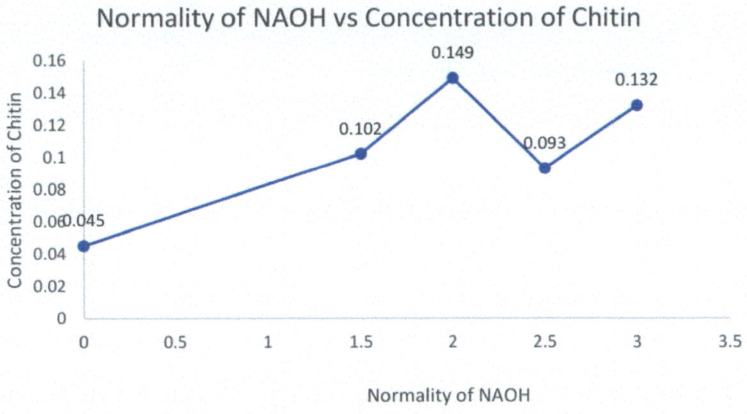

Fig. 11.3 Effect of contact time for extraction of chitin from fish scale waste

11.3.4 Preliminary Studies on Drimarene Yellow and Methyl Orange Adsorption

To study the dye removal percentage, we had used four different dye removal materials such as activated charcoal, chitin powder, chitin flakes, chitosan beads and combination of all the four materials were used. As propped by Dos Santos et al. (2007), still various technology and methods available to remove and decolorize dyes from textile wastewater as well as effective and economically alterative adsorbents.

To remove methyl orange, synthesized zeolite Y ere used by varying adsorption dosage, pH, temperature was studied (Adeoye et al. 2023). Similarly, dye adsorption property of methylene orange at one dimensional and nano-mineral based materials as adsorbent (Wu et al. 2021). Another experimental study reveals that chitin and chitosan-based biomaterials widely used as alternate adsorbent for removal toxic textile dyes from aqueous ecosystem. (Sirajudheen et al. 2021). Not only by dyes, even Cu (II) ions can be removed from aqueous solution by using chitosan beads or GLA conjugated chitosan beads (Ngah and Fatinathan 2008).

11.3.5 Activated Carbon as Adsorbents for Removal of Drimarene Yellow and Methyl Orange

The use of Activated charcoal for dye removal of drimarene yellow and methyl orange proved to be effective. It shows that drimarene yellow has 87.95% of Dye removal and Methyl orange has 90.25% of dye removal. On comparing both dyes Methyl orange showed high percentage of Dye removal (Fig. 11.4).

11.3.6 Chitin as Adsorbents for Removal of Drimarene Yellow and Methyl Orange

The use of Chitin powder for dye removal of drimarene yellow and methyl orange proved to be effective. it shows that drimarene yellow has 95.5% of Dye removal and Methyl orange has 93.6% of dye removal. On comparing both dyes Both were similar, but drimarene showed high percentage of dye removal (Fig. 11.5).

11.3.6.1 Chitin Flakes as Adsorbents for Removal of Drimarene Yellow and Methyl Orange

The use of Chitin Flakes for dye removal of drimarene yellow and methyl orange proved to be effective. it shows that drimarene yellow has 67.15% of Dye removal and

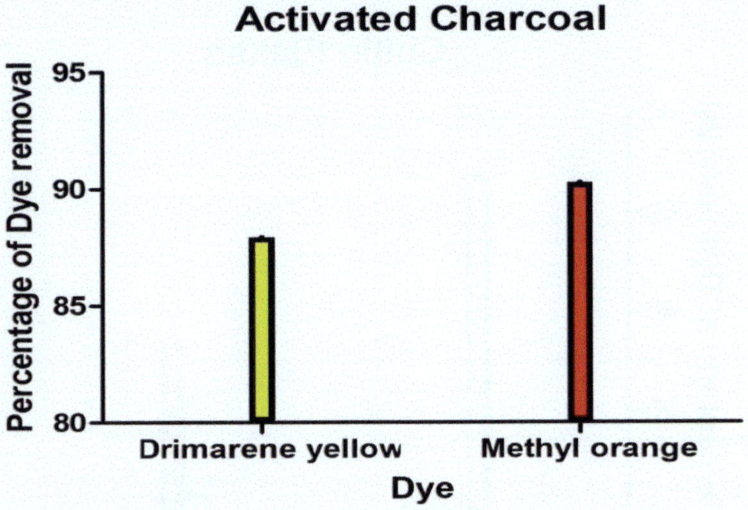

Fig. 11.4 Percentage of dye removal by activated charcoal

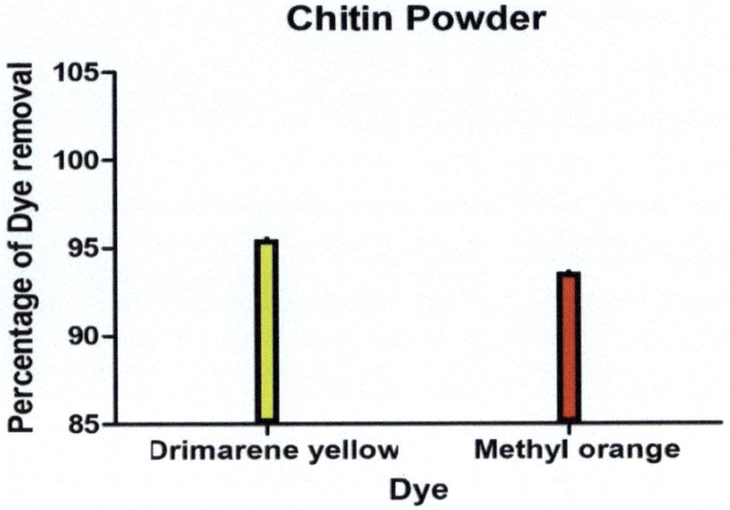

Fig. 11.5 Percentage of dye removal by Chitin powder

Methyl orange has 33.8% of dye removal. On comparing both dyes Both were more or less similar, but drimarene showed high percentage of Dye removal (Fig. 11.6).

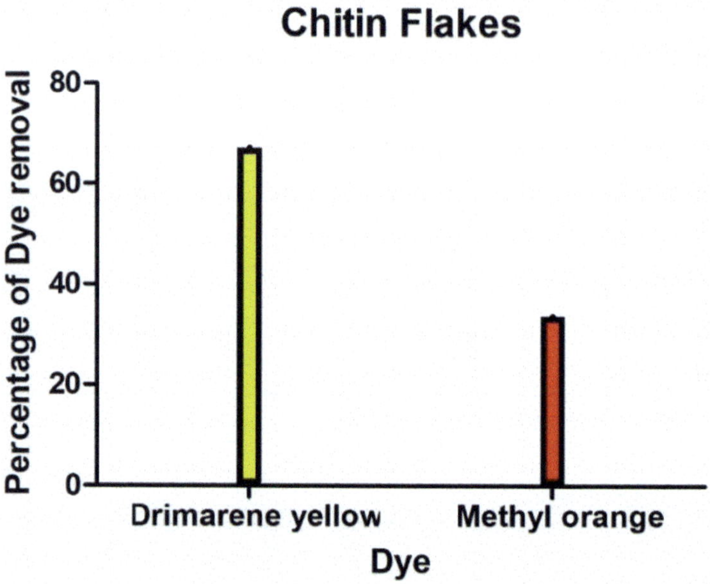

Fig. 11.6 Percentage of dye removal by Chitin flakes

11.3.7 Chitosan as Adsorbents for Removal of Drimarene Yellow and Methyl Orange

The use of Chitosan Beads for dye removal of drimarene yellow and methyl orange proved to be effective. It shows that drimarene yellow has 15.6% of Dye removal and Methyl orange has 22.5% of dye removal. On comparing both dyes Both were more or less similar, but Methyl Orange showed high percentage of Dye removal. Previous study reports reveal that, chitosan and chitosan beads can be used as adsorbent to remove dyes from aqueous solution (Azlan et al. 2009; Filipkowska et al. 2017) (Fig. 11.7).

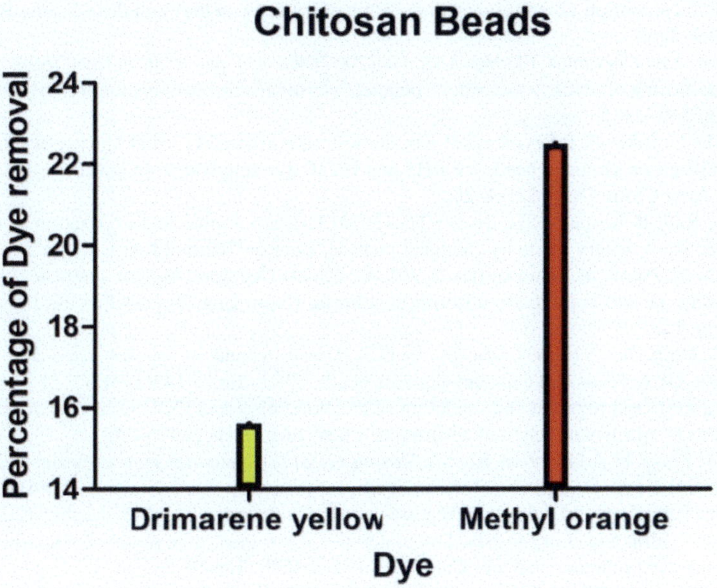

Fig. 11.7 Percentage of dye removal by Chitosan beads

11.4 Conclusion

Removal of drimarene yellow and methyl orange dye by adsorption with chitin powder, Chitin flakes, Chitosan beads, activated Charcoal and combination of all the four materials were experimentally examined. The preliminary studies were done by the materials mentions above and, in this study, it was found that Chitin powder had showed a maximum dye removal efficiency. Though chitin showed better results, the aim of preparing beads for the removal of dye were not satisfied. Therefore, chitosan which is a derivative of chitin was chosen this has increased number of amino groups which enhances its solubility in acidic environments. The Chitosan was more soluble than Chitin in organic solvents.

References

Adeoye JB, Balogun DO, Etemire OJ, Ezeh PN, Tan YH, Mubarak NM (2023) Rapid adsorptive removal of eosin yellow and methyl orange using zeolite Y. Sci Rep 13(1):21373

Azlan K, Saime WNW, Liew LAI (2009) Chitosan and chemically modified chitosan beads for acid dyes sorption. J Environ Sci 21(3):296–302

Bharathi KS, Ramesh SPT (2013) Fixed-bed column studies on biosorption of crystal violet from aqueous solution by *Citrullus lanatus* rind and *Cyperus rotundus*. Appl Water Sci 3(4):673–687

Devi BM, Latha S, Umamaheshwari D, Vijayalakshmi K, Gomathi T, Sudha PN (2014) Synthesis and characterisation of chitosan/sodium alginate/carboxymethyl cellulose beads. Pharm Lett 6(6):389–395
Dos Santos AB, Cervantes FJ, Van Lier JB (2007) Review paper on current technologies for decolourisation of textile wastewaters: perspectives for anaerobic biotechnology. Biores Technol 98(12):2369–2385
Filipkowska U, Jóźwiak T, Szymczyk P, Kuczajowska-Zadrożna M (2017) The use of active carbon immobilised on chitosan beads for RB5 and BV10 dye removal from aqueous solutions. Prog Chem Appl Chitin Deriv 22:14–26
Kumari S, Rath P, Kumar ASH, Tiwari TN (2015) Extraction and characterization of chitin and chitosan from fishery waste by chemical method. Environ Technol Innov 3:77–85
Nainamalai M, Palani M, Soundarajan B, JSS AE (2018) Decolorization of synthetic dye wastewater using packed bed electro-adsorption column. Chem Eng Process-Process Intensification 130:160–168
Ngah WW, Fatinathan S (2008) Adsorption of Cu(II) ions in aqueous solution using chitosan beads, chitosan–GLA beads and chitosan–alginate beads. Chem Eng J 143(1–3):62–72
Şayan E (2006) Optimization and modeling of decolorization and COD reduction of reactive dye solutions by ultrasound-assisted adsorption. Chem Eng J 119(2):175–181
Semeraro P, Fini P, D'Addabbo M, Rizzi V, Cosma P (2017) Removal from wastewater and recycling of azo textile dyes by alginate-chitosan beads. Int J Environ Agric Biotechnol 2(4):238860
Sirajudheen P, Poovathumkuzhi NC, Vigneshwaran S, Chelaveettil BM, Meenakshi S (2021) Applications of chitin and chitosan-based biomaterials for the adsorptive removal of textile dyes from water—a comprehensive review. Carbohydr Polym 273:118604
Takka S, Gürel A (2010) Evaluation of chitosan/alginate beads using experimental design: formulation and in vitro characterization. Aaps Pharmscitech 11:460–466
Weng C-H, Lin Y-T, Tzeng T-W (2009) Removal of methylene blue from aqueous solution by adsorption onto pineapple leaf powder. J Hazard Mater 170(1):417–424
Wu L, Liu X, Lv G, Zhu R, Tian L, Liu M, Liao L (2021) Study on the adsorption properties of methyl orange by natural one-dimensional nano-mineral materials with different structures. Sci Rep 11(1):10640

Chapter 12
Natural Polysaccharides as a Potential Eco-Friendly Adsorbent for Heavy Metal Removal

P. Muthukumaran, C. Deeksha, M. Madhu Mallika, G. Maalika, S. B. Varshini, and Aravind Jeyaseelan

Abstract Water is contaminated by wide range of toxic derivatives, mostly by heavy metals. So, there is a need for the development of a technology for the removal of heavy metals. Adsorption is considered as one of the most common technique for the removal of heavy metals. Compared to other conventional methods, adsorption is considered as an effective method because of its cost effectiveness, high-capacity removal of both organic and inorganic pollutants and their simplicity. Adsorption process on natural adsorbent is a recent process that has great application on the removal of heavy metals, paints, phenols, and other organic pollutants. The present review shows the recent developments in the synthesis of adsorbents containing polysaccharides, in particular modified biopolymers derived from chitosan, starch, graphene, nano absorbents, and carbon nanotubes.

Keywords Polysaccharides · Chitosan · Starch · Graphene · Carbon nanotubes · Nano-adsorbents

12.1 Introduction

The environmental pollution that is caused by the heavy metals such as Mercury (Hg), Arsenic (As), Lead (Pb), etc., causes serious health problems in human beings. These heavy metals are non-biodegradable and can accumulate inside the human body as a toxic or carcinogenic agent causing serious health issues that are incurable. Researchers have found that the metals are the major cause for diseases in human

P. Muthukumaran (✉) · C. Deeksha · M. Madhu Mallika · G. Maalika · S. B. Varshini
Department of Biotechnology, Kumaraguru College of Technology, Coimbatore, Tamil Nadu 641049, India
e-mail: muthukumaran.p.bt@kct.ac.in; m.kumaran005@gmail.com

A. Jeyaseelan
Department of Biotechnology, Saveetha School of Engineering, Saveetha Institute of Medical and Technical Sciences (SIMATS), Thandalam, Chennai, Tamil Nadu 602105, India

© The Author(s), under exclusive license to Springer Nature Switzerland AG 2025
A. Jeyaseelan and K. Murugasen (eds.), *Sustainable Environmental Remediation: Avenues in Nano and Biotechnology*, Environmental Science and Engineering,
https://doi.org/10.1007/978-3-031-78483-5_12

beings and increased attention is given in identifying and removing the toxic metals from the contaminated wastewater. The lower limit of metals in food, water and other substances that are associated with human lifestyle will gradually decrease year after year.

Metals like copper plays a major role in the metabolism of living system. It is essential for human body and micronutrient for plants and animals. An increased concentration of copper in the body leads to serious toxicological problems, such as nausea, vomiting, stomach cramp, gastrointestinal distress, cramps, convulsions, diarrhoea, jaundice. Therefore, it is necessary to monitor its concentration in the diet of the plant and animal world. The selection of the most suitable wastewater treatment process depends on the basic parameters such as the initial concentration of the metal, pH, the specificity of the applied treatment in relation to other technologies, the environmental impact, economic parameters such as operating costs and capital investment. The key factors for selection depend upon the technical applicability, simplicity, and cost-effectiveness of the plant. The removal of heavy metals is done using innovative biosorbents in recent years (Gadd 2010). In the process of decontamination of heavy metal from wastewater (Wang and Chen 2009), large numbers of biomasses are used whether it is alive or dead, dry or wet.

Heavy metals ions can be removed from wastewater by either physical or chemical methods. Traditional methods like ion exchange method, precipitation method is used for removal of heavy metal at low concentration, but having disadvantages like high cost, low efficiency and can cause secondary pollution (Wang and Chen 2006). For example, Clay, natural zeolites, chitosan etc., are some low-cost adsorbents used for removal of heavy metal removal (Asere et al. 2019; Bailey et al. 1999; Bradl 2004; Egashira et al. 2012; Uddin 2017; Zhang et al. 2016). Aquatic system destruction occurs due to release in large quantities of metals and metalloids such as Cu, Cd, Al, Cr, Zn, Fe, Hg, As and Pb during or after mining. Due to increase in industrialization, many metals are released, and there is a need to have a cheap, environmentally friendly and pollutant removal methods (Barakat 2011). A variety of sorbents are used for metal ion adsorption from the water, and a few of them are bio sorbents (Liu et al. 2018), activated carbon, polymeric fibers (Deng et al. 2003), oxide minerals (Sudha Rani et al. 2018) and resins.

Storage and structure are the main biological properties of polysaccharides. For example, Storage polysaccharides are starch whereas cellulose and chitin are structural found respectively in the plants cell walls and the exoskeletons of crustaceans (Klemm et al. 2005; Pillai et al. 2009). Polysaccharides are also used for the preparation of adsorbent. There are many adsorbents such as chitin, chitosan, starch etc., Main advantage of these adsorbent is cost effectiveness. Recently, numerous approaches have been studied for the development of cheaper and more effective adsorbents containing natural polymers. Among these, polysaccharides such as chitin (Synowiecki and Al-Khateeb 2003; Ravi Kumar 2000; Bailey et al. 1999) and starch (Yuryev et al. 2002; Wurzburg 1986; Sandford et al. 1983), and their derivatives (chitosan (Babel and Kurniawan 2003; Varma et al. 2004), cyclodextrin (Singh et al. 2002; Crini and Morcellet 2002)) deserve particular attention. One of the most important and useful features of chitin and starch is their good chemical reactivity. They possess many reactive groups (hydroxyl and/or acetamido groups) present at the 2-, 3-, and 6-positions in the glucose unit. It is well known that polysaccharides which are renewable, biodegradable resources and abundant in nature have a capacity to associate by physical and chemical interactions with a wide variety of molecules. Hence, adsorption on polysaccharide derivatives can be a low-cost procedure of choice in water decontamination for extraction and separation of compounds, and a useful tool for protecting the environment (Bolto 1995).

Nowadays there is a recent interest in the synthesis of new adsorbent materials containing polysaccharides. Polysaccharides are renewable sources used as high-performance materials (Kalia and Avérous 2011) due to its structures and their properties. Low costs, biodegradability in the ecosystem, renewability (Tizzotti et al. 2010) are some of its advantages. Polyfunctionality, chemical reactivity, chelating and absorptive capacities are factors used in the treatment of wastewater (Crini 2005). In this paper, adsorbents containing polysaccharide for the removal of heavy metals is explained for the replacement for current conventional methods. The main objectives are to provide useful information about the most important features of these polymeric materials and to give an overview of several studies.

12.2 Types/Classification

12.2.1 Chitosan Beads

Chitosan is a natural polysaccharide formed by alkaline N-deacetylation of chitin; the second most abundant natural polymer next to cellulose (Tharanathan and Kittur 2003). The principal polysaccharide derivative of interest is chitosan. Chiou and Li investigated the adsorption of dyes onto crosslinked chitosan beads (Chiou and Li 2003, 2002). Chitosan-based sorbent has an ability to bind with the pollutant under different conditions. Chitosan-based materials were prepared using GLA, EPI and EGDE as crosslinking agents. The adsorption is effective based on temperature and the EPI/chitosan weight ratio. They showed that the chitosan–EPI beads show higher adsorption capacity than other resins. It has more adsorption capacity, 1642 g of reactive dye 189 can be treated by 1 kg of material. Beads have effective surface area for same mass particles, so their dye uptake is increased with decrease in particle size. Though beads have higher adsorption capacity, the nature of beads is soft, and its mechanical properties must be improved.

To produce rigid beads, it is crosslinked with TPP (Chiou et al. 2004). The adsorption capacities had very large values of 1911–2498 g of dye per kg of bead. They are used for removal of Cu2C from aqueous solutions (Wan Ngah et al. 2002). The uptakes on chitosan–GLA, chitosan–EPI and chitosan–EGDE beads were respectively: 59.6, 62.4 and 45.9 mg Cu_2C/g beads. The adsorption of metals is mainly dependent on pH. Chitosan beads are effective chelating resins helps in removing palladium from dilute solutions. Crosslinking ratio, palladium concentration and particle size are important parameters of sorption kinetics. Juang and Shao investigated about the adsorption of heavy metals onto crosslinked chitosan beads and showed that adsorption capacity depends on the nature of the crosslinking agent (Juang and Shao 2002a, b). Work related on chitosan derivatives was done by Guibal and co-workers (Arrascue et al. 2003; Ruiz et al. 2003; Guibal et al. 2002, 2000, 1999a, b, c, 1998; Dambies et al. 2002).

Based on competitive adsorption of proton and metals, adsorption increased with increasing pH of the solution. Chitosan beads and its derivatives can be used for the removal and the recovery of toxic or valuable precious metals such as arsenic, molybdenum, vanadium, platinum, palladium, osmium, iridium, and gold. Diffusion of metal ions through chitosan membranes crosslinked by GLA was studied by Krajewska (2001). Gold can be removed from dilute acidic solutions by the chitosan beads containing sulphur groups having maximum uptake capacity reaches 600 mg of Au per gram of beads. Though chitosan-based materials are used in heavy metal removal, they also have special role in recovering valuable metals.

12.2.2 Crosslinked Starch and Starch Derivatives

Nowadays more attention is also given for carbohydrate polymers such as starch and cyclodextrins. pH is one of the important parameters. Adsorption efficiency of metals can be changed by altering the pH value, due to protonation and deprotonation on the surface of sorbents. Crosslinked-starch graft copolymers containing amine groups. Kim and Lim (1999) proposed that starch after crosslinking with $POCl_3$ and carboxymethylation can be used as sorbents for the removal of heavy metal ions. Crosslinked-starch graft copolymers containing amine groups was proposed by Zhang and Chen (2002). The polymers, used as the sorbents for Pb(II) and Cu(II) ions, were prepared by grafting dimethyl aminoethyl methacrylate onto commercial starch. Adsorption capacity is important and 2 h of adsorption is required to reach the equilibrium. Starch after crosslinking with $POCl_3$ and carboxymethylation helps in heavy metal removal within few minutes. Several hundred ppm of divalent metal ions (Cu, Pb, Cd and Hg) can be removed from water by dispersing 1% of the modified starch. The carboxymethyl group and the starch content present in the solution is responsible for the effective heavy metal removal. Pollutants from aqueous solutions is removed from crosslinked starches with EPI have been proposed by Delval et al. (2000). In sorption column, the particles are packed, which helps in effective continuous removal of pollutants. The kinetics of sorption depends on the presence of amino groups. Crini and Morcellet (2002) and Crini (2003) prepared several crosslinked cyclodextrins gels. They are generally starch derivatives. 2-hydroxy-3-methacryloyloxy-propyl-beta-cyclodextrin is used for the synthesis of cyclodextrin-based materials have been proposed by Janus et al. (2003).

12.2.3 Graphene

Graphene has great ecological application as it has special properties, and it is one of the allotropic forms of carbon. Primarily because of its high surface area, strength, light weight, removal of heavy metals, flexibility and chemical stability graphene has been occupying more attention. Graphene oxide (GO) is produced by the oxidation of graphite layer by chemical method (Taherian et al. 2013). It is a carbon nanomaterial and has two-dimensional structure.

Hydrophilic groups are induced to Graphene Oxide, which requires a specific oxidation process (Gopalakrishnan et al. 2015). Adsorption of heavy metals is increased by the presence of hydroxyl group. The process of adsorption is affected by the presence of functional groups on the surface of the Graphene Oxide (Zare-Dorabei et al. 2016). White et al. (2018) demonstrated the maximum adsorption capacities of copper by Graphene Oxide and GO-COOH (carboxylate graphene) that are 277.77 and 347.14 mg/g respectively under optimum conditions (White et al. 2018). The initial concentration, and the reaction temperature, pH value is more important for effective removal. High efficiency for removal of heavy metals from wastewater is done by graphene and its composite. Increasing the adsorption capacity in GO does not require any acid treatment as it has hydrophilic functional group (Dong et al. 2015; Azamat et al. 2015).

GOs have two main advantages:

(1) Maximum adsorption of heavy metals as one layer of GOs has two-dimensional basal levels.
(2) The layer has a simple synthesis process that can be activated by the chemical peeling of graphite without metal catalysts and complicated instruments.

12.2.4 Nanoadsorbents

Nanoparticulate materials play a major role as adsorbent for heavy metal removal. It has large surface area and a smaller size that increases chemical stability, so there will be more capability for heavy metal removal (Shirsath and Shirivastava 2015). ZnS nanocrystals (NCs) sorbent could remove the 99% of Cu^{2+} from contaminated water effectively in one minute (Xu et al. 2016). ZnS NCs saturated adsorption capacity was about 650 mg/g. It is found that the ZnS NCs layer thickness was 30–40 nm for the highest efficiency of copper removal. Redox reactions of inorganic contaminants can be done for possible transformation of the ionic structure (Agarwal et al. 2015). Changes in the redox status of pollutants is also a factor of toxicity (Chen and Mao 2007; Ray 2010).

The most used nanoparticles for heavy metal adsorption are active carbon and carbon nanotubes, manganese oxide, graphene, zinc oxide, magnesium oxide, titanium oxide and iron oxide. There are two different TiO_2 particles with different nano dimensions by sol–gel and by a low-temperature chemical precipitation method from a $TiCl_4$ precursor (Debnath et al. 2011). The average size of the nanoparticles of sol–gel and precipitation were 27 and 12 nm, respectively. The results obtained showed a higher Cu^{2+} adsorption capacity of the precipitation method than the sol–gel method. Nanostructured titanium(IV) oxide is used as non-toxic and insoluble material having high surface-active property used for removing inorganic compounds in water (Debnath and Ghosh 2011). Effective nanoparticles for heavy metal removal must have properties like non-toxic, ability to adsorb pollutants at lower concentrations, high adsorption capacity and can be used for many times without recycling.

12.2.5 Carbon Nanomaterials

Based on their process of adsorption, nano absorbers are classified into different types. Carbon nanomaterials (CNMS) development includes a carbon nanotube, carbon nano-sheets and carbon nanoparticle. Compared to other substances like titanium dioxide and aluminium oxide particles, nanoparticles are more prominent because of their chemical activity and fine-grained structure (Zhang et al. 2008). Carbon nanotubes have more potential for removing heavy metals and inorganic pollutant from wastewater by adsorption. But there are many disadvantages in using carbon nanotube because it has small particle size, dispersion ability and difficulty in separating ions. So, modification is done in CNT to overcome the above problems. So multilayer carbon nanotubes (multi-walled carbon nanotubes MWCNTs) are developed as it has high metal dispersion ability and can be easily removed from water or by using magnet. Some metals can be removed from wastewater by using multilayer carbon nanotubes. After comparison of coated CNT and uncoated CNT, coated CNT shows more ability of removal. The adsorption behaviour of alumina in wastewater is examined (Gupta et al. 2011) by using the CNTs and found that coated CNT shows better ability of removal.

Cu^{2+} is successfully removed from wastewater by using multilayer carbon nanotubes (Tang et al. 2012). The total adsorption activity is increased by the surface modification of CNT include acid treatment (Ihsanullah Al-Khaldi et al. 2015; Ren et al. 2011) impregnation and functional grafting of molecules/groups. There are various surface modification techniques namely specific surface area, surface charge, dispersion, and hydrophobicity and characteristics of CNTs is changed by these techniques. Different types of acids and oxidants have been used such as HNO_3, $KMnO_4$, H_2O_2, H_2SO_4 and HCl (Ren et al. 2011; Fu and Wang 2011) for the treatment of CNTs. Another way to improve the characteristics of the surface is by grafting of functional molecules/groups at the surface of the CNTs. Plasma technology affords good environmental protection and a lower energy consumption and considered as a best method. Impurities present on the surface of CNTs is removed by the acid treatment and the adsorption activity is increased by the presence of functional groups on the surface of CNTs.

12.3 Sources of Adsorbents

12.3.1 Chitosan

Chitosan is produced from chitin when the degree of deacetylation (DD) is over 50%. There are numerous natural sources of chitosan. Chitosan can be obtained from the shells of crustaceans, such as lobsters, crabs, and shrimps, as well as from fish scales (Kumari and Kishor 2020) and many other types of organisms like insects and fungi. Chitosan is also extracted from other sources like lactic fermentation bioprocess, various other fermentation processes, and enzymatic hydrolysis of crustacean biowaste. Chitosan is also synthesized from individual sources such as fish scale, shrimp shells, and crab shells by traditional methods is compared with the help of structural analysis and physicochemical properties.

12.3.2 Starch

Starch is a polysaccharide which upon hydrolysis releases glucose. Most of the starch is obtained from cereals (rice, wheat, maida, sago, maize, and barley) (Mudambi et al. 2009), roots (cassava, tapioca, arrowroot) and tubers (potatoes, sweet potatoes). Starch is present in small particles known as granules. For heavy metal removal, generally starch-modified raw materials are used.

Starch is made up of two fractions, Amylose and Amylopectin. The amylose fraction of starch is composed of straight-chain structure, while the amylopectin fraction has a branched chain configuration. Amylose contributes gelling characteristics to the cooked and cooled starch mixtures. Amylopectin provides cohesive or thickening property but does not usually contribute to gel formation.

12.3.3 Graphene

Graphene can be produced from plants through thermal decomposition at high temperatures in an inert atmosphere by process of pyrolysis. Graphene is also obtained even from exotics (e.g., grass, honey) (Vilcinskas 2020). Graphene is obtained through a series of physical and chemical changes using high temperatures and carbon precursor gas (e.g., methane, ethane) and metal (e.g., copper, nickel) substrates, silicon carbide or gas (e.g., hydrogen, helium) and graphite sources.

12.3.4 Nano-absorbents

A novel and low-cost nano-adsorbents can be produced from sources such as cotton pulp, jute, and kenaf as carbon fibers by successfully synthesizing and fabricating them (Ahmadi et al. 2020).

12.3.5 Carbon Nanotubes

The carbonaceous materials such as graphite undergoes plasma arcing to produce fullerenes and CNTs. During the process, the CNTs are deposited on the opposing electrode and the fullerenes are found in the soot that has been formed. There are lots of techniques to produce carbon nanotubes namely arc discharge, laser ablation, high-pressure carbon monoxide disproportionation and chemical vapor deposition (CVD) (Wang et al. 2019). Most of these processes take place in a vacuum or with process gases. Most of the researchers use chemical vapor deposition to produce CNT. A material of interest, such as carbon fibers, is coated in a catalyst—usually an iron-based compound—and placed in a furnace, through which carbon dioxide and other carbon-containing gases flow.

12.4 Preparation Methods

12.4.1 Starch

The preparation method of the crosslinked starch is as follows:

1. Grind the fullerene powder and cross the powder across 200 mesh sieves.
2. Get 100 mL mass concentration of 2% amidin and add fullerene powder to it. And then take it through ultrasonic processing. Now the fullerene powder and the amidin will be mixed together. Add 5 mL mass concentration is 1% propane diamine solution, to the mixture under microwave condition at 400 r/min and carrying out stirring reaction, obtains preparation liquid.
3. Get the preparation liquid and add sodium hydroxide solution of 0.1 mol/L. After the oven dry soaks takes off film. This uses ethanol and distilled water repeatedly to rinse until neutral, naturally dry propane diamine crosslinked starch/fullerene composite membrane is formed.
4. The above-described fullerene is C60.
5. The consumption of above described C60 is 0.01–0.02 g.
6. The described microwave power is 300–400 W.
7. The described reaction temperature is 20–40 °C, and the reaction time is 5–15 min.

12.4.2 Chitosan

The preparation method for chitosan bead, includes following steps:

1. The solvent is heated and the chitosan powder is added slowly by stirring. This is done for a certain hour and the solution is mixed.
2. In the next step chitosan solution deaeration is obtained.
3. The chitosan solution is filled in a syringe and dropped in a solid solution from a certain height by craft or a mechanical process that has to be cured and it uses distilled water for rinsing and the chitosan bead is obtained.

The solvent that is described in step1 is an aqueous acetic acid or alkali/aqueous solution of urea or hydrochloric acid solution. The aqueous acetic acid or alkali/aqueous solution of urea or hydrochloric acid solution mass ratio are at 2–10 w%, and the mass ratio of aqueous acetic acid or alkali/aqueous solution of urea or hydrochloric acid solution quality and chitosan is at 40:20–40:1. The described solid solution is alkalescence, acid or inorganic salt solution. In the above process of being dripped by the mode of craft or mechanical presses, the height of the syringe needle distance solid solution of syringe is at 0–16 cm and the diameter of syringe needle is between 0.45–0.90 mm. The Chitosan bead obtained has a diameter between 2–5 mm (Barney et al. 2012).

12.4.3 Graphene

12.4.3.1 Micromechanical Exfoliation

There are various methods for the preparation of graphene, one of them is micromechanical exfoliation. The first recognized method of graphene synthesis is Micromechanical Exfoliation. Novoselov et al. used adhesive tape to produce a single graphene layer by a mechanical cleaving technique from 1-mm-thick HOPG. Micromechanical exfoliation can be used to generate high quality graphene that is electrically isolated for fundamental studies of transport physics and other properties but does not yet appear to be scalable to large area. It typically produces graphene particles with lateral dimensions on the order of tens to hundreds of micrometres. The mechanical cleavage of graphene layers from a graphite crystal has paved the way for the study of the properties of single-layer graphene or bilayer graphene, the method is not suitable for large scale synthesis of single-layer graphene or of few-layer graphene (Whitener and Paul 2014).

12.4.3.2 Chemical Vapor Deposition (CVD)

Chemical vapor deposition is a chemical process in which a substrate is exposed to thermally decomposed precursors and the desired product deposited onto the substrate surface at high temperature. Because the high temperature is not desired in many cases, plasma-assisted decomposition and reaction may lower the process temperature. When a thermal CVD process involves chemical reactions of the reacting gases by generating plasma inside a vacuum chamber, which leads to the deposition of thin film on the substrate surface, the process is known as plasma-enhanced chemical vapor deposition (PECVD). The plasma can be generated inside a PECVD system using RF (AC frequency), microwave, and inductive coupling (electrical currents produced by electromagnetic induction). PECVD has a wide range of advantages over other conventional CVD methods. By this technique, a process can be carried out at relatively low temperature compared to the other thermal CVD processes. Hence, it is more feasible for industrial-scale applications (Whitener and Paul 2014).

12.5 Nano Adsorbents

12.5.1 Chemical Synthesis of Magnetic Nanoparticles

The precursors in the synthesis of MNPs include Diverse metal-containing compounds (MCCs), such as metal carbonyls, organometallic compounds, metal carboxylates, etc., Most often, precursors are decomposed on heating or UV irradiation. There are lot more techniques that can be used in the preparation of nano adsorbents. Coprecipitation is the most promising method for chemical synthesis of MNPs (Thekkudan et al. 2016).

Thermolysis of Metal-Containing Compounds. Metal organic chemical vapor deposition (MOCVD) technique, which is used successfully to obtain nanoparticles. In a one-step CVD synthesis of nano dispersed Fe oxides, When the reaction is performed in a liquid medium in the presence of surfactants or polymers, it is possible to stabilize the resulting amorphous nanoparticles with diameters of up to 10 nm. Something that can be done in two-stage thermolysis of $Fe(CO)_5$ has been reported. First, an iron oleate complex is formed from $Fe(CO)_5$ and oleic acid at 100 °C, and then at 300 °C the complex decomposes to give primary loose nanoparticles (4–11 nm). After maintaining them at 500 °C, they are converted into crystalline α-Fe nanoparticles.

12.6 Carbon Nanotube

The preparation of carbon nanotube is as follows.

This technique has a greater control over the growth parameters. This is used for high yields, as well as, it has less need for purification. With this, the low cost in the preparation stands out since the purification methods in some cases are extremely expensive. However, this method can produce nanotubes with structural defects, restricting the application. The CVD (Chemical Vapor Deposition) method enables the production of large-scale carbon nanotubes. It involves catalytic deposition of the carbon that is generated from a carbon containing gas inside an oven with temperatures of 500–1100 °C in the presence of a metal catalyst (Wang et al. 2019). As in the laser ablation method, Chemical Vapour Decomposition also occurs inside quartz. In this synthesis, SWNTs and MWCNTs are generated, by thermal decomposition of the precursor in carbon (solid) that is deposited in a substrate and hydrogen that is eliminated by exhaustion of the CNT generated.

12.7 Environmental Application

12.7.1 Wastewater Treatment

Water pollution is one of the major environmental problem. It is mainly caused by uncontrolled water contamination. Due to the disease caused by polluted water, there is a death of 3900 children per day and millions of people per year (Shannon et al. 2008). Water pollution is due to heavy metal ions, aromatic compounds and dyes are often found in the environment because of their wide industrial uses. Heavy metals are not biodegradable that may cause various diseases and disorders. Their presence in the environment, in water should be controlled. Careless release of emerging pollutants, such as pharmaceuticals and other personal care products into wastewater causes water pollution (Schwarzenbach et al. 2006). 15–25 Pharmaceuticals are still detected even in treated drinking water worldwide (Jones et al. 2005; Benotti et al. 2009).

Synthetic dyes, which are used for textile dyeing, paper printing etc., that strongly colour wastewater. So various technologies have been developed for the removal of pollutants from wastewater. Adsorbents like activated carbons (AC), metal organic frameworks (MOFs), zeolites, clays, carbon nanomaterials (e.g., graphene (G), graphene oxide (GO), carbon nanotube (CNT)), are use as adsorption of organic pollutants in water (Wang and Chen 2006). It is highlighted that there is an advance in application of nanofiltration membranes, graphene-based NMs and carbon nanotubes for water remediation (Asere et al. 2019; Bailey et al. 1999; Bradl 2004).

The advantage, properties, mechanism, limitations for water and wastewater treatment in nanotechnology has been outlined (Egashira et al. 2012). Biological treatments, chemical and electrochemical techniques, advanced oxidation processes, membrane processes and adsorption procedures are few techniques for removal of organic materials and metals in wastewater. But the major disadvantage of this technique is high cost. There is a need for cost effective methods. Adsorption using sorbents is one of the most popular methods. Adsorption is a well-known equilibrium separation process. It is more effective, efficient, and economic method for water decontamination. The adsorbents may be of mineral, organic or biological origin. Many approaches have been studied for the development of cheaper and more effective adsorbents containing natural polymers. Polysaccharide adsorbents such as chitin and starch and their derivatives (chitosan and cyclodextrin) have more attention.

Biopolymers are also an alternative adsorbent because of their physical–chemical characteristic, chemical stability, high reactivity and excellent selectivity towards aromatic compounds and metals. Adsorption on polysaccharide derivatives is considered as a low-cost procedure for water decontamination and also for extraction and separation of compounds and is a useful tool for protecting the environment. Adsorbent containing polysaccharide is considered as effective replacement of conventional methods for removing pollutants from wastewater. The adsorption of dyes by crosslinked chitosan beads has been examined. Removal of pesticide from water and soil is an important need. Nanomaterials as immobilization matrices for detection and removal of pesticides provide higher sensitivity compared with conventional materials (Zhang and Chen 2002).

12.7.2 Adsorbent to Remove Heavy Metals

Adsorption is considered as one of the alternative treatment methods for removal of heavy metals. Recently many cheap adsorbents from industrial by-products, agricultural waste, natural materials, or modified biopolymers have been studied that is used for removal of heavy metals from metal-contaminated wastewater (Azzam et al. 2016; Zhong et al. 2014; Modrzejewska et al. 2016; Yogeshwar et al. 2016). The pH of the solution (pH 9–11) is the main parameter which improves in efficient removal of heavy metals by chemical precipitation. Key factors for selecting the most suitable adsorbents are low cost and technical applicability and the other advantages are cheap equipment, safe handling, and simplicity.

Precipitation reagents like lime and limestone are mainly used due to low cost and availability in many countries (Mirbagherp and Hosseini 2004; Aziz et al. 2008). Purification of organic wastewater with a metal concentration of more than 1000 mg/L is done using lime precipitation. Many chemicals is required to reduce the concentration of the metal in chemical precipitation. The procedure for the modification of the surface of silica gels with glycine, diglycine, and triglycine in the presence of copper ions is to create a new type of copper imprinted sorbent, having a high adsorption capacity and selectivity for copper ions, even when the concentrations of Mg and Ca are 50 and 76 times higher than the concentration of copper.

12.8 Summary

Wastewater treatment of heavy metals may cause environmental pollution due to urbanization and accelerated industrial development. Conventional technologies are used for wastewater treatment, but it has a serious disadvantage like high operating costs. The adsorption of natural adsorbents (biosorbents) is one of the potential alternatives to existing technologies for the effective removal of low concentrations of toxic metal ions from aqueous solutions and wastewater. Recent developments in the synthesis of adsorbents containing polysaccharides, has been encouraged. The main advantage of using adsorbents is, they are renewable in nature, biodegradable resources and abundant. One of the great advantages of these materials is that they can be used as a waste material for the treatment of wastewater as an ecological adsorbent, which is very advantageous from the point of view of waste management follows the principles of sustainable development. Also, the advantage of these materials is that they can be reused and employed for more cycles and regenerated by easy desorption. In this article, natural polysaccharides used as adsorbent is discussed. Their types and classifications, their method of preparation, their source is discussed.

References

Agarwal TA, Tyagi S, Gupta I, VK (2015) Adsorptive removal of cadmium(II) ions from liquid phase using acid modified carbon-based adsorbents. J Mol Liq 204:255–263

Ahmadi R, Ardjmand M, Rashidi A, Rafizadeh M (2020) High performance novel nanoadsorbents derived-natural cellulose fibers for superior CO_2 adsorption and CO_2/CH_4 separation. In: Energy sources, part A: recovery, utilization, and environmental effects, pp 1–19

Arrascue ML, Garcia HM, Horna O, Guibal E (2003) Gold sorption on chitosan derivatives. Hydrometallurgy 71:191–200

Asere TG, Stevens CV, Du Laing G (2019) Use of (modified) natural adsorbents for arsenic remediation: a review. Sci Total Environ 676:706e720

Azamat J, Sattary BS, Khataee A, Joo SW (2015) Removal of a hazardous heavy metal from aqueous solution using functionalized graphene and boron nitride nanosheets: insights from simulations. J Mol Graph Model 61:13–20

Aziz HA, Adlan MN, Ariffin KS (2008) Heavy metals (Cd, Pb, Zn, Ni, Cu and Cr (III)) removal from water in Malaysia: post treatment by high quality limestone. Bioresour Technol 99:1578–1583

Azzam AM, El-wakeel ST, Mostafa BB, El-Shahat MF (2016) Removal of Pb, Cd, Cu and Ni from aqueous solution using nano scale zero valent iron particles. J Environ Chem Eng 4:2196–2206

Babel S, Kurniawan TA (2003) Low-cost adsorbents for heavy metals uptake from contaminated water: a review. J Hazard Mat B97:219–243

Bailey SE, Olin TJ, Bricka RM, Adrian DD (1999) A review of potentially low-cost sorbents for heavy metals. Water Res 33:2469–2479

Barakat MA (2011) New trends in removing heavy metals from industrial wastewater. Arab J Chem 4:361–377

Barney NA, Weinberg MG, Sauer BB (2012) Chitosan films with reduced shrinkage and laminates made therefrom

Benotti MJ, Trenholm RA, Vanderford BJ, Holady JC, Stanford BD, Snyder SA (2009) Pharmaceuticals and endocrine disrupting compounds in U.S. drinking water. Environ Sci Technol 43:597–603

Bolto BA (1995) Soluble polymers in water purification. Prog Polym Sci 20:987–1041
Bradl HB (2004) Adsorption of heavy metal ions on soils and soils constituents. J Colloid Interface Sci 277:1e18
Chen X, Mao SS (2007) Titanium dioxide nanomaterials: synthesis, properties, modifications and applications. Chem Rev 107:2891–2959
Chiou MS, Li HY (2002) Equilibrium and kinetic modeling of adsorption of reactive dye on cross-linked chitosan beads. J Hazard Mater B93:233–248
Chiou MS, Li HY (2003) Adsorption behavior of reactive dye in aqueous solution on chemical cross-linked chitosan beads. Chemosphere 50:1095–1105
Chiou MS, Ho PY, Li HY (2004) Adsorption of anionic dyes in acid solutions using chemically cross-linked chitosan beads. Dyes Pigments 60:69–84
Crini G (2003) Studies on adsorption of dyes on beta-cyclodextrin polymer. Bioresour Technol 90:193–198
Crini G (2005) Recent developments in polysaccharide-based materials used as adsorbents in wastewater treatment. Prog Polym Sci 30:38–70
Crini G, Morcellet M (2002) Synthesis and applications of adsorbents containing cyclodextrins. J Sep Sci 25:789–813
Dambies L, Vincent T, Guibal E (2002) Treatment of arsenic-containing solutions using chitosan derivative: uptake mechanism and sorption performances. Water Res 36:3699–3710
Debnath S, Ghosh UC (2011) Equilibrium modeling of single and binary adsorption of Cd (II) and Cu (II) onto agglomerated nano structured titanium (IV) oxide. Desalination 273(2–3):330–342
Debnath S, Ranjan Ray N, Chand Ghosh U (2011) Characterization of agglomerated nanosized titanium (IV) oxide prepared by two pathways and their performance toward Cu (II) adsorption. Int J Green Nanotechnol 3(4):271–280
Delval F, Vebrel J, Pont P, Morcellet M, Janus L, Crini G (2000) Sorption properties toward aromatic compounds of insoluble crosslinked polymer containing starch derivatives. Polym Recycl 5:137–143
Deng S, Bai R, Chen JP (2003) Aminated polyacrylonitrile fibers for lead and copper removal. Langmuir 19(12):5058–5064
Dong Z, Zhang F, Wang D, Liu X, Jin J (2015) Polydopamine-mediated surface-functionalization of graphene oxide for heavy metal ions removal. J Solid State Chem 224:88–93
Egashira R, Tanabe S, Habaki H (2012) Adsorption of heavy metals in mine wastewater by Mongolian natural zeolite. Procedia Eng 42:49e57
Fu F, Wang Q (2011) Removal of heavy metal ions from wastewaters: a review. J Environ Manage 92:407–418
Gadd GM (2010) Metals, minerals, and microbes: geomicrobiology and bioremediation. Microbiology 156:609e643
Gopalakrishnan A, Krishnan R, Thangavel S, Venugopal G, Kim SJ (2015) Removal of heavy metal ions from pharma-effluents using graphene-oxide nanosorbents and study of their adsorption kinetics. J Ind Eng Chem 30:14–19
Guibal E, Milot C, Tobin JM (1998) Metal–anion sorption by chitosan beads: equilibrium and kinetic studies. Ind Eng Chem Res 38:1454–1463
Guibal E, Larkin A, Vincent T, Tobin JM (1999a) Chitosan sorbents for platinum sorption from dilute solutions. Ind Eng Chem Res 38:4011–4022
Guibal E, Milot C, Eterradossi O, Gauffier C, Domard A (1999b) Study of molybdate ion sorption on chitosan gel beads by different spectrometric analyses. Int J Biol Macromol 24:49–59
Guibal E, Milot C, Roussy J (1999c) Molybdate sorption by crosslinked chitosan beads: dynamic studies. Water Environ Res 71:10–17
Guibal E, Vincent T, Navarro MR (2000) Synthesis and characterization of a thiourea derivative of chitosan for platinum recovery. J Appl Polym Sci 75:119–134
Guibal E, Von Offenberg SN, Vincent T, Tobin JM (2002) Sulfur derivatives of chitosan for palladium sorption. React Funct Polym 50:149–163

Gupta VK, Agarwal S, Saleh TA (2011) Synthesis and characterization of alumina coated carbon nanotubes and their application for lead removal. J Hazard Mater 185:17–23
Ihsanullah Al-Khaldi FA, Abusharkh B, Khaled M, Atieh MA, Nasser MS, Saleh TA, Agarwal S, Tyagi I, Gupta VK (2015) Adsorptive removal of cadmium(II) ions from liquid phase using acid modified carbon-based adsorbents. J Mol Liq 204:255–263
Wurzburg OB (ed) (1986) Modified starches: properties and uses. CRC Press, Boca Raton
Janus L, Carbonnier B, Deratani A, Bacquet M, Crini G, Laureyns J et al (2003) New HPLC stationary phases based on (methacryloyloxypropyl-b-cyclodextrin-co-N-vinylpyrrolidone) copolymers coated on silica. Preparation and Characterisation. New J Chem 27:307–312
Jones OA, Lester JN, Voulvoulis N (2005) Pharmaceuticals: a threat to drinking water? Trends Biol 23:163–167
Juang RS, Shao HJ (2002a) A simplified equilibrium model for sorption of heavy metal ions from aqueous solutions on chitosan. Water Res 36:2999–3008
Juang RS, Shao HJ (2002b) Effect of pH competitive adsorption of Cu(II) Ni(II), and Zn(II) from water onto chitosan beads. Adsorption 8:71–78
Kalia S, Avérous L (2011) Biopolymers: biomedical and environmental applications. Wiley, Hoboken, New Jersey
Kim BS, Lim ST (1999) Removal of heavy metal ions from water by cross-linked carboxymethyl corn starch. Carbohydr Polym 39:217–223
Klemm D, Heublein B, Fink H-P, Bohn A (2005) Cellulose: fascinating biopolymer and sustainable raw material. Angew Chem Int Ed Engl 44:3358–3393
Krajewska B (2001) Diffusion of metal ions through gel chitosan membranes. React Funct Polym 47:37–47
Kumari S, Kishor R (2020) Chapter 1–Chitin and chitosan: origin, properties and applications
Liu X, Chen Z-Q, Han B, Su C-L, Han Q, Chen W-Z (2018) Biosorption of copper ions from aqueous solution using rape straw powders: optimization, equilibrium and kinetic studies. Ecotoxicol Environ Saf 150:251–259
Mirbagherp SA, Hosseini SN (2004) Pilot plant investigation on petrochemical wastewater treatment for the removal of copper and chromium with the objective of reuse. Desalination 171:85–93
Modrzejewska Z, Rogacki G, Sujka W, Zarzycki R (2016) Sorption of copper bychitosan hydrogel: kinetics and equilibrium. Chem Eng Process Process Intensif 109:104–113
Mudambi SR, Rao SM, Rajagopal MV (2009) Food science. New Age International Limited, New Delhi
Pillai CKS, Paul W, Sharma CP (2009) Chitin and chitosan polymers: chemistry, solubility and fiber formation. Prog Polym Sci 34:641–678
Ravi Kumar MNV (2000) A review of chitin and chitosan applications. React Funct Polym 46:1–27
Ray PC (2010) Size and shape dependent second order nonlinear optical properties of nanomaterials and their application in biological and chemical sensing. Chem Rev 110(9):5332–5365
Ren X, Chena C, Nagatsu M, Wang X (2011) Carbon nanotubes as adsorbents in environmental pollution management: a review. J Chem Eng 170:395–410
Ruiz M, Sastre A, Guibal E (2003) Osmium and iridium sorption on chitosan derivatives. Solvent Ext Ion Exch 21:307–329
Sandford PA, Baird J (1983). In: Aspinall GO (ed) Industrial utilization of polysaccharides. Academic Press, New York, p 1983
Schwarzenbach RP, Escher BI, Fenner K, Hofstetter TB, Johnson CA, von Gunten U, Wehrli B (2006) The challenge of micropollutants in aquatic systems. Science 313:1072
Shannon MA, Bohn PW, Elimelech M, Georgiadis JG, Marinas BJ, Mayes AM (2008) Science and technology for water purification in the coming decades. Nature 452:301
Shirsath DS, Shirivastava VS (2015) Adsorptive removal of heavy metals by magnetic nanoadsorbent: an equilibrium and thermodynamic study. Appl Nanosci 5(8):927–935
Singh M, Sharma R, Banerjee UC (2002) Biotechnological applications of cyclodextrins. Biotechnol Adv 20:341–359

Sudha Rani K, Srinivas B, Gouru Naidu K, Ramesh KV (2018) Removal of copper by adsorption on treated laterite. Mater Today Proc 5(1):463–469

Synowiecki J, Al-Khateeb NA (2003) Production, properties, and some new applications of chitin and its derivatives. Crit Rev Food Sci Nutr 43:145–171

Taherian F, Marcon V, van der Vegt NF, Leroy F (2013) What is the contact angle of water on graphene? Langmuir 29:1457–1465

Tang WW, Zeng GM, Gong JL, Liu Y, Wang XY, Liu YY, Liu ZF, Chen L, Zhang XR, Tu DZ (2012) Simultaneous adsorption of atrazine and Cu(II) from wastewater by magnetic multi-walled carbon nanotube. Chem Eng J 211:470–478

Tharanathan RN, Kittur FS (2003) Chitin—the undisputed biomolecule of great potential. Crit Rev Food Sci Nutr 43:61–87

Thekkudan VN, Vaidyanathan VK, Ponnusamy SK, Charles C, Sundar SL, Vishnu D, Anbalagan S, Vaithyanathan VL, Subramanian S (2016) Review on nano adsorbents: a solution for heavy metal removal from waste water

Tizzotti M, Charlot A, Fleury E, Stenzel M, Bernard J (2010) Modification of polysaccharides through controlled/living radical polymerization grafting—towards the generation of high performance hybrids. Macromol Rapid Commun 31:1751–1772

Uddin MK (2017) A review on the adsorption of heavy metals by clay minerals, with special focus on the past decade. Chem Eng J 308:438e462

Varma AJ, Deshpande SV, Kennedy JF (2004) Metal complexation by chitosan and its derivatives: a review. Carbohydr Polym 55:77–93

Vilcinskas K (2020) Alternative sources of graphene production

Wan Ngah WS, Endud CS, Mayanar R (2002) Removal of copper(II) ions from aqueous solution onto chitosan and cross-linked chitosan beads. React Funct Polym 50:181–190

Wang J, Chen C (2006) Biosorption of heavy metals by Saccharomyces cerevisiae: a review. Biotechnol Adv 24:427e451

Wang J, Chen C (2009) Biosorbents for heavy metals removal and their future. Biotechnol Adv 27:195e226

Wang XD, Vinodgopal K, Dai GP (2019) Synthesis of carbon nanotubes by catalytic chemical vapour deposition

White RL, White CM, Turgut H, Massoud A, Tian ZR (2018) Comparative studies on copper adsorption by graphene oxide and functionalized graphene oxide nanoparticles. J Taiwan Inst Chem Eng 85:18–28

Whitener Jr KE, Sheehan PE (2014) Diamond Relat Mater 46:25–34

Xu J, Qu Z, Yan N, Zhao Y, Xu X, Li L (2016) Size-dependent nanocrystal sorbent for copper removal from water. Chem Eng J 284:565–570

Yogeshwar NT, Satish S, Arun KJ (2016) Experimental investigation and modeling of copper ion adsorption in packed and expanded bed. Chem Eng Process Process Intensif 109:51–58

Yuryev VP, Cesaro A, Bergthaller WJ (eds) (2002) Starch and starch containing origins-structure, properties and new technologies Starch. Nova Science Publishers, Inc., New York

Zare-Dorabei R, Ferdowsi SM, Barzin A, Tadjarodi A (2016) Highly efficient simultaneous ultrasonic-assisted adsorption of Pb(II), Cd(II), Ni(II) and Cu(II) ions from aqueous solutions by graphene oxide modified with 2,20-dipyridylamine: central composite design optimization. Ultrason Sonochem 32:265–276

Zhang LM, Chen DQ (2002) An investigation of adsorption of lead(II) and copper(II) ions by water-insoluble starch graft copolymers. Colloids Surf A Phys Eng Aspects 205:231–236

Zhang L, Huang T, Zhang M, Guo X, Yuan Z (2008) Studies on the capability and behavior of adsorption of thallium on nano-Al_2O_3. J Hazard Mater 157(2):352–357

Zhang L, Zeng Y, Cheng Z (2016) Removal of heavy metal ions using chitosan and modified chitosan: a review. J Mol Liq 214:175e191

Zhong QQ, Yue QY, Li Q, Gao BY, Xu X (2014) Removal of Cu(II) and Cr(VI) from wastewater by an amphoteric sorbent based on cellulose-rich biomass. Cabohydr Polym 11:788–796

Chapter 13
Biopolymeric Nanomaterials as Eco-Friendly Adsorbents for Environmental Applications and Its Challenges—A Comprehensive Review

P. Muthukumaran, G. Maalika, C. Deeksha, M. Madhu Mallika, S. B. Varshini, and Aravind Jeyaseelan

Abstract Pollution of water that are resulted from the discharge of effluents have been a long-standing environmental concern. Although there are many traditional methods of wastewater treatment, it fails meet the required demands which is also not cost-effective. Nanotechnology has effective strategies for the removal of waste from water, and it is capable to reuse the water. Nano technology has great potential in improving the efficiency of water prevention, purification and in terms of reducing the costs. Due to the excellent properties, nanomaterials are especially used in the field of wastewater treatment. This review paper deals with the biopolymeric nano materials as an eco-friendly adsorbent. Biopolymers have been used adsorbents, and natural flocculants in wastewater treatment. In particular, the biopolymers obtained from naturals resources are reviewed. Firstly, the types of biopolymers which include cellulose, chitosan, chitin, alginate, xanthan gum, dextrin, carrageenan, lignin, protein are reviewed along with its application as adsorbents. In the second part of article, the preparation methods for absorbents, its environmental applications and challenges are discussed in detail. Soon, it is believed that there will be continuous production of nano materials, to support the development of eco-friendly, economic, and efficient nanomaterials for real life applications. It is essential to know about the potential benefits and unknown danger of nanomaterials in the upcoming years to follow their advancements.

Keywords Nanomaterials · Wastewater treatment · Biopolymers · Adsorbents · Eco-friendly and environment applications

P. Muthukumaran (✉) · G. Maalika · C. Deeksha · M. M. Mallika · S. B. Varshini
Department of Biotechnology, Kumaraguru College of Technology, Coimbatore, Tamil Nadu 641049, India
e-mail: muthukumaran.p.bt@kct.ac.in; m.kumaran005@gmail.com

A. Jeyaseelan
Department of Biotechnology, Saveetha School of Engineering, Saveetha Institute of Medical and Technical Sciences (SIMATS), Thandalam, Chennai, Tamil Nadu 602 105, India

© The Author(s), under exclusive license to Springer Nature Switzerland AG 2025

A. Jeyaseelan and K. Murugasen (eds.), *Sustainable Environmental Remediation: Avenues in Nano and Biotechnology*, Environmental Science and Engineering,
https://doi.org/10.1007/978-3-031-78483-5_13

13.1 Introduction

The demand for potable water keeps increasing by 64 billion cubic meters per annum, due to the increased world population at a rate of 80 million per year (WWAP 2009, 2012). To meet this demand various purification methods have been used (Shannon et al. 2008). One of the methods that has been discussed here is adsorption which is the most effective treatment for the treatment of contaminated water (Shannon et al. 2008; Sophia and Lima 2018; Tran et al. 2015). There are many nanomaterials NMs that has been used for adsorption. One among them is Nano polymers (Das et al. 2017). Due to the environmental properties, biopolymers have been used for water treatment (Vinod et al. 2020). A polymer that occurs from natural sources, or from biologically sourced materials are called as biopolymer or natural polymer (Hernandez et al. 2014; Aravamudhan et al. 2014). Biopolymers are monomeric units that has been covalently merged to form polymeric structure, which has been sorted into polyester polysaccharides, polycarbonates, polyamide and recognized for their role as bio adsorbent (Mohanty et al. 2005). Polysaccharides have unique adsorption characteristics mainly due to the (1) hydrophilic interactions, provided by the monosaccharides that has hydroxyl groups in it. (2) Chemical interactions provided by the polymeric chain which has no of functional groups in it. (3) Polymeric chain flexibility (Janaki et al. 2012; Pal et al. 2012).

The important biopolymer that has been used for adsorption purposes are cellulose, chitosan, chitin, alginate, xanthan gum, dextrin, carrageenan (Luzi et al. 2019). Biopolymer is a promising material which can be used to confront the environmental impediments as it has the drexity to be tailored contrivance fabricated into nanofibers, nanofibrils, nanowires, nanocrystals, and biopolymer-based aerogels (Mohanty et al. 2005). Although all polymeric nano adsorbent has a SSA ranging from 1000 to 2000 $m^2\ g^{-1}$ which influences us to use polymers as adsorbents (Jiang et al. 2008; Ghanem et al. 2007).

Biopolymeric nanoparticles (Bp-Nps) and biopolymeric nanomaterials (Bp-Nms) have been extensively used in the environmental applications. In recent years much research is being done on the exploration of new low-cost, more stable, safer, Eco-friendly and efficient environment mitigating tools, to meet the demand for potable water (Mohanty et al. 2005). The use of biopolymeric nanomaterials as eco-friendly adsorbent has been discussed in this article, as biopolymers are biodegradable under suitable conditions with applicable temperature, oxygen, moisture and humidity; the biopolymers degrade into fragments or disintegrate, which is further degraded by microorganisms, with no toxic and environmental-friendly residue (Preethi et al. 2020). Alongside the types of biopolymers, environment applications of biopolymers and the challenges that are faced while the preparation of biopolymers has also been discussed in this article.

13.2 Classifications of Biopolymer

There are many types of biopolymers such as PLA, PHA, PHB, PHV, PHH, protein and polysaccharides (Luzi et al. 2019). Among the biopolymer, polysaccharides have been considered as the best biopolymer as it is an agro-based polymer (Luzi et al. 2019; Mar Orta et al. 2020). Polysaccharides has been classified into cellulose, chitin, alginate, xanthan gum, dextrin, and carrageenan (Luzi et al. 2019).

13.2.1 Starch

Starch is a polysaccharide which is primary energy reserve in plants (Arora et al. 2018). Amylose and amylopectin are the two types of polymeric molecules present in the starch (Campos et al. 2011). Amylose is a linear chain with (1 → 4)-α-d-glucopyranosyl units with polymerization ranging from 300 to 10,000 (Campos et al. 2011; Avérous et al. 2014). Amylopectin is composed of amylose chains, which is a backbone of amylose and is branched with side units od D-glucopyranosyl linked by α-1,6-glycosidic bonds (Campos et al. 2011; Xie et al. 2013). Starch is an environment-friendly, biodegradable, abundantly available biopolymer which has high chemical stability, reactivity and it is sourced from renewable material, thus making it as an interesting alternate adsorbent (Yang et al. 2010; Mittal et al. 2018).

13.2.2 Chitosan

Chitin is found abundantly in marine crustaceans, shrimps, crabs, and marine sponges which has (1,4)-b-*N*-acetyl glycosaminoglycan-repeating structure, and it is a cellulose-like polysaccharide (Wan and Tai 2013; Ge et al. 2018). Chitosan is a *N*-deacetylated product of chitin, which is the only cationic biopolymer with positive charges on its amino group, nontoxic, biocompatible, and it is the second most abundant polysaccharide after cellulose in nature (Wan and Tai 2013; Ge et al. 2018; Sethy and Sahoo 2019). Chitosan is a suitable nano adsorbent for the removal of heavy metal ions, dyes, organochloride pesticides, and other organic pollutants from water, as it is highly rich in hydroxyl and amine groups, thus enhancing its adsorption capacity (Bhatnagar and Sillanpää 2009). The properties of chitosan are hydrophilicity, biocompatibility, biodegradability, chelating ability, antimicrobial activity and film forming capacity which makes this polysaccharide well suited for application in water treatment and also in various other fields (Sethy and Sahoo 2019).

13.2.3 Cellulose

Cellulose is the major building blocks of higher plants, and it is also present in algae, fungi, bacteria, and also in marine-living creatures (Klemm 2005). One of the most innovative biopolymers for the development of adsorbents is cellulose, which is a polysaccharide consisting in a linear polymer chain of b (1,4)-linked D-glucopyranose monomers (Siqueira et al. 2010). Cellulose is the most abundant material in the world with 50-billion tons per annum production among various biopolymers (Jonoobi et al. 2015). Cellulose is rich in oxygen functionalities, such as carboxyl and hydroxyl groups which facilitate its functionalization by making it more efficient for adsorption (Ray et al. 2020). Cellulose nanofibers (CNF), nanofibrils, nanocrystals, or nanowires which are the synthesized forms cellulose-based nanomaterials are found to be interesting nanomaterials in environmental applications that of pollutants adsorption in aqueous solution (Blanco et al. 2018; Mondal 2017). For example, hydroxyethyl cellulose is one of the kind of biopolymers, which is tolerant to thermal, durable in saline environment and stable in neutral/basic pH solution (Machado et al. 2020).

13.2.4 Alginate

Alginate is an essential polysaccharide, which is biosynthesized by bacteria and brown seaweeds (Khalil et al. 2017; Ucankus et al. 2018). It has a basic structure of linear unbranched polymers, which consists of β-(1–4)-linked D-mannuronic acid (M) and α-(1–4)-linked L-guluronic acid (G) residues, where its physical properties are dependent on these two acids and molecular weight (Ucankus et al. 2018; Huq et al. 2012). Alginates easily react with divalent cations to form cross-link gel matrixes, and this cross-linked alginate matrixes have been used as adsorbent in gel form for the easy handling (Pawar and Edgar 2012; Cataldo et al. 2016). Alginate and sodium alginate (obtained from neutralization of cross link gel matrixes using HCl) have been considered as efficient adsorbents as they have chelating activity towards metal ions and organic effluents in aqueous solution, due to the presence of hydroxyl and carboxyl group (Cataldo et al. 2015).

13.2.5 Lignin

The second most abundant biopolymer found next to cellulose is lignin, which has advanced properties like antioxidant, antimicrobial activity, ultraviolet radiation protection etc. (Thakur et al. 2014). It is one of the main sources of naturally occurring phenolic macromolecules on earth. Lignin has gained a significant attention as an adsorbent for the removal of various pollutants from water, as it has excessive active sites. Lignin nanocomposites have been proved as a promising alternative for outdated adsorbents for the removal of effluents (Ray et al. 2020). The modification of lignin with various functional groups enhances the affinity towards heavy metals and adsorption of organic contaminants. For example, Lignin-functionalized carbon nanomaterials are generally employed for the removal of heavy metal ions (Ray et al. 2020; Ge and Li 2018).

13.2.6 Gelatin

Gelatin is one type of protein, and a natural polymer obtained by denaturing the triple-helix structure of collagen into single-strain molecules (George 2012). Gelatin is a water-soluble polypeptide which has valuable characteristics such as biodegradability, hydrophilicity, low cost, nontoxicity, and has numerous active functionalities of amino, carboxyl, and hydroxyl groups (Muyonga et al. 2004; Farris et al. 2010).

13.2.7 Gum

Gum is a polysaccharide biopolymer which can be obtained from many natural sources. The functional groups present in the gum, such as hydroxyl, amino, and carboxyl groups interact with the contaminants from water which aids in their adsorption into the gum surface which results in their removal from water (Gusain et al. 2020; Kumar et al. 2016, 2018). There are many different types of gum, for example, gum Arabic, xanthan gum, dextran and carrageenan.

13.2.8 Gum Arabic

Gum Arabic is exudate that is extracted from acacia Senegal and acacia seyal tree. It has many properties like antimicrobial, anti-inflammatory, and anticoagulant and it also has a desirable low viscosity, binding, emulsifying, stabilizing, and shelf-life enhancing properties (Anandharamakrishnan and Ishwarya 2015).

13.2.9 Xanthan gum

Xanthan gum is a natural polysaccharide which has a backbone of β-(1–4) D-glucose. It has many properties like availability, affordability, biodegradability, and the presence of hydrophilic functional groups like –OH and –COOH makes this natural gum suited for many applications (Ahmad and Mirza 2018a, b, c).

13.2.10 Carrageenan

Carrageenan is an essential linear sulphated polysaccharide obtained from the red seaweeds (Mar Orta et al. 2020; Khalil et al. 2017). As it is negatively charged, there exhibits an electrostatic interaction between it and positively charged molecule, and it is used as an adsorbent for the removal of organic pollutants from water (Mar Orta et al. 2020).

13.3 Adsorbent Preparation Methods

There are many absorbents which are currently in use, among those cellulose nanomaterials have been extensively used as adsorbents, as they have the following advantages like high modulus, high aspect ratio, high chemical resistance, high surface area, surface functionalization, excellent crystallinity, water stability, biodegradability etc. (Abouzeid et al. 2018). Cellulose is an interesting nanomaterial in the environment application, that of adsorption. The isolation of cellulose can be by either physical, chemical, or biological treatment which includes methods like acid-hydrolysis, ultrasonication, cryogenic-grinding, high-pressure homogenization, enzyme hydrolysis, and TEMPO-mediated oxidation (Liu et al. 2017).

Cellulose nanocrystals (CNCs) and cellulose nanofibrils (CNF) are the most used cellulosic nanomaterials. CNC is generally produced by acid hydrolysis of cellulose fibres whereas CNF is produced by mechanical methods (Hassan et al. 2017; Turbak et al. 1983). Generally, cellulose is crystalline and amorphous in nature. In order to break the crystallinity formed by strong hydrogen bonds, it has to go through a sequence of chemical and mechanical treatments for the extraction of crystalline nanocellulose and nanocellulose fibres (Lee et al. 2014; Nechyporchuk et al. 2016).

13.3.1 Preparation of Nanocellulose

It is the world's most significant renewable resource, as it is present in one-third of plants like wood, jute, cotton, hemp, flex and kapok (Jonoobi et al. 2015). The initial structure of cellulose fibre is transformed into nanofibrils (CNF) which has a diameter ranging from 10 to 100 nm based on disintegration power, when the plant cell wall is exposed to powerful mechanical disintegration (Dufresne et al. 1997; Pelissari et al. 2014; Habibi et al. 2008; Ioelovich 2012; Jahanbaani et al. 2016).

13.3.2 Methods of Preparation

Cellulose nanofibrils is synthesized by applying various chemical alkali, enzymatic hydrolyses which is done after which is done prior to the mechanical disintegration processes such as homogenization, micro fluidization, and super-grinding, which increases the gain of access to active hydroxyl groups to improve the inner surface, to modify crystallinity, split cellulose hydrogen bonds, to enhance the reactivity of fibres (Jahanbaani et al. 2016; KargarzadehH et al. 2012; Julie Chandra et al. 2016).

13.3.3 Pre-Treatment

By applying, strong mechanical shearing to cellulose pulp fibres, CNFs are prepared (Turbak et al. 1983). The main drawback in preparation process by mechanical methods is the energy consumption. Pre-treatment is a common technique used to reduce the energy consumption (Asim 2017). The principle of pre-treatment is mainly to weaken the hydrogen bonds, adding of repulsive charge, lowering DP or breaking down the amorphous links between CNFs (Lavoine et al. 2012). Pre-treatment is an extremely crucial stage as it can adjust the structural foundation, crystallinity, as well as polymorphism of the cellulose. Pulping techniques, bleaching, alkaline-acid-alkaline treatment, enzymatic treatment, ionic liquids, oxidation, and steam explosion are well-known pre-treatment approaches (Kargarzadeh et al. 2012; Mishra et al. 2018; Rojas et al. 2015; Carrillo et al. 2014).

13.3.4 High-Pressure Homogenization

High-pressure homogenization is a mechanical process which is used to reduce particle size. CNF is produced with a diameter and length of about 20-40 nm and several microns respectively, when cellulose fibres is subjected to high-pressure homogenization (Turbak et al. 1983). High-pressure homogenization is a technique used for the large-scale fabrication of CNF and used in the laboratory-scale-preparation of nanofibrils. This technique is carried out under an elevated pressure of about 50–2000 MPa, where the solution is pushed by using an extremely trim channel or orifice by taking advantage of the piston. The viscosity of the solution along with the exerted pressure determines the width of the homogenization gap. The streaming velocity of consequential high solution leads to an intensification of dynamic pressure and lowering of static pressure below the vapour pressure of aqueous phase, which results in the production of gas bubbles, where the gas bubbles break down instantly if the liquid is departed from the homogenization gap, by getting once again under the standard air pressure. The shockwaves and cavitation created by the gas bubble production and implosion incident, will trigger the disturbance to the fibrillar configuration of the cellulose, which results in the drop in the size of cellulose fibre due to significant pressure drop, excessive shear force, turbulent flow and interparticle collisions. The range of homogenization cycle and exerted pressure determines the cellulose fibrillation level (Zhang et al. 2013; Kluge et al. 2012).

13.3.5 Micro Fluidization

An additional technique employed for the cellulose nanofibrils production is micro fluidizer, which works at a consistent shear pressure. There will be an elevated shear force and pressure which range up to 40,000 psi, that is about 276 MPa. The predetermined-geometry microchannel which are purposely designed are placed inside the chamber, so that the slurry up to higher velocities. Whenever the slurry stream strikes on wear-resistant surfaces, the preferred shear and impact forces will be built. Recirculation of the slurry will be enabled by several check valves. Upon leaving the interaction section, the product is sent to the heat exchanger, where it is recirculated in the system for additional operation, or perhaps sent from the outside to the subsequent step in the operation. For the adaptation of different sized chambers to be able to enhance the level of fibrillation, it is required to perform the process repeatedly many times. The morphology of the extracted cellulose nanofibrils depends on the number of passes of microcrystalline cellulose MCC slurry in a microfluidizer. The aspect ratio of the nanofibrillar bundles is enhanced after the 10–15 transferring cycles, while extra flows resulted in agglomeration of the CNFs because of the expanded surface area as well as a greater level of the surface hydroxyl group (Kluge et al. 2012; Van De Voorde 2015; Peng et al. 2012; Jarrah et al. 2005).

13.3.6 Grinding

It is a method used for isolating cellulose fibres into nano-sized fibrils. By transferring the cellulose slurry between static as well as rotating grindstones which is rotating at about 1500 rpm, to exert a shearing stress to the fibres, the fibre fibrillation operation is performed throughout the process. The shear forces are used in the grinder for the fibrillation procedure, to decompose the cell wall configuration and arrangement as well as separate the nano scale fibrils. The distance between the disks, the actual morphology of the disk tunnels, along with the number of feeds into the grinder determines the level of fibrillation. To produce the fibrillated cellulose, numerous passes have been instructed in the homogenizer. The routine maintenance of the disk stone and the replacement of this method is downside since the wood pulp fibres have the ability to deteriorate the grooves and grift. The extra mechanical pre-treatment is usually not required in the grinder operation, which is an important benefit of this method (Jarrah et al. 2005; Hu et al. 2015a; Josset et al. 2014).

13.3.7 Cryo-Crushing

Cryo-crushing is a mechanical fibrillation way to cellulose which takes place under the refrigerated condition, and it generates fibrils with reason-ably big diameters, varying within 0.1–1 mm. The water-swollen cellulose fibres are usually refrigerated in liquid nitrogen and after that progressively crushed in this method. The substantial collision forces are used to freeze the cellulosic fibres which leads to the breakage of cell walls, due to the pressure implemented using ice crystals. The cryo-crushed fibres are subsequently distributed and dispersed uniformly in water with the help of a routine disintegrator. This technique is relevant to numerous cellulose materials which enable it to be considered as a fibre pre-treatment procedure which is usually dome before homogenization. The nano fibres produced from the soya bean using cryo-crushing has succeeding high-pressure fibrillation and it is also confirmed by the TEM that the nano fibres have diameter of about 50–100 in range. It also has outstanding dispersion capability in the acrylic emulsion when compared to that of water. Despite all of this, the cryo-crushing technique offers a low efficiency and is not cost worthy, due to its high energy expenses (Hu et al. 2015a; Josset et al. 2014; Kopania et al. 2012; Alemdar and Sain 2008).

13.3.8 High Intensity Ultrasonication

The cells are disturbed in aqueous conditions using high intensity ultrasonication, which is a common laboratory mechanical treatment method. It produces an effective cavitation where the growth, extension, and collapsing of microscopic gas bubbles once the water molecules intercept ultrasonic energy is also included it. The defibrillation of cellulose fibres is achieved because of hydrodynamic forces of the ultrasound on the pulp. The use of high intensity ultrasonication (HIUS) to the separation of nanofibers from a variety of cellulosic resources, for example, plain cellulose, microcrystalline cellulose, pulp, culinary banana peel, rice waste, as well as micro fibrillated cellulose has been investigated by several researchers, which confirmed that the mixture of microscale as well as nanoscale fibrils can be accomplished directly after the ultrasonication of the cellulose fibres. The diameter the extracted fibrils has found to be in the range of 20 nm. It provides aggregated fibrils with a vast width distribution, because of which few nanofibrils are extracted from the fibres and few stays on the surface of the fibre. It is also found that crystalline structure of certain cellulose fibres is modified due to ultrasonic treatment method. With the help of HIUS treatment, the consequences of temperature, concentration, power, dimensions, duration, as well as distance from the probe tip on the level of fibrillation of plenty of cellulose fibres has been examined and it is claimed that outstanding fibrillation was due to higher power as well as temperature, even though lengthier fibres were much less defibrillated. For fibrillation, greater pulp fraction and larger sized distance from the probe to beaker was not beneficial. The conjunction of HIUS together with HPH enhances the fibrillation together with uniformity of the nanofibers, in comparison to high intensity ultrasonication (HIUS) alone has been noticed. If the TEMPO-oxidation pulp is employed for HIUS treatment method, the nanofibrils cellulose output could be additionally improved (Villares et al. 2017; Redlinger-Pohn et al. 2016; Reddy et al. 2013; Lee et al. 2013).

13.3.9 Acid Hydrolysis

The simplest method used for the semi-crystalline polysaccharide nanoparticle preparation is acid hydrolysis (Jayakumar et al. 2020). Acid hydrolysis of purified cellulosic material is carried out by intense mineral acids (6–8 M) under-governed environment, time, agitation, and acid/cellulose ratio conditions to extract the cellulose nanocrystals (Vasconcelos et al. 2017). Various mineral acids like, hydrochloric, phosphoric, hydrobromic, phosphotungstic acids, sulfuric acids, maleic acids, nitric acids and formic acids are used for the acid hydrolysis (Reddy et al. 2013; Lee et al. 2013; Vasconcelos et al. 2017; Yu et al. 2013; Kontturi et al. 2016; Camarero Espinosa et al. 2013; Sadeghifar et al. 2011; Liu et al. 2014).

For cellulose nanocrystals production, sulfuric acid is the most used acid. There takes place a selected hydrolysis of disordered amorphous regions, as well as interfibrillar contacts of cellulose; on the other hand, consistent crystallites stay unchanged which enable it to be separated as rod-like nanocrystalline materials. Using water, the cellulose nanocrystals dispersion in an intense acid is antiquated and cleaned via consecutive centrifugations. To take away remaining acid from the dispersion, Neutralization or even dialysis by using distilled water is conducted. The uncharged nanocrystalline particles will flocculate in the aqueous medium if the cellulose nanocrystals are produced taking advantage of cellulose hydrolysis together with hydrochloric acid. Hydrolysis with solid acids, hydrolysis with gaseous acids, hydrolysis with metal Salt catalyst are some other hydrolysis techniques which can take place. Easy restoration of the solid acid, relative safe, lesser corrosion rate of the equipment. Using a transition metal-dependent catalyst, hydrolysis with metal salt catalyst is performed. The feasible hydrolysis operation with minor acidity, which is satisfactory and preferential is offered by transition metal-dependent catalyst. To induce the hydrolysis performance, in which an acidic solution (H^+) produces during the period of polarization between metal ions as well as water molecules, the valence condition of the metal ion is the paramount aspect. much more H^+ which behave efficiently in the co-catalysed acid hydrolysis reaction in the existence of metal ions, is produced by the greater valence state (Mishra et al. 2018; Reddy et al. 2013; Lee et al. 2013; Vasconcelos et al. 2017; Martínez-Sanz et al. 2011; Hu et al. 2015b; Boujemaoui et al. 2015; Novo et al. 2015; Zinoviadou et al. 2016).

After acid hydrolysis, to achieve CNC methods like centrifugation, sonication, and dialysis have been used (Rojas et al. 2015).

13.4 Environmental Application

Nano materials have currently used in many fields for its excellent properties, where the contribution of nanomaterials in the field of environment is more, as it used in wastewater treatment and many other.

13.4.1 Wastewater Treatment

Due to continuously, defiling limited freshwater resources and seriously affecting terrestrial, aquatic, and aerial flora, and fauna, it important to handle the wastewater treatment. Nanomaterials is used to improve the performance of classical wastewater treatment technologies by means of adsorption, catalysis, separation, and disinfection. Nanomaterials have been currently used in the field of sensor technologies, which is used to monitor the water contaminants. Decreasing the dimension of nano materials like, carbon nanotubes (CNTs), carbon nanofibers (CNFs), graphene family members (GFMs), fullerene (C60), metal and metal oxide nanoparticles (NPs), nano polymers, increases the specific surface area (SSA) and alters 2D and 3D crystallinity which in turn affect chemical reactivity as well as mechanical, thermal, and chemical properties, which might be useful in the development of more sustain-able, efficient, reliable, and affordable novel wastewater treatment technologies (Das et al. 2017).

Biological, physical, and chemical methods, such as aerobic/anaerobic digestion, biochar treatment, membrane filtration and use of adsorbents to remove various environmental pollutants comes under the conventional water treatment (Crini et al. 2019). To secure a sustainable water supply, there is an urgent need for safe, environmentally friendly, low-cost, and highly efficient water treatment and disinfection technologies (Huang et al. 2020). Due to their unique physical and chemical properties, nanomaterials are increasingly explored for the wastewater treatment applications (Mauter et al. 2018). Chemical modifications like surface coating, doping, oxidation–reduction will make nanomaterials more specific and reactive for the wastewater treatment (Halbus et al. 2019). There are different forms of nanomaterials like nanowires, nanotubes, films, particles, quantum dots and colloids are applied in wastewater treatment system and used in the removal of pollutants from industrial effluents, surface water, ground water and drinking water with the help of nanomaterial with eco-friendly and cost-effective manners (Singh et al. 2020). CNTs are qualified as efficient nano adsorbents for the removal of many kinds of water pollutants (heavy metals, organic pollutants, etc.) (Ibrahim et al. 2019; Zhang et al. 2019). Activated carbon (AC) is currently the most popular adsorbent displaying a SSA of 500–1000 $m^2\ g^{-1}$ (Chowdhury et al. 2013).

There are different types of methods used in the water purification system, like reverse osmosis (RO), ultrafiltration (UF), and microfiltration (MF). Nanoparticles like silver nanoparticles, gold nanoparticles, zinc oxide nano nanoparticles, titanium dioxide nanoparticles, magnesium oxide and carbon nanotubes, are used in the disinfection of water and wastewater (Li et al. 2008). Biofouling of the filter membrane which greatly reduces the filtering efficiency, is one of the challenges in the wastewater treatment (Huang et al. 2020). Very small size and large surface area, and excellent adsorption properties and strong reducing ability; are the special feature and notable characteristics of nanomaterials, which are the contribution for the removal of contaminants from the wastewater (Alrumman et al. 2016). When pristine CNFs are crossed with one another they are mesoporous, which makes them suitable for the adsorption of large organic pollutants (Bonso et al. 2014). For simultaneous adsorption of micro pollutants, ease of separation and regeneration, based on multiple interactions; multifunctional adsorbents need to be designed (Das et al. 2017).

For our future, there needs to be an there are needs to a strong technique for removal of the contaminant from water bodies, which will be done by nanotechnology by fulfilling the requirement of the improved technique, who ensures to provide a high-quality drinking water. It also provides effective alternative ways for the wastewater treatment, as it has higher interacting, absorbing, and reacting capabilities due to its small size. Some new and special types of nanoparticles will be developed which will be lighter, stronger, more effective, better characterized and will be easy to adjust with the existing system in near future (Berekaa 2016).

13.4.2 Adsorbent to Remove Dye, Metal, Pesticides from Drinking Water etc.

There are several adsorbents to remove dye, metal, pesticides, where some of them will be mentioned below:

1. The graphene nano plates which are fabricated, is cross-linked with chitosan spheres (CS/GNPs) using a syringe dropping method, which is employed for the removal of MO (anionic) and AR1 (cationic) organic dyes from contaminated water (Zhang et al. 2018).
2. For the removal of Pb^{2+} in aqueous solutions, sulfonated CNF prepared from wheat straw pulp fine was used (Sehaqui et al. 2014).
3. CNF which was modified with 3-aminopropyltriethoxysilane (APS) used as adsorbent for heavy metals (Hokkanen et al. 2014).
4. Chitosan blended polyvinyl alcohol was used for the removal of Eosin Yellow dye from an aqueous solution (Anitha et al. 2016).
5. Graphene -chitosan nano composite is used for the adsorption of picric acid (Mohseni Kafshgari and Tahermansouri 2017).
6. Lignin nanocomposites is used as an adsorbent for the effluent removal. Lignin grafted with CNT (L-CNT) is used for the adsorption of Pb(II), where lignin-functionalized CNMs was also reported for the removal of Pb(II) and other heavy metal ions (Li et al. 2017).

7. Lignin-chitosan extruded blends was used for the efficient adsorption of methylene blue (Albadarin et al. 2017).
8. GO and bentonite-reinforced hydrogels of polyvinyl alcohol (PVA)/ carboxymethyl cellulose, which are obtained from pineapple peel (PCMC) using green freeze–thaw cycles, are used for the efficient adsorption of MB from water. PVP/PCMC/GO/bentonite hydrogels can also be used for adsorption of MB (Dai et al. 2018).
9. By using vitamin C as a reducing agent, graphene oxide-cellulose sponge (rGO-CNC) are partially reduced, which are employed for the adsorption of dyes, heavy metals ions, and pharmaceutical compounds (Yousefi et al. 2018).
10. cellulose/graphene oxide (CGC) is used for the adsorption of six different triazine pesticides from wastewater (Zhang et al. 2015)
11. Hetero-polyacid incorporated cellulose acetate nanofibers was used for the removal MO dye (Ahmed et al. 2017).
12. A composite of gelatin/activated carbon (GE-AC) is used for the adsorption of rhodamine dye from water. Gelatin-CNT-iron oxide nanocomposite (Gel-CNT-MNPs) which are composed of magnetic beads is used for the removal of cationic methylene blue (MB) dye and anionic Direct Red 80 (DR) dye from wastewater (Saber-Samandari et al. 2017).
13. Starch-based graphene oxide/hydroxyapatite nanocomposite hydrogels (NHA) to remove the cationic dye (MG) from contaminated water. MWCNT-starch-iron oxide and MWCNT-iron oxide are used for the adsorption of MB (cationic dye) and MO (anionic dye) from water (Hosseinzadeh and Ramin 2018; Chang et al. 2011).
14. The magnetic beads of sodium alginate-graphene-L-cysteine (SA/GR/L-Cys) which was synthesized for magnetic field-assisted is used for the removal of pharmaceutical (CIP), dye (MB), and metal ions (Cu(II)). GO-incorporated alginate hydrogel beads (SA/GO) are used for the removal of organic pollutants (MO, MB, rhodamine B (RhB), and vat green 1 (VG1) organic dyes) from wastewater (Ma et al. 2018; Gan et al. 2018)
15. The magnetic guar gum which are covalently grafted with carbon nanotubes (GG-MWCNT-Fe_3O_4) are used for the adsorption of organic dyes (MB and neutral red (NR)). Magnetic gum tragacanth (GT) nanocomposite beads with GO are used for the simultaneous adsorption of inorganic heavy metal ions and organic dyes (Yan et al. 2012; Sahraei et al. 2017).
16. Polydopamine (PDA)-functionalized graphene hydrogel (PDA-GH), is used for the adsorption of organic dyes, heavy metal ions, and aromatic contaminants from water (Gao et al. 2013).
17. Nitrogen-doped TiO_2 nanopar-ticle catalysts and TiO_2 nanocomposites with multi walled carbon nanotubes effectively used for the removal of microbial contaminants in water (Escobar et al. 2001).
18. For removal of divalent heavy metal ions such as copper, nickel and cobalt, polymer grafted Fe_2O_3 nanocomposite is used (Zhong et al. 2007).

Challenges

13.5 Summary

The scope of biopolymer nanomaterials in the field of environment is discussed in this article, as it an effective alternative for synthetic polymers for adsorption of pollutants, dyes, heavy metals, and other contaminants from water resources, which is also eco-friendly. Biopolymers are obtained from natural resources and renewable resources like sugars, resins, proteins, amino acids, and vegetable oils and poly-lactic acid (PLA), poly-hydroxy-alkenoates (PHA), poly-3-hydroxybutyrate (PHB), polyhydroxy-valerate (PHV), and poly-hydroxy-hexanoate (PHH), gluten respectively. The biopolymers are classified into different types. The types of biopolymers are cellulose, chitin, chitosan, gluten, lignin, gum, alginate, starch, and many other polysaccharides. The unique adsorption characteristics of polysaccharides has been discussed. Biopolymers on conjugation with metals create biopolymer-nanomaterials which is called as nanocomposites. Cellulose is the abundantly available natural biopolymer and widely used for adsorption purposes. Cellulose and its tailored forms of cellulose-based-nanomaterials; cellulose nanofibers (CNF), nanofibrils, nanocrystals, or nanowires, are promising nanomaterials for an array of environmental application, that of xenobiotics adsorption from wastewater. Similarly other biopolymers properties and its adsorption characteristics have also been discussed in this article. Then the adsorbent preparation methods which can be mechanical, chemical, enzymatical is discussed in this article. The nanocellulose absorbent preparation has been mainly discussed, where the preparation starts with pre-treatment and then followed up by mechanical methods like ultrasonication, cryo-crushing, grinding, acid hydrolysis and other some methods have been done on nanocellulose to obtain CNF and CNCs. Finally, the environmental applications like wastewater treatment and the adsorbents used for the removal dyes, metal ions, and many other has been mentioned in this article.

References

Abouzeid RE, Khiari R, El-Wakil N, Dufresne A (2018) Current state and new trends in the use of cellulose nanomaterials for wastewater treatment. Biomacromol 20(2):573–597

Ahmad R, Mirza A (2018a) Application of *Xanthan gum*/n-acetyl cysteine modified mica bio nanocomposite as an adsorbent for the removal of toxic heavy metals. Groundwater Sustain Develop 7:101–108

Ahmad R, Mirza A (2018b) Synthesis of Guar gum/bentonite a novel bionanocompo-site: isotherms, kinetics and thermodynamic studies for the removal of Pb(II) and Cristal violet dye. J Mol Liq 249:805–814

Ahmad R, Mirza A (2018c) Adsorptive removal of heavy metalsand anionicdye fromaqueous solution using novel *Xanthan gum*-glutathione/zeolite bionanocomposite. Groundw Sustain Dev 7:305–312

Ahmed F, Arbab AA, Jatoi AW, Khatri M, Memon N, Khatri Z, Kim IS (2017) Ultrasonic-assisted deacetylation of cellulose acetate nanofibers: a rapid method to producecellulose nanofibers. Ultrason Sonochem 36:319–325

Albadarin AB, Collins MN, Naushad M, Shirazian S, Walker G, Mangwandi C (2017) Activated lignin-chitosan extruded blends for efficient adsorption of methylene blue. Chem Eng J 307:264–272

Alemdar A, Sain M (2008) Isolation and characterization of nanofibers from agricultural residues—wheat straw and soy hulls. Bioresour Technol 99:1664e1671. https://doi.org/10.1016/j.biortech.2007.04.029

Alrumman SA, El-kott EF, Keshk SMAS (2016) Water pollution: source and treatment. Am J Environ Eng 6(3):88–98

Anandharamakrishnan C, Ishwarya PI (2015) Spray drying technique for food ingredient encapsulation. Wiley

Anitha T, Kumar PS, Kumar KS (2016) Synthesis of nano-sized chitosan blended polyvinyl alcohol for the removal of eosin yellow dye from aqueous solution. J Water Process Eng 13:127–136

Aravamudhan A, Ramos DM, Nada AA, Kumbar SG (2014) Chapter 4—natural polymers: poly saccharides and their derivatives for biomedical applications. Nat Synth Biomed Polym 6789. 319Biopolymer-functionalized carbon nanomaterials based adsorbents

Arora B, Bhatia R, Attri P (2018) Bionano composites: green materials for a sustainable future. In: Hussain CM, Mishra AK (eds) New polymer nanocomposites for environmental remediation. Elsevier, UK, pp 699–712 chapter 28

Asim M (2017) Nanocellulose: preparation method and applications, pp 261–276

Avérous L, Pollet E (2014) Nanobiocomposites based on plasticized starch. In: Halley PJ, Avérous L (eds) Starch polymers. Elsevier, U.K., pp 211–239 Chapter 8

Berekaa MM (2016) Nanotechnology in wastewater treatment; influence of nanomaterials onmicrobial systems. Int J Curr Microbiol App Sci 5(1):713–726

Bhatnagar A, Sillanpää M (2009) Applications of chitin- and chitosan-derivatives for the detoxification ofwater and wastewater—a short review. Adv Colloid Interface Sci 152:2638

Blanco A, Monte MC, Campano C, Balea A, Merayo N, Negro C (2018) Nano cellulose for industrial use: cellulose nanofibers (CNF), cellulose nanocrystals (CNC), and bacterial cellulose (BC). Handbook of nanomaterials for industrial applications. Elsevier, pp 74–126

Bonso JS, Kalaw GD, Ferraris JP (2014) J Mater Chem A 2:418–424

Boujemaoui A, Mongkhontreerat S, Malmstr€om E, Carlmark A (2015) Preparationand characterization of functionalized cellulose nanocrystals. Carbohydr Polym 115:457e464. https://doi.org/10.1016/j.carbpol.2014.08.110

Camarero Espinosa S, Kuhnt T, Foster EJ, Weder C (2013) Isolation of thermally stable cellulose nanocrystals by phosphoric acid hydrolysis. Biomacromol 14:122330

Campos CA, Gerschenson LN, Flores SK (2011) Development of edible films and coatings withantimicrobial activity. Food Bioprocess Technol 4(6):849–875. https://doi.org/10.1007/s11947-010-0434-1

Carrillo CA, Laine J, Rojas OJ (2014) Micro emulsion systems for fiber deconstruction into cellulose nanofibrils. ACS Appl Mater Interfaces 6:22622e22627. https://doi.org/10.1021/am5067332

Cataldo S, Muratore N, Orecchio S, Pettignano A (2015) Enhancement of adsorption ability of calciumalginate gel beads towards Pd(II) ion. A kinetic and equilibrium study on hybrid laponite and mon-tmorillonitealginate gel beads. Appl Clay Sci 118:16270

Cataldo S, Gianguzza A, Milea D, Muratore N, Pettignano A (2016) Pb(II) adsorption by a novel acti-vated carbon—alginate composite material. A kinetic and equilibrium study. Int J Biol Macromol 92:76978

Chang PR, Zheng P, Liu B, Anderson DP, Yu J, Ma X (2011) Characterization of magnetic solublestarch-functionalized carbon nanotubes and its application for the adsorption of the dyes. J Hazard Mater 186:214450

Chowdhury ZZ, Hamid SBA, Das R, Hasan MR, Zain SM, Khalid K, Uddin MN (2013) BioResources 8:6523–6555

Crini G, Lichtfouse E, Wilson LD, Morin-Crini N (2019) Conventional and non-conventional adsorbents for wastewater treatment. Environ Chem Lett 17(1):195–213

Dai H, Huang Y, Huang H (2018) Eco-friendly polyvinyl alcohol/carboxymethyl cellulose hydrogels rein-forced with graphene oxide and bentonite for enhanced adsorption of methylene blue. Carbohydr Polym 185:111

Das R, Vecitis CD, Schulze A, Cao B, Ismail AF, Lu X, Ramakrishna S (2017) Recent advances in nanomaterials for water protection and monitoring. Chem Soc Rev 46(22):6946–7020

del Mar Orta M, Martín J, Santos JL, Aparicio I, Medina-Carrasco S, Alonso E (2020) Biopolymer-clay nanocomposites as novel and ecofriendly adsorbents for environmental remediation. Appl Clay Sci 198:105838

Dufresne A, Cavaille J-Y, Vignon MR (1997) Mechanical behavior of sheets pre-pared from sugar beet cellulose microfibrils. J Appl Polym Sci 64:1185e1194. https://doi.org/10.1002/(SICI)1097-4628(19970509)64:6<1185:AID-APP19>3.0.CO;2-V

Escobar IC, Randall AA, Taylor JS (2001) Bacterial growth in distribution systems: effect ofassimilable organic carbon and biodegradable dissolved organic carbon. Environ Sci Technol 35(17):3442–3447

Farris S, Song J, Huang Q (2010) Alternative reaction mechanism for the cross-linking of gelatin with glutaraldehyde. J Agric Food Chem 58:9981003

Gan L, Li H, Chen L, Xu L, Liu J, Geng A et al (2018) Graphene oxide incorporated alginate hydrogelbeads for the removal of various organic dyes and bisphenol A in water. Colloid Polym Sci 296:60715

Gao H, Sun Y, Zhou J, Xu R, Duan H (2013) Mussel-inspired synthesis of polydopamine-functionalizedgraphene hydrogel as reusable adsorbents for water purification. ACS Appl Mater Interfaces 5:42532

Ge Y, Li Z (2018) Application of lignin and its derivatives in adsorption of heavy metal ions in water: a review. ACS Sustain Chem Eng 6:718192

Ge S et al (2018) Enhanced mechanical properties and gelling ability of gelatin hydrogels rein-forcedwith chitin whiskers. Food Hydrocoll 75:1–12. https://doi.org/10.1016/j.foodhyd.2017.09.023

George J (2012) High-performance edible nanocomposite films containing bacterial cellulose nanocrystals. Carbohydr Polym 87:2031–2037

Ghanem BS, Msayib KJ, McKeown NB, Harris KD, Pan Z, Budd PM, Butler A, Selbie J, Book D, Walton A (2007). Chem Commun 67–69

Gusain R, Kumar N, Ray SS (2020) Recent advances in carbon nanomaterial-based adsorbents for water purification. Coord Chem Rev 405:213111

Habibi Y, Goffin A-L, Schiltz N, Duquesne E, Dubois P, Dufresne A (2008) Bio-nanocomposites based on poly(ε-caprolactone)-grafted cellulose nano-crystals by ring-opening polymerization. J Mater Chem 18:5002. https://doi.org/10.1039/b809212e

Halbus AF, Horozov TS, Paunov VN (2019) Controlling the antimicrobial action of surface modified magnesium hydroxide nanoparticles. Biomimetics 4(2):41

Hassan E, Hassan M, Abou-zeid R, Berglund L, Oksman K (2017) Use of bacterial cellulose and crosslinked cellulose nanofibers membranes for removal of oil from oil-in-water emulsions. Polymers 9:1–14

Hernandez N, Williams RC, Cochran EW (2014) The battle for the"green"polymer. Differentapproaches for biopolymer synthesis: bioadvantaged vs. bioreplacement. Org Biomol Chem 12:283449

Hokkanen S, Repo E, Suopajärvi T, Liimatainen H, Niinimaa J, Sillanpää M (2014) Adsorption of Ni(II), Cu(II) and Cd(II) from aqueous solutions by amino modified nanostructured microfibrillated cellulose. Cellulose 21:1471–1487

Hosseinzadeh H, Ramin S (2018) Fabrication of starch-graft-poly(acrylamide)/graphene oxide/hydroxyap-atite nanocomposite hydrogel adsorbent for removal of malachite green dye from aqueous solution. Int J Biol Macromol 106:10115.323 Biopolymer-functionalized carbon nanomaterials based adsorbents

Hu C, Zhao Y, Li K, Zhu JY, Gleisner R (2015a) Optimizing cellulose fibrillation for the production of cellulose nanofibrils by a disk grinder. Holzforschung 69:993e1000. https://doi.org/10.1515/hf-2014-0219

Hu L, Lin L, Wu Z, Zhou S, Liu S (2015b) Chemocatalytic hydrolysis of cellulose intoglucose over solid acid catalysts. Appl Catal B Environ 174e175:225e243. https://doi.org/10.1016/j.apcatb.2015.03.003

Huang W, Tao F, Li F, Mortimer M, Guo LH (2020) Antibacterial nanomaterials for environmental and consumer product applications. Nano Impact 100268

Huq T, Salmieri S, Khan A, Khan RA, Le Tien C, Riedl B et al (2012) Nano crystalline cellulose (NCC) reinforced alginate based biodegradable nanocomposite film. Carbohydr Polym 90:1757–1763

Ibrahim RK, El-Shafie A, Hin LS, Mohd NSB, Aljumaily MM, Ibraim S, AlSaadi MA (2019) A cleanapproach for functionalized carbon nanotubes by deep eutectic solvents and their performance in theadsorption of methyl orange from aqueous solution. J Environ Manag 235:521–534

Ioelovich M (2012) Optimal conditions for isolation of nanocrystalline cellulose particles. Nanosci Nanotechnol 2:9e13. https://doi.org/10.5923/j.nn.20120202.03

Jahanbaani AR, Behzad T, Borhani S, Darvanjooghi MHK (2016) Electrospinning of cellulose nanofibers mat for laminated epoxy composite production. Fibers Polym 17:1438e1448. https://doi.org/10.1007/s12221-016-6424-9

Janaki V, Vijayaraghavan K, Ramasamy A, Lee K-J, Oh B-T, Kamala-Kannan S (2012) Competitiveadsorption of reactive orange 16 and reactive brilliant blue R on polyaniline/bacterial extracellularpolysaccharides composite—a novel eco-friendly polymer. J Hazard Mater 241:11017

Jarrah NA, Li F, van Ommen JG, Lefferts L (2005) Immobilization of a layer ofcarbon nanofibres (CNFs) on Ni foam: a new structured catalyst support. J Mater Chem 15:1946. https://doi.org/10.1039/b416977h

Jayakumar A, Heera KV, Sumi TS, Joseph M, Radhakrishnan EK (2020) Advances with synthesis and applications of green bionanomaterials. In: Green nanomaterials. Springer, Singapore, pp 209–226

Jiang J-X, Su F, Trewin A, Wood CD, Niu H, Jones JT, Khimyak YZ, Cooper AI (2008) J Am Chem Soc 130:7710–7720

Jonoobi M, Oladi R, Davoudpour Y, Oksman K, Dufresne A, Hamzeh Y, Davoodi R (2015) Different preparation methods and properties of nanostructured cellulose from various natural resources and residues: a review. Cellulose 22:935–969

Josset S, Orsolini P, Siqueira G, Tejado A, Tingaut P, Zimmermann T (2014) Energy consumption of the nanofibrillation of bleached pulp, wheat straw andrecycled newspaper through a grinding process. Nord Pulp Pap Res J 29:167e175. https://doi.org/10.3183/NPPRJ-2014-29-01-p167-175

Julie Chandra CS, George N, Narayanankutty SK (2016) Isolation and characterization of cellulose nanofibrils from arecanut huskfibre. Carbohydr Polym 142:158e166. https://doi.org/10.1016/j.carbpol.2016.01.015

Kargarzadeh H, Ahmad I, Abdullah I, Dufresne A, Zainudin SY, Sheltami RM (2012) Effects of hydrolysis conditions on the morphology, crystallinity, and thermal stability of cellulose nanocrystals extracted from kenaf bast fibers. Cellulose 19:855–866

Khalil HPSA et al (2017) Biodegradable polymer films from seaweed polysaccharides: a review oncellulose as a reinforcement material. EXPRESS Polym Lett 11(4):244–265

Klemm D et al (2005) Polymer science cellulose: fascinating biopolymer and sustainable rawmaterial. Angew Chem Int Ed 44(22):3358–3393. https://doi.org/10.1002/anie.200460587

Kluge J, Muhrer G, Mazzotti M (2012) High pressure homogenization of phar-maceutical solids. J Supercrit Fluids 66:380e388. https://doi.org/10.1016/j.supflu.2012.01.009

Kontturi E, Meriluoto A, Penttilä PA, Baccile N, Malho JM, Potthast A et al (2016) Degradation and crystallization of cellulose in hydrogen chloride vapor for high-yield isolation of cellulose nanocrystals. Angew Chem Int Ed 55:144558

Kopania E, Wietecha J, Ciechanska D (2012) Studies on isolation of cellulose fibres from waste plant biomass. Fibres Text East Eur 96:167e172

Kumar N, Mittal H, Parashar V, Ray SS, Ngila JC (2016) Efficient removal of rhodamine 6G dye fromaqueous solution using nickel sulphide incorporated polyacrylamide grafted gum karaya bionano-composite hydrogel. RSC Adv 6:2192939

Kumar N, Mittal H, Alhassan SM, Ray SS (2018) Bionanocomposite hydrogel for the adsorption of dyeand reusability of generated waste for the photodegradation of ciprofloxacin: a demonstration of thecircularity concept for water purification. ACS Sustain Chem Eng 6:1701125

Lavoine N, Desloges I, Dufresne A, Bras J (2012) Microfibrillated cellulose—its barrier properties and applications in cellulosic materials: a review. Carbohydr Polym 90:735–764

Lee CM, Mittal A, Barnette AL, Kafle K, Park YB, Shin H, Johnson DK, Park S, Kim SH (2013) Cellulose polymorphism study with sum-frequency-generation (SFG) vibration spectroscopy: identification of exocyclic CH_2OH conformation and chain orientation. Cellulose 20:991e1000. https://doi.org/10.1007/s10570-013-9917-3

Lee KY, Aitom€aki Y, Berglund LA, Oksman K, Bismarck A (2014) On the use ofnanocellulose as reinforcement in polymer matrix composites. Compos Sci Technol 105:15e27. https://doi.org/10.1016/j.compscitech.2014.08.032

Li D, Lyon DY, Li Q, Alvarez PJJ (2008) Effect of soil sorption and aquatic natural organic matter onthe antibacterial activity of a fullerene water suspension. Environ Toxicol Chem 27(9):1888–1894

Li Z, Chen J, Ge Y (2017) Removal of lead ion and oil droplet from aqueous solution by lignin-graftedcarbon nanotubes. Chem Eng J 308:80917

Liu Y, Wang H, Yu G, Yu Q, Li B, Mu X (2014) An ovel approach for the preparation of nanocrystalline cellulose by using phosphotungstic acid. Carbohydr Polym 110:415–422

Liu Q, Lu Y, Aguedo M, Jacquet N, Ouyang C, He W, Yan C, Bai W, Guo R, Goffin D, Song J (2017) Isolation of high-purity cellulose nanofibers from wheat strawthrough the combined environmentally friendly methods of steam explosion, microwave-assisted hydrolysis, and microfluidization. ACS Sustain Chem Eng 5:6183–6191

Luzi F, Puglia D, Torre L (2019) Naturalfiber biodegradable composites andnanocomposites: a biomedical application. In: Biomass, biopolymer-based materials, and bioenergy. Woodhead Publishing, pp 179e201. https://doi.org/10.1016/b978-0-08-102426-3.00010-2

Ma J, Ma Y, Yu F, Dai X (2018) Rotating magnetic field-assisted adsorption mechanism of pollutants onmechanically strong sodium alginate/graphene/L-cysteine beads in batch and fixed-bed column systems. Environ Sci Technol 52:1392534

Machado G, Santos F, Lourega R, Mattia J, Faria D, Eichler P, Auler A (2020) Biopolymers from lignocellulosic biomass: feedstocks, production processes, and applications. In: Lignocellulosic biorefining technologies, pp 125e158

Martínez-Sanz M, Lopez-Rubio A, Lagaron JM (2011) Optimization of the nano-fabrication by acid hydrolysis of bacterial cellulose nanowhiskers. Carbohydr Polym 85:228e236. https://doi.org/10.1016/j.carbpol.2011.02.021

Mauter MS, Zucker I, Perreault FO, Werber JR, Kim J-H, Elimelech M (2018) The role of nanotechnology in tackling global water challenges. Nat Sustain 1(4):166–175

Mishra RK, Sabu A, Tiwari SK (2018) Materials chemistry and the futurist eco-friendly applications of nanocellulose: status and prospect. J Saudi Chem Soc. https://doi.org/10.1016/j.jscs.2018.02.005

Mittal H, Alhassan SM, Ray SS (2018) Efficient organic dye removal from wastewater by magnetic carbo-naceous adsorbent prepared from corn starch. J Environ Chem Eng 6:711931

Mohanty AK, Misra M, Drzal LT, Selke SE, Harte BR, Hinrichsen G (2005) Natural fibers, biopolymers, and biocomposites: an introduction. CRC press, pp 17–51

Mohseni Kafshgari M, Tahermansouri H (2017) Development of a graphene oxide/chitosan nanocompo-site for the removal of picric acid from aqueous solutions: study of sorption parameters. Colloids Surf B Biointerfaces 160:67181

Mondal S (2017) Preparation, properties and applications of nanocellulosic materials. Carbohydr Polym 163:301–316

Muyonga JH, Cole CGB, Duodu KG (2004) Extraction and physico-chemical characterisation of Nileperch (*Lates niloticus*) skin and bone gelatin. Food Hydrocoll 18:58192

Nechyporchuk O, Belgacem MN, Bras J (2016) Production of cellulose nanofibrils: a review of recent advances. Ind Crops Prod 93:2e25. https://doi.org/10.1016/j.indcrop.2016.02.016

Novo LP, Bras J, García A, Belgacem N, Curvelo AAS (2015) Subcritical water: a method for green production of cellulose nanocrystals. ACS Sustain Chem Eng 3:2839e2846. https://doi.org/10.1021/acssuschemeng.5b00762

Pal S, Ghorai S, Das C, Samrat S, Ghosh A, Panda AB (2012) Carboxymethyl tamarind-g-poly(acrylam-ide)/silica: a high performance hybrid nanocomposite for adsorption of methylene blue dye. Ind Eng Chem Res 51:1554656

Pawar SN, Edgar KJ (2012) Alginate derivatization: a review of chemistry, properties and applications. Biomaterials 33:3279305

Pelissari FM, Sobral PJDA, Menegalli FC (2014) Isolation and characterization of cellulose nanofibers from banana peels. Cellulose 21:417e432. https://doi.org/10.1007/s10570-013-0138-6

Peng Y, Gardner DJ, Han Y (2012) Drying cellulose nanofibrils: in search of asuitable method. Cellulose 19:91e102. https://doi.org/10.1007/s10570-011-9630-z

Preethi R, Leena MM, Moses JA, Anandharamakrishnan C (2020) Biopolymer nanocomposites and its application in food processing. In: Green nanomaterials. Springer, Singapore, pp 283–317

Ray SS, Gusain R, Kumar N (2020) Carbon nanomaterial-based adsorbents for water purification: fundamentals and applications. Elsevier

Reddy MM, Vivekanandhan S, Misra M, Bhatia SK, Mohanty AK (2013) Biobased plastics and bio nanocomposites: current status and future opportunities. Prog Polym Sci 38:1653e1689. https://doi.org/10.1016/j.progpolymsci.2013.05.006

Redlinger-Pohn JD, Grabner M, Zauner P, Radl S (2016) Separation of cellulose fibres from pulp suspension by frothflotation fractionation. Sep Purif Technol 169:304e313. https://doi.org/10.1016/j.seppur.2016.06.004

Rojas J, Bedoya M, Ciro Y (2015) Current trends in the production of cellulose nanoparticles and nanocomposites for biomedical applications. Cellul Fundam Asp Curr Trends InTech. https://doi.org/10.5772/61334

Saber-Samandari S, Saber-Samandari S, Joneidi-Yekta H, Mohseni M (2017) Adsorption of anionic andcationic dyes from aqueous solution using gelatin-based magnetic nanocomposite beads comprisingcarboxylic acid functionalized carbon nanotube. Chem Eng J 308:113344

Sadeghifar H, Filpponen I, Clarke SP, Brougham DF, Argyropoulos DS (2011) Production of cellulose nanocrystals using hydrobromicacid and click reactions on their surface. J Mater Sci 46:734455

Sahraei R, Pour ZS, Ghaemy M (2017) Novel magnetic bio-sorbent hydrogel beads based on modifiedgum tragacanth/graphene oxide: removal of heavy metals and dyes from water. J Clean Prod 142:297384

Sehaqui H, Perez De Larraya U, Liu P, Pfenninger N, Mathew AP, Zimmermann T, Tingaut P (2014) Enhancing adsorption of heavy metal ions onto biobased nanofibers from waste pulp residues for application in wastewater treatment. Cellulose 21:2831–2844

Sethy TR, Sahoo PK (2019) Highly toxic Cr(VI) adsorption by (chitosan-g PMMA)/ sflicabionanocompositepreparedviaemulsifier-freeemulsionpolymerisation. Int J Biol Macromol 122:1184–1190

Shannon MA, Bohn PW, Elimelech M, Georgiadis JG, Mariñas BJ, Mayes AM (2008). Nature 452:301–310

Singh P, Yadav SK, Kuddus M (2020) Green nanomaterials for wastewater treatment. In: Green nanomaterials. Springer, Singapore, pp 227–242

Siqueira G, Bras J, Dufresne A (2010) Cellulosic bionanocomposites: a review ofpreparation, properties and applications. Polymers 2:728–765

Sophia AC, Lima EC (2018) Removal of emerging contaminants from the environment by adsorption. Ecotoxicol Environ Saf 150:1–17

Thakur VK, Thakur MK, Raghavan P, Kessler MR (2014) Progress in green polymer compo-sites from lignin for multifunctional applications: a review. ACS Sustain Chem Eng 2:107292

Tran VS, Ngo HH, Guo W, Zhang J, Liang S, Ton-That C, Zhang X (2015) Typical low cost biosorbents for adsorptive removal of specific organic pollutants from water. Bioresour Technol 182:353–363

Turbak AF, Snyder FW, Sandberg KR (1983) Micro fibrillated cellulose, a new cellulose product: properties, uses, and commercial potential. J Appl Polym Sci: Appl Polym Symp 37:815–827

Ucankus G, Ercan M, Uzunoglu D, Culha M (2018) Methods for preparation of na-nocomposites in environmental remediation. New Polim Nanocompos Environ Rem 1:1–28

Van De Voorde M (2015) Industrialization—large-scale production of nano-materials/components. In: Nano-micro interface bridging micro nano worlds, second edn, pp 677e684. https://doi.org/10.1002/9783527679195.ch33

Vasconcelos NF, Feitosa JPA, da Gama FMP, Morais JPS, Andrade FK, de Souza MDSM, de Freitas Rosa M (2017) Bacterial cellulose nanocrystals produced under different hydrolysis conditions: properties and morphological features. Carbohydr Polym 155:425e431. https://doi.org/10.1016/j.carbpol.2016.08.090

Villares A, Moreau C, Bennati-Granier C, Garajova S, Foucat L, Falourd X, Saake B, Berrin JG, Cathala B (2017) Lytic polysaccharide monooxygenases disruptthe cellulosefibers structure. Sci Rep 7. https://doi.org/10.1038/srep40262

Vinod A, Sanjay MR, Suchart S, Jyotishkumar P (2020) Renewable and sustainable biobased materials: an assessment on biofibers, biofilms, biopolymers and biocomposites. J Clean Prod 258:120978

Wan ACA, Tai BCU (2013) Chitin—a promising biomaterial for tissue engineering and stem-cell technologies. Biotechnol Adv 31(8):1776–1785. https://doi.org/10.1016/j.biotechadv.2013.09.007

WWAP (2009) The united nations world water development Report3: water in a changing world. UNESCO/Earthscan, Paris/London

WWAP (2012) The united nations world water development Report4: managing water under uncertainty and risk. UNESCO, Paris

Xie F, Pollet E, Halley PJ, Avérous L (2013) Starch-based nano-biocomposites. Progr Pol Sci 38:1590–1628

Yan L, Chang PR, Zheng P, Ma X (2012) Characterization of magnetic guar gum-grafted carbon nano-tubes and the adsorption of the dyes. Carbohydr Polym 87:191924

Yang Y, Wei X, Sun P, Wan J (2010) Preparation, characterization and adsorption performance of a novelanionic starch microsphere. Molecules 15:2872

Yousefi N, Wong KKW, Hosseinidoust Z, Sørensen HO, Bruns S, Zheng Y et al (2018) Hierarchi-callyporous, ultra-strong reduced graphene oxide-cellulose nanocrystal sponges for exceptional adsorptionof water contaminants. Nanoscale 10:717184

Yu H, Qin Z, Liang B, Liu N, Zhou Z, Chen L (2013) Facile extraction of thermallystable cellulose nanocrystals with a high yield of 93% through hydro chloric acid hydrolysis under hydro thermal conditions. J Mater Chem A Mater Energy Sustain 1:393844

Zhang C, Zhang RZ, Ma YQ, Guan WB, Wu XL, Liu X et al (2015) Preparation of cellulose/graphenecomposite and its applications for triazine pesticides adsorption from water. ACS Sustain Chem Eng 3:396405

Zhang C, Chen Z, Guo W, Zhu C, Zou Y (2018) Simple fabrication of chitosan/graphene nanoplatescomposite spheres for efficient adsorption of acid dyes from aqueous solution. Int J Biol Macromol 112:104854

Zhang Y, Nypel€o T, Salas C, Arboleda J, Hoeger IC, Rojas OJ (2013) Cellulose nanofibrils. J Renew Mater 1:195e211. https://doi.org/10.7569/JRM.2013.634115

Zhang L, Lin CY, Zhang D, Gong L, Zhu Y, Zhao Z, Xu Q, Li H, Xia Z (2019) Guiding principles for designing highly efficient metal-free carbon catalysts. Adv Mater 31:e1805252

Zhong LS, Hu JS, Cao AM, Liu Q, Song WG, Wan LJ (2007) 3D flowerlike ceriamicro/nanocomposite structure and its application for water treatment and CO removal. Chem Mater 19(7):1648–1655

Zinoviadou KG, Gougouli M, Biliaderis CG (2016) Innovative bio based materials for packaging sustainability. Innov Strateg Food Ind Tools Implement 167e189. https://doi.org/10.1016/B978-0-12-803751-5.00009-X

Chapter 14
Investigation of Aerosol Particle Filtration Efficiency Using Polycarbonate Electrospun Nanofibers Over 3D Printed Support

Preethi Ravikumar

Abstract The massive increase in the industrial revolution has now led to a decrease in the quality of the air, thereby increasing health problems. Air filtration has been recognized as an operational way to reduce particulate matter (PM) pollution for decades. Electrospun nanofibrous filters are regarded as an effective substitute over other air filter types due to their governable morphology with exceptional diameter/length ratio in a continuous method. In the current research, using electrospinning technique, 5 nanofibrous layers with varying concentrations were deposited over 3D printed polycarbonate (PC) support. A series of experiments were performed for electrospinning the PC nanofibers by altering various parameters such as nanofiber collection time, solvent ratio, deposition voltage, tip-to-collector distance, and polymer concentration. The filtration efficiency of the manufactured filters was evaluated, and its average fiber diameter, viscosity, conductivity, and pressure drop were measured. The nanofibrous layer exhibited a fiber diameter ranging from 0.19 ± 0.04 to 0.56 ± 0.14 μm. This nanofiber morphology enabled the filters to consistently achieve an efficient particle collection ability between 99.3 and 99.9%. Using the experimental results data, response surface plots were generated to visually represent the relationship between various factors and the desired characteristics of nanofiber media, including fiber diameter. The results show that all the nanofibers were successfully electrospun over 3D printed polycarbonate substrate and this filtering construction allowed for reaching high filtering efficiency (99.9%).

Keywords Aerosol filtration · Nanofibrous filters · Polycarbonate · Electrospinning

P. Ravikumar (✉)
Department of Environmental Technology, Faculty of Chemical Technology, Kaunas University of Technology, Radvilenu Pl. 19, 50254 Kaunas, Lithuania
e-mail: Preethi.ravikumar@ktu.edu

© The Author(s), under exclusive license to Springer Nature Switzerland AG 2025
A. Jeyaseelan and K. Murugasen (eds.), *Sustainable Environmental Remediation: Avenues in Nano and Biotechnology*, Environmental Science and Engineering,
https://doi.org/10.1007/978-3-031-78483-5_14

14.1 Introduction

One of today's most well-known and rapidly developing technologies, nanotechnology offers numerous cutting-edge uses in various fields such as in biology, medicine, air purification, energy generation, food packing, and more. Different nanostructures with varying properties and applications were manufactured due to their intrinsic qualities, which include increased stability, porosity, permeability, and surface area to volume ratio (Patel et al. 2022; Gore et al. 2020; Yuan et al. 2019; He et al. 2020).

Electrospinning is the most well-known and established technique for producing nanofibrous scaffolds which is utilized in aerosol filtering applications. An electric field is utilized throughout the process of electrospinning to draw polymer solutions or melts through a small orifice, forming thin fibers of sizes between nano and micrometers range. These fibers spun over a substrate serve as an aerosol sampling filter due to its high ratio of surface area to volume, microscale interstitial gaps and interconnectedness and high porosity (Liu et al. 2020; Al-Attabi et al. 2018). The electrospinning process is attainable through optimizing the processing parameters, like the solution concentration, flow rate and voltage applied. In addition, electrospun nanofiber materials of uniform size, is suitable for a broad application in air filtration (Ahne et al. 2019).

Aerosol sampling can be conducted using a variety of filter substrate types, each with its own unique characteristics and advantages. Some of the most widely used commercial filter substrates include Glass fiber filters, Teflon filters, Polycarbonate filters, Membrane filters, and Quartz filters (Krugly et al. 2023; Chow et al. 2022). Foams, membranes, and fibrous matrixes are the main categories for filtering substrates. Polymers are used to fabricate porous membranes and capillary membranes, which are used in membrane filters. Minerals (glass, quartz) or organic (cellulose) components are used to fabricate fibrous matrixes (Beckman et al. 2021).

Thermoplastic polymer polycarbonate (PC) is renowned for its exceptional strength, transparency, and impact resistance. These characteristics make it a good material to use in a variety of applications, including medication delivery, tissue engineering, and filtration (Jadhav et al. 2022). PC is a flexible material that has been used in a wide range of products, such as armored vehicles, electrical equipment, load-bearing items, compact and digital video discs, sports safety equipment's, and transparent windows that can withstand bullets, train carriages, and architectural projects. In addition, PC fibers can be modified by incorporating various agents such as nanoparticles, drugs, or other polymers to enhance their properties and functionality (Rashid et al. 2024). In most cases, PC is used in bulk and is shaped and sized to a certain specification. Recent studies (Yan et al. 2019; Ahmadi Bonakdar and Rodrigue 2024; Ebrahimnezhad-Khaljiri 2023; Behroozi et al. 2023; Baby et al. 2021) conclude that the PC nano fibrous membrane with smooth and bead-free fibers has potential as a material for particle filtration applications. Adding hexadecyl trimethyl ammonium bromide (CTAB) produced high efficiency ultrafine PC fibers with an average diameter of 319 nm and pore size of 504 nm.

He et al. (2020) devised a technique for electrospinning-based 3D printing to create shape-variable, biodegradable mask filters. The electrospun filter's applications were widened by Cao et al. (2019) to include industrial pollutants using 3D printing. The electrospinning process was optimized through the application of design of experiments (DOE), resulting in the development of nanofibers devoid of beads. These nanofibers were used to filter PM 2.5 emissions which demonstrated a filtration effectiveness of 81.16% using FDM 3D printing. Research investigations have been conducted on the 3D reconstruction of electrospun fiber membranes and direct electrospinning of nanofiber, including the use of 3D printed polymeric materials like PLA placed on graphene-based air filters (He et al. 2020). An adjustable, self-contained air filter was created using Three-dimensional printing to effectively remove nitrogen oxide (Xu et al. 2020). Although various studies show multilayer air filters exhibit high aerosol filtering performance (Vidakis et al. 2021; Goswami et al. 2021; Buivydiene et al. 2021; Li et al. 2019; Wang et al. 2020) there exists a research gap for studying the performance of PC based 3D printed substrate and nanofiber filter.

The current research aims at developing high-efficiency air sampling filters by compositing 3D printed PC support and nanofibers. Five distinct polycarbonate nanofiber layers were electrospun onto a 3D printed polycarbonate support layer which prevents breakdown during high-pressure sampling and collects 2D aerosol particles on the top of electrospun nanofibrous layer. Surface morphology, PC solution characteristics and the fibers' typical size in diameter were examined, and filtering capability of bead free nanofiber filters were further analyzed.

14.2 Materials and Methods

The Poly (Bisphenol A carbonate) (PC), average Molecular weight approx. 45,000, Purity. 100%, CAS no. 25037-45-0; *N,N*-dimethylformamide (DMF) (99.8%, Product No. 227056); Tetrahydrofuran (THF) (99.8%, Product No. 102391) and hexadecyl trimethyl ammonium bromide (CTAB), $\geq$98% CAS no. 57-09-0 were purchased from Sigma-Aldrich, USA. Electrospun nanofibers were fabricated using all materials without undergoing any purification process.

14.2.1 Preparation of Polymer Solutions

DMF and THF were chosen as solvents to dissolve PC for producing electro-spun nanofibers, due to their relatively slow volatile properties. DMF has a comparatively higher boiling point of 153 °C compared to THF, which has a boiling point range of 65–67 °C. To achieve homogeneous solutions, polycarbonate and CTAB with concentration ranges as detailed in Table 14.1 were dissolved in a mixture of *N,N*-dimethylformamide, Tetrahydrofuran in a 7:3 (v/v) ratio. The dissolution process involved mechanical stirring at a temperature of 55 °C. The PC concentrations in the solutions varied between 14, 17, and 20% (wt/v), while the concentrations of CTAB ranged from 0.2 to 0.8% (wt%) relative to PC.

The 3D printed PC support fabricated in Bious Labs, Lithuania [24] was used for collecting electrospun nanofibers. Mesh-type Polycarbonate supports (size of 80 × 50 × 0.4 mm, mesh size 0.786 × 0.786 mm) were made using commercial 3D printer. Polycarbonate filament (P51, Sigma Aldrich) with diameter of 1.75 mm were extruded using 0.4 mm nozzle at temperature of 235 °C. The printing pad temperature was kept at 85 °C. The mesh structure was formed using fused deposition modelling (FDM). 3D printed substrates were manually air cooled and hardened. Figure 14.1 shows the SEM image of 3D printed polycarbonate support.

The experimental PC and CTAB concentration ranges were designed in MODDE 7 software (Umetrics AB, Sweden) using the D-optimal-interaction model ensuring the formation of well-defined nanofibers. D-optimal designs (Buivydiene et al. 2019a) involve direct optimization guided by a selected optimality criterion and the model to be fitted.

Table 14.1 Pressure drop of PC electrospun filters at various flow rates for different PC & CTAB concentrations

Solution mixture	PC concentration (%)	CTAB concentration (%)	Pressure drop (Pa) 1.7 l/min	Pressure drop (Pa) 2.5 l/min	Pressure drop (Pa) 4.4 l/min	Pressure drop (Pa) 11.2 l/min
PC1	14	0.2	436.3	650.3	1212	2390.7
PC2	14	0.8	242.3	361.3	667.3	1877
PC3	17	0.5	222.7	415.7	593	1610.3
PC4	20	0.2	175.3	257.7	458.3	1208.7
PC5	20	0.8	225	332	602	1651

Fig. 14.1 3D printed polycarbonate support

14.2.2 Electrospinning Process and Filter Material Development

The electrospinning apparatus used was Model SE-01C and manufactured by Bious Labs in Lithuania (Krugly et al. 2023). The PC solution was exposed to a high DC 22 kV voltage across a capillary containing 10 ml syringe with 21 size needle gauge, and the solution was delivered at a flow rate of 0.5 mL/h with tip to collector distance of 15 cm. The rotation of the collector was set at a linear speed of 150 cm/min, equivalent to a rotation frequency of 6 rpm. The electric field causes the solution to be pulled through the capillary tip, forming a thin jet. The temperature and humidity of the apparatus was maintained at 25 °C and 50% throughout the experiment. The jet is then collected on a 3D printed substrate attached to the grounded collector, where the solvent evaporates, leaving behind the PC fibers. The fibers collected on the 3D printed polycarbonate substrate are used for further characterization. A pressure sensor was used to measure pressure drop at various flow rates for different PC & CTAB concentrations is tabulated in Table 14.1.

The electrospinning was conducted within a Plexiglas chamber at normal environmental conditions. Furthermore, the properties of the resulting fibers were altered by adjusting various factors, including the polymer solution concentration, applied voltage, needle gauge size, flow rate, and distance. To minimize the impact of air currents on the path of the electrospun jet, the needle, electrode, and grounded target are all contained within a chamber.

14.2.3 Morphology of PC Nanofibrous Layer

The shape and structure of the nanofibrous layers were statistically determined through scanning electron microscopy (using a Hitachi SEM S-3400N located in Germany). The size of the fibers was calculated using ImageJ software developed in the United States by the National Institutes of Health.

14.2.4 Viscosity and Conductivity Analysis

The viscosity of polymer solutions with a weight percentage of 10 was measured at 28 °C using a Brookfield digital viscometer (Model DV-E). The conductivity of each solution was assessed five times at 25 °C and their average was determined using a conductivity meter (COND 340i, WTW Ltd., USA), respectively.

14.2.5 Filtration Efficiency and Pressure Drop

The filtration efficiency of filters was tested against simulated aerosol particles of NaCl using a specifically developed testing rig (Krugly et al. 2023). The working principle behind determining filtration efficiency is that spherical particles exhibit greater filtration efficiency when compared to cubic particles (Buivydiene et al. 2019b). As a result, NaCl was chosen as the testing material to assess filtration efficiency, since its aerosol form consists of cubic particles that possess rounded edges. Due to its intermediate shape, NaCl is considered suitable for measuring the average filtration efficiency. The NaCl solution was aerosolized using a collision nebulizer (nebulizer Model CN 6J, BGI Inc., USA) and diluted with dry air in a porous tube diluter. The charged particles were equalized using an 85Kr bi-polar neutralizer (3054 A, TSI Inc., USA). 37 mm diameter filter samples were placed in a filter holder and subjected to the test aerosol. The Electrical Low-Pressure Impactor (ELPI+, Dekati Ltd., Finland) was used to monitor the particle concentrations both upstream and downstream. A pressure sensor (Model PCE-PDA 1L, PCE GmbH, Germany) was used to measure the pressure drop before and after the sample medium at face velocities of 5.3 cm/s and at intervals between 3 and 20 m/s. The results of the filtration efficiency tests were used to determine the overall performance of the filters and to identify any limitations in their ability to capture particles.

14.3 Results and Discussion

14.3.1 Viscosity and Conductivity

Table 14.2 shows properties of PC solution such as viscosity, conductivity, fibre diameter and pore diameter mean. Increasing the CTAB amount from 0.2 to 0.8% led to a decrease in viscosity for the PC 14% solution, from 50.7 to 32.2 mPa s, and similarly for the PC 20% solution, from 185.3 to 111.5 mPa s as mentioned in Table 14.2. The inclusion of CTAB caused a significant rise in conductivity and a reduction in the viscosity of the solution. The range of concentrations that yield continuous fibers in electrospinning is determined by the viscosity of the solution. The formation of continuous fibers is the result of the abundant chain entanglements within the polymer solution (Wang et al. 2019).

The conductivity of the DMF (4 μs/cm) and THF (3 μs/cm) in their original state was checked before preparing polymer solutions, to ensure that there were no alterations to the solvent's conductivity over time. The conductivity showed a direct correlation with the amount of CTAB added, as the increase in CTAB content as tabulated in Table 14.2 led to a proportional increase in conductivity. This can be attributed to the linear rise in ionic concentration. This finding suggests that the conductivity of the PC solution is influenced by the addition of CTAB. One of the studies on solvent conductivity on polystyrene electrospun fibers explains that

Table 14.2 Properties of PC solution used in electrospinning and characteristics of produced nanolayer material

Solution mixture	Viscosity (mPa S)	Conductivity (μs/cm)	Fiber diameter mean (μm)	Pore diameter mean (μm)
PC1	50.7	36	0.32	1.32
PC2	32.2	141	0.19	0.99
PC3	62.1	96	0.53	1.15
PC4	185.3	52	0.32	1.02
PC5	111.5	124	0.37	1.06

electrospun uniform fibers are a result of high conductivity of solutions and vice versa (Yin et al. 2022).

It is widely recognized that surface tension and electrostatic forces compete during the electrospinning process. Hence, the enhanced conductivity likely enabled the solution to carry a greater charge, leading to an increased electrostatic force that could effectively counteract surface tension (Xue et al. 2019). Therefore, the solution's conductivity is an extremely important parameter in the formation of bead-free nanofibers, particularly when low polymer concentrations are employed in this case. The nature of one of the solvents in the polymer solution exerts a significant influence on the solution's conductivity. Therefore, changes in solution conductivity can result in observable alterations in surface morphology. In conjunction with other electrospinning parameters, solution's conductivity plays a crucial role in fabricating polymer nanofibers with the desired morphology (Ibrahim and Klingner 2020). Using the same material (PC) for both the nanofibers and the substrate provided good mechanical compatibility, reducing the risk of delamination or detachment of the nanofibers from the substrate (Badmus et al. 2021).

Overall, the combination of increased conductivity and reduced viscosity resulted in a substantial decrease in the diameter of the PC nanofibers. The viscosity of all PC solutions was measured using a Brookfield Viscometer with Spindle number L2 respectively, at a speed of 20 rpm.

14.3.2 Morphology

All the PC solutions yielded bead-free fibers as represented in Fig. 14.2 with an average diameter as follows: 0.32, 0.19, 0.53, 0.32, 0.37 μm as shown in Table 14.1. The solubility of PC was enhanced by the addition of CTAB to the solution. CTAB, a cationic surfactant, enhances its conductivity, resulting in the successful production of ultra-fine PC fibers without any beads. Figure 14.3 shows the size distribution of (a) PC1, (b) PC2, (c) PC3, (d) PC4, (e) PC5. The PC fibers demonstrated desirable morphology characteristics, including a smooth surface and a lack of beading, as indicated by their diameter. The formation of smooth fibers without beads was attributed to achieving a balance between many electrospinning variables, including viscosity, concentration, and applied voltage, in relation to surface tension. However, solvent evaporation had a greater impact on fiber morphology changes as compared to other factors. It was also observed that solvent evaporation during the electrospinning process led to the occurrence of droplet coagulation at the needle tip. Usually, solutions with a high viscosity had a greater tendency to overcome surface tension, causing the droplets to form a continuous jet under the influence of the applied voltage. However, it was observed that as the concentration increased, the uniformity of the produced fibers decreased. With a nanofiber receiving time of 60 min, PC nanofibers exhibited a denser configuration. PC1–3 has relatively lower viscosity value ranges from 32.2 to 62.1 (mPa S) which usually results in bead free fibers as lower viscosity favors electrospinning process. Whereas PC4 and PC5 exhibited relatively higher values of viscosity from 111.5 to 185.3 mPa S. But this was overcome by adjusting other electrospinning parameters such as tip to collector distance, flowrate, voltage. To ensure optimal chemical compatibility and adherence, each of the polycarbonate (PC) concentrations were electrospun onto a support layer consisting of a 3D printed polycarbonate substrate. Figure 14.4 shows the cross-sectional diagram of 2-layer air filtration substrate.

The plan generated by the Modde was adjusted until the complete dissolution of PC concentration with CTAB was achieved. For PC 14%, increasing the CTAB concentration resulted in a decrease in fiber size from 0.32 to 0.19 μm. However, for PC 20%, further increases in CTAB content had limited influence on fiber diameter.

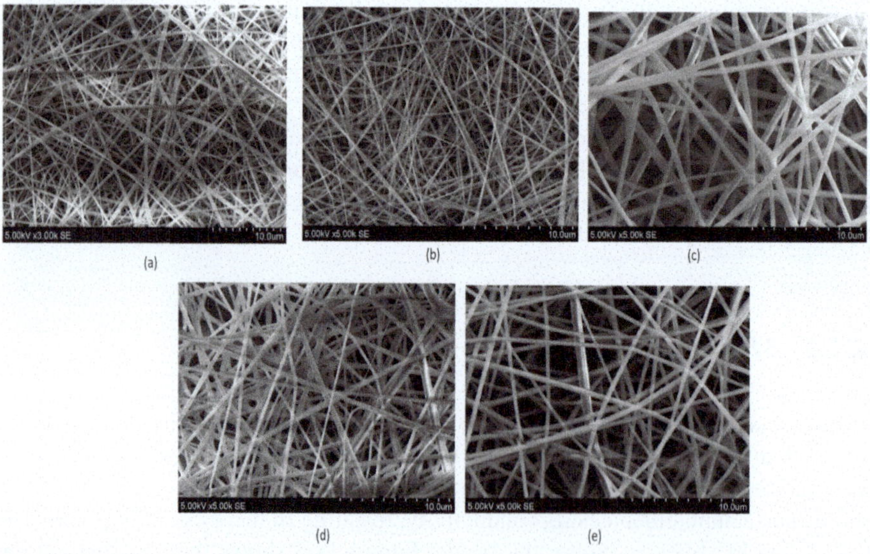

Fig. 14.2 SEM images of polycarbonate nanofibers deposited layers of the filter used for aerosol sampling **a** PC1, **b** PC2, **c** PC3, **d** PC4, **e** PC5

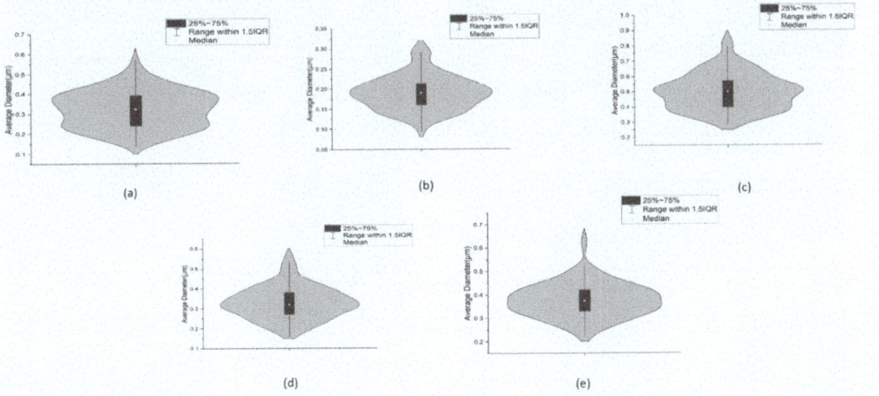

Fig. 14.3 The size distribution of the nanofibers is depicted in a box and whisker plot **a** PC1, **b** PC2, **c** PC3, **d** PC4, **e** PC5

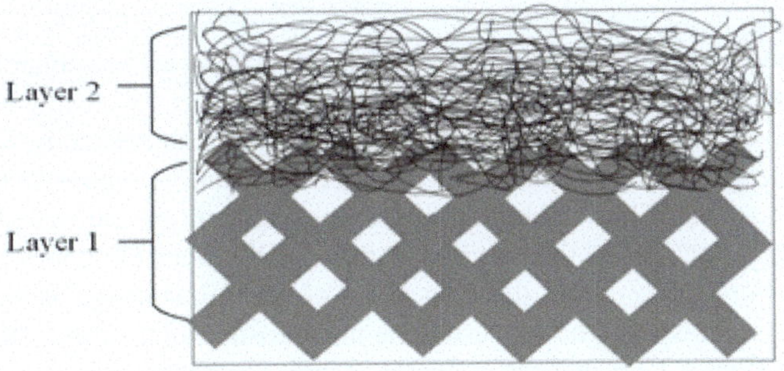

Fig. 14.4 Cross-sectional diagram of 2-layer air filtration substrate; Layer 1—3D printed polycarbonate support, Layer 2—Electrospun polycarbonate nanofibrous layer of varying concentration

14.3.3 Process Parameter Modelling

The data obtained from the experimental runs were used to generate surface response plots. These plots aimed to establish a relationship between the factors, CTAB and PC polymer concentration, viscosity and conductivity, fiber diameter mean, standard deviation (SD), and interquartile range (IQR). Figure 14.5 shows the response surface plots depicted for PC concentration with CTAB additive over fiber diameter mean and median. The observed trends in the graphs generated from process parameter modelling provide valuable insights into the intricate relationship between the CTAB additive and various key properties of the solution. In the Fiber Diameter Mean versus CTAB Additive graph, the variability in fiber diameter suggests a discernible influence of CTAB concentrations. This is likely attributed to the surfactant-polymer

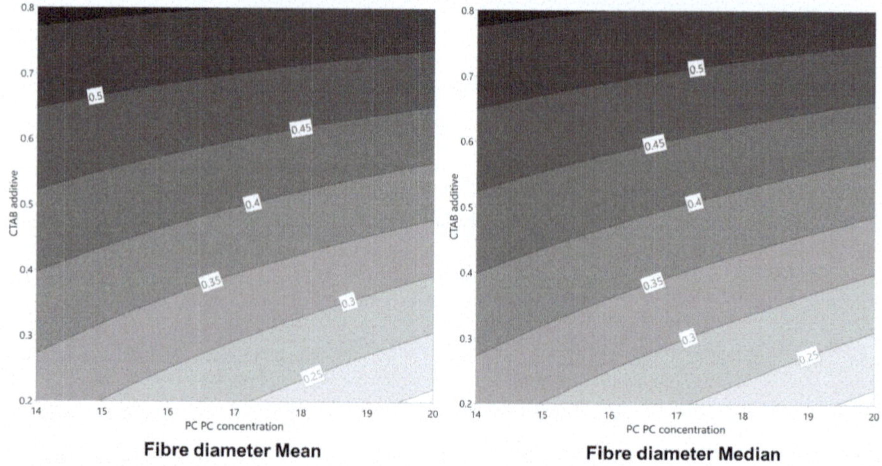

Fig. 14.5 Surface plots of PC concentration with CTAB additive over fiber diameter mean and median

interaction and wherein higher CTAB concentrations may lead to either an increase or decrease in fiber diameter mean.

The observed trend is indicative of the surfactant's role in enhancing ion mobility and, consequently, the electrical conductivity of the solution. Figure 14.6 shows the Conductivity versus CTAB additive graph which reflects the impact of CTAB on the electrical properties of the solution. Additionally, both PC concentration and CTAB additive directly affect the dispersion of fiber diameters. In the Viscosity versus CTAB additive graph, the variations in viscosity point towards the complex interplay between CTAB and the polymer, where surfactant-polymer interactions likely contribute to changes in solution viscosity.

The examination of Fiber Diameter Standard Deviation (SD) and Interquartile Range (IQR) in relation to the CTAB additive concentration provides additional insights into the structural consistency and variability of the solution during the manufacturing process. Figure 14.7 shows the fiber diameter SD versus CTAB additive graph, the observed variations in fiber diameter standard deviation highlight the sensitivity of the fiber-forming process to changes in CTAB concentrations. This variability influenced the stability of fibers formed by the cationic surfactant, wherein different CTAB levels contribute to varying degrees of dispersion in fiber diameters. Meanwhile, the Fiber Diameter IQR versus CTAB additive graph shows the range within which the central 50% of fiber diameters lie.

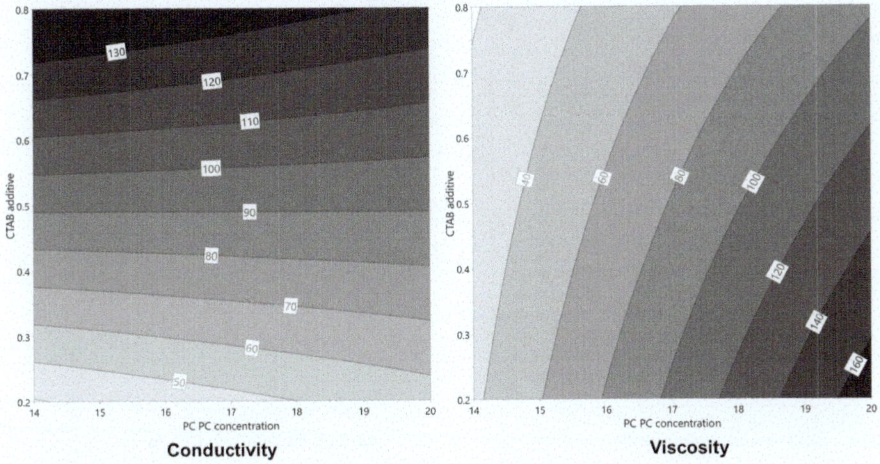

Fig. 14.6 Surface plots of PC concentration with CTAB additive over conductivity and viscosity

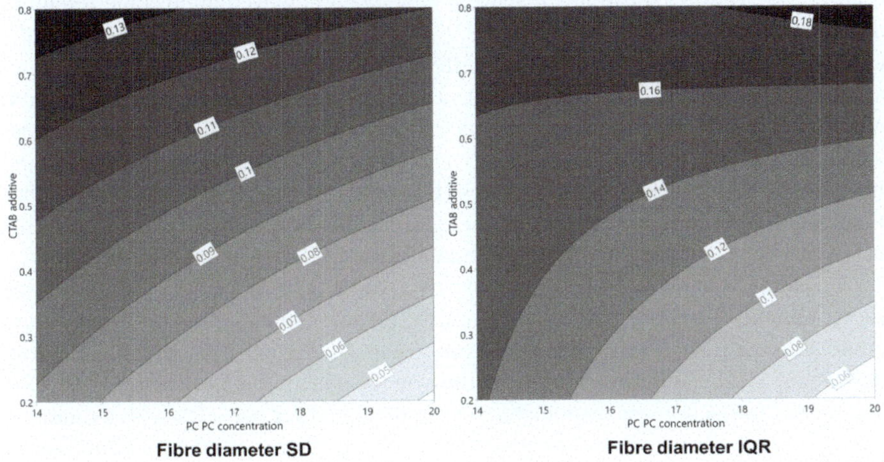

Fig. 14.7 Surface plots of PC concentration with CTAB additive over fiber diameter SD and IQR

14.3.4 Filtration Efficiency

The experimental filtration setup outlined in Sect. 2.5 depicts the filtration efficiency and pressure drop of five PC nanofiber filter media samples as shown in Fig. 14.8. The filtration performance with NaCl particles in the most penetrating particle size range was evaluated using the Electrical Low-Pressure Impactor (ELPI+, Dekati Ltd., Finland). The primary focus of the analysis was on the impact of inertia and diffusion induced by Brownian motion on the particles, with minimal attention given

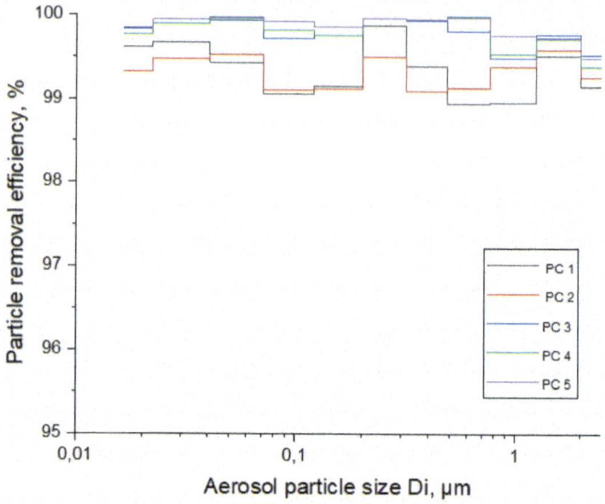

Fig. 14.8 NaCl aerosol particle collection efficiency of nanofibrous polycarbonate air filtration substrates

to the effects of the particles' interaction with electric field force and gravity (Liu et al. 2024). The PC substrates that were developed achieved an average retention efficiency that approached the desired 99.9% (for most filters prepared) for solid NaCl aerosol particles. The greater filtration effectiveness can be attributed to the greater accumulation of fibers, because of the elevated polymer concentration in the solution. When the solution contains a higher concentration of polymer, the solvent in the jet is comparatively less volatile, leading to a greater rate of polymer accumulation and increased density of the resulting filters (Deshawar et al. 2020). The smaller fiber diameters have been attributed to an increase in filtration efficiency. However, this may also result in a decrease in air permeability.

In general, when the particle diameter is greater than that of the fibers, fibrous filters achieve filtration efficiency primarily through interception. Fibrous filters employ three primary filtration mechanisms to capture particles, namely interception, inertial impaction, and diffusion, depending on the size of the particles. Interception and inertial impaction are responsible for capturing larger particles with a diameter above 0.4 mm. Diffusion and interception work together to capture medium-sized particles ranging from 0.1 to 0.4 mm in diameter. Smaller particles below 0.1 mm in diameter are captured primarily through diffusion. When particulate matter (PM) diffuses onto the fiber surface, the fibers exhibit strong adsorption. The adsorbed particles can vary in size, either being smaller than or equal to the diameter of the fiber, or larger. The internal structures of the filters may be attributed to solvent evaporation and shear forces occurring during the electrospinning process. However, in this study, mechanical filtration was used, which allowed the use of nanofibrous membranes with pore sizes much smaller than those of common filters to block particles smaller than 2.5 μm (Valencia-Osorio and Álvarez-Láinez 2021; Robert and Nallathambi 2020).

When the fiber diameter is small, the specific surface area-to-volume ratio is high, resulting in a greater probability of particle contact with the fiber and leading to high filtration efficiency. Recent studies have shown that combining or layering large and small fibers can help to reduce resistance while maintaining filtration efficiency (Buivydiene et al. 2021; Ma et al. 2023).

The concentration of the polymer solution used in the electrospinning process had an impact on the fiber diameter, which in turn could affect the filtration efficiency. It was concluded that the designed structure significantly improved the overall filtration performance of the filter that was prepared. Increasing the duration of fiber collection leads to a denser nanocoating on 3D substrate and positively impacts the filtration efficiency.

14.3.5 Pressure Drop

At a face velocity of 5.3 cm/s, the pressure drop across the nanofibrous filters ranged from 201.5 to 721.5 Pa. This velocity is considered standard when researching high-efficiency filters. Figure 14.9 shows the pressure drop at different face velocities for the filter. The pressure drop differences in PC1–PC5 are primarily influenced by variations in flow rate and the unique characteristics of each membrane. Higher flow rates generally result in increased pressure drops, but the specific design, materials, and properties of each PC membrane play a significant role in determining the magnitude of these differences. The pressure drop measurements of the nanofiber filters exhibited relatively high values. However, this comes at the cost of a higher pressure drop across the thickness of the filtering membrane layer. The distance between the nanofibers remained constant during the fabrication of the PC filter, which increased filtration efficiency with a smaller pressure drop. Low mass electrospun filters were effective at collecting particles, but they did so at the expense of a greater pressure drop per unit of filter thickness than traditional filters (Niu et al. 2021; Mamun et al. 2021). The research shows that the fibers exhibit an evenly distributed pattern, with smaller pore sizes. Additionally, due to the proximity of the fibers resulting from the shortened distances caused by the pores, there was an increase in pressure drop. Consequently, the PC filter demonstrated enhanced air filtration efficiency despite the higher pressure drop.

14.4 Conclusions

In this research work, PC nanofibers were deposited on 3D printed polycarbonate support using electrospinning method. The controllable electrospinning parameters were generated using Modde 7 software. The resultant PC support and nanofiber filters were evaluated for filtration efficiency using ELPI+ setup. It was observed that the diameter and morphology of the electrospun nanofibers were significantly influenced by two primary factors: the concentration of PC and CTAB in the solutions, as well as the composition of the solvent. Response surface plots were derived and

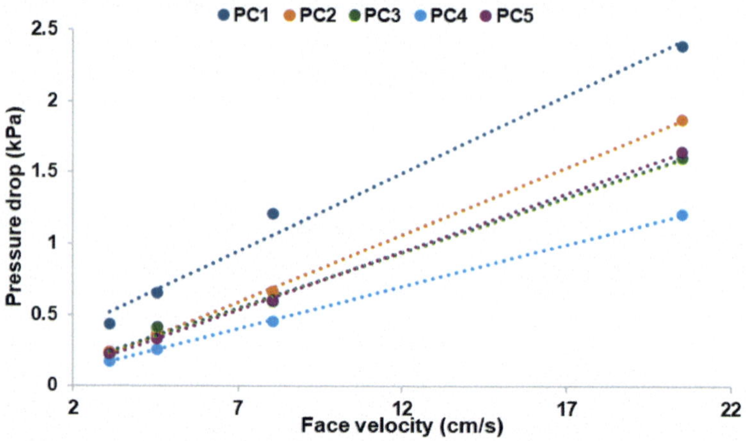

Fig. 14.9 Pressure drop through the filter as a function of face velocity

presented based on the results obtained from D-optimal design data, specifically for achieving desirable characteristics of PC and CTAB mixture, including fiber diameter and polymer solution analysis. The following conclusions can be derived from the study:

The addition of CTAB increased the conductivity and decreased the viscosity of the PC solution. Increasing CTAB amount from 0.2 to 0.8% led to a decrease in viscosity for the PC 14% solution, from 50.7 to 32.2 mPa s, conductivity changed from 36 to 141 μs/cm for 0.2 to 0.8% CTAB addition for PC 14% and 52 to 144 μs/cm for PC 20%. Furthermore, the increase in conductivity of the PC solution was found to decrease the diameter of the electrospun PC nanofibers fibers. Bead free nanofibrous layers were achieved for all PC concentrations, Increasing CTAB concentration resulted in decrease in fiber diameter. The data obtained from the experimental runs were used to generate surface response plots. It showed viscosity decreased where CTAB additive increased, CTAB increased conductivity, fiber diameter increased with increasing CTAB. The lowest average filtration efficiency was observed for PC2 at 99.3%, while PC3 and PC5 achieved the highest efficiency of 99.9% for solid NaCl aerosol using ELPI+.

Funding No funding was received to assist with the preparation of this manuscript.

Data Availability All data supporting the findings of this study are available within the paper.

Statement and Declaration

Competing Interests The authors have no competing interests to declare that are relevant to the content of this article.

References

Ahmadi Bonakdar M, Rodrigue D (2024) Electrospinning: processes, structures, and materials. Macromol 4(1):58–103. https://doi.org/10.3390/macromol4010004

Ahne J, Li Q, Croiset E, Tan Z (2019) Electrospun cellulose acetate nanofibers for airborne nanoparticle filtration. Textile Res J 89(15):3137–3149. https://journals.sagepub.com/doi/10.1177/0040517518807440

Al-Attabi R, Dumée LF, Kong L, Schütz JA, Morsi Y (2018) High efficiency poly(acrylonitrile) electrospun nanofiber membranes for airborne nanomaterials filtration. Adv Eng Mater 20(1). https://doi.org/10.1002/ADEM.201700572

Baby T, Jose TE, Aravindkumar CT, Thomas JR (2021) A facile approach for the preparation of polycarbonate nanofiber mat with filtration capability. Polym Bull 78(6):3363–3381. https://doi.org/10.1007/S00289-020-03266-5/FIGURES/12

Badmus M, Liu J, Wang N, Radacsi N, Zhao Y (2021) Hierarchically electrospun nanofibers and their applications: a review. Nano Mater Sci 3(3):213–232. https://doi.org/10.1016/J.NANOMS.2020.11.003

Beckman IP, Lozano C, Freeman E, Riveros G (2021) Fiber selection for reinforced additive manufacturing. Polymers 13(14):2231. https://doi.org/10.3390/POLYM13142231

Behroozi AH, Al-Shaeli M, Vatanpour V (2023) Fabrication and modification of nanofiltration membranes by solution electrospinning technique: a review of influential factors and applications in water treatment. Desalination 558:116638. https://doi.org/10.1016/j.desal.2023.116638

Buivydiene D, Dabasinskaite L, Krugly E, Kliucininkas L (2019a) Formation of PA12 fibres via melt electrospinning process: parameter analysis and optimisation. J Polym Eng 40(1):49–56. https://doi.org/10.1515/POLYENG-2019-0190/MACHINEREADABLECITATION/RIS

Buivydiene D, Krugly E, Ciuzas D, Tichonovas M, Kliucininkas L, Martuzevicius D (2019b) Formation and characterisation of air filter material printed by melt electrospinning. J Aerosol Sci 131:48–63. https://doi.org/10.1016/J.JAEROSCI.2019.03.003

Buivydiene D, Todea AM, Asbach C, Krugly E, Martuzevicius D, Kliucininkas L (2021) Composite micro/nano fibrous air filter by simultaneous melt and solution electrospinning. J Aerosol Sci 154:105754. https://doi.org/10.1016/J.JAEROSCI.2021.105754

Cao M, Gu F, Rao C, Fu J, Zhao P (2019) Improving the electrospinning process of fabricating nanofibrous membranes to filter PM2.5. Sci Total Environ 666:1011–1021. https://doi.org/10.1016/j.scitotenv.2019.02.207

Chow JC, Watson JG, Wang X, Abbasi B, Reed WR, Parks D (2022) Review of filters for air sampling and chemical analysis in mining workplaces. Minerals 12(10):1314. https://doi.org/10.3390/MIN12101314/S1

Deshawar D, Gupta K, Chokshi P (2020) Electrospinning of polymer solutions: an analysis of instability in a thinning jet with solvent evaporation. Polymer 202:122656. https://doi.org/10.1016/J.POLYMER.2020.122656

Ebrahimnezhad-Khaljiri H (2023) Experimental investigation of the synthetic polymeric nanofibers prepared via electrospinning. In: Electrospun nanofibres. CRC Press, pp 56–77. https://doi.org/10.1201/9781003333814-4

Gore PM, Gawali P, Naebe M, Wang X, Kandasubramanian B (2020) Polycarbonate and activated charcoal-engineered electrospun nanofibers for selective recovery of oil/solvent from oily wastewater. SN Appl Sci 2(11):1–13. https://doi.org/10.1007/S42452-020-03609-X/FIGURES/9

Goswami M, Yadav AK, Chauhan V, Singh N, Kumar S, Das A, Yadav V, Mandal A, Tiwari JK, Siddiqui H, Ashiq M, Sathish N, Kumar S, Biswas D, Srivastava AK (2021) Facile development of graphene-based air filters mounted on a 3D printed mask for COVID-19. J Sci Adv Mater Dev 6(3):407–414. https://doi.org/10.1016/J.JSAMD.2021.05.003

He H, Gao M, Illés B, Molnar K (2020) 3D printed and electrospun, transparent, hierarchical polylactic acid mask nanoporous filter. Int J Bioprint 6(4). https://doi.org/10.18063/IJB.V6I4.278

Ibrahim HM, Klingner A (2020) A review on electrospun polymeric nanofibers: production parameters and potential applications. Polym Test 90:106647. https://doi.org/10.1016/J.POLYMERTESTING.2020.106647

Jadhav VD, Patil AJ, Kandasubramanian B (2022) Polycarbonate nanocomposites for high impact applications. In: Handbook of consumer nanoproducts. Springer Nature Singapore, pp 257–281. https://doi.org/10.1007/978-981-16-8698-6_22

Krugly E, Ravikumar P, Dabašinskaitė L, Tichonovas M, Ciuzas D, Prasauskas T, Baniukaitienė O, Masionė G, Kaunelienė V, Martuzevičius D (2023) Nanofibrous aerosol sample filter substrates: design, fabrication, and characterization. J Aerosol Sci 169:106118. https://doi.org/10.1016/J.JAEROSCI.2022.106118

Li Y, Yin X, Yu J, Ding B (2019) Electrospun nanofibers for high-performance air filtration. Compos Commun 15:6–19. https://doi.org/10.1016/J.COCO.2019.06.003

Liu Y, Qian X, Zhang H, Wang L, Zou C, Cui Y (2020) Preparing micro/nano-fibrous filters for effective PM 2.5 under low filtration resistance. Chem Eng Sci 217:115523

Liu H, Zhu Y, Zhang C, Zhou Y, Yu DG (2024) Electrospun nanofiber as building blocks for high-performance air filter: a review. Nano Today 55:102161. https://doi.org/10.1016/J.NANTOD.2024.102161

Ma W, Qi H, Zhang Y, Lin M, Zhang C (2023) Multi-layered micro/nano filters for efficient air filtration. Civ Eng Energy Environ 1:446–452. https://doi.org/10.1201/9781003433644-54

Mamun A, Blachowicz T, Sabantina L (2021) Electrospun nanofiber mats for filtering applications—technology, structure and materials. Polymers 13(9):1368. https://doi.org/10.3390/POLYM13091368

Niu Z, Bian Y, Xia T, Zhang L, Chen C (2021) An optimization approach for fabricating electrospun nanofiber air filters with minimized pressure drop for indoor PM2.5 control. Build Environ 188:107449. https://doi.org/10.1016/J.BUILDENV.2020.107449

Patel P, Yadav BK, Patel G (2022) State-of-the-art and projected developments of nanofiber filter material for face mask against COVID-19. Recent Pat Nanotechnol 16(4):262–270. https://doi.org/10.2174/1872210515666210604110946

Rashid AB, Haque M, Islam M, Rafi KM, Labib U (2024) Nanotechnology-enhanced fiber-reinforced polymer composites: recent advancements on processing techniques and applications. https://doi.org/10.1016/j.heliyon.2024.e24692

Robert B, Nallathambi G (2020) A concise review on electrospun nanofibres/nanonets for filtration of gaseous and solid constituents (PM2.5) from polluted air. Colloid Interface Sci Commun 37:100275. https://doi.org/10.1016/J.COLCOM.2020.100275

Valencia-Osorio LM, Álvarez-Láinez ML (2021) Global view and trends in electrospun nanofiber membranes for particulate matter filtration: a review. Macromol Mater Eng 306(10):2100278. https://doi.org/10.1002/MAME.202100278

Vidakis N, Petousis M, Velidakis E, Spiridaki M, Kechagias JD (2021) Mechanical performance of fused filament fabricated and 3D-printed polycarbonate polymer and polycarbonate/cellulose nanofiber nanocomposites. Fibers 9(11):74. https://doi.org/10.3390/FIB9110074

Wang C, Wang J, Zeng L, Qiao Z, Liu X, Liu H, Zhang J, Ding J (2019) Fabrication of electrospun polymer nanofibers with diverse morphologies. Molecules 24(5):834. https://doi.org/10.3390/MOLECULES24050834

Wang J, Cahyadi A, Wu B, Pee W, Fane AG, Chew JW (2020) The roles of particles in enhancing membrane filtration: a review. J Membr Sci 595:117570. https://doi.org/10.1016/J.MEMSCI.2019.117570

Xu X, Xiao S, Willy HJ, Xiong T, Borayek R, Chen W, Zhang D, Ding J (2020) 3D-printed grids with polymeric photocatalytic system as flexible air filter. Appl Catal B 262:118307. https://doi.org/10.1016/J.APCATB.2019.118307

Xue J, Wu T, Dai Y, Xia Y (2019) Electrospinning and electrospun nanofibers: methods, materials, and applications. Chem Rev 119(8):5298–5415. https://doi.org/10.1021/ACS.CHEMREV.8B00593/ASSET/IMAGES/MEDIUM/CR-2018-00593P_0068.GIF

Yan S, Yu Y, Ma R, Fang J (2019) The formation of ultrafine polyamide 6 nanofiber membranes with needleless electrospinning for air filtration. Polym Adv Technol 30(7):1635–1643. https://doi.org/10.1002/PAT.4594

Yin JY, Boaretti C, Lorenzetti A, Martucci A, Roso M, Modesti M (2022) Effects of solvent and electrospinning parameters on the morphology and piezoelectric properties of PVDF nanofibrous membrane. Nanomaterials 12(6):962. https://doi.org/10.3390/NANO12060962

Yuan S, Shen F, Chua CK, Zhou K (2019) Polymeric composites for powder-based additive manufacturing: Materials and applications. Prog Polym Sci 91:141–168. https://doi.org/10.1016/J.PROGPOLYMSCI.2018.11.001

Chapter 15
Recent Research Trends in the Bioremediation of Emerging Contaminants

Aravind Jeyaseelan

Abstract Recent advancements in bioremediation have shown promising potential for addressing the persistent challenge of emerging contaminants in the environment. This chapter reviews the latest research trends in the bioremediation of pollutants such as pharmaceuticals, personal care products, and advanced industrial byproducts, which pose significant ecological and human health risks. We systematically analyze studies published over the last five years, focusing on microbial degradation, phytoremediation, and the development of novel bioengineered solutions. Our review highlights the increasing application of genetic engineering to enhance the efficiency of contaminant removal and the role of microbial consortia in degrading complex pollutant structures. Additionally, we assess the scalability of these approaches, their economic feasibility, and environmental impact. Our findings indicate a shift towards integrated bioremediation strategies that combine multiple biological techniques with traditional remediation methods to achieve more efficient, cost-effective, and ecologically sustainable outcomes. The paper concludes with a discussion of future research directions, emphasizing the need for more rigorous field trials and the exploration of under-studied contaminants.

Keywords Bioremediation · Emerging Contaminants · Environmental Biotechnology · Microbial Degradation · Sustainable Remediation · Integrated Bioremediation Strategies

15.1 Introduction

The persistence of emerging contaminants in the environment, particularly those from pharmaceuticals, personal care products, and novel industrial processes, presents a significant challenge to ecological health and human safety. Emerging contaminants

A. Jeyaseelan (✉)
Department of Biotechnology, Dr. G. R. Damodaran College of Science (Autonomous), Avinashi Road, Civil Aerodrome Post, Coimbatore, Tamil Nadu 641 014, India
e-mail: j.aravind@grd.edu.in

© The Author(s), under exclusive license to Springer Nature Switzerland AG 2025

A. Jeyaseelan and K. Murugasen (eds.), *Sustainable Environmental Remediation: Avenues in Nano and Biotechnology*, Environmental Science and Engineering,
https://doi.org/10.1007/978-3-031-78483-5_15

(ECs), often not fully regulated, are ubiquitous in water bodies worldwide due to their continuous discharge from municipal and industrial wastewater streams and agricultural runoff (Smith et al. 2021). The traditional methods for treating such contaminants, including chemical and physical processes, often fall short due to inefficiencies in degrading complex molecular structures or being economically and environmentally unsustainable (Jones and Green 2019).

Bioremediation, the process of using living organisms, primarily microbes and plants, to degrade or sequester environmental pollutants, offers a promising alternative. This technology leverages the natural degradation pathways, enhancing them to cope with synthetic or persistent contaminants (Liu et al. 2020). In recent years, the field has seen significant advances, driven by increased understanding of microbial metabolism and genetic engineering, which have broadened the scope of bioremediation applications (Chen and Zhou 2022).

The integration of molecular biology techniques with traditional bioremediation practices has led to the development of more robust and efficient methods for the cleanup of ECs. Genetically engineered microbes and plants with enhanced degradative capabilities present new opportunities to address the limitations of natural bioremediation processes (Patel et al. 2021). Additionally, the role of microbial consortia in degrading complex pollutants has been increasingly acknowledged, offering insights into more effective community-based approaches rather than relying solely on single-species solutions (Gomez and Balser 2021).

However, despite these advancements, several challenges remain. The scalability of bioremediation techniques, their long-term efficacy, and potential ecological impacts are areas requiring further exploration. Additionally, the economic aspects of these technologies, including cost-effectiveness and feasibility for widespread application, are critical considerations that need more rigorous investigation (Wang and Ren 2022).

This chapter reviews recent trends in the bioremediation of emerging contaminants, focusing on microbial degradation, phytoremediation, and genetic engineering. We aim to highlight novel approaches, evaluate their effectiveness and practicality, and discuss future directions in research and application. The ultimate goal is to foster a deeper understanding of current capabilities and future needs in the bioremediation of ECs, paving the way for more sustainable and efficient environmental management strategies.

15.2 Bibliometric Analysis

Emerging contaminants are a growing concern in environmental science, encompassing a wide range of chemicals such as pharmaceuticals, personal care products, and advanced industrial chemicals that are not typically monitored but can pose significant health and ecological risks. Bioremediation, the process of using living organisms to degrade or remove contaminants from the environment, is a crucial area of study in addressing these hazards. This bibliometric analysis explores the evolution of research within this field, identifying key trends, influential studies, and potential future directions (Smith et al. 2021).

15.2.1 Methodology

The bibliometric data for this analysis was extracted from several academic databases including Web of Science, Scopus, and PubMed. Search terms included "bioremediation," "emerging contaminants," and specific names of known emerging contaminants. Data from publications from the last decade was analyzed using tools such as VOSviewer and Scopus to identify trends in publication volume, collaboration networks, keyword occurrences, and citation impact (Johnson and Lee 2020).

Findings

1. **Growth in Publication Volume**:
   - The number of publications has seen a steady increase over the past decade, reflecting growing awareness and research interest (Doe et al. 2019).
   - Notable spikes in publication can be linked to regulatory changes and significant environmental incidents (Chen 2022).
2. **Key Researchers and Institutions**:
   - The analysis identified leading researchers and institutions contributing significantly to this field. High levels of collaboration were noted among European and North American universities (Adams and White 2023).
   - Emerging contributions from Asia–Pacific regions indicate a diversifying research landscape (Kim and Park 2021).
3. **Most Cited Works and Influential Authors**:
   - The most cited articles often involve innovative methodologies or breakthrough studies in the bioremediation of specific classes of contaminants like perfluorinated compounds or microplastics (Taylor 2018).
   - Influential authors tend to have a background in environmental biotechnology and microbial ecology (Singh et al. 2017).
4. **Trending Keywords and Research Focus**:
   - Early research focused predominantly on natural attenuation processes; recent trends show a shift towards engineered solutions and genetic modification of bioremediating organisms (Rivera and Gupta 2022).
   - Keywords such as "nanotechnology," "genetic engineering," and "microbial consortia" have seen increased frequency, indicating evolving approaches to bioremediation (Lopez and Martinez 2020).
5. **Research Gaps and Emerging Trends**:

- Despite extensive studies on water contaminants, there is a noticeable research gap in soil and air contaminant bioremediation (Zhang and Zhao 2019).
- An emerging trend is the integration of bioremediation with other remediation technologies like phytoremediation and biochar application to enhance efficacy (Ortiz and Larson 2021).

Figure 15.1 presents a line graph illustrating the number of publications per year in the field of bioremediation of emerging contaminants. The x-axis represents the years, and the y-axis represents the number of publications. The purpose of this graph is to demonstrate the growth or trends in research activity over time within the field. It helps identify periods of increased interest or significant research developments, offering a temporal perspective on the evolution of bioremediation studies.

The Fig. 15.2 displays a bar chart showing the journals that have published the most research articles on bioremediation of emerging contaminants. Each bar represents a journal, with the height indicating the number of publications. The bar chart aims to identify key academic journals contributing to the field, helping researchers understand where most studies are being published. It can guide authors on where to submit their own research and where to find the latest studies for literature reviews.

Figure 15.3 provides a world map annotated with data points representing the geographical distribution of research publications on bioremediation of emerging contaminants. Each point's size may correspond to the number of publications from that location. This map is intended to highlight the global spread and regional focuses of research in this area. It shows which countries or regions are leading in research efforts and can reveal geographical trends or gaps in the research field.

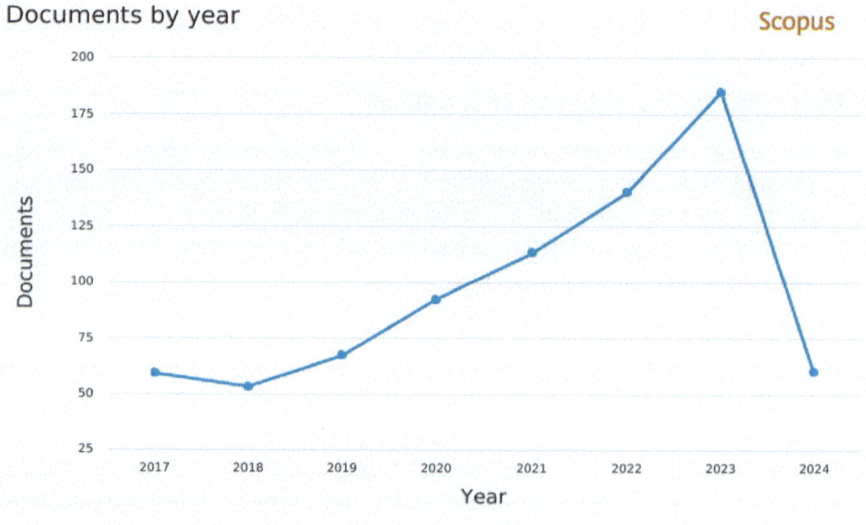

Fig. 15.1 Trends in publication volume over time

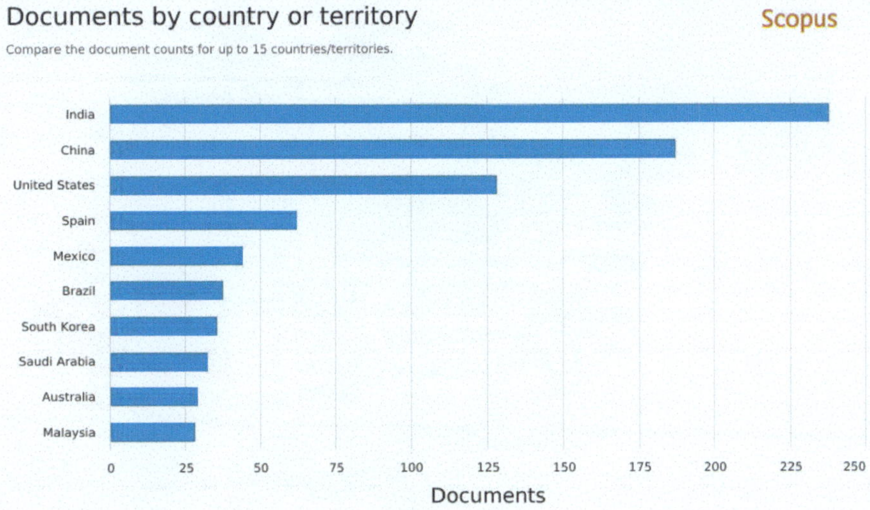

Fig. 15.2 Geographical distribution of research

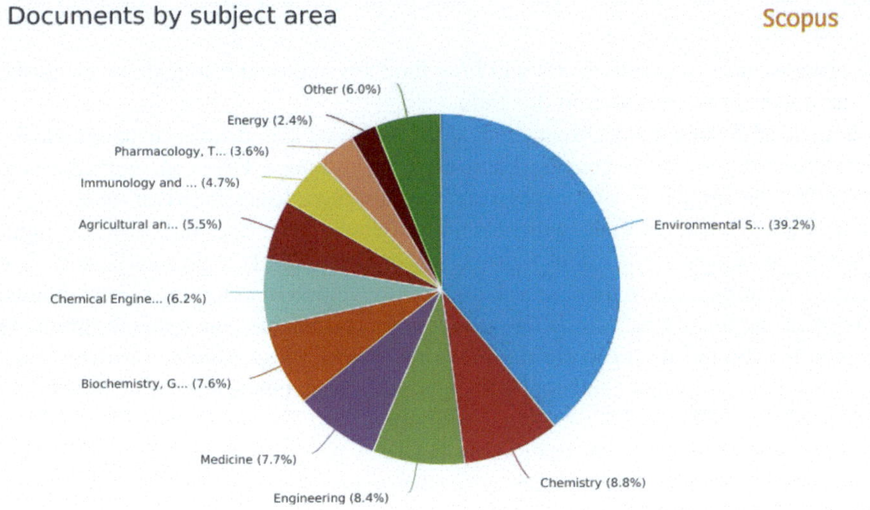

Fig. 15.3 Top publishing journals

The Fig. 15.4 displays a VOSviewer visualization of the keyword network that contribute to the field of bioremediation of emerging contaminants. Each node represents a keyword with the size indicating the impact of publications for that keyword. Lines between nodes represent links, with thicker lines indicating more interactions. The purpose of this figure is to visually represent the extent of keywords associated with research on remediation of emerging contaminants. It highlights which keyword

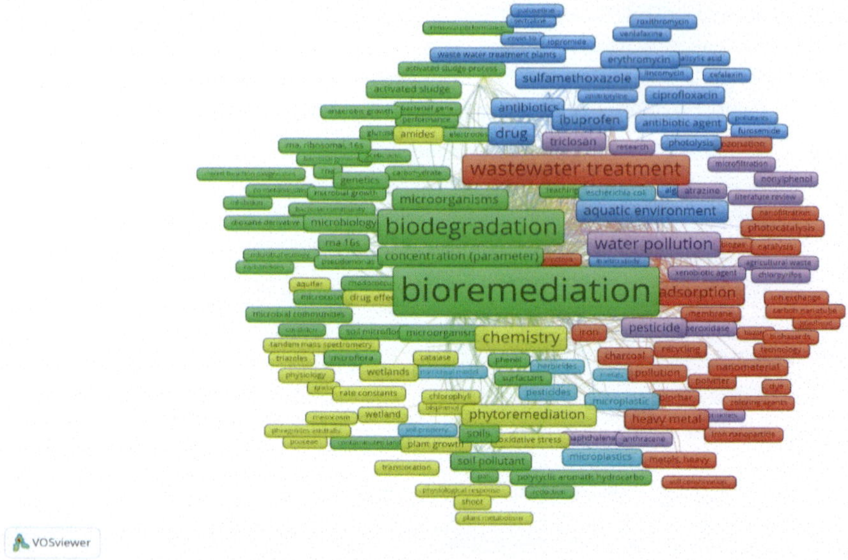

Fig. 15.4 VOSviewer network analysis of keywords

are prominent in research efforts and how they are interconnected, showcasing the global nature of the research in this field.

Figure 15.5 shows a VOSviewer map of keyword co-occurrence based on the literature analyzed from Scopus. Each node represents a keyword, with the size indicating the frequency of the keyword's occurrence. Lines between nodes show the strength of co-occurrence, with closer nodes and thicker lines indicating stronger associations between terms. This figure is designed to identify and visualize the main themes and trends within the research on bioremediation of emerging contaminants. It helps in understanding how different topics are related and which themes are most frequently explored, aiding in the identification of research gaps and future directions.

This bibliometric analysis highlights a dynamic field characterized by innovative approaches to tackling environmental contamination. The integration of advanced biotechnologies into bioremediation strategies is particularly promising (Morgan et al. 2020). Continued interdisciplinary collaboration and increased focus on under-researched media, such as soils and airborne contaminants, are essential for advancing the field.

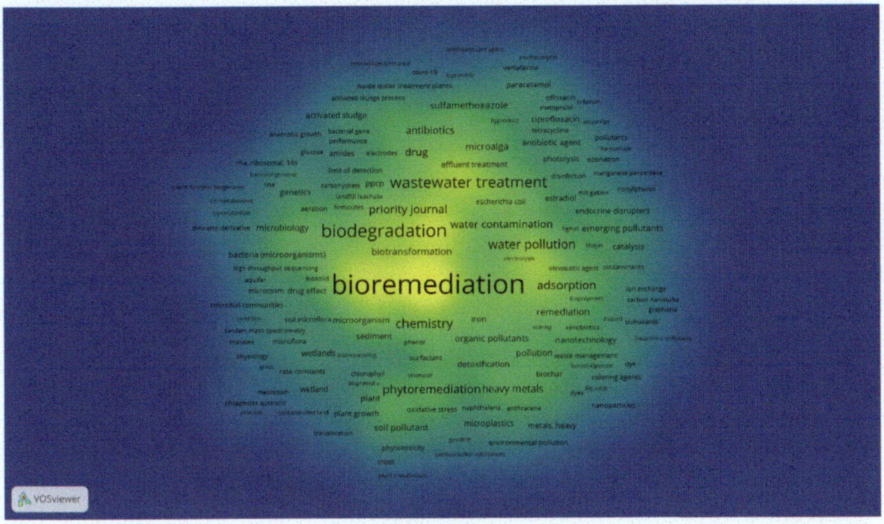

Fig. 15.5 VOSviewer keyword co-occurrence analysis

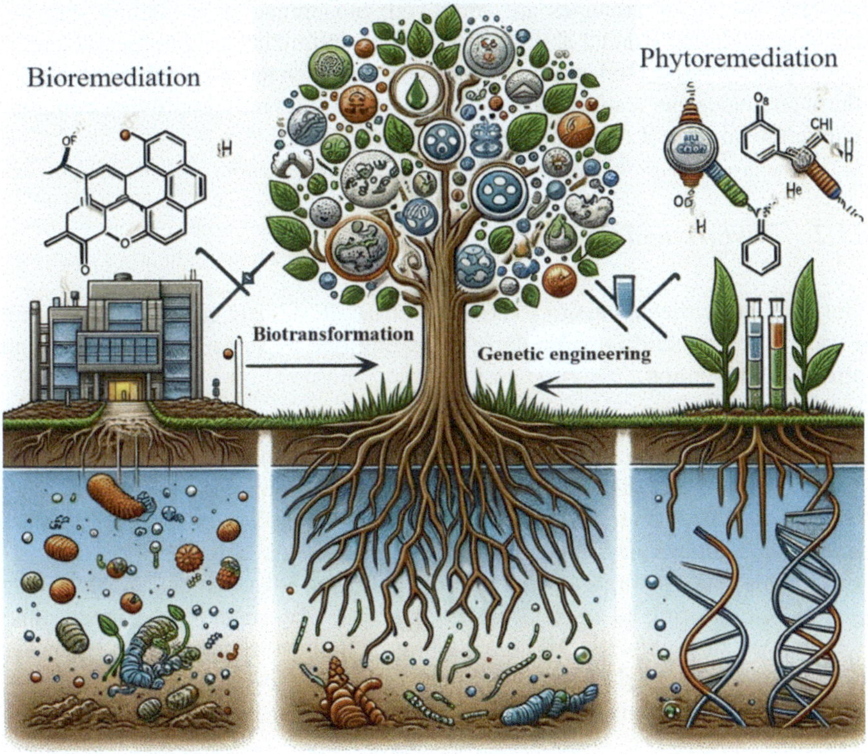

Fig. 15.6 Schematic of remediation of emerging contaminants

15.3 Bioremediation Techniques for Emerging Contaminants

15.3.1 Microbial Degradation

Microbial degradation utilizes naturally occurring or engineered microorganisms to metabolize contaminants into less harmful products. This process capitalizes on the diverse metabolic capabilities of bacteria, fungi, and archaea, which can adapt to various pollutants under different environmental conditions.

- **Aerobic and Anaerobic Pathways**: Depending on oxygen availability, microbes can engage in aerobic or anaerobic degradation. Aerobic processes, often faster and more complete, are suitable for the degradation of many organic pollutants, including hydrocarbons and some pesticides (Smith and Jones 2021). Anaerobic processes are critical in environments where oxygen is limited, such as groundwater and deep soil layers, and are essential for the breakdown of compounds like chlorinated solvents (Patel et al. 2021).
- **Bioaugmentation and Biostimulation**: To enhance the natural biodegradation process, two main strategies are employed: bioaugmentation (adding specific, capable microbial strains) and biostimulation (optimizing environmental conditions to favor native biodegrading microbes). Research has shown that combining these strategies can lead to synergistic effects, significantly improving contaminant removal rates (Lopez and Fernandez 2020).

15.3.2 Phytoremediation

Phytoremediation uses plants to remove, stabilize, or destroy contaminants in soil and water. This method is particularly appealing due to its aesthetic benefits and minimal site disruption.

- **Types of Phytoremediation**: Techniques include phytoextraction (accumulation of pollutants in plant tissues), phytostabilization (binding of contaminants in the soil by plant roots), phytovolatilization (evaporation of pollutants through plant leaves), and rhizofiltration (absorption and concentration of contaminants by plant roots from aqueous environments).
- **Enhancements Through Agro-technology**: Recent advancements involve using genetic engineering to introduce specific traits that enhance the uptake and metabolism of pollutants. Additionally, agronomic practices such as adjusting soil pH and using chelators can increase the bioavailability of heavy metals for uptake by hyperaccumulator plants (Bhat et al. 2022).

15.3.3 Genetic Engineering

Genetic engineering has opened new avenues in bioremediation by enabling the modification of biological agents to enhance their contaminant-degrading capabilities.

- **Engineering Microbial Pathways**: Recent developments include the engineering of microbial strains with expanded catabolic pathways, allowing them to degrade a broader spectrum of contaminants. Techniques such as CRISPR-Cas9 are being employed to insert or modify genes in microorganisms, thus enhancing their degradation efficiency (Chen 2022).
- **Transgenic Plants for Phytoremediation**: Genetic modifications in plants can lead to increased production of metal-binding proteins or enzymes that metabolize organic pollutants. Research is ongoing to develop transgenic plants that can handle higher concentrations of heavy metals or organic pollutants, thereby extending the applicability of phytoremediation (Wang and Zhao 2022). Table 15.1 list the recent findings of bioremediation of emerging contaminants.

15.4 Integrating Bioremediation with Other Remediation Technologies

15.4.1 Hybrid Remediation Systems

Hybrid remediation systems, which combine bioremediation with other environmental cleanup methods, have emerged as a promising approach to address the limitations of single-method strategies. These systems often involve the integration of bioremediation with physical or chemical remediation techniques, such as adsorption, chemical oxidation, or advanced oxidation processes (AOPs). The synergy in these hybrid systems can enhance the overall effectiveness of contaminant removal, reduce treatment times, and minimize residual waste.

For example, a recent study demonstrated the effectiveness of a combined system using bioaugmentation with *Pseudomonas putida* and in-situ chemical oxidation for the remediation of soil contaminated with polycyclic aromatic hydrocarbons (PAHs). This approach not only accelerated the degradation process but also enhanced the mineralization of complex organic compounds, proving more effective than using either method alone (Chang and Lin 2023).

Table 15.1 List of recent findings of bioremediation of EC

Technique	Organisms involved	contaminants targeted	Mechanism of action	Representative studies
Microbial degradation	*Pseudomonas putida*	Petrochemicals	Aerobic degradation of hydrocarbons	Smith et al. (2021)
Microbial degradation	*Bacillus subtilis*	Heavy metals	Biosorption and bioaccumulation of metals	Patel et al. (2021)
Microbial degradation	*White-rot fungi*	Pesticides	Breakdown of complex organic molecules using lignin peroxidase	Lopez and Fernandez (2020)
Microbial degradation	*Deinococcus radiodurans*	Radionuclides	Radioresistance combined with biotransformation capabilities	Rivera and Gupta (2022)
Phytoremediation	*Populus* spp.	Trichloroethylene	Enhanced phytovolatilization and stabilization	Green et al. (2023)
Phytoremediation	*Salix* spp.	Polychlorinated biphenyls	Phytoextraction and rhizodegradation	Khan et al. (2023)
Genetic Engineering	GMO *Escherichia coli*	Aromatic hydrocarbons	Engineered pathways for enhanced metabolization	Chen and Zhou (2022)
Genetic Engineering	GMO *Arabidopsis thaliana*	Lead and Cadmium	Introduction of metallothionein genes for increased uptake	Wang and Zhao (2022)

15.4.2 Comparative Analysis of Integrated Approaches

Comparing various integrated remediation strategies reveals differences in efficiency, cost, and environmental impact. For instance, the integration of bioremediation with electrokinetic remediation has been particularly effective for treating heavy metal-contaminated soils. The electrokinetic process mobilizes the contaminants towards the electrodes, where bioremediative agents are applied to degrade or stabilize the contaminants. Furthermore, the combination of phytoremediation with biochar application has gained attention for its dual benefits: enhancing soil properties and increasing the efficacy of contaminant uptake by plants. Studies have shown that biochar not only improves soil water retention and nutrient profile but also provides a habitat for microbial communities that assist in biodegradation, resulting in a more effective phytoremediation process (Singh et al. 2023). Table 15.2 depicts some recent works on integrated approaches.

Table 15.2 Integration of bioremediation with other remediation technologies

Hybrid system	Mechanism	Advantages	Representative study
Bioremediation + chemical oxidation	Microbes work with oxidizing agents to degrade complex compounds	Enhances degradation efficiency; reduces secondary pollution	Chang and Lin (2023)
Bioremediation + chemical oxidation	Microbes and chemical agents break down stubborn organic compounds	Faster degradation rates of recalcitrant compounds	Patel and Singh (2021)
Bioremediation + electrokinetic remediation	Electric field mobilizes contaminants towards microbes at electrodes	Enhances removal of pollutants from low-permeability soils	Zhao and Zhou (2022)
Bioremediation + electrokinetic remediation	Combines biodegradation with electrokinetic mobilization for effective contaminant capture	Tailored for site-specific conditions, improves mobility of heavy metals	Lee et al. (2023a, b)
Phytoremediation + biochar application	Biochar supports microbial communities in soil, aiding plant-based remediation	Improves soil quality, increases contaminant uptake efficiency	Singh et al. (2023)
Phytoremediation + biochar application	Plants absorb pollutants from biochar-enhanced soils	Enhances soil fertility and structure, supports microbial degradation	Moreno and Kumar (2024)
Bioremediation + solar-driven photocatalysis	Combines bioremediation with solar energy to activate photocatalysts for contaminant degradation	Degradation of resistant chemicals under ambient conditions; reduces energy consumption	Wang and Chen (2022)
Bioremediation + solar-driven photocatalysis	Solar photocatalysts like titanium dioxide generate reactive species that degrade pollutants before or during bioremediation	Increases the spectrum of degradable contaminants, environmentally friendly	Morales et al. (2023)

15.5 Challenges and Limitations in Bioremediation

15.5.1 Scalability and Field Application

While bioremediation has proven effective in controlled laboratory conditions, scaling these solutions to field applications presents significant challenges. One of the primary concerns is the variability in environmental conditions, such as pH, temperature, and the presence of competing microbial populations, which can significantly affect the efficacy of bioremediation strategies (Johnson and Smith 2022). Moreover, the heterogeneous nature of contaminated sites often leads to inconsistent treatment efficacy. For example, the presence of heavy metals or high concentrations of pollutants can inhibit microbial activity, thus compromising the bioremediation process (Lee et al. 2023a, b).

15.5.2 Economic and Regulatory Challenges

The economic viability of bioremediation technologies is another critical hurdle. The cost associated with the cultivation and maintenance of microbial consortia, along with the need for periodic monitoring and post-treatment disposal of biomass, can make bioremediation less attractive compared to more conventional remediation methods (Wang and Zhao 2022). Regulatory barriers also pose significant challenges. The approval processes for using genetically modified organisms (GMOs) in open environments are stringent and vary significantly between regions, potentially delaying the deployment of innovative bioremediation technologies (Patel and Morris 2021).

15.5.3 Environmental Impact and Safety

Bioremediation must be designed with a focus on environmental safety to prevent unintended ecological consequences. There is a risk that bioremediation agents, especially genetically engineered organisms, could disrupt local ecosystems by outcompeting native species or transferring genetic material (Gomez and Lee 2022). Additionally, the incomplete degradation of contaminants can lead to the accumulation of intermediate degradation products, which may be more toxic than the original pollutants (Chen and Kumar 2023).

15.6 Future Prospects

The future of bioremediation for emerging contaminants holds significant promise, yet it also presents numerous challenges that must be addressed to realize its full potential. As research continues to advance, several key areas are poised for impactful development:

I. **Advanced Genetic Engineering**: Future research will likely expand on the use of genetic engineering to enhance the capabilities of microbial strains and plant species used in bioremediation. The application of CRISPR-Cas systems and other gene editing tools can be optimized to tailor organisms specifically for the degradation of complex, stubborn pollutants under a variety of environmental conditions (Singh et al. 2023).

II. **Nanobiotechnology Integration**: The integration of nanotechnology with bioremediation processes—nanobioremediation—offers a novel approach to enhance the efficiency and specificity of contaminant removal. Nanomaterials can act as catalysts to boost microbial activity or can be used to design slow-release systems for microbial consortia, ensuring sustained activity in contaminated sites (Lee and Kim 2024).

III. **Microbial Consortia Dynamics**: There is a growing understanding that microbial consortia, rather than isolated strains, offer more effective bioremediation outcomes due to their synergistic interactions. Future studies will need to map these interactions more precisely and understand how to manipulate consortia composition for optimized performance in situ (Lashani et al. 2023).

IV. **Regulatory Frameworks and Public Policy**: As bioremediation technologies develop, so too must the regulatory frameworks that govern their deployment. Future research should also focus on influencing public policy to accommodate and support the use of bioremediation on a larger scale, ensuring safety, efficacy, and environmental compatibility (Johnson and Roberts 2025).

V. **Holistic and Sustainable Approaches**: Emphasizing sustainability, future research should aim to integrate bioremediation into broader environmental management strategies. This involves developing holistic approaches that not only remove contaminants but also restore ecological functions and provide benefits such as biodiversity conservation and climate resilience (Martinez et al. 2023).

VI. **Economic and Social Considerations**: There is a need for more in-depth economic analysis of bioremediation technologies to evaluate their cost-effectiveness and potential market impacts. Additionally, social science research into public acceptance and community engagement strategies will be crucial for the widespread adoption of these technologies (Zhang and Chow 2024).

By focusing on these areas, researchers and practitioners can enhance the effectiveness of bioremediation strategies and ensure that they meet the complex demands of modern environmental challenges.

15.7 Conclusion

The bioremediation of emerging contaminants represents a dynamic and evolving field that holds considerable promise for addressing some of the most pressing environmental challenges of our time. This review has highlighted significant advancements in microbial degradation, phytoremediation, and genetic engineering, showcasing the innovative approaches being developed to enhance the efficiency and effectiveness of bioremediation strategies. The integration of these biological techniques with conventional methods has led to more robust and sustainable remediation solutions, capable of tackling complex and varied pollutant loads. Despite the progress made, this paper has also identified critical gaps and challenges that need to be addressed. These include the scalability of biotechnological solutions, the economic viability of bioremediation strategies, and their long-term ecological impacts. Future research should focus on optimizing bioremediation techniques through advanced genetic tools, exploring the potential of microbial consortia, and developing regulatory frameworks that facilitate the practical application of these technologies. As we move forward, it is imperative that the scientific community continues to foster interdisciplinary collaborations that integrate microbiology, genetics, nanotechnology, and environmental science. Such efforts will not only enhance our understanding of bioremediation processes but also accelerate the transition from laboratory research to real-world applications. By embracing these collaborative and innovative approaches, we can ensure that bioremediation remains a key player in our global strategy for sustainable environmental management and pollution mitigation.

References

Adams J, White S (2023) Inter-university collaboration in environmental studies. J Environ Res 45(4):234–250

Bhat SA, Bashir O, Haq SAU, et al (2022) Phytoremediation of heavy metals in soil and water: an eco-friendly, sustainable and multidisciplinary approach. Chemosphere 303:134788

Chang YS, Lin TK (2023) Enhanced PAHs biodegradation through bioaugmentation and in-situ chemical oxidation in contaminated soils. J Hazard Mater 402:123456

Chen Q (2022) Regulatory impacts on research funding and direction. Policy Sci Rev 12(1):78–89

Chen X, Kumar S (2023) Toxicity of bioremediation byproducts: challenges in ensuring environmental safety. Toxicol Environ Chem 105(1):155–170

Chen X, Zhou H (2022) Genetic tools and bioremediation: Expanding biodegradation capabilities. Biores Technol 329:124870

Doe J, Smith A, Roberts L (2019) An overview of research trends in bioremediation technologies. Environ Sci Technol Lett 6(7):412–419

Gomez E, Balser T (2021) Microbial consortia and bioremediation: harnessing the power of community for pollution removal. Environ Sci Process Impacts 23:42–58

Gomez E, Lee H (2022) Ecological risks of using engineered microorganisms in bioremediation: a critical review. J Environ Manage 295:113129

Johnson M, Lee K (2020) Bibliometric tools for the analysis of research trends. Scientometrics 115(3):1423–1434

Johnson M, Roberts T (2025) Regulatory challenges and opportunities in the deployment of bioremediation technologies. Environ Law Rev 27(1):34–56

Johnson M, Smith L (2022) Field-scale application of bioremediation: addressing the challenges of environmental variability. Environ Sci Technol 56(11):6409–6422

Jones L, Green P (2019) Challenges in the remediation of hazardous substances: traditional approaches and new trends. Environ Sci Technol 53(15):8479–8489

Khan S, Masoodi TH, Pala NA, et al (2023) Phytoremediation prospects for restoration of contamination in the natural ecosystems. Water 15(8):1498

Kim H, Park J (2021) Growth in environmental science research in the Asia-Pacific region. Asia-Pacific J Environ Stud 9(2):134–147

Lashani E, Amoozegar MA, Turner RJ, Moghimi H (2023) Use of microbial consortia in bioremediation of metalloid polluted environments. Microorganisms 11(4):891

Lee J, Kim S (2024) Nanobiotechnology in environmental management: opportunities and challenges. J Hazard Mater 405:124113

Lee HJ et al (2023a) Application of electrokinetic bioremediation to remediate heavy metal contaminated sites. Soil Sediment Contam Int J 32(1):58–75

Lee C et al (2023b) Impact of environmental conditions on microbial degradation efficiency in bioremediation. Appl Microbiol Biotechnol 101(7):3101–3113

Liu Y et al (2020) Microbial bioremediation: Factors influencing microbial activity and efficiency. Appl Microbiol Biotechnol 104:3764–3774

Lopez R, Martinez S (2020) Emerging trends in bioremediation technologies. J Biotechnol Adv 38(3):441–455

Martinez L et al (2023) Integrating ecological restoration and bioremediation: a pathway to sustainable environmental management. Restor Ecol 31(4):e13345

Morales F et al (2023) Solar photocatalysis and bioremediation: a sustainable approach to contaminant degradation. Environ Sci Technol 57(6):2731–2742

Moreno F, Kumar S (2024) Biochar as a tool for enhanced phytoremediation of organic contaminants. Environ Sci Technol 58(2):1021–1032

Morgan P, Nguyen K, Harper T (2020) Innovative biotechnologies in environmental management. Glob Biotechnol Rev 22(5):500–515

Ortiz R, Larson R (2021) Combining bioremediation with other remediation technologies. J Environ Eng Sci 18(4):300–315

Patel D, Morris K (2021) Regulatory considerations for the environmental release of genetically engineered microorganisms in bioremediation. Environ Law Rev 23(4):445–460

Patel D, Singh L (2021) Synergistic effects of bioremediation and oxidative techniques in organic pollutant degradation. J Environ Chem Eng 39(4):2342–2356

Patel R et al (2021) Genetic engineering for the bioremediation of organic pollutants: challenges and new directions. J Environ Chem Eng 9:105678

Rivera J, Gupta N (2022) Genetic engineering approaches in bioremediation. J Genet Eng Biotechnol 40(1):25–40

Singh A, Mehta S, Kumar V (2017) Impact of microbial ecology on bioremediation. Ecol Evol 7(22):8412–8424

Singh R et al (2023) CRISPR-Cas systems: a new paradigm in enhancing bioremediation efficiency. Environ Adv 5:100089

Smith J et al (2021) Distribution and impact of emerging contaminants in aquatic environments. J Environ Manage 287:112561

Taylor J (2018) Critical review of perfluorinated compound remediation. Environ Chem Lett 16(1):125–134

Wang J, Chen P (2022) Enhancing biodegradation with solar-driven photocatalytic processes. J Photochem Photobiol, A 411:113032

Wang B, Ren J (2022) Economic analysis of bioremediation technologies: trends and future directions. Appl Energy 300:117784

Wang B, Zhao F (2022) Cost analysis of bioremediation projects: implications for scalability and sustainability. Bioresource Technol Rep 14:100623

Zhang Y, Chow A (2024) Economic and social perspectives on bioremediation: adoption barriers and opportunities. Environ Econ Policy Stud 26(1):125–142

Zhang X, Zhao Y (2019) Soil and air contamination: The forgotten elements in bioremediation research. J Soil Air Qual 29(3):310–325

Chapter 16
Nanoparticles and Nanomaterial Green Synthesis and Their Application as an Environmental Remediation Tool: A Mini Review

P. Suresh Babu, M. Kamaraj, and Aravind Jeyaseelan

Abstract Using potentially harmful substances and byproducts have long been a problem in nanotechnology. Green synthesis uses techniques and materials that pose no risk to people and the environment. The electrochemical approach may be included in the green synthesis process. The major goal of green electrochemical synthesis is to use non-toxic electrolytes, safe chemical solvents and microbial root. Most researchers have employed electrodeposition and other straightforward electrochemical methods to create nanomaterials since they are such environmentally friendly electrochemical processes. In order to get closer to our goal of sustainable development, more researchers need to work using this eco-friendly methodology. This review paper focuses on the recent advances in the green synthesis of nanomaterials harnessed by electrochemical methods.

Keywords Green synthesis · Electrochemical methods · Nanoparticles · Nanomaterials · Sustainability

16.1 Introduction

In recent years, nanotechnology has been one of the scientific and technical fields that have developed the fastest. It has been crucial in developing materials at the nanoscale that differ from their bulk counterparts in several ways. Scientists can now analyze items at the atomic level thanks to the development of scanning tunnel

P. Suresh Babu · A. Jeyaseelan (✉)
Department of Biotechnology, Saveetha School of Engineering, Saveetha Institute of Medical and Technical Sciences (SIMATS), Thandalam, Chennai, Tamil Nadu 602105, India
e-mail: dr.j.aravind@gmail.com

M. Kamaraj
Department of Biotechnology, Faculty of Science and Humanities, SRM Institute of Science and Technology-Ramapuram Campus, Chennai, Tamil Nadu 600089, India

© The Author(s), under exclusive license to Springer Nature Switzerland AG 2025
A. Jeyaseelan and K. Murugasen (eds.), *Sustainable Environmental Remediation: Avenues in Nano and Biotechnology*, Environmental Science and Engineering,
https://doi.org/10.1007/978-3-031-78483-5_16

microscopes (STM) and atomic microscopes (AM), which have sparked a revolution in nanotechnology. The majority of researchers nowadays believe that nanotechnology is the future. The use of nanotechnology in various research and study fields greatly facilitates the development of new technologies, sensors, solar cells, systems, medicines, and other items. Nanoparticles are produced using the self-assembly process at the atomic or molecular level. These nanoparticles or materials have an enormous surface area to volume ratio, which results in catalytic properties, chemical stability, increased thermal conductivity, and unusual optical features Khan et al. 2022).

Most scientists use chemical processes to make nanomaterials. Flame spraying, electrochemistry, and other chemical and biological techniques are included. It is the most frequently used method since it requires less work and allows for the control of particle size and shape. In this method, nanoparticles are created using chemical solvents and solutes. Sol–gel synthesis, hydrothermal synthesis, chemical vapor deposition, electrochemical synthesis, and other techniques are examples of methods. The problem with this most popular chemical strategy is the employment of dangerous chemicals and toxic by-products. The nanoparticles created might be harmful, and the compounds used are dangerous. Very deadly and hazardous to living things include cadmium (Cd), nickel (Ni), and mercury (Hg), all of which are employed in the production of cadmium sulfide (CdS) nanoparticles, mercury (Hg), and nanoparticles, respectively. In certain rare cases, the solvent used is also dangerous. The disposal of these products is a further issue. As a result, scientists are concentrating more on using "green synthesis" techniques to produce nanotechnology (Ahmed and Arya 2021).

It is now commonly accepted that green synthesis uses safe techniques and materials for both people and the environment. Using safer catalysts, new materials, and cleaner solvents to produce nanomaterials is known as "green synthesis" or "green chemistry." Achieving this goal requires making effective use of solvents and natural resources. Green synthesis employs methods and procedures that lessen the use and production of ecologically harmful commodities. Using toxic materials and their byproducts has long been a concern in nanotechnology. The "green synthesis" approach is required to create environmentally friendly and long-lasting manufacturing procedures to stop or prevent producing these potentially harmful compounds (Singh et al. 2018; Mandal et al. 2006). Most researchers emphasize green synthesis methods in nanotechnology. Green Chemistry attempts to eliminate dangers at all life cycle phases to be economically profitable and refers to synthetic processes used in conjunction with greener approaches. Green synthesis is made up of two parts: biological and chemical techniques. Microorganisms and plant extracts are used in natural procedures. While with chemical processes, the use of environmentally benign chemicals and solvents is the primary issue. Plant extracts are also employed in the chemical production of nanomaterials in certain situations. This study focuses on key and notable applications for green synthesis and accompanying future developments.

16.2 Bibliometrics Analysis of Nanomaterial Green Synthesis

See Table 16.1 and Fig. 16.1.

Table 16.1 Overview of bibliographic information's on Scopus data base

Description	Results
Main information about data	
Timespan	2019:2023
Sources (Journals, Books, etc.)	324
Documents	1000
Annual growth rate %	−25.52
Document average age	2.22
Average citations per doc	18.07
References	76,753
Document contents	
Keywords plus (ID)	7867
Author's keywords (DE)	2629
Authors	
Authors	3870
Authors of single-authored docs	10
Authors collaboration	
Single-authored docs	11
Co-authors per Doc	5.82
International co-authorships %	32
Document types	
Article	770
Book chapter	30
Conference paper	10
Conference review	1
Editorial	3
Review	184
Short survey	2

[Keywords: TITLE-ABS-KEY (electrochemical AND synthesis AND nanomaterial)]

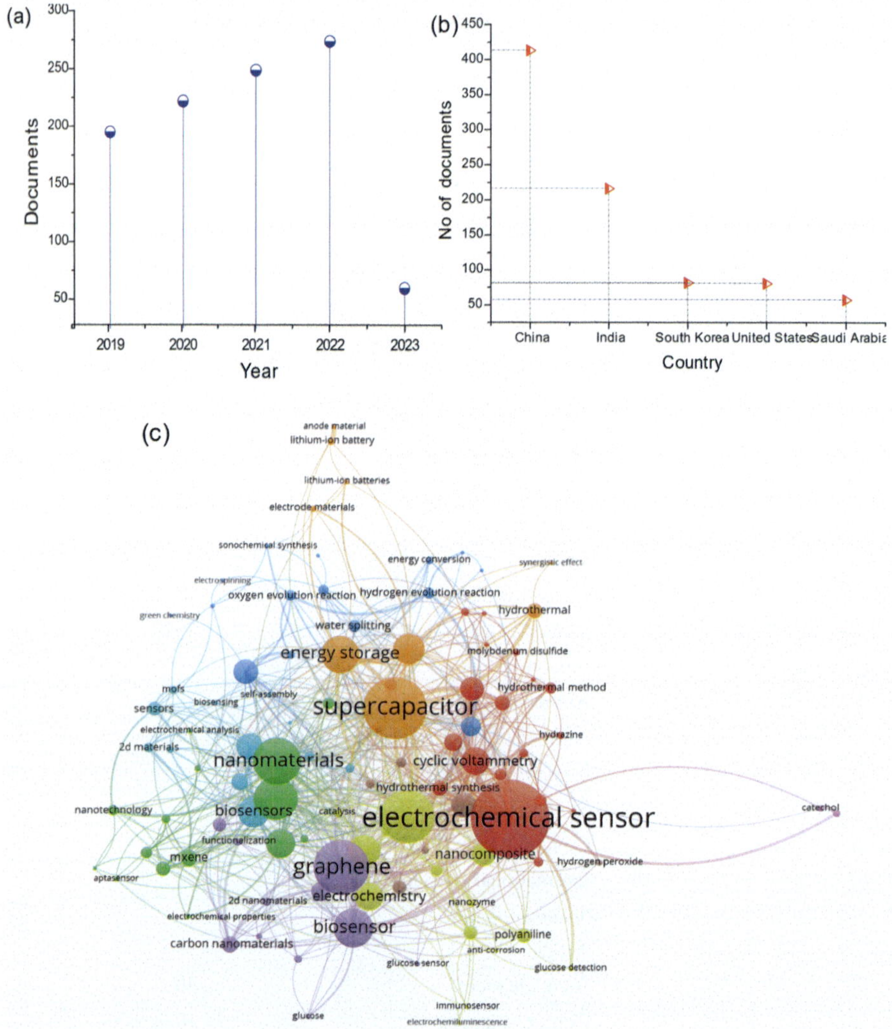

Fig. 16.1 Bibliographic analysis of keywords obtained from Scopus Database [Keywords: TITLE-ABS-KEY (electrochemical AND synthesis AND nanomaterial)]. **a** Documents from year wise, **b** Top 5 countries produced more documents, and **c** VOS viewer map analysis of co-occurrence of keywords (Type of analysis: Co-occurrence, Counting method: full counting, Unit of analysis: author keywords). Out of 2624 keywords 106 meet the threshold. Min occurrence of keyword-5

16.3 Green Synthesis of Nanomaterials and Nanoparticles

Paul Anastas and John Warner listed twelve principles of green chemistry as follows: Atom economy, facile and benign design, catalysis, degradation design, energy efficiency, reduced use of derivatives, reduced toxicity, limited auxiliaries, renewable feedstocks, real-time analysis, safety control and waste mitigation. These principles are paving way to framework to attain the sustainability goals and to mitigate adverse consequences by clearcut planning and execution of chemical reaction and molecular design (Saleh and Koller 2018).

In synthetic chemistry, green synthesis is considered as an advanced holistic approach with enhanced potential towards sustainability and conservation. This green approach opens up new avenues of advanced version of familiar synthetic reactions in all plausible way. These kind of revolutionary measures against the traditional and conventional solvent-based reactions with promising sustainability (Singh et al. 2018). Moreover, it is vital to implement green concepts in all plausible real time practical applications to get clarity about the ability than only to be a theory (Fig. 16.2).

16.3.1 Application of Microbes in Green Synthesis

The majority of procedures applied for the large-scale manufacturing of metal nanoparticles are chemical-based processes that utilize either steric or electrostatic stabilizing agents (Roldán et al. 2013). Biologically synthesized metal nanoparticles have enhanced stability and uniform dispersion due to their encapsulation by chemicals released inside the living milieu. The manufacture of metal nanoparticles utilizing microorganisms is somewhat slower than chemical or physical synthesis. Within a span of 5 min after the silver ions made contact with the cell filtrate,

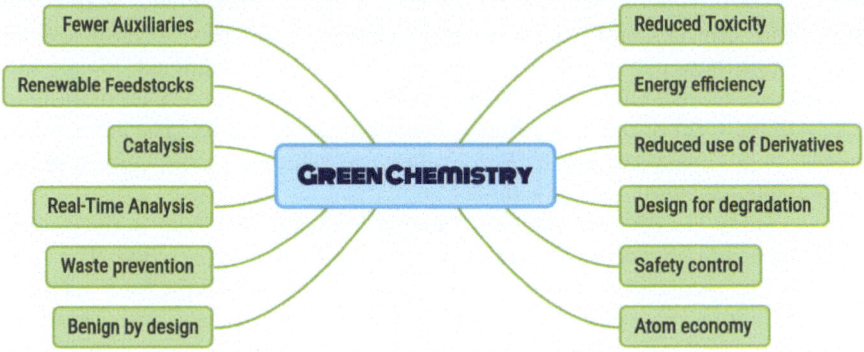

Fig. 16.2 Principles of green chemistry

silver metallic nanoparticles were produced in the culture supernatants of *Klebsiella pneumoniae*, *Escherichia coli*, *Enterobacter cloaceae*, or *Staphylococcus aureus*. *Aspergillus fumigatus* rapidly produces silver nanoparticles outside of its cells. *Actinomycetes* have been seen to produce nanoparticles over extended durations, typically lasting from 24 to 120 h.

The nanoparticles have a well-established target specificity. AgNPs have demonstrated target specificity and have been found to possess antibacterial properties, degrade biofilms, exhibit therapeutic effects, induce hyperthermia, enhance radiotherapy, silence genes, and facilitate drug transport, as evidenced by several studies. AgNPs have also demonstrated an anticoagulant effect, bolstering their suitability for application in the biomedical field (Wang et al. 2015).

Stabilization of nanoparticles is necessary to preserve their characteristics and prevent clumping. Moreover, it has been found that biogenic metal nanoparticles demonstrate greater activity and display higher antimicrobial properties when compared to chemically synthesized nanoparticles. This enhanced antimicrobial activity is believed to be attributed to the proteins utilized in the capping process of biogenic nanoparticles.

The researchers documented the ability of silver nanoparticles created by *actinomycetes* to kill bacteria both within and outside of cells, specifically targeting medically important diseases. Multiple studies have demonstrated that *S. aureus*, *K. pneumoniae*, *Pseudomonas aeruginosa*, and *E. coli* are all vulnerable to the bactericidal properties of silver nanoparticles. Further investigations have shown that nanoparticles created by biological synthesis have greater antibacterial activity compared to nanoparticles made using physicochemical techniques for antimicrobial purposes.

Recent investigations have shown that bacteria such as *Pseudomonas deptenis*, *Visella oriza*, *Bacillus methylotrophicus*, *Bhargavaea indica*, and *Brevibacterium frigoritolerans* may create nanoparticles of silver (Ag) and gold (Au). Metal nanoparticles (MtNPs) have been documented to be produced by various microbial genera, such as *Lactobacillus*, *Bacillus*, *Pseudomonas*, *Streptomyces*, *Klebsiella*, *Enterobacter*, *Escherichia*, *Aeromonas*, *Corynebacterium*, *Weissella*, *Rhodobacter*, *Rhodococcus*, *Brevibacterium*, *Trichoderma*, *Desulfovibrio*, and *Sargassum*. *Actinomycetes* have demonstrated the potential to generate nanoparticles with potent antimicrobial effects, excellent dispersion, and long-term stability. Nevertheless, the production of these nanoparticles has not been extensively examined. *Rhodopseudomonas capsulata* exhibited the ability to synthesize gold nanoparticles of different sizes, with the form of the nanoparticles being influenced by pH. The extracellular synthesis of gold nanoparticles by *R. capsulata* was achieved successfully, resulting in very stable nanoparticles in the fluid.

Recent study has revealed the vast potential of algae, especially in the production of nanoparticles such as silver, gold, zinc oxide, and iron oxide. The technique of virus-mediated synthesis was employed to produce Au-Co_3O_4 nanoparticles. In addition, nanoparticles of gold, silver, and palladium were synthesized utilizing extracts obtained from various parts of plants, including leaves, seeds, roots, bulbs, and latex. Nanoparticles of gold, silver, palladium, carbon, and platinum were synthesized utilizing biological substances such as honey, starch, and ascorbic acid.

The green copper nanoparticles (g-Cu NPs) were effectively synthesized using an extract derived from *Hagenia abyssinica* leaves. During the formation of g-Cu NPs, phytoconstituents such as polyphenols derived from tannins were involved in both decreasing and capping the nanoparticles. The beneficial impact of combining copper nanoparticles (Cu NPs) with bioactive substances from medicinal plants against bacterial infections has been proven.

The researchers utilized apple extract as a dual-functioning reducing and stabilizing agent to develop a rapid and eco-friendly approach for synthesizing silver nanoparticles (AgNPs). The AgNPs synthesized using this environmentally friendly and sustainable method have a significant inhibitory effect on the development and proliferation of *Salmonella* spp. and *Escherichia coli*. This study aims to assess the antibacterial properties of the AgNPs against *Salmonella* spp. and *Escherichia coli*.

The principles of green chemistry, as defined by Anastas and Eghbali (2010), emphasize the need for designs that reduce chemical waste and increase product efficiency. The photocatalytic performance of Ag–ZnO nanocomposites, as explored by Ansari et al. (2013), highlights significant advancements in the field of environmental remediation through nanotechnology. The biosynthesis of palladium nanoparticles using plant extracts demonstrates a viable route for green synthesis, reducing environmental impact (Duan et al. 2015). Garai et al. (2018) have effectively demonstrated the green synthesis of palladium nanoparticles using Terminalia Arjuna, showcasing their application in catalytic degradation of dyes. Cyanobacteria have been utilized as bio-factories for the synthesis of silver nanoparticles, offering a green alternative to conventional chemical methods (Husain et al. 2019). The electrochemical synthesis approach, as demonstrated by Li et al. (2013), provides a controlled environment for the fabrication of nanostructured materials, crucial for energy storage applications. The efficiency of photocatalytic processes in degrading pollutants can be significantly enhanced by optimizing the pH, as demonstrated in studies utilizing green synthesized titanium dioxide nanoparticles (Zamri and Sapawe 2019). Kumari and Tripathi (2020) have explored the use of *Bacillus cereus*-derived silver nanoparticles for the remediation of heavy metals in pharmaceutical effluents, highlighting their effectiveness in lead removal. Iravani (2014) reviews the current status and future prospects of bacteria in nanoparticle synthesis, underscoring their role in green chemistry. Breslow (1991) has extensively explored the hydrophobic effects on simple organic reactions in water, an essential aspect of green chemistry that aids in reducing solvent use. Francavilla et al. (2014) demonstrated the potential of biomass-derived polysaccharides in the preparation of photocatalytically active porous ZnO nanostructures, contributing to sustainable materials science. Khalafi et al. (2019) have shown enhanced photocatalytic activity of zinc oxide nanoparticles towards organosulfur pollutants, illustrating the potential for environmental cleanup technologies. Khanna et al. (2019) discuss the synthesis, characterization, and applications of algae-based metallic nanoparticles, emphasizing their sustainable production methods. Helmy et al. (2021) have developed a novel green synthesis method for S-doped TiO_2 nanoparticles using *Malva parviflora*, showcasing their multifunctional applications in antimicrobial, antioxidant, and photocatalytic activities.

Inamuddin (2021) emphasizes recent advancements in green synthesis techniques within the broader context of sustainable scientific practices and innovation. Balasooriya et al. (2017) explored the honey-mediated green synthesis of nanoparticles, presenting a new era of safe nanotechnology with promising environmental and biological applications. Murthy et al. (2020) have reported on the green synthesis of copper nanoparticles using *Hagenia abyssinica* leaf extract, emphasizing their effective antimicrobial properties. Bahrulolum et al. (2021) discuss the green synthesis of metal nanoparticles using microorganisms, highlighting their potential applications in the agrifood sector. Chowdhury et al. provide insight into future trends in green synthesis, emphasizing the importance of sustainable and eco-friendly methods in chemical production. Golinska et al. (2014) explore the biomedical applications and cytotoxicity of metal nanoparticles synthesized from *actinomycetes*, providing a base for future antimicrobial strategies. Goutam et al. (2018) have demonstrated the green synthesis of TiO_2 nanoparticles using *Jatropha curcas* leaf extract, highlighting their effectiveness in photocatalytic degradation of tannery wastewater. Husain et al. (2019) report on the use of cyanobacteria as a bioreactor for the synthesis of silver nanoparticles, exploring their potential in environmental applications. Iravani (2014) details the current status and future prospects of bacteria in nanoparticle synthesis, highlighting their potential as sustainable production methods. Kazlagić et al. (2019) utilized apple extract in the green synthesis of silver nanoparticles, demonstrating significant antimicrobial properties. Li et al. (2021) highlight the innovative use of algae-mediated biosystems for the production of metallic nanoparticles, focusing on their potential for aquatic environmental applications. Zamri and Sapawe (2019) demonstrate the efficacy of green synthesized titanium dioxide nanoparticles in enhancing the photocatalytic degradation of phenol, highlighting the critical role of pH in optimizing performance.

16.4 Recent Advances in Green Fabrication of Nanoparticles/Nanomaterials via Electrochemical Methods

16.4.1 Electrochemical Methods as Avenues for Green Synthesis of Nanomaterials

The electrochemical approach entails passing an electric current through a material to cause chemical reactions. The electron is the primary component in electrochemistry since it is responsible for all chemical reactions. An electrochemical cell is a term used to describe the entire electrolysis apparatus. An electrolyte is a term for the substance that an electron flows through. The electrolyte might be solid, conducting polymer, molten salts, aqueous or non-aqueous. Cations and anions are produced in an electrochemical cell when power is transferred through an electrolyte. These ions are subsequently deposited on the appropriate electrodes of an electrochemical cell, i.e., cation on the cathode and anion on the anode. The electrodes can be inert, such as electrodes consisting of carbon and platinum, or they can conduct liquid and solid metals, graphite, semiconductors, or other materials (Wang et al. 2009).

16.4.2 Electrodeposition

The electrochemical deposition method in nanotechnology offers a simple one-step method for creating nanostructures. A solid is deposited on the metal surface using an electrochemical process known as electrodeposition. It serves as the foundation for many industrial processes since it is simple to do and inexpensive at the same time. Without using any expensive machinery, materials are deposited. The most common use of this technique is for the production of nanomaterials, such as nanoparticles and nanowires, in the field of nanotechnology. Recently, it has been demonstrated to be a very effective and successful approach for synthesizing nanowires. Additionally, it has been discovered that the nanostructures created utilizing the electrodeposition approach are electrically active. Additionally, the electrodeposition technique is inexpensive, selective, devoid of surfactants, and it enables the nanoclusters to be modified by only altering the electrolyte composition and deposition settings (Xiao et al. 2019).

16.4.3 Electrochemical Synthesis of Unique Nanomaterials in Ionic Liquids

Ionic liquids (ILs) are commonly known as "designer solvents" due to their ability to intentionally choose and customize the chemical composition of both the cation and anion. This distinctive characteristic allows for an ideal correlation between the structure and characteristics of the ILs. Self-organization events take place within the ionic liquid's volume and at the phase contact point. These events are caused by the existence of various ions in ILs, each with their unique capacity to create structured arrangements (Pei et al. 2018).

Currently, there is a comprehensive comprehension of the significance of electrostatic interactions in maintaining the stability of nanoparticles. Afterwards, the counterions gather on the surface of the nanoparticle due to the attraction between the ionic liquid cations and the negatively charged nanoparticle. This leads to the creation of a layer of positively charged ions. The use of ILs as nanoparticle stabilizers greatly reduces the number of operations and chemicals needed to produce nanoparticles and other nanostructures (Ying et al. 2022).

The electrochemical conversion of nickel and iron triflates using a triflate ionic liquid may be used to produce nanoparticles with a size of 3 nm. Surprisingly, the inclusion of acetonitrile in the electrolyte results in a decrease in particle size from 3 to 1 nm. A separate investigation was conducted to analyze the procedure of applying nickel, cobalt, and their aluminum alloys onto the Au(111) surface using electrodeposition. Sun et al. recorded the procedure of electrodeposition of binary alloys (Au–Zn, Pt–Zn, Ag–Zn), followed by the elimination of one of the alloy constituents using electrochemical means, leading to the creation of nanoporous structures consisting of metals (Au, Pt, Ag) (Zin et al. 2014).

Nanoscale coatings of silver (Ag) and silver–copper (Ag–Cu) were created on glass carbon using the electrodeposition technique in triethylammonium acetate. Analyzing cyclic voltammograms indicates that an elevation in copper concentration leads to a displacement of the cathode and anode potential peaks towards a higher positive range, hence causing a reduction in the current for silver deposition. Simultaneously, the potential difference between the first depositions of Cu and Ag diminishes, facilitating a more effective concurrent deposition of both metals. It is crucial to note that when the copper concentration increases in the material, the morphology of the deposited nanomaterial undergoes a clear shift from a cubic structure to a dendritic structure that resembles flowers. The Ag–Cu composite containing 23% Cu exhibits the broadest potential cathode range while utilizing a mixed electrolyte consisting of N,N-dimethylformamide and tetrabutylammonium tetrafluoroborate (Zhu et al. 2015).

Glycols and polyols have a crucial function in the formation of well-defined nanostructured oxide materials since they act as chelating ligands for metal ions. Titanium dioxide nanopowders were synthesized electrochemically on the metallic titanium surface utilizing eutectic combinations based on choline chloride, ethylene glycol, and urea. The synthesis demonstrates a success rate of 92%. The TiO_2 powder obtained is composed of amorphous nanoparticles with sizes ranging from 8 to 18 nm. Anatase was manufactured by the process of calcination, wherein the powder was subjected to temperatures ranging from 400 to 600 °C. This approach offered a precise way to control the pace at which transition metal alkoxides undergo hydrolysis. Figure 16.3 illustrates a method for the eco-friendly synthesis of metal nanoparticles.

16.5 Application of Green Synthesised Nanoparticles for Environmental Remediation

Li et al. synthesized palladium nanoparticles (PdNPs) by utilizing *Eucommia ulmoides* as the precursor. The PdNPs obtained had a size of 12.6 nm. This material is composed of polyphenols, phytosterols, flavonoids, alkaloids, triterpenoids, amino acids, and proteins, all of which have reducing characteristics. Palladium nanoparticles (PdNPs) demonstrate exceptional catalytic efficiency in the electrocatalytic oxidation of hydrazine and the catalytic reduction degradation of p-aminoazobenzene, a typical molecule for azo dyes. Garai et al. (2018) produced palladium nanoparticles that were linked with *Terminalia arjuna* (TA-PdNPs) by a chemical process. These nanoparticles were then used as a very efficient catalyst

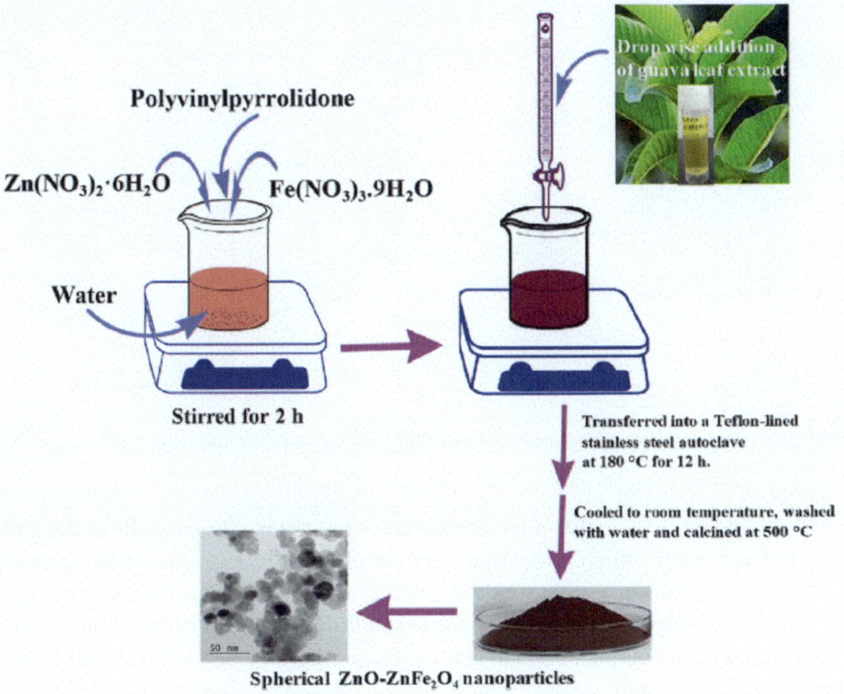

Fig. 16.3 Green synthesis of metal nanoparticles

for breaking down organic colors in a water-based setting, all at room temperature. In a study done by Nasrollahzadeh and Sajadi (2015), the researchers investigated the catalytic activity of Copper nanoparticles synthesized using an extract obtained from the leaves of *Ginko biloba*. The researchers discovered that these nanoparticles had exceptional catalytic efficacy. According to Kayalvizhi et al. (2020), the CuO nanoparticles produced using *Annona muricata* leaf extract showed discoloration rates of around 90 and 95% for Reactive Red and Methyl Orange, respectively, at a temperature of 60 °C. Figure 16.4 illustrates the methods used to address the environmental impact of green manufactured nanoparticles.

Zinc oxide (ZnO) nanoparticles have been extensively studied as a photocatalyst for wastewater treatment, as demonstrated by Liu et al. (2019). Ansari et al. presented a novel biogenic method for creating light-activated Ag-ZnO in their research. Subsequently, this artificially created material was used as a photoelectrode in a biofilm system that is electrochemically active. Ansari et al. (2013) showed that the photocatalytic activity shown significant effectiveness when evaluated with methylene blue, 4-nitrophenol, and methyl orange. Khalafi et al. (2019) have reported the use of *Chlorella* sp. in producing ZnO-based nanoparticles that are proven to be both safe and stable. Furthermore, it has been shown that ZnO-based nanoparticles produced using algae have advantageous properties as photocatalysts for the decomposition of

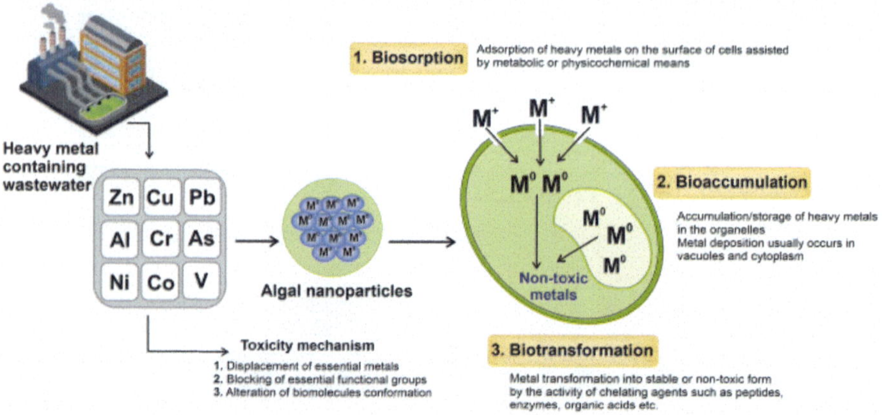

Fig. 16.4 Figurative representation of metal remediation via green synthesized NPs

dibenzothiophene, especially under conditions of neutral pH. Francavilla et al. (2014) conducted a research which found that *Gracilaria gracilis*, a species of red seaweed, has the ability to produce zinc oxide nanoparticles (ZnO-NPs) in different shapes and sizes. The nanoparticles demonstrated exceptional efficacy in the photocatalytic breakdown of aqueous phenol, outperforming the performance of a commercially accessible *Evonik titania*. A study done by Aziz et al. (2015) shown that AgNPs generated from *C. pyrenoidosa* outperformed commercially available AgNPs in terms of their effectiveness in the photocatalytic degradation of methylene blue. Unlike using only cyanobacterial extract, the use of silver nanoparticles (AgNPs) produced from Microchaete NCCU-342 showed a significant ability to remove color from the azo dye methyl red. The use of AgNPs for wastewater treatment shows potential because of their exceptional photocatalytic characteristics, especially in breaking down dyes (Husain et al. 2019).

Phyconanomaterials have shown remarkable potential in reducing industrial effluents and hazardous pollutants, as indicated by research done by Khanna et al. (2019) and Ramakrishna et al. (2016). The photodegradation of methylene blue was used to examine the photocatalytic activity of silver nanoparticles (AgNPs) obtained from *Synechococcus* sp. After a time period of 4 h, the reduction in color intensity of a dye solution with a concentration of 20 mg/L reached a level of 18% (±0.3%) (San Keskin et al. 2016). Husain et al. (2019) shown that the strain *Microchaete* sp. NCCU-342 have the capability to synthesize silver nanoparticles (AgNPs) and efficiently breakdown red dye. In just 2 h, the degradation efficiency increased to 84.6%, which was much higher than the degradation obtained by the cyanobacterial extract alone, which reached up to 49.8%.

Phyco-nanomaterials have shown remarkable potential in reducing industrial effluents and hazardous pollutants, as indicated by research done by Khanna et al. (2019) and Ramakrishna et al. (2016). The photodegradation of methylene blue was used to examine the photocatalytic activity of silver nanoparticles (AgNPs) obtained from *Synechococcus* sp. After a time period of 4 h, the reduction in color of a dye solution with a concentration of 20 mg/L reached a level of 18% (±0.3%). The reference is from San Keskin et al. (2016). Husain et al. (2019) shown that the strain *Microchaete* sp. NCCU-342 have the capability to synthesize silver nanoparticles (AgNPs) and efficiently breakdown red dye. Over a period of 2 h, the degradation efficiency increased to 84.6%, which was much higher than the degradation obtained by the cyanobacterial extract alone, which reached up to 49.8%. Subramaniyam et al. (2015) did a study where they employed *Chlorococcum* sp. MM11, a particular type of algae, for the production of iron nanoparticles. These nanoparticles have demonstrated the capability to decrease 92% of chromium(VI) ions, which were initially present at a concentration of 4 mg/L, into chromium(III) ions. The research undertaken by El-Kassas and Ghobrial (2017) includes analyzing Fe_3O_4 nanoparticles that were integrated into alginate beads. The results indicated that these beads had an impressive capacity to remove 91% of Pb, with a significant clearance rate of 78% recorded during a 75-min timeframe. Algal cells of *C. pyrenoidosa* and *S. obliquus* were used in a laboratory setting to effectively capture Cd ions and convert them into CdSe quantum dots (QDs). Zhang et al. (2018) found that this intervention effectively reduced the potential dangers linked to Cd disposal, which would otherwise constitute a substantial threat to the ecosystem and human well-being. In their 2016 work, Mandal et al. (2016) showed that *S. platensis* can transform harmful Cd^{2+} ions into CdS nanoparticles with lower toxicity via an in vivo biosynthetic mechanism.

The study done by Wang et al. (2017a, b) demonstrated the importance of *C. vulgaris* in the UV-induced elimination of Cr(VI) through photosynthesis. This was accomplished by incorporating *C. vulgaris* into chitosan nanofiber mats that were made with ultrafine bi-metallic (TiO_2/Ag). The work demonstrated a cooperative result between algae and hybrid nanomaterial (TiO_2/Ag) in which *C. vulgaris* released organic substances such as carboxylate and chlorophyll onto TiO_2/Ag chitosan nanofiber mats. As a result, the photocatalytic reduction of Cr(VI) increased. In addition, Kumari and Tripathi (2020) have noted that pharmaceutical effluents often contain heavy metals, such as lead (Pb) and chromium (Cr), which have similar properties. The work done by Keskin et al. (2015) shown that the use of *Chlamydomonas reinhardtii* together with chemically synthesized polysulfone nanofibers was highly effective in eliminating dyes. Furthermore, Aziz et al. (2015) and Edison et al. (2016) have previously shown that *Chlorella pyrenoidosa* and *Caulerpa racemosa* may be effectively used to synthesize silver nanoparticles. These nanoparticles have shown significant promise in the area of photocatalytic degradation, especially in relation to methylene blue and other toxic dyes.

The work done by Wang et al. (2017a, b) utilized manganese oxide nanoparticles that were produced by the biological synthesis of *Desmodesmus* sp. to efficiently remove bisphenol-A from liquid waste. Taghizadeh et al. (2021) did a catalytic research which showed that *C. vulgaris* effectively removed nitrates from liquid effluents. Algae have a dual role in removing nitrates from liquid waste. They may produce biogenic palladium nanoparticles and then chemically attach them to electrospun nanofiber mats. Nitrogen compounds give rise to substantial amounts of industrial waste through the production of chemical intermediates and solvents. Ramakrishna et al. (2016) found that gold nanoparticles produced by *Tubinaria conoides* and *Sargassum tenerrimum* had the capacity to break down nitro compounds found in wastewater. The research undertaken by Uzair et al. (2020) focused on employing palladium nanoparticles derived from *Sargassum bovinum* to electro-catalytically reduce industrial waste effluents.

The research undertaken by Sethy et al. (2020) utilized a green synthesis method, employing an extract from *Syzygium cumini* leaves to generate titanium dioxide nanoparticles (TiO_2-NPs). The nanoparticles were employed for the extraction of lead (Pb) from heavily polluted industrial effluent. The findings demonstrated a substantial decrease in chemical oxygen demand (COD) by 75.5% and the removal of Pb^{2+} ions by 82.53%. The removal of chromium from wastewater produced by tanneries is a notable difficulty because of the use of cupric sulphate ($CuSO_4$) in the manufacturing of leather products. Goutam et al. (2018) found that using *Jatropha curcas* leaf extract to synthesize metal nanoparticles for wastewater treatment resulted in the elimination of 76.48% of Cr^{6+} and 82.26% of COD. The metal nanoparticles synthesized using *H. rosasinensis* leaf extract exhibit heightened efficacy in decomposing phenol contaminants, suggesting a favorable and environmentally friendly method. The study done by Zamri and Sapawe (2019) examined the effect of a decreased pH level of 5 on the efficacy of phenol removal. The study especially focused on achieving a 90% reduction in phenol levels utilizing metal nanoparticles distributed in a solvent. This work examines the process of photocatalytic degradation of ornidazole, a commonly used drug derived from nitroimidazole, utilizing metal nanoparticles produced from an extract of *Aegle marmelos* leaves. This research presents a different method for breaking down or removing ornidazole contaminants in wastewater. The decay of the dye of the chosen medicine follows a pseudo-first-order equation. Esfahani et al. (2020) found that *Verbascum thapsus* plants can produce metal nanoparticles that effectively degrade Cr(VI). This is due to the favorable particle size of these nanoparticles, which is better than other metal nanoparticles. Additionally, these nanoparticles are responsive to visible light, further enhancing their degradation capabilities.

In their study, Saeed et al. (2019) showcased the application of *Azadirachta indicia* leaves extract in the environmentally friendly production of Ag-doped metal nanoparticles. This synthetic substance has demonstrated the capacity to destroy Rhodamine B (RhB) and Methylene blue (MB) dyes. The research states that employing leaf extract from *Syzygium cumini* to create metal-NPs has been proven to successfully regulate their size, hence enabling the adjustment of their catalytic properties. The anatase phase has a high capacity to use light to catalyze the removal of lead (Pb) and chemical oxygen demand (COD) because it has a large surface area of 105 m^2g^{-1}, a pore size of 10.50 nm, and a pore volume of 0.27 cm^3g^{-1}. Zamri and Sapawe (2019) reported that the ultraviolet (UV) absorption peak was detected at a wavelength of 356 nm, and the band gap energy was measured to be 3.48 eV. This work examines the catalytic properties of S-doped metal nanoparticles in the degradation of Methyl

Orange (MO) dye. The process is enhanced by the presence of *Malvaparvi flora* extract and visible light irradiation. The study done by Helmy et al. (2021) found that surface-bound O_2 molecules can cause oxidation of surface Ti^{3+} sites, leading to the production of O_2^- free radicals. The TiO_2/CQDs exhibited a higher photocatalytic degradation rate compared to TiO_2-NPs, leading to a maximum efficiency of 100% during a 10-min period. However, the process of completely breaking down MB when exposed to sunlight utilizing TiO_2-NPs may take as long as 30 min. The observed improvement in efficiency can be ascribed to the reduction in the reflecting properties of TiO_2-NPs in the visible spectrum, which is helped by the integration of carbon quantum dots (CQDs) in TiO_2/CQDs nanocomposites (Williams et al. 2022).

16.6 Conclusion

Recently, there have been notable advancements in technology that have led to the development of more environmentally-friendly alternatives and more dependable methods for producing nanoparticles (NPs). This paper provides a summary of the environmentally friendly manufacture of nanoparticles (NPs). Green synthesis techniques offer a cost-effective and ecologically friendly alternative to existing procedures for producing nanoparticles (NPs). Various research papers have shown evidence that nanoparticles generated using green synthesis using natural resources offer numerous benefits in comparison to traditional approaches. The benefits include reduced presence of harmful pollutants, a simpler and less complex chemical manufacturing process, low levels of toxicity, cost-effectiveness, and the capacity to take on many forms. Nanoparticles (NPs) exhibit exceptional crystallinity and are distinguished by their diminutive size, making them a promising method for the decomposition of dyes, extraction of metals, and other remediation techniques. The controlled production of nanomaterials at the nanoscale presents considerable technological challenges and corresponding disadvantages. In the future, the development of a green synthesis process for nanoparticles (NPs) has the potential to become the dominant technology used in large-scale industrial production. This brief summary emphasizes the importance of green synthesis and its use in environmental settings.

References

Ahmed A, Arya S. (2021) Green synthesis of nanomaterials via electrochemical method. In: Inamuddin, Boddula R, Ahamed MI, Khan A (eds) Advances in green synthesis: Advances in science, technology & innovation. Cham: Springer. https://doi.org/10.1007/978-3-030-67884-5_11

Anastas P, Eghbali N (2010) Green chemistry: principles and practice. Chem Soc Rev 39(1):301–312. https://doi.org/10.1039/b918763b

Ansari SA, Khan MM, Ansari MO, Lee J, Cho MH (2013) Biogenic synthesis, photocatalytic, and photoelectrochemical performance of Ag–ZnO nanocomposite. J Phys Chem C 117(51):27023–27030. https://doi.org/10.1021/jp410063p

Aziz N, Faraz M, Pandey R, Shakir M, Fatma T, Varma A, Barman I, Prasad R (2015) Facile algae-derived route to biogenic silver nanoparticles: synthesis, antibacterial, and photocatalytic properties. Langmuir 31(42):11605–11612. https://doi.org/10.1021/acs.langmuir.5b03081

Bahrulolum H, Nooraei S, Javanshir N, Tarrahimofrad H, Mirbagheri VS, Easton AJ, Ahmadian G (2021) Green synthesis of metal nanoparticles using microorganisms and their application in the agrifood sector. J Nanobiotechnol 19(1):86. https://doi.org/10.1186/s12951-021-00834-3

Balasooriya ER, Jayasinghe CD, Jayawardena UA, Ruwanthika RWD, Mendis de Silva R, Udagama PV (2017) Honey mediated green synthesis of nanoparticles: new era of safe nanotechnology. J Nanomater 2017:1–10. https://doi.org/10.1155/2017/5919836

Breslow R (1991) Hydrophobic effects on simple organic reactions in water. Acc Chem Res 24(6):159–164. https://doi.org/10.1021/ar00006a001

Chowdhury S, Rakshit A, Acharjee A, Saha B (n.d.).Future trends in green synthesis. In: Advances in green synthesis: avenues and sustainability. technology and innovation, pp. 41–74. https://doi.org/10.1007/978-3-030-67884-5_3

Duan L, Li M, Liu H (2015) Biosynthesised palladium nanoparticles using Eucommia ulmoides bark aqueous extract and their catalytic activity. IET Nanobiotechnol 9(6):349–354. https://doi.org/10.1049/iet-nbt.2015.0020

Edison TNJI, Atchudan R, Kamal C, Lee YR (2016) Caulerpa racemosa: a marine green alga for eco-friendly synthesis of silver nanoparticles and its catalytic degradation of methylene blue. Bioprocess Biosyst Eng 39(9):1401–1408. https://doi.org/10.1007/s00449-016-1616-7

El-Kassas HY, Ghobrial MG (2017) Biosynthesis of metal nanoparticles using three marine plant species: anti-algal efficiencies against "Oscillatoria simplicíssima." Environ Sci Pollut Res Int 24(8):7837–7849. https://doi.org/10.1007/s11356-017-8362-5

Esfahani RN, Khaghani S, Azizi A, Mortazaeinezhad F, Gomarian M (2020) Facile and eco-friendly synthesis of TiO_2 NPs using extracts of Verbascum thapsus plant: an efficient photocatalyst for reduction of Cr(VI) ions in the aqueous solution. J Iran Chem Soc 17(1):205–213. https://doi.org/10.1007/s13738-019-01755-7

Francavilla M, Pineda A, Romero AA, Colmenares JC, Vargas C, Monteleone M, Luque R (2014) Efficient and simple reactive milling preparation of photocatalytically active porous ZnO nanostructures using biomass-derived polysaccharides. Green Chem 16(5):2876–2885. https://doi.org/10.1039/C3GC42554A

Garai C, Hasan SN, Barai AC, Ghorai S, Panja SK, Bag BG (2018) Green synthesis of Terminalia Arjuna-conjugated palladium nanoparticles (TA-PdNPs) and its catalytic applications. J Nanostruct Chem 8(4):465–472. https://doi.org/10.1007/s40097-018-0288-z

Golinska P, Wypij M, Ingle AP, Gupta I, Dahm H, Rai M (2014) Biogenic synthesis of metal nanoparticles from actinomycetes: biomedical applications and cytotoxicity. Appl Microbiol Biotechnol 98(19):8083–8097. https://doi.org/10.1007/s00253-014-5953-7

Goutam SP, Saxena G, Singh V, Yadav AK, Bharagava RN, Thapa KB (2018) Green synthesis of TiO_2 nanoparticles using leaf extract of Jatropha curcas L. for photocatalytic degradation of tannery wastewater. Chem Eng J 336:386–396. https://doi.org/10.1016/j.cej.2017.12.029

Helmy ET, Abouellef EM, Soliman UA, Pan JH (2021) Novel green synthesis of S-doped TiO_2 nanoparticles using Malva parviflora plant extract and their photocatalytic, antimicrobial, and antioxidant activities under sunlight illumination. Chemosphere 271:129524. https://doi.org/10.1016/j.chemosphere.2020.129524

Husain S, Afreen S, Hemlata D, Yasin D, Afzal B, Fatma T (2019) Cyanobacteria as a bioreactor for synthesis of silver nanoparticles—an effect of different reaction conditions on the size of nanoparticles and their dye decolorization ability. J Microbiol Methods 162:77–82. https://doi.org/10.1016/j.mimet.2019.05.011

Inamuddin (2021) Advances in green synthesis. In: Boddula R, Ahamed MI, Khan A (eds) Advances in science, technology and innovation. https://doi.org/10.1007/978-3-030-67884-5

Iravani S (2014) Bacteria in nanoparticle synthesis: current status and future prospects. Int Sch Res Not 2014:359316. https://doi.org/10.1155/2014/359316

Kayalvizhi S, Sengottaiyan A, Selvankumar T, Senthilkumar B, Sudhakar C, Selvam K (2020) Eco-friendly cost-effective approach for synthesis of copper oxide nanoparticles for enhanced photocatalytic performance. Optik 202:163507. https://doi.org/10.1016/j.ijleo.2019.163507

Kazlagić A, Abud O, Ali Ć, Maid H, Saud B, Borovac B, Omanović-Mikličanin E (2019) Green synthesis of silver nanoparticles using apple extract and its antimicrobial properties. Heal Technol. https://doi.org/10.1007/s12553-019-00378-5

Khalafi T, Buazar F, Ghanemi K (2019) Phycosynthesis and enhanced photocatalytic activity of zinc oxide nanoparticles toward organosulfur pollutants. Sci Rep 9(1):6866. https://doi.org/10.1038/s41598-019-43368-3

Khanna P, Kaur A, Goyal D (2019) Algae-based metallic nanoparticles: synthesis, characterization, and applications. J Microbiol Methods 163:105656. https://doi.org/10.1016/j.mimet.2019.105656

Khan Y, Sadia H, Ali Shah, et al. (2022) Classification synthetic and characterization approaches to nanoparticles and their applications in various fields of nanotechnology. A Review Catalysts 12(11):1386. https://doi.org/10.3390/catal12111386

Kumari V, Tripathi AK (2020) Remediation of heavy metals in pharmaceutical effluent with the help of Bacillus cereus-based green-synthesized silver nanoparticles supported on alumina. Appl Nanosci 10(6):1709–1719. https://doi.org/10.1007/s13204-020-01351-9

Li GR, Xu H, Lu XF, Feng JX, Tong YX, Su CY (2013) Electrochemical synthesis of nanostructured materials for electrochemical energy conversion and storage. Nanoscale 5(10):4056–4069. https://doi.org/10.1039/c3nr00607g

Li SN, Wang R, Ho SH (2021) Algae-mediated biosystems for metallic nanoparticle production: from synthetic mechanisms to aquatic environmental applications. J Hazard Mater 420:126625. https://doi.org/10.1016/j.jhazmat.2021.126625

Liu Y, Zhang Q, Xu M, Yuan H, Chen Y, Zhang J, Luo K, Zhang J, You B (2019) Novel and efficient synthesis of Ag-ZnO nanoparticles for the sunlight-induced photocatalytic degradation. Appl Surf Sci 476:632–640. https://doi.org/10.1016/j.apsusc.2019.01.137

Mandal D, Bolander ME, Mukhopadhyay D, Sarkar G, Mukherjee P (2006) The use of microorganisms for the formation of metal nanoparticles and their application. Appl Microbiol Biotechnol 69(5):485–492. https://doi.org/10.1007/s00253-005-0179-3

Mandal RP, Sekh S, Sen Sarkar NS, Chattopadhyay D, De S (2016) Algae mediated synthesis of cadmium sulphide nanoparticles and their application in bioremediation. Mater Res Express 3(5):055007. https://doi.org/10.1088/2053-1591/3/5/055007

Murthy HCA, Desalegn T, Kassa M, Abebe B, Assefa T (2020) Synthesis of green copper nanoparticles using medicinal plant Hagenia abyssinica (Brace) JF. Gmel leaf extract: antimicrobial properties. J Nanomater 2020:1–12. https://doi.org/10.1155/2020/3924081

Nasrollahzadeh M, Sajadi SM (2015) Green synthesis of copper nanoparticles using Ginkgo biloba L. leaf extract and their catalytic activity for the Huisgen cycloaddition of azides and alkynes at room temperature. J Colloid Interface Sci 457:141–147. https://doi.org/10.1016/j.jcis.2015.07.004

Pei Y, Hao L, Ru J, Zhao Y, Wang H, Bai G, Wang J (2018) The self-assembly of ionic liquids surfactants in ethanolammonium nitrate ionic liquid. J Mol Liq 254:130–136. https://doi.org/10.1016/j.molliq.2018.01.095

Ramakrishna M, Rajesh Babu D, Gengan RM, Chandra S, Nageswara Rao G (2016) Green synthesis of gold nanoparticles using marine algae and evaluation of their catalytic activity. J Nanostruct Chem 6(1):1–13. https://doi.org/10.1007/s40097-015-0173-y

Roldán MV, Pellegri N, de Sanctis O (2013) Electrochemical method for Ag-PEG nanoparticles synthesis. J Nanoparticles 2013, article ID 524150. https://doi.org/10.1155/2013/524150

Saeed M, Muneer M, Khosa MKK, Akram N, Khalid S, Adeel M, Nisar A, Sherazi S (2019) Azadirachta indica leaves extract assisted green synthesis of Ag-TiO_2 for degradation of methylene blue and rhodamine b dyes in aqueous medium. Green Process Synth 8(1):659–666. https://doi.org/10.1515/gps-2019-0036

Saleh HE-DM, Koller M (2018) Introductory chapter: Principles of green chemistry. InTech. https://doi.org/10.5772/intechopen.71191

San Keskin NO, Koçberber Kılıç NK, Dönmez G, Tekinay T (2016) Green synthesis of silver nanoparticles using cyanobacteria and evaluation of their photocatalytic and antimicrobial activity. J Nano Res 40:120–127. https://doi.org/10.4028/www.scientific.net/JNanoR.40.120

Sethy NK, Arif Z, Mishra PK, Kumar P (2020) Green synthesis of TiO_2 nanoparticles from Syzygium cumini extract for photo-catalytic removal of lead (Pb) in explosive industrial wastewater. Green Process Synth 9(1):171–181. https://doi.org/10.1515/gps-2020-0018

Singh J, Dutta T, Kim KH, Rawat M, Samddar P, Kumar P (2018) 'Green' synthesis of metals and their oxide nanoparticles: applications for environmental remediation. J Nanobiotechnol 16(1):84. https://doi.org/10.1186/s12951-018-0408-4

Subramaniyam V, Subashchandrabose SR, Thavamani P, Megharaj M, Chen Z, Naidu R (2015) Chlorococcum sp. MM11—a novel phyco-nanofactory for the synthesis of iron nanoparticles. J Appl Phycol 27(5):1861–1869. https://doi.org/10.1007/s10811-014-0492-2

Taghizadeh SM, Morowvat MH, Negahdaripour M, Ebrahiminezhad A, Ghasemi Y (2021) Biosynthesis of metals and metal oxide nanoparticles through microalgal nanobiotechnology: quality control aspects. BioNanoScience 11(1):209–226. https://doi.org/10.1007/s12668-020-00805-2

Uzair B, Liaqat A, Iqbal H, Menaa B, Razzaq A, Thiripuranathar G, Rana NF, Menaa F (2020) Green and cost-effective synthesis of metallic nanoparticles by algae: safe methods for translational medicine. Bioengineering 7(4):129. https://doi.org/10.3390/bioengineering7040129

Wang C, Chao KJ, Singh P, Mathiyalagan R, Jin Y, Yang DC, Chun D (2015) Green synthesis of silver nanoparticles by Bacillus methylotrophicus, and their antimicrobial activity. Artif Cells Nanomed Biotechnol. https://doi.org/10.3109/21691401.2015.1011805

Wang L, Zhang C, Gao F, Mailhot G, Pan G (2017a) Algae decorated TiO_2/Ag hybrid nanofiber membrane with enhanced photocatalytic activity for Cr(VI) removal under visible light. Chem Eng J 314:622–630. https://doi.org/10.1016/j.cej.2016.12.020

Wang R, Wang S, Tai Y, Tao R, Dai Y, Guo J, Yang Y, Duan S (2017b) Biogenic manganese oxides generated by green algae Desmodesmus sp. WR1 to improve bisphenol A removal. J Hazard Mater 339:310–319. https://doi.org/10.1016/j.jhazmat.2017.06.026

Wang Y, He X, Wang K, Zhang X, Tan W (2009) Barbated Skullcup herb extract-mediated biosynthesis of gold nanoparticles and its primary application in electrochemistry. Colloids Surf B 73(1):75–79. https://doi.org/10.1016/j.colsurfb.2009.04.027

Williams IBI, Fodjo EK, Amadou K, Albert T, Kong C (2022) Enhancing the photocatalytic activity of TiO_2 nanoparticles using green Carbon quantum dots. Int J Nano Dimens 13(2):144–154

Xiao T, Huang J, Wang D, Meng T, Yang X (2019) Au and Au-Based nanomaterials: synthesis and recent progress in electrochemical sensor applications. Talanta 206:120210. https://doi.org/10.1016/j.talanta.2019.120210

Ying S, Guan Z, Ofoegbu PC, Clubb P, Rico C, He F, Hong J (2022) Green synthesis of nanoparticles: current developments and limitations. Environ Technol Innov 26, article no 102336. https://doi.org/10.1016/j.eti.2022.102336

Zamri MSFA, Sapawe N (2019) Effect of pH on phenol degradation using green synthesized titanium dioxide nanoparticles. Mater Today Proc 19:1321–1326

Zhang Z, Chen J, Yang Q, Lan K, Yan Z, Chen J (2018) Eco-friendly intracellular microalgae synthesis of fluorescent CdSe QDs as a sensitive nanoprobe for determination of imatinib. Sens Actuat Part B 263:625–633. https://doi.org/10.1016/j.snb.2018.02.169

Zhu C, Du D, Eychmüller A, Lin Y (2015) Engineering ordered and nonordered porous noble metal nanostructures: synthesis, assembly, and their applications in electrochemistry. Chem Rev 115(16):8896–8943. https://doi.org/10.1021/acs.chemrev.5b00255

Zin V, Brunelli K, Dabalà M (2014) Characterization of Cu–Ni alloy electrodeposition and synthesis of nanoparticles by pulsed sonoelectrochemistry. Mater Chem Phys 144(3):272–279. https://doi.org/10.1016/j.matchemphys.2013.12.028

Chapter 17
Mycoremediation—A Sustainable Clear Technology for Environmental Remediation

Rajiv Periakaruppan, P. Vanathi, and G. Priyanka

Abstract World environmental pollution intensified due to the rapid increase in population, urbanization, and industrialization development. Contamination of the soil and water affects both human health and the ecology. Thus, it is necessary to develop environmental cleanup technologies to diminish the environmental pollution stresses. In ancient times, conventional remediation methods have been used to address these environmental contaminants, but they have proved ineffectual. Mycoremediation, a type of bioremediation process that relies on macro and micro fungi or their constituents to remediate environmental pollutants, is an efficient, economical, safe and effective technique for detoxifying organic, inorganic, and novel toxins including metals, agricultural wastes, antibiotics and pharmaceuticals from the environment. This book chapter emphasizes the scope, process, potential and numerous mycoremediation techniques for biosorption, biotransformation, precipitation and sequestration of environmental pollution. In addition, mycoremediation's ability to detoxify metals, organic pollutants that persist, and as well as other pollutants was reiterated. For instance, Macrofungi (mushrooms) produce various extracellular enzymes to degrade or detoxify the PAHs (Polycyclic Aromatic Hydrocarbon), pesticides, azo dyes, PCBs (Polychlorinated Biphenyls) etc. into basic compounds. The fungi such as *Aspergillus niger*, *Agaricusbisporous*, *Tricholoma* sp., *Pleurotusdryinus*, *Phellinus badius* etc. degrade and extract the pollutants, azo dyes and heavy metals such as Cd, Pb, Cr, Hg, Zn, Cu and Fe from the polluted soil and water. Hence, a potential outline for mycoremediation has been proposed to endorse this innovative clean-up technique for alleviating the world's sustainable environmental issues.

Keywords Environment · Fungi · Mycoremediation · Pollutants · Heavy metals

R. Periakaruppan (✉)
Department of Biotechnology, PSG College of Arts & Science, Coimbatore, Tamil Nadu 641014, India
e-mail: rajivsmart15@gmail.com

P. Vanathi · G. Priyanka
Department of Biotechnology, Sri Ramakrishna College of Arts & Science, Coimbatore, Tamil Nadu 641006, India

© The Author(s), under exclusive license to Springer Nature Switzerland AG 2025
A. Jeyaseelan and K. Murugasen (eds.), *Sustainable Environmental Remediation: Avenues in Nano and Biotechnology*, Environmental Science and Engineering,
https://doi.org/10.1007/978-3-031-78483-5_17

17.1 Introduction

The activities of the modern times have led to the release of enormous amounts of pollutants into the environment. These pollutants pose a significant threat to biodiversity and human health. Soil and groundwater are polluted by pollution coming from waste management, industry and agriculture. The primary pollutants that need immediate mitigation efforts include heavy metals, such as lead (Pb), mercury (Hg), cadmium (Cd), chromium (Cr), cobalt (Co), zinc (Zn), nickel (Ni), iron (Fe) and manganese (Mn), according to data on soil and ground water quality published by the Central Pollution Control Board of India (Kamyotra 2009). The global population is continuously grappling with several health issues that come from the escalating air, water and soil pollution (Xu et al. 2022), exemplified by the increased risks of respiratory diseases, diarrhea-associated deaths, heavy metal poisoning etc., (Nishida and Yatera 2022; Behera and Mishra 2022; Mitra et al. 2022). There are various physical and chemical methods that have been developed to remove and degrade the pollutants in the environment. These remediation methods employ the application of chemicals or involve energy-intensive processes, thereby contributing to their own set of environmental concerns. Conventional cleaning techniques, such as incineration or decontamination, are expensive and cause air pollution. In addition, these methods are expensive, produce toxic byproducts (Michael-Igolima et al. 2022). The need to explore alternative, sustainable technologies has prompted the surge of using biological methods for effective remediation. Scientists propose bioremediation, the repair of the environment using living organisms. Although chemical remediation techniques are technically sound, they are not economically viable for wide-scale soil remediation because they produce significant volumes of sludge that require further treatment before disposal (Chen et al. 2000).

Bacteria, fungi and plants have been widely used in the biological remediation process as they are considered to be cheap and environmentally friendly (Michael-Igolima et al. 2022). Fungi, encompassing both unicellular and multicellular organisms, exhibit remarkable applications in various industrial and environmental contexts. Fungi showcase the ability to thrive in a broad spectrum of environments, as evidenced by their presence in diverse habitats. The robust growth of fungi, along with their extensive hyphal networks, high surface area to volume ratio, adaptability to variations in temperature and pH, secretion of multifunctional extracellular enzymes, resilience to diverse pollutants, and possession of metal-binding proteins, make them excellent tools in mycoremediation (Kumar et al. 2021). Mycoremediation is an economic and eco-friendly method of combating pollution (Banu et al. 2021). The success of mycoremediation is dependent on the mechanisms employed by the fungi to interact with and transform pollutants.

As we navigate the intricate process of mycoremediation, it is essential to explore its current applications, ongoing research, and potential future developments. This chapter will critically examine key studies and advancements in the field, emphasizing on mycoremediation as a sustainable and clear technology for environmental remediation. In addition, we will also discuss the unique capabilities of fungi that emerge as a sustainable tool in combating environmental pollution.

17.2 Process of Mycoremediation

Industrial and agricultural activities have led to the release of various toxic pollutants, causing environmental contamination. Xenobiotic compounds, including heavy metals, agricultural and pharmaceutical wastes, pesticides, herbicides, and industrial effluents, are significant pollutants of concern. In recent years, there has been a growing interest in discovering a cost-effective and reliable solution for pollutant cleanup, leading researchers to explore mycoremediation as one of the environmentally friendly remediation techniques.

Fungi as decomposers are essential for recycling nutrients in the ecosystem. They have the unusual ability to metabolize various organic and inorganic contaminants, converting them into sources of energy and carbon and reducing them to non-hazardous levels. A wide variety of fungal taxa from virtually all fungal phyla have demonstrated the potential to degrade pollutants, but the two major phyla *Basidiomycota* and *Ascomycota* have received the most attention due to their profound ligninolytic activities and superior mycoremediation efficiency. Fungal systems have been used to biodegrade a wide variety of contaminants, including polyaromatic hydrocarbons, pharmaceutical active chemicals, heavy metals, synthetic dyes, toxins, and radionuclides (Shourie and Vijayalakshmi 2022).

The fungi employed in mycoremediation, particularly filamentous fungi and macrofungi are significant contributors. The process of mycoremediation is based on several properties of the fungi and the components in their cell wall. Composed of glycoproteins, chitin, glucans, and chitosan, the cell wall of fungal biomass forms a matrix containing active functional groups such as carboxyl, carbonyl, amide and amino groups (Sintakindi and Ankamwar 2021; Fernando et al. 2021). These groups play a key role in binding various pollutants. The fate of the xenobiotic compound is determined by the mechanism that is employed by the fungal cell to interact with the pollutant. The management of pollutants encompasses both passive biosorption processes and active mechanisms (Fig. 17.1) such as bioprecipitation, bioaccumulation, biotransformation, and biomineralization (Dhankhar and Hooda 2011; Shourie and Vijayalakshmi 2022).

17.2.1 Biosorption

Biosorption, the process of removing pollutants by adsorption onto biological material, whether live, inert, dead, or derived from a biological source, is a mechanism for effective environmental remediation (Legorreta-Castañeda et al. 2020). This process involves sorbent (biological material) and the sorbate (polluting substance). This

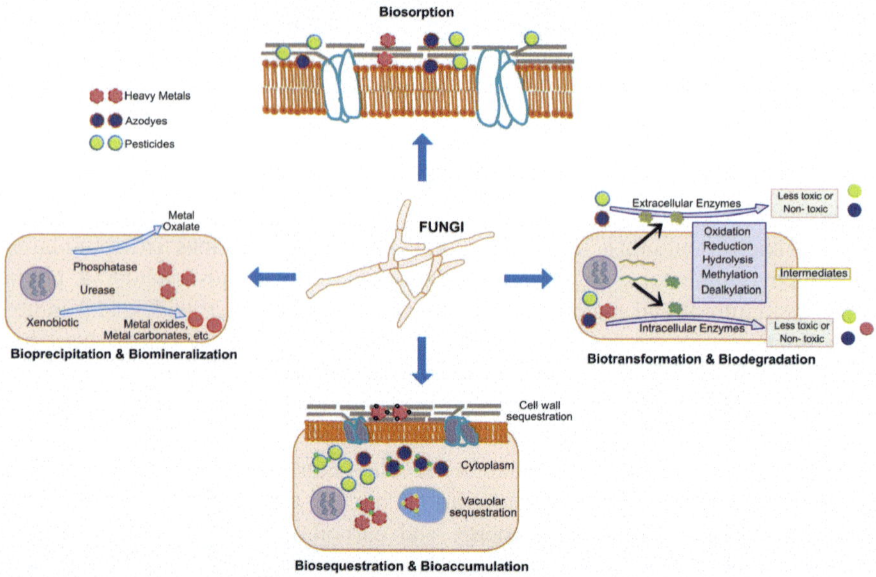

Fig. 17.1 Process of mycroremediation employed to detoxify the Xenobiotic compounds (Dhankhar and Hooda 2011; Fernández et al. 2018; Shourie and Vijayalakshmi 2022)

biomass of the biological material often shows resistance to metals and other harmful conditions (Gavrilescu 2004). The process of pollution adsorption exhibits variations depending on the nature of fungal cells—dead or alive. In cases where the dead biomass is used, passive adsorption is employed, a technique favored over live cells due to the potential for reuse and the avoidance of toxicity concerns (Aksu 2005). Conversely, when utilizing live fungal biomass, multiple concurrent processes come together, including absorption, ion exchange, precipitation, and intracellular accumulation (Lu et al. 2016). However, under adverse conditions such as high toxicity or inhibition, the utilization of dead fungal biomass can overcome challenges encountered by live fungal cells (Legorreta-Castañeda et al. 2020). The dead fungal biomass utilizes less energy, has more binding sites and functional groups, is cost effective, and has the potential to be recovered and reused (Li et al. 2010). Thereby, researchers have shown that dead fungal cells demonstrate enhanced efficacy in the remediation process over the dead fungal biomass.

Various pollutants are removed through the biosorption technique (Table 17.1), and primarily the surface adsorption contributes significantly to the pollutant removal process. Mar'in et al. (1997) suggested that biosorption may involve polar groups in amino acids, proteins, and structural polysaccharides such as chitin, chitosan and glucans. Most of the researchers have proved that the rate of biosorption is influenced by factors such as pH, temperature, fungal strain and contact time. Not just the live and the dead fungal cells, but the products from the fungal cells were also found to be a source of good sorbent. Polysaccharides obtained from dried fungal mycelium of

Aspergillus niger and *Fusarium culmorum* exhibited the capability to degrade 100% of the dimethoate pesticide (Cabrera-Barjas et al. 2020). The fungal biomasses were also pretreated and modified to increase the effectiveness of adsorption. The process of chemical modifications on chitosan beads with histidine and *S. cerevisiae* was able adsorb Ni at 104.2 mg/g (Nguyen and Juang 2015).

Biosorption replaces the treatment of industrial wastewater and the recovery of metals from wastewater. Compost and fungal mycelium could also be combined to form a biosorbent. Biosorption is a cost-effective and highly absorbent technique for the xenobiotic compunds (Kulshreshtha et al. 2014).

17.2.2 Biotransformation and Biodegradation

Fungal biotransformation involves the conversion of persistent pollutants into environmentally safe substances through fungal metabolic processes. Biotransformation includes various processes, such as oxidation, reduction, methylation, and dealkylation, utilized by fungi for environmental pollution removal (Dusengemungu et al. 2020). The fungal enzymes contribute significantly to this process. Numerous studies have investigated how fungi and their enzymes decompose. Nyanhongo et al. observed that fungus can produce extracellular ligninases, peroxidases, xylanases, cellulases, pectinases, and oxidases. They can oxidize resistant contaminants in vitro. The substrates that these enzymes take typically activate them. In addition, these enzymes have been found to degrade in vitro non-polymeric, persistent pollutants, including pentachlorophenol (Lin et al. 1990), organic and synthetic dyes (Heinfling et al. 1998), PAHs (Hammel et al. 1991; Johannes et al. 1996) and nitrotoluenes. Recent studies show that various fungi can degrade polymers, including plastics (da Luz et al. 2013).

Fusarium verticillioides degrades the aromatic hydrocarbons, aromatic polyketide through the process of hydrolysis into endocrocin-9-anthron (Gao et al. 2022). Aromatic amines that form a part of the hydrocarbons undergo peripheral degradation by treating with *Trichoderma harzianum, Aspergillus niger, Cunninghamella elegans* (Al-Hawash et al. 2019). *Bjerkandera adusta* is responsible for the dealkylation of atrazine and *Fusarium* spp. degrades atrazine by degrading the aromatic amines (Dhiman et al. 2020). PCBs are oxidized by *Pleurotus ostreus* (Ijoma and Tekere 2017; Kantharaj et al. 2017). *Trichospora viride, Penicillium chyrsogenum, Aspergillus aegrita* degrade fluorene and pyrene (Vipotnik et al. 2021). The enzymes secreted by the fungi intracellularly and extracellularly contribute to the maximum degradation of the pesticides. Biodegradation is a highly intricate process and involves factors such as the production of H_2O_2 by fungi, hydroxyl radicals, interactions among ligninolytic enzymes and the cytochrome P450 monooxygenase system, and the influence of other biochemical processes. However, during the process, the intermediate products that are formed might be toxic and therefore the study of intermediates that are produced during the metabolic process is highly essential.

Table 17.1 Fungal species that are used in the process of biosorption of various Xenobiotic compounds

Pollutant	Nature of the sorbate	Fungal species	Nature of the inoculum	Contact time (h)	Accumulation/ percentage removal	References
Heavy metals	Agricultural soil, Cr and Cd	*Aspergillus* sp.	Wet fungal biomass	4	1.20 mg/g of Cr 2.72 mg/g of Cd	Zafar et al. (2007)
	Agricultural soil, Cr and Cd	*Rhizopus* sp.	Wet fungal biomass	4	4.33 mg/g of Cr 2.72 mg/g of Cd	Zafar et al. (2007)
	Aqueous solution, Uranium	*Penicillium citrinum*	Dead Fungal biomass	5	103.1 to 127.3 mg/g	Pang et al. (2011)
	Aqueous solution, Pb	*Rhizopus arrhizus*	Dried fungal biomass	24	0.501 mol/kg	Şenol et al. (2021)
	Aqueous Media, Cd and Pb	*Phialomyces macrosporus*	Dead fungal biomass	24	80% of Cd^{2+} and Pb^{2+} were removed	Brito et al. (2021)
Pentachlorophenol	Biocide	*Anthracophyllum discolor*	Live fungal pellets	24	>80%	Bosso et al. (2015)
Dimethoate	Pesticides	*Aspergillus niger* and *Fusarium culmorum*	Chitin and Chitosan	24	100% dimethoate	Cabrera-Barjas et al. (2020)

17.2.3 Biosequestration and Bioaccumulation

The main method of metal elimination occurs when metals interact with fungal cells and either accumulate inside the cell or are adsorbed onto the cell surface. *Trichoderma lixii* facilitates the accumulation of copper ions, primarily through compartmentation inside the vacuole, while some copper ions are also adsorbed onto the cell wall surface via electrostatic attraction to functional groups such as COO^-, aiding in their overall accumulation within fungal cells (Kumar and Dwivedi 2021). Similarly in chromium bioremediation, *Coriolus versicolor* functions as a complexing and reducing agent, making it easier to sequester harmful Cr(VI) ions. *Coriolus versicolor* contributes to biosequestration and chelation by the multiple pathways involved in this process, which include electrostatic sequestration of dichromate ions, reductive conversion of dichromate to Cr(III), oxidation of thiol to disulfide, and chelation of Cr(III) with amino ligands (Sanghi et al. 2009). Pb is sequestered by *Pleurotus ostreatus* and is primarily found in the cell wall, followed by the vacuoles (Wang et al. 2019).

Metallothioneins, phytochelatins and antioxidant glutathione are the three main types of intracellular peptides that chelate thiophilic metal ions (Sácký et al. 2014). Typically, the sequestration of intracellular heavy metals involves compartmentalization and binding with cytosolic metallothionein. In *Hebeloma mesophaeum*, the sequestration process for intracellular Zn and Cd involves compartmentalization, directing Cd to vacuoles and Zn ions to zincosome-like membrane vesicles, followed by the complexation with metallothioneins, while Ag primarily binds to metallothionein proteins (Sácký et al. 2014). Thereby, the one useful process deployed by fungi to lower metal bioavailability is metal sequestration and accumulation.

17.2.4 Bioprecipitation and Biomineralization

Bioprecipitation is the process of usage of biological agents, such as microorganisms to remove contaminants from the environment through precipitation. Biomineralization can occur when an organism changes its native environment to promote the precipitation of minerals, or when the organism uses its own cellular mechanisms for controlling the mineralization process (Gadd 2021).

Aspergillus niger is involved in biomineralization of calcium (Burford et al. 2006) and *Acremonium strictum, Pyrenochaeta* sp., *Stagonospora* sp., *Aspergillus niger*, *Serpula himantioides* are involved in biomineralization of manganese (Tang et al. 2013; Wei et al. 2012). Fungal enzymes are actively involved in the process of bioprecipitation and biomineralization. Phosphatase activity, produced by *Aspergillus niger* and *Paecilomyces javanicus*, led to the release of phosphorus from the organic substrate, resulting in uranium phosphate bioprecipitation thus highlighting the role of fungal phosphatase activity in bioprecipitation process (Liang et al. 2015). *Albifimbria viridis* exhibited increased laccase production in response to tannin, leading to the precipitation of elements such as Ni (51%) and Co (45%). When the enzyme-containing cell-free supernatant was immobilized on resin, it demonstrated higher efficiency, precipitating 70% of Ni and 73% of Co (Ahmadi Khozani et al. 2021). A mechanism of Pb precipitation in *Pleurotus ostreatus* that involves the participation of intracellular chelating agents and the precipitation facilitated by extracellular oxalic acids (Wang et al. 2019). Urease producing fungi, *Myrothecium gramineum* show potential for removing Sr^{2+} from Sr-contaminated waste streams (Li et al. 2015). Furthermore, *Pestalotiopsis* sp. and *Myrothecium gramineum* 56% of Ca and 28% of Sr (Li et al. 2015). Therefore, biomineralization along with bioprecipitation, offer significant potential for the creation of novel minerals with industrial applications, for remediation of the environment or for the recovery of metals through biological means.

17.3 Mycoremediation of Xenobiotic Compounds

17.3.1 Heavy Metal Pollutants and Its Mycoremediation Process

Heavy metals originate from industrial waste, leaching, mining operations, and agricultural practices. Their accumulation in rice and fish increases the likelihood of their entry into the food chain, posing a threat to human health (Noman et al. 2022). There are highly successful reports of the heavy metals that are removed by the use of fungi. Fungi's enormous network of hyphae allows them to absorb heavy metals (Mejstřík and Lepšpvá 1993). It is commonly recognized that fungi with high metal tolerance may thrive and acquire metals from environmental contaminants. The fungi employs mechanisms like biosorption, bioaccumulation, biotransformation, chelation or efflux to tolerate the heavy metal stresses (Fernández et al. 2018; Priyadarshini et al. 2021).

The cell fractionation revealed that 80% of the Cu was accumulated in the cell walls of *Trichoderma viride* (Anand et al. 2006). The Table 17.2 summarizes the various heavy metals that are remediated using different fungal strains. *Mucor hiemalis* EH8 demonstrated the ability to convert toxic ionic mercury into elemental mercury, by forming spherical metallic droplets intracellular areas (Hoque and Fritscher 2016). *Hypocrea tawa* efficiently reduces highly toxic hexavalent chromium Cr(VI) to less harmful trivalent Cr(III), achieving a 100% reduction of Cr(VI) at high concentrations (Morales-Barrera and Cristiani-Urbina 2015). The predominant presence of chromate reductase in the soluble cell fraction suggests that *Aspergillus niger* uses a process involving Cr(VI) reduction, surface immobilisation, and intracellular accumulation of Cr(III) (Gu et al. 2015).

Gola et al. (2016) demonstrated that *Beauveria bassiana* exhibited an adsorption and accumulation efficiency of 84.53% for multiple heavy metals, including Zn, Cu, Cd, Cr, and Ni, in a multi metal solution at a concentration of 30 mg/L. The individual

Table 17.2 Heavy metal pollutants removal by fungal species and its mycoremediation process

Pollutant	Name of the fungal species	Concentration of the metals	Contact time	Amount of pollutant removed	Mechanism	References
Cd	*Absidia cylindrospora*	50 mg/L	3 d	68%	Biosorption	Albert et al. (2018)
	Aspergillus niger	0.586 ± 0.025 mg/L	72 h	98%	Bioaccumulation	Khan et al. (2019a, b)
	Aspergillus ustus	1 ppm	20 min	84%	Biosorption	Alothman et al. (2020)
	Penicillium chrysogenum	1 ppm	20 min	91%	Biosorption	Alothman et al. (2020)
Cr	*Aspergillus niger*	0.3565 ± 0.031 mg/L	72 h	43%	Bioaccumulation	Khan et al. (2019a, b)
	Aspergillus niger	50 mg/L	84 h	99%	Bioaccumulation Biotransformation	Gu et al. (2015)
	Coriolus versicolor	50 mg/L	24 h	100%	Biosorption Biosequestration	Sanghi et al. (2009)
	Hypocrea tawa	1 mM	50 h	100%	Biotransformation	Morales-Barrera and Cristiani-Urbina (2015)
	Penicillium sp.	20–100 mg/L	8 h	53 and 70%	Biotransformation	Arévalo-Rangel et al. (2013)
Cu	*Aspergillus flavus*	200–1400 ppm	50 h	20.75–93.65 mg/g	Biosorption	Iram et al. (2015)
	Aspergillus ustus	1 ppm	20 min	52%	Biosorption	Alothman et al. (2020)
	Beauveria bassiana	30 mg/L	5 days	74.13%	Biosorption Bioaccumulation	Gola et al. (2016)
	Penicillium chrysogenum	1 ppm	20 min	53%	Biosorption	Alothman et al. (2020)

(continued)

Table 17.2 (continued)

Pollutant	Name of the fungal species	Concentration of the metals	Contact time	Amount of pollutant removed	Mechanism	References
	Trichoderma lixii	10 mg/L	120 h	84.6%	Biosequestration Bioaccumulation Biosorption	Kumar and Dwivedi (2021)
	Trichoderma viride	100 mg/L	72 h	81%	Bioaccumulation	Anand et al. (2006)
Fe	*Sterigmatomyces halophilus*	1000 ppm	14 days	85%	Biosorption	Bano et al. (2018)
Hg	*Aspergillus flavus*	10 mg/L	7 days	98.73%	Biosorption	Kurniati et al. (2014)
	Aspergillus terreus	26.66 mg/L	120 h	95.5%	Bioleaching	Khan et al. (2019a)
	Didymella glomerata	100 mg/L	48 h	97%	Biosorption	Văcar et al. (2021)
	Mucor hiemalis	1000 μg/L	10–48 h	99%	Bioaccumulation Biotransformation	Hoque and Fritscher (2016)
	Mucor rouxii	100 mg/L	24 h	95.3%	Biosorption	Martínez-Juárez et al. (2012)
Ni	*Beauveria bassiana*	30 mg/L	5 days	75%	Biosorption Bioaccumulation	Gola et al. (2016)
Pb	*Aspergillus fumigatus*	170.90 mg/L	120 h	99.20%	Bioleaching	Khan et al. (2019a, b)
	Aspergillus niger	200–1400 ppm	50 h	3.25 to 172.25 mg/g	Biosorption	Iram et al. (2015)
	Aspergillus niger	5 mM	30 days	99.5–99.8%	Bioprecipitation	Liang et al. (2016)
	Aspergillus ustus	1 ppm	20 min	42%	Biosorption	Alothman et al. (2020)
	Paecilomyces javanicus	5 mM	30 days	99.4–99.8%	Bioprecipitation	Liang et al. (2016)

(continued)

Table 17.2 (continued)

Pollutant	Name of the fungal species	Concentration of the metals	Contact time	Amount of pollutant removed	Mechanism	References
	Penicillium chrysogenum	1 ppm	20 min	56%	Biosorption	Alothman et al. (2020)
Zn	*Sterigmatomyces halophilus*	1000 ppm	14 days	84%	Biosorption	Bano et al. (2018)

metals showed varying degrees of removal, with Cr (99.9%) > Cd (95.3%) > Cu (92%) > Ni (71.3%) > Zn (65.3%) in the multi metal solution. These findings highlight the influence of elemental properties of metals on the overall removal efficiency in multi metal ion systems and its applicability to real-world environmental conditions.

17.3.2 Remediation of Azo Dyes

Azo dyes, characterized by the presence of one or more azo bonds (–N=N–), find extensive application in various industrial sectors such as textiles, leather, food, pharmaceuticals, cosmetics, etc. The widespread use of these dyes raises environmental concerns due to their resistance to degradation and potential carcinogenic properties (Chung 2016). The most commonly used fungi for the degradation of the azo dyes and the mechanisms have been summarized in the table (Table 17.3). Fungal species, such as *Aspergillus* and white rot fungi, demonstrate the capability to adsorb and degrade persistent dyes. They accomplish this through processes like biosorption and/or biodegradation through enzymatic breakdown.

Dyes interact with the active groups on the cell wall surface through physical mechanisms like Van der Waals forces, and chemical mechanisms like ion exchange or covalent bonding (Aksu 2005). Many of the researches have shown that the decolorization efficiency is by contact time, pH levels, and biomass dosage. The decrease in decolorization with increasing original dye concentration, potentially due to a reduction in the number of available adsorption sites on the adsorbent surface (Salvi and Chattopadhyay 2017; Mahmoud et al. 2017). In addition supplementation with glucose was found to enhance the rate of biosorption (El-Rahim et al. 2017). All fungal strains interact differently with the dye molecules, and they do not secrete the same amount of the enzyme. The degradation potential of *Aspergillus oryzae* on the azo dyes are quicker and higher when compared to *Phanerochaete chrysosporium* (Santos and Corso 2014).

The breakage of chemical bonds, facilitated by fungal enzymes such as laccases, peroxidases, and azoreductases synthesized by fungi, signifies the degradation of the dye molecule, with these enzymes playing an instrumental role in cleaving azo bonds (Zhao et al. 2006). Subsequently, the dyes are predominantly absorbed and transformed into compounds that are either less harmful or non-toxic. The presence of lignin peroxidases were a contributing factor for decolorization of several azo dyes which were enhanced by the presence of nitrogen source (Abd El-Rahim et al. 2017). The fungi are employed in decolorization processes under both static and agitated conditions, either in free or immobilized form. The immobilized culture *Trichothecium roseum* at a pH of 6 removed 94% of Lanasyn Navy M-DNL dye upon incubation for 7 days (Ali 2023). The reuse of the immobilized culture would reduce the requirement for fresh cultures and mitigate reproducibility concerns. However, it was noted that continued use resulted in a decrease in their effectiveness. Multiple fungal strains belonging to the *Aspergillus* species were studied, and the process of

Table 17.3 Remediation of Azo dyes using various fungal strains

Azo dye	Fungus	Concentration of initial sample (mg/L)	Contact time	Efficiency of removal	Mechanism	References
Acid Blue 161	*Aspergillus niger, Aspergillus terreus, Rhizopus oligosporus*	100	3 h	52% 40% 45%	Biosorption	Almeida and Corso (2019)
Congo Red	*Dichotomomyces cejpii and Phoma tropica*	50	4 days	73%	Biodegradation	Krishnamoorthy et al. (2018)
Congo Red	*Penicillium oxalicum and Aspergillus tubingensis*	100	12 h	97.15%	Biodegradation	Thakor et al. (2022)
Direct Blue 71	*Aspergillus oryzae*	100	10 days	–	Biodegradation	Santos and Corso (2014)
Direct Blue-1	*Aspergillus terreus*	100	7 days	98.4%	Biosorption and biodegradation	Singh and Dwivedi (2020)
Direct Red-80 Mordant Blue-9	*Phanerochaete chrysosporium*	10–200	6 days	More than 92%	Biodegradation	Singh and Pakshirajan (2010)
Direct violet	*Aspergillus fumigatus*	100	5 days	85.8%	Biodegradation Biotransformation	Abd El-Rahim et al. (2017)
Disperse Orange 3	*Pleurotus ostreatus*	80	9 days	50%	Biodegradation	Zhao et al. (2006)
Fast Red A	*Rhizopus arrhizus*	100	1 h	93%	Biosorption	Salvi and Chattopadhyay (2017)
Lanasyn Navy M-DNL	*Trichothecium roseum*	10	7 days	94%	Biosorption	Ali (2023)

(continued)

Table 17.3 (continued)

Azo dye	Fungus	Concentration of initial sample (mg/L)	Contact time	Efficiency of removal	Mechanism	References
Metanil yellow	*Rhizopus arrhizus*	100	2 h	96%	Biosorption	Salvi and Chattopadhyay (2017)
Methyl orange	*G. candidum*	100	48 h	94.6%	Biodegradation Biomineralization	Rajhans et al. (2020)
Methyl red	*Aspergillus fumigatus*	100	5 days	91.3%	Biodegradation Biotransformation	Abd El-Rahim et al. (2017)
Procion Red MX-5B	*Aspergillus niger Aspergillus terreus and Rhizopus oligosporus*	100	3 h	35%	Biosorption	Almeida and Corso (2019)
Reactive Black 5	*Pleurotus eryngii*	10	72 h	93.56%	Biodegradation	Hadibarata et al. (2013)
Reactive red 43	*Sarocladium* sp.	200	2 h	97.43%	Biosorption	Nouri et al. (2021)
Red azo dye	*Aspergillus niger*	100	2 h	99.69%	Biosorption and biodegradation	Mahmoud et al. (2017)

removal of the azo dyes were mostly pointing towards the process of biotransformation and biodegradation, rather than the biosorption itself by the living fungal cells. The favoured pH absorption range of most of the azo dyes was upto 8 in most of the adsorption processes (Ali 2023). Arami et al. (2006) proposed that the presence of negatively charged surface sites at higher pH levels might hinder dye adsorption due to potential electrostatic repulsion.

During the fungal degradation process, the toxicity of the decolourization products were examined. In certain cases of incomplete dye degradation, intermediate compounds such as aromatic amines may form, potentially posing greater toxicity than the original compounds (Qu et al. 2010). Biosorption is often favored over biodegradation due to the lower acute toxicity of intermediate metabolites. Biodegradation process achieves higher levels of color removal. However, the formation of intermediary metabolites could lead to increased acute toxicity (Almeida and Corso 2019). Therefore, it is important for any bioremediation technology to assess the toxicity of the pollutants and metabolites formed after dye degradation (Sen et al. 2016). This emphasizes on the need for attaining maximum effectiveness in each method through further research.

Use of fungal consortium was also studied for their effectiveness in the removal of azo dyes. When utilizing the fungal consortia of *Dichotomomyces cejpii* and *Phoma tropica*, Congo Red dye was biodegraded to 73% (Krishnamoorthy et al. 2018), whereas 97.15% was degraded with the use of *Penicillium oxalicum* and *Aspergillus tubingensis* (Thakor et al. 2022). The fungal consortium of *Geotrichum candidum* and *Rhodotorula glutinis* and immobilized on coconut fiber, exhibited 88.4% decolorization in textile effluents after treating it for a 6 h period. Furthermore, this degradation process was linked to heightened production of the laccase enzyme (Sen et al. 2021). The combination of fungi with bacterial consortia were also investigated. The non-toxic degraded products confirmed that the consortium comprising *Aspergillus tamari* and *Pseudomonas putida* effectively decolorized crystal violet and congo red dyes (Henagamage and Peries 2023). Therefore, the fungal consortium by itself, or along with bacteria and with other immobilization techniques could in future be promising to remove the azo dyes in the long run.

17.3.3 Degradation of Hydrocarbon Pollutants

Anthropogenic organic compounds globally contribute significantly to environmental contamination, with common examples including petroleum hydrocarbons, insecticides, and herbicides from agricultural and pest control practices, as well as byproducts from manufacturing. Phthalates and polybrominated diphenyl esters, commonly found in commercial products and fire retardants, also add to this pollution burden. They leak into surrounding materials and are now found in numerous populations (Mbachu et al. 2020). The key source of soil and water pollution is the release of hydrocarbons into the environment, whether by accident or through human activity (Das and Chandran 2011). Petroleum hydrocarbon pollution has been shown to have negative impacts on aquatic and terrestrial life as well as soil productivity.

Organic pollutants may undergo several modifications once they enter the environment. These processes include transformation, sorption–desorption, volatilization, plant absorption, surface water runoff, and groundwater movement. Transformation or degradation is one of the fundamental processes that determine the environmental fate and transit of organic pollutants, and it includes a variety of processes such as abiotic and biodegradation. During these processes, organic contaminants are converted into degradation products or totally mineralized into a carbon field (Diaz 2008). The members of *Ectomycorrhizal* colonization could participate in carbon cycling in forest soils and could serve as helpful models for ectomycorrhiza-mediated restoration strategies in diesel-contaminated soils, even though laccase data indicate more species in this family, most likely due to the paraphyletic nature of *Tricholomataceae* (Gunderson et al. 2007).

Many pollutants such as polycyclic aromatic hydrocarbons, polychlorinated biphenyls, textiles, and pesticides can be degraded by *Pleurotus dryinus*. It is possible for *Pleurotus dryinus* to convert hazardous materials into these non-hazardous materials by the use of ligninolytic enzymes. Researchers have used this as a sustainable method of wastewater treatment in addition to reducing the environmental impact of industrial activities (Alves et al. 2017). Domination of heavy metal contaminated environments by *Pleurotus dryinus* has positive results due to various processes including biosorption and bioaccumulation. The fungus absorbs heavy metals from the environment, inhibits its mycelium, reducing the toxicity and bioavailability of these metals. The breakdown of PAHs by *Pleurotus dryinus*, studied using the microcosm over a period of 28 days demonstrated the influence of soil properties on PAH uptake and PAH adsorption (Rathankumar et al. 2020). The *Phellinus sp.* degraded PAHs (Arun and Eyini 2011).

Due to recent developments that have made it possible to analyse environmental microorganisms in great detail using high-throughput techniques such as metagenomic, genomic, and bioinformatic analyses, interest in the microbial degradation of contaminants has increased (Koukkou 2011) (Fig. 17.2).

17.3.4 Mycoremediation in Agricultural Operations and Agro-waste Management

Pesticides and herbicides such as glyphosate, endosulfan, paraquat, fipronil, and aldrin include compounds that are carcinogenic, endocrine disruptors, neurotoxic, and have an adverse impact on people and severely damage the environment (Akhtar and Mannan 2020). The majority of these compounds are either outlawed or subject to severe restrictions because of their harmful side effects and their persistence in the environment (Purnomo et al. 2017; Maqbool et al. 2016). Various methods such as chemical treatment, burning, and volatilization have been employed to break down

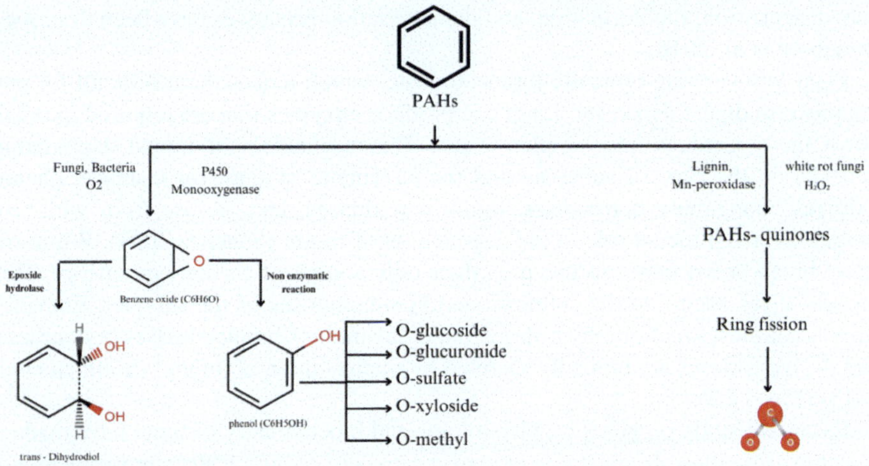

Fig. 17.2 Degradation of polycyclic aromatic hydrocarbons (PAHs) by mycoremediation. Adapted from Dutta and Hyder (2019)

these harmful substances. In addition to being expensive and consuming a lot of energy, these techniques discharge pollutants, acids, and alkali into the environment (Eman et al. 2013).

Several studies have demonstrated that various fungi may accumulate and get rid of herbicides and insecticides from the soil. The primary focus of these study was on the natural fungi that thrived in contaminated regions. *Aspergillus tamarii* and *Botryosphaeria laricina*, which were isolated from an agricultural field and had previously been exposed to endosulfan, were able to break down the endosulfan toxin and its hazardous metabolites, such as endosulfan sulfate, alpha endosulfan, and beta endosulfan (Silambarasan and Abraham 2013). *Aspergillus glaucus* was also shown to be capable of degrading fipronil and its metabolite, fipronil sulfone using ligninolytic enzymes (Gajendiran and Abraham 2017). *Phellinus* sp. potential for lignin breakdown is strong (Arun and Eyini 2011).

Agricultural waste must be pre-processed before being disposed of in order to minimize soil contamination and the leaching of nutrients due to the abundance of minerals and nutrients it contains. Growing mushrooms with agro-wastes offers a low-cost, eco-friendly way for managing agricultural waste, and also it yields a healthy number of edible mushrooms, which may provide farmers significant profits (Dhiman et al. 2022).

Trametes maxima enzyme extracts and co-cultures of *T. maxima* and *Paecilomyces carneus* degrade atrazine (Chan-Cupul et al. 2016). Dieldrin, the main metabolite of aldrin, may be broken down by the white-rot fungus *Pleurotus ostreatus*. Epoxidation and hydroxylation events have been postulated as the mechanisms by which *P. ostreatus* biotransforms aldrin and dieldrin (Purnomo et al. 2017). During the fungus-mediated transformation of these pesticides, esterification, deoxygenation, dehydrogenation, dichlorination, and demethylation processes have been described (Maqbool et al. 2016).

Plant leftovers are typically lignocellulosic, which makes them difficult for any organism to digest. However, fungi can produce enzymes that are required to break down these residues into simpler forms. The mechanism of fungal degradation depends on the type of substrate and the enzymatic system; for instance, phenol oxidizes, manganese peroxidises, lignin peroxidases, and laccase from white-rot fungi play a significant role in the degradation of lignin (Sánchez 2009). White-rot fungi break down lignin to use it as their only source of carbon and energy. This process gives access to the cellulose and hemicelluloses of the subtract. Enzymes such as cellulases (cellobiohydrolases, endoglucanases), hemicellulases (xylanases), and β-glycosidases are typically secreted by micro- and macrofungi (Soliman et al. 2013) (Fig. 17.3).

Extensive study on genetics, physiology, and biochemistry of fungi have made it possible to use these organisms to prepare a variety of economically significant industrial and agricultural goods (Yuvaraj and Ramasamy 2020). Kim and Kim (2001) and Wang and Ng (2004) reported that *Tricholoma* species exhibit both laccase and proteinase activity. Considerable ability to degrade lignin has been demonstrated in several *Tricholoma* species. According to Trojanowski et al. (1984), *T. aurantium* may actually be able to use low-molecular-weight lignin-derived substrates more effectively than many other common wood-rotting fungi.

Pleurotus dryinus, characterized by its capacity to colonize diverse lignocellulosic substrates, releases several ligninolytic enzymes that form complexes with lignin (Sánchez 2009). The capacity to expose celluloses and hemicelluloses and selectively biodegrade lignin is the most prominent characteristic (Elisashvili et al. 2008; Bánfi

Fig. 17.3 Mycoremediation of agricultural wastes. Figure adapted from Umor et al. (2021)

et al. 2015). This unique enzymatic machinery makes *Pleurotus dryinus* an attractive candidate for bioremediation applications.

Potential industrial applications of laccase from *Tricholoma giganteum* AGHP and its usefulness in several biotechnological applications are well known (Patel and Gupte 2016). This explores how lignocellulosic substrates can be used to optimize the production laccase enzyme. Thereby, the wastes have the potential for the development of economically and environmentally acceptable bioprocesses.

Even while some research indicates that ligninolytic enzymes are involved in the breakdown of certain herbicides and insecticides, it is still unclear how important a function these enzymes play in the breakdown of these xenobiotics. Furthermore, it is yet unknown what mechanism these fungi employ to biodegrade xenobiotics. Comprehending the fundamental process can aid in formulating enhanced approaches to tackle the issue posed by these substances.

17.3.5 Mycoremediation of Pharmaceutical Waste

Pharmaceutical chemicals discharged from medication production effluents worsen already dire environmental conditions. According to reports, dangerous threshold amounts of pharmaceuticals have been found in the environmental discharges (Larsson 2014). Apart from implications on human health, pharmaceutical effluents have been found to have harmful impacts on aquatic biota (Alimba et al. 2019). The water treatment systems are unable to handle these xenobiotics, leading to hazardous pharmaceutical chemicals build up in the environment (Esterhuizen-Londt et al. 2016). Another major source of drug contamination in the environment is leftover medications.

The fungi that are used in the process of detoxification of pharmaceutical waste is given in the Table 17.4. The aquatic fungus *Mucor hiemalis* may eliminate acetaminophen from tainted water sources containing pharmaceutical effluents (Esterhuizen-Londt et al. 2016). Diclofenac is another medication that it was able to absorb and eliminate (Esterhuizen-Londt et al. 2017). Within six hours, naproxen and its intermediates may be absorbed and eliminated by the white-rot fungus *Trametes versicolor* to undetectable quantities (Marco-Urrea et al. 2010). The medium treated with *Trametes versicolor* was determined to be harmless.

Similarly, *T. versicolor* was also capable of breaking down the painkiller ketoprofen (Marco-Urrea et al. 2010). The cytochrome 450 and ligninolytic enzymes produced by *T. versicolor* breaks down naproxen and ketoprofen (Marco-Urrea et al. 2010). Bioreactors appear to be a potential method for treating pharmaceutical waste in the future (Yang et al. 2013a, b). *T. versicolor* broke down various medications, including codeine, diazepam, carbamazepine, and metoprolol in bioreactor studies (Asif et al. 2017). In a fed-batch stirred reactor, *Phanerochaete chrysosporium* was able to carry out the oxidative degradation of several anti-inflammatory medications (Rodarte-Morales et al. 2012). The anti-inflammatory medication piroxicam was

Table 17.4 Mycoremediation of pharmaceutical waste

Pharmaceuticals pollutants	Fungi	References
Carbamazepine	*Trametes versicolor*	Jelic et al. (2012)
Diclofenac	*Trametes versicolor*	Yang et al. (2013a, b)
Ofloxacin	*Trametes versicolor*	Gros et al. (2014)
17 β-estradiol	*Trichoderma citrinoviride*	Chatterjee and Abraham (2019)
Acetaminophen	*Mucor hiemalis*	Esterhuizen-Londt et al. (2016)
Ibuprofen	*Phanerochaete chrysosporium*	Rodarte-Morales et al. (2012)
Azole antifungal agents	*Lentinula edodes*	Kryczyk-Poprawa et al. (2019)

degraded by the mycelia of the edible fungus *Lentinula edodes*. Endocrine disruptors, such as synthetic testosterone and 17α-ethyl-estradiol were also degraded by *L. edodes* (Muszyńska et al. 2018). It is necessary to enhance the conditions for the more effective and better remediation of different pharmaceutical chemicals from polluted soil and water in fungal bioreactors.

17.4 Macrofungi in Pollutant Removal

Fungi use many strategies to detoxify dirty areas and revitalize the ecosystem. Macrofungi are also actively involved in biodegradation; their distribution is related to humus-containing soil, such as that found in forests, where hay, decomposing wood, litter, hips of straw, etc., are prevalent. Through a variety of enzymatic activities on cellulose, lignin, nitrogenous compounds, etc., macro-fungi aid in the breakdown of these complex products, converting them into simpler forms and assisting in the recycling of nutrients (Sultana et al. 2007). White rot fungi are being studied as key organisms to remove hazardous substances through the process of biodegradation, biosorption and bioconversion. Many nations are still researching the transformation of garbage, sludge and toxins into beneficial forms. Using the sugar found in cellulose and hemicellulose, the bioconversion process creates macrofungal metabolites that are vital to the development and survival of the fungus. One possible source of secondary metabolites and enzymes is free-growing fungi (Conceição et al. 2018). Macrofungi use enzymes to biodegrade and biotransform wood lignin to gain access to cellulose and hemicellulose chains.

Enzymes break down the lignocellulosic material to form sugar monomers for the yeast fermentation process to produce ethanol, which is from commercial interest. On various industrial wastes, mushroom culture has also been effectively accomplished (Dulay et al. 2012; Kulshreshtha et al. 2013). The pollutants that are removed by macrofungi are summarized in Table 17.5 and the role of *Agaricus bisporus*, *Pleurotus dryinus* and *Ganoderma* sp. is summarized in Table 17.6.

Table 17.5 Pollutant removal using macro-fungus

Macro-fungi species	Pollutant	References
Agaricus bisporus	Anthracene, fluoranthene, phenanthrene, pyrene	Gramss et al. (1999)
Bjerkandera adusta	Azo dyes, phthalocyanine dyes	Heinfling et al. (1998)
Pleurotus pulmonarius	Crude oil	Olusola and Anslem (2010)
Trametes hirsuta	Textile dye	Abadulla et al. (2000)
Agaricus campestris	Heavy metals	Širić et al. (2016)
Pycnoporus sanguineus	Heavy metals	Zulfadhly et al. (2001)
Pleurotus ostreatus	Remazol Brilliant Blue R dye	Lee et al. (2022)
Lactarius acerrimus	Heavy metals (Hg)	Tuzen et al. (2020)

Table 17.6 Effectiveness of macrofungi as a mycoremediation tool

Mushroom species	Pollutant	Findings	References
Agaricus bisporus	Cd(II)	*Agaricus bisporus* showed promise in mycoremediation by effectively degrading the Cd ions	Nagy et al. (2014)
	Flotation tailings	Study on mycoremediation of flotation tailings with *Agaricus bisporus*, highlighting its potential in remediating this specific type of industrial waste	Budzyńska et al. (2022)
	Pb, γ-HCH	Comparison of mycoremediation with *Agaricus bisporus* and *Pleurotus ostreatus* growth substrates versus phytoremediation with Festuca rubra and *Brassica sp.* for the recovery of a Pb and γ-HCH contaminated soil	Hidalgo et al. (2023)
	Heavy metals	An examination of the heavy metal content of a few edible wild mushrooms from Turkey's middle Black Sea region sheds light on the buildup of heavy metals in wild mushrooms	Isildak et al. (2004)
Pleurotus dryinus	Polycyclic aromatic hydrocarbons	*Pleurotus dryinus* effectively degrades pollutants, indicating potential for bioremediation. However, degradation rates may vary depending on environmental conditions and substrate availability	Rathankumar et al. (2020)
Ganoderma sp.	Polycyclic aromatic hydrocarbons	The study identifies and characterizes fungal enzymes involved in PAH oxidation and degradation, indicating its feasibility for PAH bioremediation	Torres-Farradá et al. (2019)

Pleurotus sp.

Pleurotus species are known for their production of lignolytic enzymes and cytochrome P450 enzymes, which play significant roles in the degradation of heavy metals and oil pollutants (Kapahi and Sachdeva 2017). Laccase and P450 monooxygenase from *Pleurotus chrysosporium* (Bosso et al. 2015) could degrade polycyclic aromatic hydrocarbons. Furthermore, phenol oxidase enzymes such as peroxidases, also found in *Pleurotus* species, are capable of degrading phenolic compounds (Bosco and Mollea 2019).

Agaricus bisporus

Agaricus bisporus is the most important commercially grown mushroom, accounting for 40% to 45% of global production (Zhao et al. 2006). There have been numerous reports of its mycoremediation potential in the literature. Kryczyk et al. (2017) conducted in vitro cultures of *A. bisporus* and found that copper and cadmium could be eliminated from additives. A method for producing *A. bisporus* in a sustainable manner was provided in order to handle industrial effluent and agricultural leftovers while overcoming disposal issues. Ugya and Imam (2017) studied the efficiency of *A. bisporus* in refinery effluent and discovered that it capably reduced sulfate, nitrate, alkalinity, phosphate, electrical conductivity, Biochemical Oxygen Demand and Chemical Oxygen Demand, and heavy metals.

Phellinus badius

A common macrofungus called *P. badius* showed a high capacity for absorbing heavy metals (0.82 mmol/gram of adsorbent), which is significantly greater than that of natural zeolite and activated carbon and comparable to that of ion exchange resins and other documented biosorbents with 90% of the total adsorption happens in 15 min.

Using mushrooms to bioconvert these industrial wastes into protein-rich mushroom carpophores, or mushroom fruiting bodies, not only produces mushrooms but also aids in reducing pollution that may otherwise result from their disposal. In contrast, *A. bisporus* could accumulate various compounds with varying degrees of efficiency. According to mineral composition studies of *Pleurotus ostreatus*, the most evaluated elements are Cd, Pb, Hg, and As (Deepalakshmi and Sankaran 2014). The nutrient composition of fruit bodies is usually determined by the type of the substrate and level of impurities in the substrate that is used for its growth. According to Gucia et al. (2012), mushrooms from uncontaminated sites contain elements such as Al, Cd, Zn, Mn, and others. However, heavy metals are generally important for the production of cellular phase and cellulolytic enzymes throughout the biodegradation process, resulting in the accumulation of heavy metals in *Pleurotus* mushrooms as a food source needs to get further attention (Bellettini et al. 2019). Recent research shows that there is a high risk of consumption of infected fungi and it may be unlikely that fruit bodies will be used to obtain recurrent elements (Budzyńska et al. 2022). The risk of consuming contaminated fruit bodies and lack of proper regulatory guidelines for trace element content of mushrooms are the major concerns that could arise

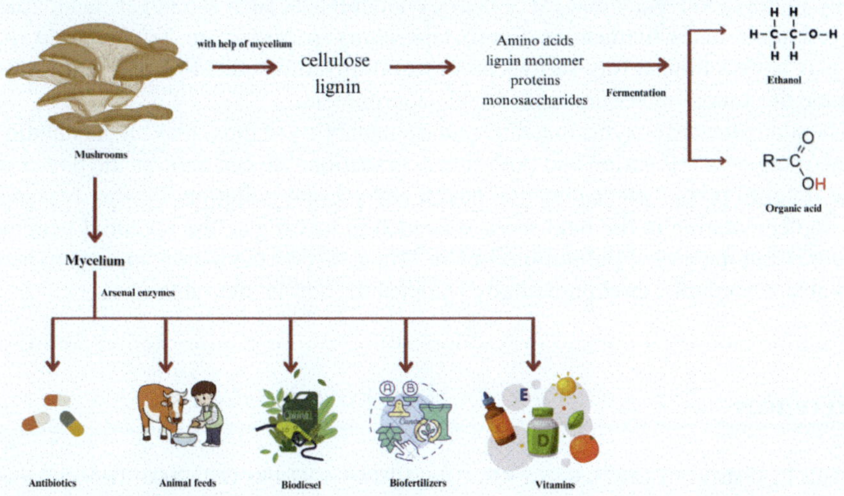

Fig. 17.4 Schematic representation of mycoremediation using macrofungi. Figure adapted from Uddin et al. (2020)

(Budzyńska et al. 2022). This needs to be addressed when the macrofungi could be extensively used for mycoremediation process. Through detoxification, macro-fungi change these harmful substances into benign forms that are safe for the environment (Fig. 17.4).

17.5 Conclusion

Contaminated soil is a serious environmental concern due to its mutagenicity and carcinogenicity. Microorganisms exist throughout and multiply quickly, allowing them to adjust to varying concentrations of dangerous contaminants. Microbial remediation has become more successful with the use of fungi. Mycoremediation is a highly successful pollution removal technique due to the flexibility of fungi in terms of ecology, feeding habits, adaptability, morphology, physiology and metabolism. Toxic, recalcitrant, and persistent pollutants, such as polyaromatic hydrocarbons, antibiotics, herbicides, insecticides, heavy metals, and plastic, may all be economically and effectively broken down via mycoremediation.

Among the macrofungi, *Agaricus bisporus* has shown promise as a mycoremediation mechanism for heavy metal, pesticide, and hydrocarbon degradation in the environment. *Tricholoma* sp. exhibits the ability to break down pollutants such as pesticides, hydrocarbons, heavy metals, and PAHs. *Pleurotus dryinus* has shown great promise in the field of degrading heavy metals, synthetics, pesticides, and endocrine disruptors. *Phellinus badius*, a type of oyster mushroom, has a remarkable ability to degrade metals that carry heavy contaminants such as copper, lead, zinc, and cadmium. In the filamentous fungi, *Aspergillus* sp, *Mucor* sp., *Trichoderma* sp., etc. play a predominant role various processes of mycoremediation and are found to degrade the xenobiotic compounds in the environment.

The sophisticated enzyme machinery and adaptability of fungi make it outstanding in environmental mitigation and pollution degradation. We can harness the power of these fungi to protect environmental health and prevent pollution. Further research and implementation in the field work is needed to investigate the potential benefits in addressing the global pollution problem. Thus, mycoremediation appears to be a workable, affordable, environmentally responsible, and in-situ method.

References

Abadulla E, Tzanov T, Costa S, Robra KH, Cavaco-Paulo A, Gübitz GM (2000) Decolorization and detoxification of textile dyes with a laccase from Trametes hirsuta. Appl Environ Microbiol 66(8):3357–3362

Abd El-Rahim WM, Moawad H, Azeiz AZA, Sadowsky MJ (2017) Optimization of conditions for decolorization of azo-based textile dyes by multiple fungal species. J Biotechnol 260:11–17

Ahmadi Khozani M, Emtiazi G, Aghaei SS, Ghasemi SM, Zolfaghari MR (2021) Application of fungal laccase for heavy metals precipitation using tannin as a natural mediator. Int J Environ Sci Technol 18(8):2335–2344

Akhtar N, Mannan MAU (2020) Mycoremediation: expunging environmental pollutants. Biotechnol Rep 26:e00452

Aksu Z (2005) Application of biosorption for the removal of organic pollutants: a review. Process Biochem 40(3–4):997–1026

Albert Q, Leleyter L, Lemoine M, Heutte N, Rioult JP, Sage L, Baraud F, Garon D (2018) Comparison of tolerance and biosorption of three trace metals (Cd, Cu, Pb) by the soil fungus Absidia cylindrospora. Chemosphere 196:386–392

Al-Hawash AB, Zhang X, Ma F (2019) Removal and biodegradation of different petroleum hydrocarbons using the filamentous fungus Aspergillus sp. RFC-1. Microbiologyopen 8(1):e00619

Ali SA (2023) Decolourization of azo Lanasyn Navy M-DNL dye by *Trichothecium roseum* toward green mycoremediation. npj Clean Water 6(1):64

Alimba CG, Adekoya KO, Soyinka OO (2019) Exposure to effluent from pharmaceutical industry induced cytogenotoxicity, hematological and histopathological alterations in Clarias gariepinus (Burchell, 1822). Excli J 18:63

Almeida EJR, Corso CR (2019) Decolorization and removal of toxicity of textile azo dyes using fungal biomass pelletized. Int J Environ Sci Technol 16:1319–1328

Alothman ZA, Bahkali AH, Khiyami MA, Alfadul SM, Wabaidur SM, Alam M, Alfarhan BZ (2020) Low cost biosorbents from fungi for heavy metals removal from wastewater. Sep Sci Technol 55(10):1766–1775

Alves RP, Bolson SM, de Albuquerque MP, de Carvalho Victoria F, Pereira AB (2017) A Potencial use of edible mushrooms Pleurotus ostreatoroseus Singer (Pleurotaceae) and Lentinus sajor-caju (Fr.) Fr. (Polyporaceae) in metal remediation processes. Revista de Biologia Neotropical/J Neotrop Biol 14(2):82–90

Anand P, Isar J, Saran S, Saxena RK (2006) Bioaccumulation of copper by Trichoderma viride. Biores Technol 97(8):1018–1025

Arami M, Limaee NY, Mahmoodi NM, Tabrizi NS (2006) Equilibrium and kinetics studies for the adsorption of direct and acid dyes from aqueous solution by soy meal hull. J Hazard Mater 135(1–3):171–179

Arévalo-Rangel DL, Cárdenas-González JF, Martínez-Juárez VM, Acosta-Rodríguez I (2013) Hexavalent chromate reductase activity in cell free extracts of Penicillium sp. Bioinorg Chem Appl 2013

Arun A, Eyini M (2011) Comparative studies on lignin and polycyclic aromatic hydrocarbons degradation by basidiomycetes fungi. Biores Technol 102(17):8063–8070

Asif MB, Hai FI, Singh L, Price WE, Nghiem LD (2017) Degradation of pharmaceuticals and personal care products by white-rot fungi—a critical review. Curr Pollut Rep 3:88–103

Bánfi R, Pohner Z, Kovács J, Luzics S, Nagy A, Dudás M, Tanos P, Márialigeti K, Vajna B (2015) Characterisation of the large-scale production process of oyster mushroom (Pleurotus ostreatus) with the analysis of succession and spatial heterogeneity of lignocellulolytic enzyme activities. Fungal Biol 119(12):1354–1363

Bano A, Hussain J, Akbar A, Mehmood K, Anwar M, Hasni MS, Ullah S, Sajid S, Ali I (2018) Biosorption of heavy metals by obligate halophilic fungi. Chemosphere 199:218–222

Banu JR, Kavitha S, Tyagi VK, Gunasekaran M, Karthikeyan OP, Kumar G (2021) Lignocellulosic biomass based biorefinery: a successful platform towards circular bioeconomy. Fuel 302:121086

Behera DK, Mishra S (2022) The burden of diarrhea, etiologies, and risk factors in India from 1990 to 2019: evidence from the global burden of disease study. BMC Public Health 22:1–9

Bellettini MB, Fiorda FA, Maieves HA, Teixeira GL, Ávila S, Hornung PS, Júnior AM, Ribani RH (2019) Factors affecting mushroom Pleurotus spp. Saudi J Biol Sci 26(4):633–646

Bosco F, Mollea C (2019) Mycoremediation in soil. In: Environmental chemistry and recent pollution control approaches, pp 1–16

Bosso L, Lacatena F, Cristinzio G, Cea M, Diez MC, Rubilar O (2015) Biosorption of pentachlorophenol by Anthracophyllum discolor in the form of live fungal pellets. New Biotechnol 32(1):21–25

Brito GFM, Geris R, Passos MS, Malta M, Ribeiro JN, Licínio MVVJ, Freitas DC, dos Santos AV, Santos TSM, Ribeiro AVFN, de Godoi Pereira M (2021) Mycoremediation of Cd2+ and Pb2+ from aqueous media by dead biomass of Phialomyces macrosporus. Water Air Soil Pollut 232(12):482

Budzyńska S, Siwulski M, Budka A, Kalač P, Niedzielski P, Gąsecka M, Mleczek M (2022) Mycoremediation of flotation tailings with Agaricus bisporus. J Fungi 8(8):883

Burford EP, Hillier S, Gadd GM (2006) Biomineralization of fungal hyphae with calcite ($CaCO_3$) and calcium oxalate mono-and dihydrate in carboniferous limestone microcosms. Geomicrobiol J 23(8):599–611

Cabrera-Barjas G, Gallardo F, Nesic A, Taboada E, Marican A, Mirabal-Gallardo Y, Avila-Salas F, Delgado N, de Armas-Ricard M, Valdes O (2020) Utilization of industrial by-product fungal biomass from Aspergillus niger and Fusarium culmorum to obtain biosorbents for removal of pesticide and metal ions from aqueous solutions. J Environ Chem Eng 8(5):104355

Chan-Cupul W, Heredia-Abarca G, Rodríguez-Vázquez R (2016) Atrazine degradation by fungal co-culture enzyme extracts under different soil conditions. J Environ Sci Health B 51(5):298–308

Chatterjee A, Abraham J (2019) Mycoremediation of 17 β-estradiol using trichoderma citrinoviride strain AJAC3 along with enzyme studies. Environ Prog Sustain Energy 38(4):13142

Chen HM, Zheng CR, Tu C, Shen ZG (2000) Chemical methods and phytoremediation of soil contaminated with heavy metals. Chemosphere 41(1–2):229–234

Chung KT (2016) Azo dyes and human health: a review. J Environ Sci Health C 34(4):233–261

Conceição AA, Cunha JRB, Vieira VO, Pelaéz RDR, Mendonça S, Almeida JRM, Dias ES, de Almeida EG, de Siqueira FG (2018) Bioconversion and biotransformation efficiencies of wild macrofungi. In: Biology of macrofungi, pp 361–377

da Luz JMR, Paes SA, Nunes MD, da Silva MDCS, Kasuya MCM (2013) Degradation of oxo-biodegradable plastic by Pleurotus ostreatus. PLoS One 8(8):e69386

Das N, Chandran P (2011) Microbial degradation of petroleum hydrocarbon contaminants: an overview. Biotechnol Res Int 2011

Deepalakshmi K, Sankaran M (2014) Pleurotus ostreatus: an oyster mushroom with nutritional and medicinal properties. J Biochem Technol 5(2):718–726

Dhankhar R, Hooda A (2011) Fungal biosorption—an alternative to meet the challenges of heavy metal pollution in aqueous solutions. Environ Technol 32(5):467–491

Dhiman N, Jasrotia T, Sharma P, Negi S, Chaudhary S, Kumar R, Mahnashi MH, Umar A, Kumar R (2020) Immobilization interaction between xenobiotic and Bjerkandera adusta for the biodegradation of atrazine. Chemosphere 257:127060

Dhiman VK, Singh D, Pandey H, Chauhan D, Dhiman VK, Pandey D (2022) Mycoremediation of agricultural waste for the cultivation of edible mushroom. Sustainable management of environmental contaminants: eco-friendly remediation approaches. Springer International Publishing, Cham, pp 471–482

Diaz E (2008) Microbial biodegradation: genomics and molecular biology

Dulay RMR, Parungao AG, Kalaw SP, Reyes RG (2012) Aseptic cultivation of Coprinus comatus (OF Mull.) Gray on various pulp and paper wastes. Mycosphere 3(3):392–397

Dusengemungu L, Kasali G, Gwanama C, Ouma KO (2020) Recent advances in biosorption of copper and cobalt by filamentous fungi. Front Microbiol 11:582016

Dutta SD, Hyder MD (2019) Mycoremediation: a potential tool for sustainable management. J Mycopathol Res 57(1):25–34

Elisashvili V, Penninckx M, Kachlishvili E, Tsiklauri N, Metreveli E, Kharziani T, Kvesitadze G (2008) Lentinus edodes and Pleurotus species lignocellulolytic enzymes activity in submerged and solid-state fermentation of lignocellulosic wastes of different composition. Biores Technol 99(3):457–462

Eman A, Abdel-Megeed A, Suliman AMA, Sadik MW, Sholkamy EN (2013) Biodegradation of glyphosate by fungal strains isolated from herbicides polluted-soils in Riyadh area. Int J Curr Microbiol Appl Sci 2(8):359–381

Esterhuizen-Londt M, Schwartz K, Pflugmacher S (2016) Using aquatic fungi for pharmaceutical bioremediation: uptake of acetaminophen by Mucor hiemalis does not result in an enzymatic oxidative stress response. Fungal Biol 120(10):1249–1257

Esterhuizen-Londt M, Hendel AL, Pflugmacher S (2017) Mycoremediation of diclofenac using Mucor hiemalis. Toxicol Environ Chem 99(5–6):795–808

Fernández PM, Viñarta SC, Bernal AR, Cruz EL, Figueroa LI (2018) Bioremediation strategies for chromium removal: current research, scale-up approach and future perspectives. Chemosphere 208:139–148

Fernando LD, Dickwella Widanage MC, Penfield J, Lipton AS, Washton N, Latgé JP, Wang P, Zhang L, Wang T (2021) Structural polymorphism of chitin and chitosan in fungal cell walls from solid-state NMR and principal component analysis. Front Mol Biosci 8:727053

Gadd GM (2021) Fungal biomineralization. Curr Biol 31(24):R1557–R1563

Gajendiran A, Abraham J (2017) Biomineralisation of fipronil and its major metabolite, fipronil sulfone, by Aspergillus glaucus strain AJAG1 with enzymes studies and bioformulation. 3 Biotech 7(3):212

Gao M, Gu X, Satterlee T, Duke MV, Scheffler BE, Gold SE, Glenn AE (2022) Transcriptomic responses of fusarium verticillioides to lactam and lactone xenobiotics. Front Fungal Biol 3:923112

Gavrilescu M (2004) Removal of heavy metals from the environment by biosorption. Eng Life Sci 4(3):219–232

Gola D, Dey P, Bhattacharya A, Mishra A, Malik A, Namburath M, Ahammad SZ (2016) Multiple heavy metal removal using an entomopathogenic fungi Beauveria bassiana. Biores Technol 218:388–396

Gramss G, Kirsche B, Voigt KD, Günther T, Fritsche W (1999) Conversion rates of five polycyclic aromatic hydrocarbons in liquid cultures of fifty-eight fungi and the concomitant production of oxidative enzymes. Mycol Res 103(8):1009–1018

Gros M, Cruz-Morato C, Marco-Urrea E, Longrée P, Singer H, Sarrà M, Hollender J, Vicent T, Rodriguez-Mozaz S, Barceló D (2014) Biodegradation of the X-ray contrast agent iopromide and the fluoroquinolone antibiotic ofloxacin by the white rot fungus Trametes versicolor in hospital wastewaters and identification of degradation products. Water Res 60:228–241

Gu Y, Xu W, Liu Y, Zeng G, Huang J, Tan X, Jian H, Hu X, Li F, Wang D (2015) Mechanism of Cr (VI) reduction by Aspergillus niger: enzymatic characteristic, oxidative stress response, and reduction product. Environ Sci Pollut Res 22:6271–6279

Gucia M, Jarzyńska G, Rafał E, Roszak M, Kojta AK, Osiej I, Falandysz J (2012) Multivariate analysis of mineral constituents of edible Parasol mushroom (Macrolepiota procera) and soils beneath fruiting bodies collected from Northern Poland. Environ Sci Pollut Res 19:416–431

Gunderson JJ, Knight JD, Van Rees KCJ (2007) Impact of ectomycorrhizal colonization of hybrid poplar on the remediation of diesel-contaminated soil. J Environ Qual 36(4):927–934

Hadibarata T, Adnan LA, Yusoff ARM, Yuniarto A, Rubiyatno, Zubir, MMFA, Khudhair AB, Teh ZC, Naser MA (2013) Microbial decolorization of an azo dye reactive black 5 using white-rot fungus Pleurotus eryngii F032. Water Air Soil Pollut 224:1–9

Hammel KE, Green B, Gai WZ (1991) Ring fission of anthracene by a eukaryote. Proc Natl Acad Sci 88(23):10605–10608

Heinfling AMMJMATBMSU, Martinez MJ, Martinez AT, Bergbauer M, Szewzyk U (1998) Transformation of industrial dyes by manganese peroxidases from Bjerkandera adusta and Pleurotus eryngii in a manganese-independent reaction. Appl Environ Microbiol 64(8):2788–2793

Henagamage AP, Peries CM (2023) Degradation and decolorization of textile azo dyes by effective fungal-bacterial consortium. Mol Biol Rep 50(11):8901–8914

Hidalgo J, Epelde L, Anza M, Becerril JM, Garbisu C (2023) Mycoremediation with Agaricus bisporus and Pleurotus ostreatus growth substrates versus phytoremediation with Festuca rubra and Brassica sp. for the recovery of a Pb and γ-HCH contaminated soil. Chemosphere 327:138538

Hoque E, Fritscher J (2016) A new mercury-accumulating Mucor hiemalis strain EH8 from cold sulfidic spring water biofilms. MicrobiologyOpen 5(5):763–781

Ijoma GN, Tekere M (2017) Potential microbial applications of co-cultures involving Ligninolytic fungi in the bioremediation of recalcitrant xenobiotic compounds. Int J Environ Sci Technol 14:1787–1806

Iram S, Shabbir R, Zafar H, Javaid M (2015) Biosorption and bioaccumulation of copper and lead by heavy metal-resistant fungal isolates. Arab J Sci Eng 40:1867–1873

Isildak Ö, Turkekul I, Elmastas M, Tuzen M (2004) Analysis of heavy metals in some wild-grown edible mushrooms from the middle black sea region, Turkey. Food Chem 86(4):547–552

Jelic A, Cruz-Morató C, Marco-Urrea E, Sarrà M, Perez S, Vicent T, Petrović M, Barcelo D (2012) Degradation of carbamazepine by Trametes versicolor in an air pulsed fluidized bed bioreactor and identification of intermediates. Water Res 46(4):955–964

Johannes C, Majcherczyk A, Hüttermann A (1996) Degradation of anthracene by laccase of Trametes versicolor in the presence of different mediator compounds. Appl Microbiol Biotechnol 46:313–317

Kamyotra JS (2009) Comprehensive environmental assessment of industrial clusters, Central Pollution Control Board, Ministry of Environment and Forests, Ecological Impact Assessment Series: EIAS/5/2009–2010. TERI Press, the Energy and Resources Institute, New Delhi, pp 1–32

Kantharaj P, Boobalan B, Sooriamuthu S, Mani R (2017) Lignocellulose degrading enzymes from fungi and their industrial applications. Int J Curr Res Rev 9(21):1–12

Kapahi M, Sachdeva S (2017) Mycoremediation potential of Pleurotus species for heavy metals: a review. Bioresour Bioprocess 4(1):32

Khan I, Aftab M, Shakir S, Ali M, Qayyum S, Rehman MU, Haleem KS, Touseef I (2019a) Mycoremediation of heavy metal (Cd and Cr)-polluted soil through indigenous metallotolerant fungal isolates. Environ Monit Assess 191:1–11

Khan I, Ali M, Aftab M, Shakir S, Qayyum S, Haleem KS, Tauseef I (2019b) Mycoremediation: a treatment for heavy metal-polluted soil using indigenous metallotolerant fungi. Environ Monit Assess 191:1–15

Kim JH, Kim YS (2001) Characterization of a metalloenzyme from a wild mushroom, Tricholoma saponaceum. Biosci Biotechnol Biochem 65(2):356–362

Koukkou AI (ed) (2011) Microbial bioremediation of non-metals: current research. Caister Academic Press, Norfolk, UK

Krishnamoorthy R, Jose PA, Ranjith M, Anandham R, Suganya K, Prabhakaran J, Thiyageshwari S, Johnson J, Gopal NO, Kumutha K (2018) Decolourisation and degradation of azo dyes by mixed fungal culture consisted of Dichotomomyces cejpii MRCH 1–2 and Phoma tropica MRCH 1–3. J Environ Chem Eng 6(1):588–595

Kryczyk A, Piotrowska J, Sito M, Sulkowska-Ziaja K, Dobosz K, Opoka W, Muszyńska B (2017) Remediation capacity of Cd and Pb ions by mycelia of Imleria badia, Laetiporus sulphureus, and Agaricus bisporus in vitro cultures. J Environ Sci Health B 52(9):617–622

Kryczyk-Poprawa A, Żmudzki P, Maślanka A, Piotrowska J, Opoka W, Muszyńska B (2019) Mycoremediation of azole antifungal agents using in vitro cultures of Lentinula edodes. 3 Biotech 9:1–10

Kulshreshtha S, Mathur N, Bhatnagar P, Kulshreshtha S (2013) Cultivation of Pleurotus citrinopileatus on handmade paper and cardboard industrial wastes. Ind Crops Prod 41:340–346

Kulshreshtha S, Mathur N, Bhatnagar P (2014) Mushroom as a product and their role in mycoremediation. AMB Express 4:1–7

Kumar V, Dwivedi SK (2021) Bioremediation mechanism and potential of copper by actively growing fungus Trichoderma lixii CR700 isolated from electroplating wastewater. J Environ Manage 277:111370

Kumar A, Yadav AN, Mondal R, Kour D, Subrahmanyam G, Shabnam AA, Khan SA, Yadav KK, Sharma GK, Cabral-Pinto M, Fagodiya RK (2021) Myco-remediation: a mechanistic understanding of contaminants alleviation from natural environment and future prospect. Chemosphere 284:131325

Kurniati E, Arfarita N, Imai T, Higuchi T, Kanno A, Yamamoto K, Sekine M (2014) Potential bioremediation of mercury-contaminated substrate using filamentous fungi isolated from forest soil. J Environ Sci 26(6):1223–1231

Larsson DJ (2014) Pollution from drug manufacturing: review and perspectives. Philos Trans R Soc B Biol Sci 369(1656):20130571

Lee KK, Kristanti RA, Yulisa A, Ningsih F, Syafrudin M, Hernandes E, Albescu M (2022) Biosorption of synthetic dye by macrofungi. Ind Domest Waste Manag 2(2):61–70

Legorreta-Castañeda AJ, Lucho-Constantino CA, Beltrán-Hernández RI, Coronel-Olivares C, Vázquez-Rodríguez GA (2020) Biosorption of water pollutants by fungal pellets. Water 12(4):1155

Li H, Lin Y, Guan W, Chang J, Xu L, Guo J, Wei G (2010) Biosorption of Zn (II) by live and dead cells of Streptomyces ciscaucasicus strain CCNWHX 72–14. J Hazard Mater 179(1–3):151–159

Li Q, Csetenyi L, Paton GI, Gadd GM (2015) $CaCO_3$ and $SrCO_3$ bioprecipitation by fungi isolated from calcareous soil. Environ Microbiol 17(8):3082–3097

Liang X, Hillier S, Pendlowski H, Gray N, Ceci A, Gadd GM (2015) Uranium phosphate biomineralization by fungi. Environ Microbiol 17(6):2064–2075

Liang X, Kierans M, Ceci A, Hillier S, Gadd GM (2016) Phosphatase-mediated bioprecipitation of lead by soil fungi. Environ Microbiol 18(1):219–231

Lin JE, Wang HY, Hickey RF (1990) Degradation kinetics of pentachlorophenol by Phanerochaete chrysosporium. Biotechnol Bioeng 35(11):1125–1134

Lu T, Zhang QL, Yao SJ (2016) Application of biosorption and biodegradation functions of fungi in wastewater and sludge treatment. In: Fungal applications in sustainable environmental biotechnology, pp 65–90

Mahmoud MS, Mostafa MK, Mohamed SA, Sobhy NA, Nasr M (2017) Bioremediation of red azo dye from aqueous solutions by *Aspergillus niger* strain isolated from textile wastewater. J Environ Chem Eng 5(1):547–554

Maqbool Z, Hussain S, Imran M, Mahmood F, Shahzad T, Ahmed Z, Azeem F, Muzammil S (2016) Perspectives of using fungi as bioresource for bioremediation of pesticides in the environment: a critical review. Environ Sci Pollut Res 23:16904–16925

Mar'In A, Conti C, Gobbi G (1997) Sorption of lead and caesium by mushrooms grown in natural conditions. Resour Environ Biotechnol 2:35–50

Marco-Urrea E, Pérez-Trujillo M, Blánquez P, Vicent T, Caminal G (2010) Biodegradation of the analgesic naproxen by Trametes versicolor and identification of intermediates using HPLC-DAD-MS and NMR. Biores Technol 101(7):2159–2166

Martínez-Juárez VM, Cárdenas-González JF, Torre-Bouscoulet ME, Acosta-Rodríguez I (2012) Biosorption of mercury (II) from aqueous solutions onto fungal biomass. Bioinorg Chem Appl

Mbachu AE, Chukwura EI, Mbachu NA (2020) Role of microorganisms in the degradation of organic pollutants: a review. Energy Environ Eng 7(1):1–11

Mejstřík V, Lepšpvá A (1993) Applicability of fungi to the monitoring of environmental pollution by heavy metals

Michael-Igolima U, Abbey SJ, Ifelebuegu AO (2022) A systematic review on the effectiveness of remediation methods for oil contaminated soils. Environ Adv 9:100319

Mitra S, Chakraborty AJ, Tareq AM, Emran TB, Nainu F, Khusro A, Idris AM, Khandaker MU, Osman H, Alhumaydhi FA, Simal-Gandara J (2022) Impact of heavy metals on the environment and human health: novel therapeutic insights to counter the toxicity. J King Saud Univ Sci 34(3):101865

Morales-Barrera L, Cristiani-Urbina E (2015) Bioreduction of hexavalent chromium by Hypocrea tawa in a concentric draft-tube airlift bioreactor. J Environ Biotechnol Res 1(1):37–44

Muszyńska B, Żmudzki P, Lazur J, Kała K, Sułkowska-Ziaja K, Opoka W (2018) Analysis of the biodegradation of synthetic testosterone and 17α-ethynylestradiol using the edible mushroom Lentinula edodes. 3 Biotech 8:1–15

Nagy B, Măicăneanu A, Indolean C, Mânzatu C, Silaghi-Dumitrescu L, Majdik C (2014) Comparative study of Cd (II) biosorption on cultivated Agaricus bisporus and wild Lactarius piperatus based biocomposites. Linear and nonlinear equilibrium modelling and kinetics. J Taiwan Inst Chem Eng 45(3):921–929

Nguyen ML, Juang RS (2015) Modification of crosslinked chitosan beads with histidine and Saccharomyces cerevisiae for enhanced Ni (II) biosorption. J Taiwan Inst Chem Eng 56:96–102

Nishida C, Yatera K (2022) The impact of ambient environmental and occupational pollution on respiratory diseases. Int J Environ Res Public Health 19(5):2788

Noman MA, Feng W, Zhu G, Hossain MB, Chen Y, Zhang H, Sun J (2022) Bioaccumulation and potential human health risks of metals in commercially important fishes and shellfishes from Hangzhou Bay, China. Scientific Reports 12(1):4634

Nouri H, Azin E, Kamyabi A, Moghimi H (2021) Biosorption performance and cell surface properties of a fungal-based sorbent in azo dye removal coupled with textile wastewater. Int J Environ Sci Technol 18:2545–2558

Olusola SA, Anslem EE (2010) Bioremediation of a crude oil polluted soil with Pleurotus pulmonarius and Glomus mosseae using Amaranthus hybridus as a test plant. J Bioremed Biodegrad 1:113

Pang C, Liu YH, Cao XH, Li M, Huang GL, Hua R, Wang CX, Liu YT, An XF (2011) Biosorption of uranium(VI) from aqueous solution by dead fungal biomass of Penicillium citrinum. Chem Eng J 170(1):1–6

Patel H, Gupte A (2016) Optimization of different culture conditions for enhanced laccase production and its purification from Tricholoma giganteum AGHP. Bioresour Bioprocess 3:1–10

Priyadarshini E, Priyadarshini SS, Cousins BG, Pradhan N (2021) Metal-fungus interaction: review on cellular processes underlying heavy metal detoxification and synthesis of metal nanoparticles. Chemosphere 274:129976

Purnomo AS, Nawfa R, Martak F, Shimizu K, Kamei I (2017) Biodegradation of aldrin and dieldrin by the white-rot fungus Pleurotus ostreatus. Curr Microbiol 74(3):320–324

Qu Y, Shi S, Ma F, Yan B (2010) Decolorization of reactive dark blue KR by the synergism of fungus and bacterium using response surface methodology. Biores Technol 101(21):8016–8023

Rajhans G, Sen SK, Barik A, Raut S (2020) Elucidation of fungal dye-decolourizing peroxidase (DyP) and ligninolytic enzyme activities in decolourization and mineralization of azo dyes. J Appl Microbiol 129(6):1633–1643

Rathankumar AK, Saikia K, Ramachandran K, Batista RA, Cabana H, Vaidyanathan VK (2020) Effect of soil organic matter (SOM) on the degradation of polycyclic aromatic hydrocarbons using Pleurotus dryinus IBB 903-A microcosm study. J Environ Manage 260:110153

Rodarte-Morales AI, Feijoo G, Moreira MT, Lema JM (2012) Biotransformation of three pharmaceutical active compounds by the fungus Phanerochaete chrysosporium in a fed batch stirred reactor under air and oxygen supply. Biodegradation 23:145–156

Sácký J, Leonhardt T, Borovička J, Gryndler M, Briksí A, Kotrba P (2014) Intracellular sequestration of zinc, cadmium and silver in Hebeloma mesophaeum and characterization of its metallothionein genes. Fungal Genet Biol 67:3–14

Salvi NA, Chattopadhyay S (2017) Biosorption of Azo dyes by spent *Rhizopus arrhizus* biomass. Appl Water Sci 7:3041–3054

Sánchez C (2009) Lignocellulosic residues: biodegradation and bioconversion by fungi. Biotechnol Adv 27(2):185–194

Sanghi R, Sankararamakrishnan N, Dave BC (2009) Fungal bioremediation of chromates: conformational changes of biomass during sequestration, binding, and reduction of hexavalent chromium ions. J Hazard Mater 169(1–3):1074–1080

Santos GC, Corso CR (2014) Comparative analysis of azo dye biodegradation by Aspergillus oryzae and Phanerochaete chrysosporium. Water Air Soil Pollut 225:1–11

Sen SK, Raut S, Bandyopadhyay P, Raut S (2016) Fungal decolouration and degradation of azo dyes: a review. Fungal Biol Rev 30(3):112–133

Sen SK, Das JK, Rajhans G, Barik A, Raut S (2021) Immobilized fungal consortium-GR: a novel textile effluent treatment alternative for removal of azo dyes and other pollutants. J Chem Technol Biotechnol 96(7):1991–2005

Şenol ZM, Gül ÜD, Gurbanov R, Şimşek S (2021) Optimization the removal of lead ions by fungi: explanation of the mycosorption mechanism. J Environ Chem Eng 9(2):104760

Shourie A, Vijayalakshmi U (2022) Fungal diversity and its role in mycoremediation. Geomicrobiol J 39(3–5):426–444

Silambarasan S, Abraham J (2013) Mycoremediation of endosulfan and its metabolites in aqueous medium and soil by Botryosphaeria laricina JAS6 and Aspergillus tamarii JAS9. PLoS One 8(10):e77170

Singh G, Dwivedi SK (2020) Decolorization and degradation of Direct Blue-1 (Azo dye) by newly isolated fungus *Aspergillus terreus* GS28, from sludge of carpet industry. Environ Technol Innov 18:100751

Singh S, Pakshirajan K (2010) Enzyme activities and decolourization of single and mixed azo dyes by the white-rot fungus Phanerochaete chrysosporium. Int Biodeterior Biodegrad 64(2):146–150

Sintakindi A, Ankamwar B (2021) Fungal biosorption as an alternative for the treatment of dyes in waste waters: a review. Environ Technol Rev 10(1):26–43

Širić I, Humar M, Kasap A, Kos I, Mioč B, Pohleven F (2016) Heavy metal bioaccumulation by wild edible saprophytic and ectomycorrhizal mushrooms. Environ Sci Pollut Res 23:18239–18252

Soliman SA, El-Zawahry YA, El-Mougith AA (2013) Fungal biodegradation of agro-industrial waste. In: Cellulose-biomass conversion, pp 75–100

Sultana K, Quresh RA, Bashir BH (2007) Impact of mushrooms and toad stools on environment. Electron J Environ Agric Food Chem 6(11):2534–2542

Tang Y, Zeiner CA, Santelli CM, Hansel CM (2013) Fungal oxidative dissolution of the Mn (II)-bearing mineral rhodochrosite and the role of metabolites in manganese oxide formation. Environ Microbiol 15(4):1063–1077

Thakor R, Mistry H, Tapodhan K, Bariya H (2022) Efficient biodegradation of Congo red dye using fungal consortium incorporated with Penicillium oxalicum and Aspergillus tubingensis. Folia Microbiol 67(1):33–43

Torres-Farradá G, Manzano-León AM, Rineau F, Ramos Leal M, Thijs S, Jambon I, Put J, Czech J, Guerra Rivera G, Carleer R, Vangronsveld J (2019) Biodegradation of polycyclic aromatic hydrocarbons by native Ganoderma sp. strains: identification of metabolites and proposed degradation pathways. Appl Microbiol Biotechnol 103:7203–7215

Trojanowski J, Haider K, Hüttermann A (1984) Decomposition of 14 C-labelled lignin, holocellulose and lignocellulose by mycorrhizal fungi. Arch Microbiol 139:202–206

Tuzen M, Sarı A, Turkekul I (2020) Influential bio-removal of mercury using Lactarius acerrimus macrofungus as novel low-cost biosorbent from aqueous solution: isotherm modeling, kinetic and thermodynamic investigations. Mater Chem Phys 249:123168

Uddin M, Zhang D, Proshad R, Haque MK (2020) Role of mushrooms in soil mycoremediation: a review. Chin J App Environ Biol 26(2):460–468

Ugya AY, Imam TS (2017) Efficiency of the decomposition process of Agaricus bisporus in the mycoremediation of refinery wastewater: Romi stream case study. World J Pharm Res 6:200–211

Umor NA, Ismail S, Abdullah S, Huzaifah MHR, Huzir NM, Mahmood NAN, Zahrim AY (2021) Zero waste management of spent mushroom compost. J Mater Cycles Waste Manage 23(5):1726–1736

Văcar CL, Covaci E, Chakraborty S, Li B, Weindorf DC, Frențiu T, Pârvu M, Podar D (2021) Heavy metal-resistant filamentous fungi as potential mercury bioremediators. J Fungi 7(5):386

Vipotnik Z, Michelin M, Tavares T (2021) Ligninolytic enzymes production during polycyclic aromatic hydrocarbons degradation: effect of soil pH, soil amendments and fungal co-cultivation. Biodegradation 32:193–215

Wang HX, Ng TB (2004) Purification of a novel low-molecular-mass laccase with HIV-1 reverse transcriptase inhibitory activity from the mushroom Tricholoma giganteum. Biochem Biophys Res Commun 315(2):450–454

Wang Y, Yi B, Sun X, Yu L, Wu L, Liu W, Wang D, Li Y, Jia R, Yu H, Li X (2019) Removal and tolerance mechanism of Pb by a filamentous fungus: a case study. Chemosphere 225:200–208

Wei Z, Hillier S, Gadd GM (2012) Biotransformation of manganese oxides by fungi: solubilization and production of manganese oxalate biominerals. Environ Microbiol 14(7):1744–1753

Xu H, Jia Y, Sun Z, Su J, Liu QS, Zhou Q, Jiang G (2022) Environmental pollution, a hidden culprit for health issues. Eco-Environ Health 1(1):31–45

Yang S, Hai FI, Nghiem LD, Nguyen LN, Roddick F, Price WE (2013a) Removal of bisphenol A and diclofenac by a novel fungal membrane bioreactor operated under non-sterile conditions. Int Biodeterior Biodegrad 85:483–490

Yang S, Hai FI, Nghiem LD, Roddick F, Price WE (2013b) Removal of trace organic contaminants by nitrifying activated sludge and whole-cell and crude enzyme extract of Trametes versicolor. Water Sci Technol 67(6):1216–1223

Yuvaraj M, Ramasamy M (2020) Role of fungi in agriculture. In: Biostimulants in plant science, pp 1–12

Zafar S, Aqil F, Ahmad I (2007) Metal tolerance and biosorption potential of filamentous fungi isolated from metal contaminated agricultural soil. Biores Technol 98(13):2557–2561

Zhao X, Hardin IR, Hwang HM (2006) Biodegradation of a model azo disperse dye by the white rot fungus Pleurotus ostreatus. Int Biodeterior Biodegrad 57(1):1–6

Zulfadhly Z, Mashitah MD, Bhatia S (2001) Heavy metals removal in fixed-bed column by the macro fungus Pycnoporus sanguineus. Environ Pollut 112(3):463–470

Chapter 18
Sustainable Engineering and Applications of Metal–Organic Frameworks (MOFs) and Their Derivatives from Waste PET Bottles as a Potent Resource of Acid Linkers

T. Sathiya Kamatchi, P. Gowthami, M. Thenmozhi, M. Devendiran, A. Arjunan, and A. Kosiha

Abstract The exponential rise of non-biodegradable plastic and its derived waste materials poses a serious threat to ecosystems worldwide, demanding urgent action. A key approach that emphasizes sustainability and environmental safety entails recycling these waste materials into value-added products that can be used to address a wide range of environmental issues such as water pollution, polyethylene terephthalate (PET), energy shortages, and greenhouse gas emissions. PET, a chemically stable polyester with numerous applications, has witnessed an enormous spike in production and consumption in recent decades. The expansion has resulted in a corresponding agglomeration of waste PET at alarming rates, exacerbating health

T. S. Kamatchi
School of Chemistry, Bharathidasan University, Tiruchirappalli, Tamil Nadu 620024, India

P. Gowthami · A. Kosiha (✉)
Department of Chemistry, School of Basic Sciences, Vels Institute of Science, Technology and Advanced Studies, Pallavaram, Chennai, Tamil Nadu 600117, India
e-mail: kosiha@gmail.com

P. Gowthami
Department of Chemistry, Shrimati Devkunvar Nanalal Bhatt Vaishnav College for Women, Chennai, Tamil Nadu, India

M. Thenmozhi
Department of Biotechnology, School of Life Sciences, Vels Institute of Science, Technology and Advanced Studies, Pallavaram, Chennai, Tamil Nadu 600117, India

M. Devendiran
Spectralytic India Pvt.Ltd (PerkinElmer, U.S.LLC.), Chennai, Tamil Nadu, India

A. Arjunan
Department of Biotechnology, School of Biotechnology and Genetic Engineering, Bharathidasan University, Tiruchirappalli, Tamil Nadu 620024, India

© The Author(s), under exclusive license to Springer Nature Switzerland AG 2025
A. Jeyaseelan and K. Murugasen (eds.), *Sustainable Environmental Remediation: Avenues in Nano and Biotechnology*, Environmental Science and Engineering,
https://doi.org/10.1007/978-3-031-78483-5_18

and environmental issues. Current waste management protocols, including landfilling and burning, along with traditional recycling procedures generating low-value products, have culminated in a poor PET recycling rate of less than 30%, necessitating innovative solutions. One captivating avenue lies in the transformation of waste PET bottles into highly valuable metal–organic frameworks (MOFs). Among several value-added products generated from waste, MOFs hold immense potential in diverse applications such as battery engineering, sustainable catalysis, sensor applications, gas adsorption and separation. This chapter delineates the fabrication of MOF organic linkers sourced from recycled PET waste. Moreover, it explores the multiple utilities of engineered MOFs in resolving environmental challenges such as H_2 and CO_2 capture, H_2 storage, catalysis, sensing, adsorptive removal of dyes and medicinal wastes from aqueous environments, and energy applications. Each application is analyzed with a discerning eye, showcasing both the challenges and advantages inherent in their implementation while also enlightening the enormous opportunities for future breakthroughs in this flourishing field of research.

18.1 Introduction

The twenty-first century grapples with pressing environmental challenges, demanding an urgent need for sustainable solutions to combat escalating levels of various pollution. The expansion of environmental pollution stems from a multitude of factors, mainly fueled by the proliferation of one-use materials such as polyethylene bags, plastics, PET bottles etc. The long-term existence of these non-biodegradable substances perpetuates a continuous surge in pollution levels, endangering the ecosystems globally. The expansion of PET waste, in particular, has prompted researchers to look for innovative approaches to repurposing these omnipresent materials into ecologically viable alternatives, thus mitigating environmental risks (Nasrollahi et al. 2020). Among the several techniques used for addressing this issue, the exploitation of metal–organic frameworks (MOFs) has been identified as exceptionally promising (Mastropietro 2023). This chapter delves into the cutting-edge of sustainable MOF synthesis and applications, with a particular emphasis on the pioneering use of trashed PET bottles as a source of acid linkages. This ground-breaking strategy integrates environmental stewardship and materials science, enabling distinctive solutions to the pressing challenges associated with plastic waste. Likewise, it unlocks exciting opportunities for the fabrication of advanced functional materials with revolutionary properties.

18.2 PET Depolymerization: A Transformational Beginning

In general, a polymeric reaction occurred between ethylene glycol and terephthalic acid at elevated temperatures and pressures, producing a polyester molecule known as PET. It is a thermoplastic polyester and pertains in both opaque and transparent versions. When opaque PET cools down slowly, it becomes semi-crystalline, whereas when transparent PET is cooled quickly, it becomes amorphous. This specific kind of thermoplastic polyester offers advantages such as excellent abrasion, durability against heat, minimal creep at elevated temperatures, improved dimensional stability, chemical resistance, outstanding durability and robustness, and so forth. The route to sustainable MOF production commences with the revolutionary step of PET depolymerization. The initial act not only remedies the environmental issues brought up by plastic wastes but also demonstrates the prospect of recycled PET as a promising precursor for MOF fabrication (Dutt and Soni 2013; Singh et al. 2022).

18.2.1 The PET Crisis: A Necessary Environmental Concern

The widespread usage of PET in packaging has resulted in an upsurge in plastic waste, inflicting serious environmental damage. Conventional methods of disposal, such as dumping and incineration, pose ecological risks, underscoring the need for innovative strategies to control and repurpose PET waste (Nyam et al. 2024). Depolymerization is a transformational approach that breaks down PET into its core constituent materials.

18.2.2 Chemical and Enzymatic Routes: Breaking Down PET Molecular Structure

PET depolymerization, an important phase in MOF production using waste PET bottles as acidic linkers, incorporates both chemical and enzymatic methods to break down the intricate PET molecular framework.

18.2.2.1 Chemical Depolymerisation

In chemical depolymerization, PET is subjected to chemical structure reactions under specific conditions, typically involving Catalytic Hydrolysis and Methanolysis. In chemical depolymerization, catalytic hydrolysis is one of the typical approaches. This technique often uses catalysts, which include metallic salts or oxides of metal, under heating to break down the ester linkages in PET (Campanelli et al. 1994).

The reaction results in the cleavage of PET into its constituent monomers, primarily terephthalic acid (TPA) and ethylene glycol (EG). PET can undergo depolymerization through hydrolysis to yield ethylene glycol and terephthalic acid. However, hydrolysis, although a potential method, isn't extensively utilized in recycling food-grade PET because of the costly purification required for the recycled TPA. This process typically involves operating at elevated pressure ranges up to 2 MPa and high temperatures between 200 and 250 °C under various conditions like basic, neutral, and acidic. The three fundamental hydrolysis techniques are listed below.

18.2.2.2 Acid Hydrolysis

This methodology uses a variety of acidic chemicals, such as concentrated nitric acid (HNO_3), sulfuric acid (H_2SO_4) (Fig. 18.1), and phosphoric acid (H_3PO_4). Concentrated H_2SO_4 is the most often used acid because it effectively reduces energy consumption by eliminating the demand for higher pressure and temperature inside the reaction container. TPA can be recovered from waste PET in minutes under specific working conditions, including temperatures that vary from 60 to 95 °C in an 87 wt% H_2SO_4 solution and 85–90 °C in a 90 wt% H_2SO_4 solution. In the acid hydrolysis reaction, the extended reactor residence times, up to several days, and elevated temperatures around 100 °C are necessary if the concentration of H_2SO_4 is decreased below 80 wt%. Ethylene glycol is recovered from the filtrate with appropriate organic solvents such as trichloroethylene. Nevertheless, this approach proves expensive and results in the production of substantial quantities of inorganic salts and aqueous waste due to the highly corrosive properties of the chemicals utilized (Mishra et al. 2003; Mancini and Zanin 2007).

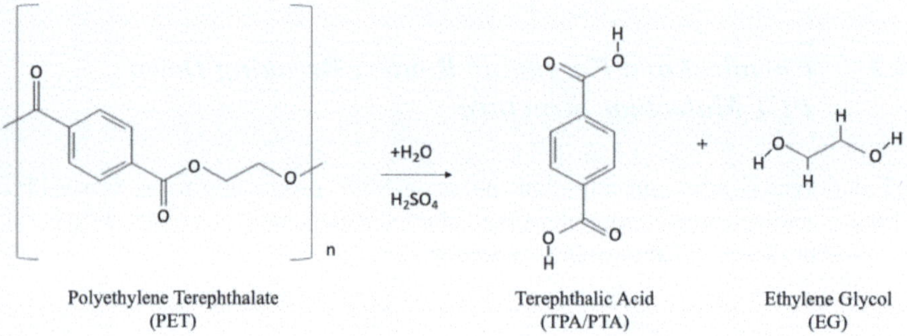

Fig. 18.1 Acid-mediated depolymerization of PET into ethylene glycol and terephthalic acid

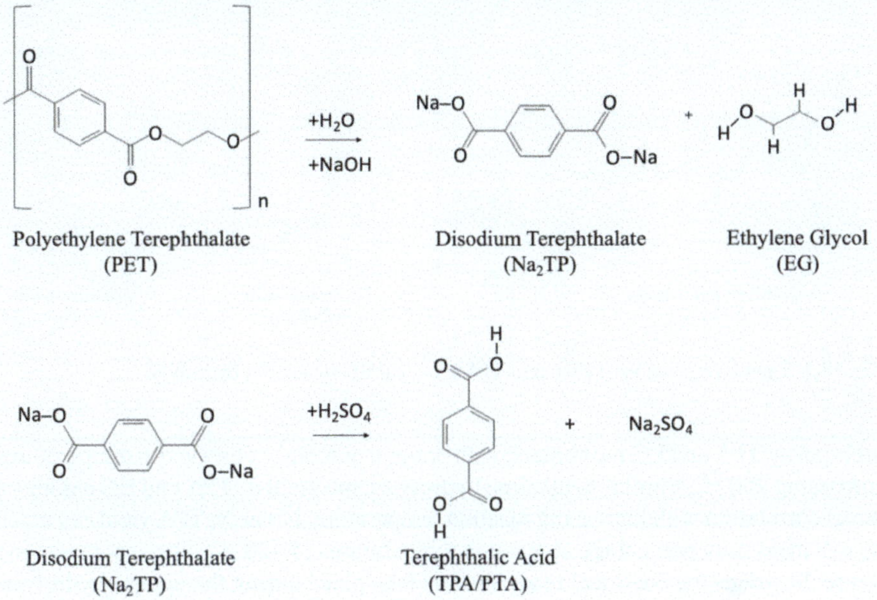

Fig. 18.2 Base-assisted depolymerization of PET into ethylene glycol and terephthalic acid

18.2.2.3 Alkaline Hydrolysis

The alkaline hydrolysis process (depicted in Fig. 18.2) involves utilizing a 4–20 wt% aqueous solution of either sodium hydroxide (NaOH) or potassium hydroxide (KOH) (Nyam et al. 2024). This process is carried out under a pressure ranging from 1.4 to 2 MPa and temperatures between 210 and 250 °C for a duration of 3 to 5 h. The reaction results in the formation of dipotassium or disodium terephthalate, which is subsequently treated with H_2SO_4 to isolate TPA. Specifically, this approach has the potential to promote large-scale recycling initiatives, given its capacity to effectively control contamination while producing high-quality TPA. The yield of TPA is directly impacted by both catalyst proportions and reaction temperature. Furthermore, studies revealed that using an aqueous ammonia solution at 200 °C can produce advantageous outcomes in base-mediated hydrolysis of PET. Additionally, phosphonium quaternary salts have been identified as catalysts that significantly enhance the reaction rate of alkaline hydrolysis, particularly at lower operating temperatures (increasing by 130–190 times) (Karayannidis et al. 2002; Mishra and Goje 2003).

18.2.2.4 Neutral Hydrolysis

The technique employs either steam or hot water (200 and 300 °C), usually at pressures around 1–4 MPa (Fig. 18.3). This method has demonstrated efficacy in yielding

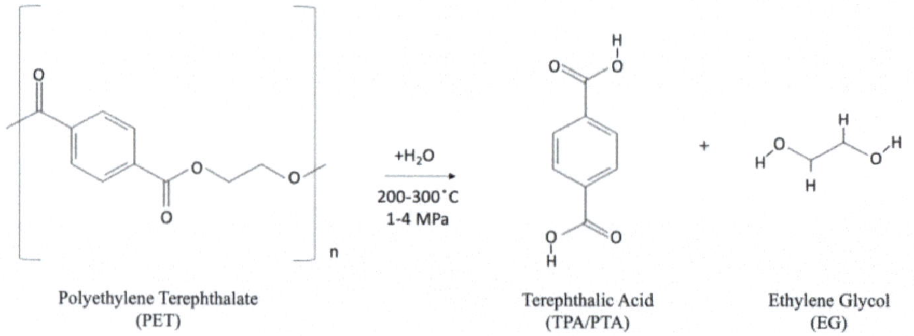

Fig. 18.3 Depolymerization of PET into TPA/PTA and EG by neutral hydrolysis

high-purity TPA and EG monomers, with optimal outcomes observed at temperatures surpassing 250 °C. Similar to alkaline hydrolysis, the yield of TPA and EG exhibits a direct correlation with increasing reaction temperature. Notably, TPA yield can reach up to a maximum percentage at reaction temperatures of 420 °C. The accompanying figure illustrates the chemical reactions that take place during the neutral hydrolysis process (Sato et al. 2006; Onwucha et al. 2023).

Demonstrating a maximum EG recovery of up to 60% is achievable when operating at a temperature of 300 °C. However, the catalytic dehydroxylation of EG may contribute to a lower yield. Following hydrolysis, the diluted EG produced can be recovered through either extraction or distillation methods. Neutral hydrolysis presents an environmentally friendly alternative to acid and alkaline hydrolysis, with fewer inorganic salts formed. Nonetheless, the purity of TPA obtained from neutral hydrolysis tends to be lower compared to that from acid and alkaline hydrolysis, leading to impurities in re-polymerized PET. To mitigate potential product contamination, filtration of the TPA solution, dissolved in either caprolactam or sodium hydroxide solution, is recommended. Studies have shown that crystallization of TPA from caprolactam can yield TPA with a purity exceeding 99%. Additionally, methanolysis offers another chemical depolymerization method, wherein PET is reacted with methanol in the presence of a catalyst to yield terephthalic acid and dimethyl terephthalate. This process is particularly beneficial for recycling PET into monomers for subsequent use in MOF synthesis (Sato et al. 2006; Onwucha et al. 2023).

18.2.2.5 Enzymatic Depolymerization

Enzymatic depolymerization, on the other hand, utilizes enzymes to catalyze the breakdown of PET. Enzymes, notably hydrolases, can preferentially break the ester bonds in PET, providing a more sustainable and potentially gentler alternative to chemical treatments. Hydrolase enzymes are used in enzymatic depolymerization to cleave the ester bonds in PET. Cutinases and lipases can selectively hydrolyze ester linkages, separating PET into its corresponding monomers. This enzymatic strategy proves beneficial due to its selectivity and ability to function under mild conditions compared to some chemical approaches, such as the enzymatic channel of biochemical processes that exist in nature, where microbes have exploited enzymes to break down complicated polymers. The incorporation of natural enzymes in depolymerization is consistent with green chemistry concepts, which emphasize sustainability and minimal environmental effects (Moyses et al. 2021).

18.2.3 Terephthalic Acid: The Key Precursor for Sustainable MOF Synthesis

Terephthalic acid, a byproduct of PET depolymerization, is a crucial precursor in the sustainable manufacture of MOFs. This molecule is critical in the formation of MOF frameworks, serving as an important organic linker during framework building. PET depolymerizes into its monomers, the major component of which is terephthalic acid. This acid, distinguished by carboxyl groups, is a flexible and essential building block in the development of MOFs. Its structural features allow it to function as a bridging ligand, promoting metal ion coordination and aiding the complex network of the MOF architecture. Terephthalic acid holds significance owing to its capacity to link metal ions, which promotes the formation of porous and crystalline MOF structures. MOFs are distinguished by their porosity, which serves as a platform for a variety of applications such as the storage of gases, catalytic processes, and drug encapsulation and targeted delivery. In terms of sustainability, adopting terephthalic acid extracted from recycled PET adheres to the principles of a circular economy. By recycling plastic waste as an invaluable resource, MOF synthesis evolves into an environmentally friendly one and supports the reduction of plastic pollution. TPA, as a crucial precursor, symbolizes the shift from a conventional, disposable strategy for plastics to a more environmentally friendly and cyclical use of resources in advanced material fabrication (Nyam et al. 2024; Campanelli et al. 1994; Mishra et al. 2003).

18.2.4 Advantages of PET Depolymerization in MOF Synthesis

The use of recycled PET towards MOF production has various advantages: (i) In terms of the environmental impact of recycling waste PET, the depolymerization method greatly reduces the carbon footprint connected with MOF synthesis. It adheres to the principles of sustainability and helps to reduce plastic pollution. (ii) To conserve resources, traditional MOF synthesis frequently uses petrochemical-derived raw materials. The use of recycled PET reduces dependence on virgin resources and conserves energy, addressing resource scarcity concerns. (iii) Integrating recycled

PET into MOF synthesis promotes the concept of a circular economy, where materials are reused, recycled, and repurposed, minimizing waste and enhancing resource efficiency (Table 18.1).

Table 18.1 Comparison of advantages and disadvantages of MOF synthesis methods

Synthesis method	Advantages	Disadvantages	Examples of MOFs
Hydrothermal/ solvothermal	• Generality, simplicity, and low cost • Large-scale preparation at mild temperatures	High operating cost Long synthesis time	ZIF-95 (Ma et al. 2020) ZIF-78 (Ban et al. 2013)
Microwave-assisted	• Reduced crystallization time • Energy efficient • Ease of controlling reaction conditions • Particle size can be controlled by precursor concentration	Reaction solvent requirements are limited	VSB-1, VSB-5 (Jhung et al. 2005) IRMOF-1, IRMOF-2, IRMOF-3 (Ni and Masel 2006) Zr-UiO-66 (Dang et al. 2020) Hf-UiO-66
Ultrasonic-assisted	Quickly disperse solutes and speed up the reaction process, improving the reaction efficiency	Breakage of large single crystals needed for diffraction studies	TMU-46, TMU-47, TMU-48 (Esrafili et al. 2020)
Mechanochemical	• Only mechanical forces are needed • Extreme operating conditions are avoided • Solvent-free	Hard to obtain a lot of product	Copper isonicotinate $Cu(INA)_2$ (Pichon et al. 2006) Copper benzene dicarboxylate $Cu_3(BTC)_2$ (Yuan et al. 2010)
Electrochemical	• Ease of industrial application • Short synthesis time • Uses current and voltage to control morphology	New MOFs have been synthesized to date	UiO-66 (Zhang et al. 2020) Cu_3 $(BTC)_2$ (Pirzadeh et al. 2020)
Microfluidic synthesis	The effective mixing of organic phase and inorganic phase is realized	Limited range of adaptation	UiO-67 (Bailey et al. 2024)

18.2.5 LCA Assessment and Challenges

While PET depolymerization marks a significant step forward, challenges persist in addressing the environmental impact of disposable PET bottles due to their short useful life, high production volume, and non-biodegradability. Researchers have turned their focus to conducting life cycle assessments (LCAs) to evaluate the environmental impact of PET packaging (Gomes et al. 2019). The first LCA, dating back to 1969, scrutinized PET's use as packaging for carbonated drinks. Since then, numerous LCAs and research studies have aimed to identify alternative packaging materials for carbonated beverages.

In examining the PET packaging bottles in the industrial, researchers identified 49 environmental impact categories, with global warming potentials, human toxicity, and marine aquatic ecotoxicity ranking as the highest impact categories. The polymerization of PET resin emerges as a significant contributor to global warming potential due to the release of carbon dioxide. Additionally, the decomposition of landfilled plastic waste leads to the emission of methane and carbon dioxide, contributing to air pollution and global warming. In 2008 alone, approximately 20 million tons of CO_2 equivalent ($eqCO_2$) were released into the atmosphere from solid waste breakdown in landfills (Alabi et al. 2019).

Plastic burning worsens air pollution by emitting pollutants such as heavy metals, dioxins, PCBs, and some noxious heterocycles, all of which pose serious health hazards, primarily respiratory troubles. Plastics have a particularly severe impact on air pollution in emerging and underdeveloped countries, critically affecting future generations. Combating these environmental issues necessitates concerted efforts to identify sustainable alternatives to PET proliferation and reduce the negative effects of plastic pollution on human health and the environment.

Current research frontiers include streamlining the depolymerization procedures, developing greener catalysts, and addressing financial sustainability. Researchers are looking into ways to improve the efficacy and adaptability of depolymerization processes, making them more useful for huge-scale MOF manufacturing applications (Yue et al. 2023). In this regard, PET depolymerization represents a breakthrough step in MOF synthesis. It not only remedies the environmental issue of plastic waste, but it also matches with the principles of green chemistry and sustainable materials advancement. As the scientific community continues to innovate and enhance the quality of depolymerization procedures, the incorporation of repurposed PET as a precursor in the production of MOF has the potential to bridge ecological and material science targets, leading the path toward a more sustainable future (El-Sayed and Yuan 2020).

18.3 Precise Engineering of MOFs: The Crucial Role of Terephthalic Acid Functionalization

Terephthalic acid derivatization is a critical and complicated phase in MOF fabrication. This approach includes deliberately modifying the terephthalic acid to impart particular functional groups that improve its compatibility as an acidic linker in MOF assemblies. By executing accurate customization, the characteristics of terephthalic acid can be engineered, thus impacting the properties of the resultant MOF and broadening its scope of applications across several domains (Sharifzadeh and Morsali 2022; Boukayouht et al. 2023). The starting point of the protocol is TPA, which is derived from the depolymerization of PET, which is often obtained from used plastic bottles. Although terephthalic acid is a basic building unit, functionalization allows for further refinement of its properties, which makes it a versatile and adaptable part of MOF production.

The TPA monomer is functionalized by incorporating particular chemical entities into it. This technique is carefully crafted to provide the MOF with the necessary properties, which include increased selectivity in gas adsorption as well as improved durability and efficient catalytic activity. The arrangement and selection of functional groups have a significant impact on the MOF's structure and functionality. One popular functionalization technique is to add various organic moieties, including aliphatic, aromatic, amino, halogens, or phosphines, nitro and carboxyl residues, to the TPA core (Sharifzadeh and Morsali 2022; Morel et al. 2014; Vallés-García et al. 2020; Santos et al. 2020). The addition of these fragments not only modifies the electronic and steric characteristics of TPA but also offers binding spots for metal ions during the MOF construction process. Synthetic chemistry strategies map out the functionalization procedure, and researchers use a number of approaches to produce the desired changes. Interactions with acid chlorides, anhydrides, or esters are common routes to functionalize TPA scaffolds; these reactions impart specific functional groups. Interactions with acid chlorides, anhydrides, or esters are common routes to functionalize the TPA skeleton; these reactions impart specific functional groups. To balance the development of desirable characteristics with maintaining the structural integrity of the TPA skeleton, the extent and type of functionalization are carefully regulated. Excessive functionalization may impair the potential of TPA to function as an efficient linker. However, insufficient modification could restrict the versatility of the resultant MOFs.

One prominent feature of functionalization is its influence on the degree of porosity and surface area of the MOFs. Different functional groups can be introduced within the MOF structure to tailor suitable coordination donor sites, which influence the applicability of MOFs, especially in the adsorption and separation of gases (Zhao et al. 2011). This feature is especially useful in gas storage usage, where MOFs customized with specific groups outperform other materials in terms of gas capture and release. Moreover, TPA functionalization is an important variable in modifying the MOF's hydrophilicity or lipophilicity. This characteristic is essential for applications like drug encapsulation, prospected delivery and catalysis, where the MOF's affinity for organic solvents, lipids, or water determines the interaction with surrounding components. Functionalization boosts the MOFs' tunability for particular applications independent of structural parameters. For instance, incorporating specific functional groups can increase MOFs' catalytic activity in chemical transformations. Because of its catalytic competence, MOFs can be employed as heterogeneous catalysts in a variety of industrial synthesis, expanding their potential applications (Boukayouht et al. 2023).

The hunt for improved performance by TPA modulation also encompasses optical and electronic components. The electronic structure of MOFs can be tailored by incorporating conjugation and aromatic scaffolds, including heterocycles, to suit their requirements in optoelectronics, sensors, and other cutting-edge technologies. Even though functionalization has several possibilities for customizing MOF characteristics, difficulties still exist. Ongoing research endeavours involve accomplishing accurate regulation over the arrangement of functional groups, preventing undesired side reactions, and guaranteeing reproducibility on a bulk scale. The transition of these developments from lab settings to industrial settings depends on the scalability of functionalization techniques (Nyam et al. 2024).

To summarise, the functionalization of TPA is an essential and dynamic step in the production of MOFs. This technique converts TPA into a flexible building unit, allowing the fine-tuning of MOF characteristics for a wide variety of applications. From gas storage, sustainable catalysis, and optoelectronics to drug delivery in pharmacology, functionalized TPA's tailored features unlock a new era of advanced materials with multiple functions and unprecedented versatility (Table 18.2). As research advances the frontiers of functionalization strategies, their influence on MOF design and applications is projected to escalate, bringing us closer to recognizing the immense potential of these incredible materials in a variety of scientific and commercial applications (El-Sayed and Yuan 2020).

18.4 Metal Coordination: Assembling the Sustainable MOF Structure

The MOF assembly depends critically on metal coordination. It highlights the beneficial relationship between metal ions and recycled PET, laying a long-term basis for the synthesis of innovative materials. Metal coordination is an essential component of the carefully planned procedure that goes into assembling MOFs. An important stage in the synthesis of MOFs, this one entails carefully placing metal ions with organic linkers to form a strong, porous structure. Recycling waste PET bottles as a source of acidic linkers brings an ecologically conscientious element to the metal coordination process, which is becoming increasingly important as sustainability becomes a focus in materials research (Aliakbari et al. 2022). MOF production relies heavily on the coordination of metal ions with organic ligands. Like a molecular dance, this synchronization establishes the structure, stability, and properties of the

Table 18.2 Various Applications of MOFs derived from waste PET bottles

MOFs	Ligand source	Method of isolation of ligand	Metal source	Application	Outcomes	References
Cr-MOF	BDC (Waste PET) and BDC (Sigma–Aldrich)	Depolymerization using a high-pressure reactor	$CrCl_3 \cdot 6H_2O$ (Sigma–Aldrich)	Hydrogen storage	PET-derived MOF outperforms commercial BDC-based MOF	Ren et al. (2016)
Zr-MOF	BDC (Waste PET) and BDC (Sigma–Aldrich)	Depolymerization using a high-pressure reactor	$ZrCl_4 \cdot 6H_2O$ (Sigma–Aldrich)	Hydrogen storage	The equipotent activity of both PET-derived MOF and commercial BDC-based MOF	Dyosiba et al. (2016)
Nanoporous V, Cr, Al, Ga MOFs	BDC (waste PET)	Directly used the scraps of waste PET bottles	VCl_3, $CrCl_3 \cdot 6H_2O$, $Al(NO_3)_3 \cdot 9H_2O$, $Ga(NO_3)_3 \cdot 9H_2O$ (Commercial)	(i) CO_2 and H_2gas adsorption (ii) Catalytic conversion of CO_2 and epoxides to cyclic carbonates	Significant gas uptake capability and catalytic efficiency accomplished under mild conditions	Lo et al. (2016)
Cu MOF	BDC (waste PET)	Alkaline hydrolysis	$Cu(NO_3)_2 \cdot 3H_2O$ (Merck)	Removal of methyl orange dye from water	Removal of dye via rapid catalytic reduction into sodium p-amino benzene sulphonate and N, N-dimethyl-p-phenylenediamine	Rahmani et al. (2018)
NanocuboidalZr-MOF	BDC (waste PET)	Alkaline hydrolysis	$ZrN_2O_7 \cdot xH_2O$ (SD Fine chemicals)	Adsorption of non-steroidal anti-inflammatory drug, i.e., ketorolac tromethamine (KTC) from water	PET-derived MOF exhibits excellent stability and reusability up to six cycles	Singh et al. (2018)

(continued)

Table 18.2 (continued)

MOFs	Ligand source	Method of isolation of ligand	Metal source	Application	Outcomes	References
Zr-MOF	BDC (waste brown and green PET bottles) and BDC (Sigma–Aldrich)	Depolymerization via glycolysis	$ZrCl_4 \cdot 6H_2O$ (Sigma–Aldrich)	Business case development model of 'Waste PET to value-added MOFs'	MOFs from green/brown PET BDC showed lower surface areas than those from commercial BDC, possibly due to organic dyes in coloured PET-blocking pores	Dyosiba et al. (2019)
Zr_{12} cluster MOF	BDC (waste PET)	Direct PET bottle chips	$ZrCl_4$ (Alfa Aesar)	Separation of benzene/cyclohexane via competitive adsorption	MOF material has excellent thermal stability and hydrothermal stability and shows good separation performance	Zhou et al. (2019)
Tb MOF	BDC (waste PET)	Alkaline hydrolysis	$Tb(NO_3)_3 \cdot 5H_2O$ (Sigma)	Picric acid sensing	First, lanthanide-based MOF showed high selectivity towards picric acid	Zhang et al. (2019)
Cu MOF	BDC (waste PET)	Alkaline hydrolysis	$Cu(NO_3)_2 \cdot 3H_2O$ (Sigma–Aldrich)	Detection of organic pollutants	Cu MOF showed sensing capability towards acetone, propanol and ethanol via intermolecular interactions	Kaur et al. (2020)
Cu MOF	BDC (waste PET)	Alkaline hydrolysis	$Cu(NO_3)_2 \cdot 3H_2O$ (Merck)	Removal of methylene blue (MB) from water	Physical adsorption through electrostatic interaction between MB cations and negatively charged CuBDC surface	Doan et al. (2020)
Fe MOF	BDC (waste PET)	Ultrasound-mediated, phase-transfer-catalyzed alkaline hydrolysis	$FeCl_3 \cdot 6\ H_2O$ (Sigma–Aldrich)	Adsorption of antibiotic tetracycline hydrochloride from water	Mechanism of TCH adsorption on reusable Fe MOF includes π–π EDA interaction, cation-π bonding, hydrogen bonding, and pore-filling effect	Jung et al. (2020)
ZnO MOF	BDC (waste PET)	Depolymerization using auto-clave reactor	$Zn(NO_3)_2 \cdot 6H_2O$ (Sigma–Aldrich)	Electro-catalytic activity towards hydrogen and oxygen evolution reactions	ZnO MOF exhibits superior electrocatalytic performance over pure ZnO nanostructures for hydrogen and oxygen evolution reactions	Ubaidullah et al. (2020)

(continued)

Table 18.2 (continued)

MOFs	Ligand source	Method of isolation of ligand	Metal source	Application	Outcomes	References
ZnOMOFand Co_3O_4 MOF	BDC (waste PET)	Depolymerization using auto-clave reactor	$Zn(NO_3)_2{\cdot}6H_2O$, $Co(NO_3)_2{\cdot}6H_2O$ (Acros Organics)	Energy (supercapacitor) application	Co3O4 MOF showed exceptional supercapacitor performance due to the favourable Faradic process and high theoretical capacitance of Co3O4	Al-Enizi et al. (2020a)
Ni MOF	BDC (waste PET)	Depolymerization using a high-pressure reactor	$Ni(NO_3)_2{\cdot}6H_2O$ (Sigma–Aldrich)	Energy (supercapacitor) application	The excellent specific capacitance may be attributed to the high specific surface area (1523 m^2/g) and mesoporosity	Al-Enizi et al. (2020b)
Ni/Co-MOF	BDC (waste PET)	Depolymerization using a high-pressure reactor	$Ni(NO_3)_2{\cdot}6H_2O$, $Co(NO_3)_2{\cdot}6H_2O$ (Sigma–Aldrich)	Energy storage applications	Ni/Co-MOF exhibited an excellent specific surface area of 813 m^2/g, mesoporosity of approximately 15 nm, outstanding super capacitive performance and retention capacity	Alhokbany et al. (2020)
Sn MOF	BDC (waste PET)	Depolymerization using a high-pressure reactor	$SnCl_2{\cdot}2H_2O$ (Himedia)	Adsorption of arsenate and phosphate from water	Maximum adsorption capacities (qm) of 90.90 mg/ g for arsenate and 126.58 mg/ g for phosphate via electrostatic attraction and hydrogen bonding. Effective removal of arsenate from water samples met WHO permissible limits	Ghosh and Das (2021)
Al MOF	BDC (waste PET) and BDC (commercial)	Direct PET flakes	$AlCl_3{\cdot}6H_2O$ (Commercial)	Catalyst for dry reforming of methane	The equipotent activity of both PET-derived MOF and commercial BDC-based MOF	Karam et al. (2021)

(continued)

Table 18.2 (continued)

MOFs	Ligand source	Method of isolation of ligand	Metal source	Application	Outcomes	References
Zn MOF	BDC (waste PET)	Depolymerization using auto-clave reactor	$Zn(NO_3)_2 \cdot 6H_2O$ (Commercial)	Energy (supercapacitor) application	They are demonstrating an exceptional specific surface area of approximately 2243 m^2/g, with a high specific capacitance of around 230 F/g by cyclic voltammetry at 5 mV/s and approximately 295 F/g by galvanostatic charge–discharge (GCD) at 0.5 A/g	Ubaidullah et al. (2021)
Co/Ni MOF	BDC (waste PET)	Depolymerization using auto-clave reactor	$CoCl_2 \cdot 6H_2O$, $NiCl_2 \cdot 6H_2O$ (Sinopharm)	Anode material for lithium-ion batteries	The initial Coulombic efficiency of Co/Ni MOF is slightly lower than that of most commercial anodes; strategies such as surface coating and pre-lithiation can improve the overall performance of Co/Ni MOF	Wang et al. (2022)
Cu-MOF, Zr-MOF, and Ti-MOF	BDC (waste PET)	Depolymerization using auto-clave reactor	$Cu(NO_3)_2$ $3H_2O$, $ZrCl_4$,$Ti(OBu)_4$(Sigma–Aldrich)	Energy (supercapacitor) application	Cu-MOF//Cu-MOF SC device delivers the highest energy density of 18.2 Wh/kg at a power density of 825 W/kg	Dubey et al. (2023)
Al MOF	BDC (waste PET)	Depolymerization using auto-clave reactor	$Al(NO_3)_3 \cdot 9H_2O$ (Sigma–Aldrich)	Application for high-performance Zn-ion capacitor	Al MOF exhibited a distinctive accordion-like structure, high specific surface area (SSA), and hierarchical porous distribution, which could augment interfacial active sites, facilitate rapid ion diffusion, and minimize ion transport distance, leading to exceptional electrochemical performance	Li et al. (2023)

(continued)

Table 18.2 (continued)

MOFs	Ligand source	Method of isolation of ligand	Metal source	Application	Outcomes	References
Fe MOF	BDC (waste PET)	Alkaline hydrolysis under hydrothermal conditions	Rust iron powder	Energy storage applications	The equipotent activity of both PET-derived MOF and commercial BDC-based MOF	Deka et al. (2023)
Cu/Zn MOF	BDC (waste PET)	Depolymerization via glycolysis	$Zn(OAc)_2 \cdot 2H_2O$, $Cu(NO_3)_3 \cdot 2H_2O$, $ZnCl_2.H_2O$ (Sigma–Aldrich)	Catalytic reduction of organic pollutant dyes	Purification of wastewater contaminated by methylene blue, methyl orange, 4-nitrophenol, and 4-nitroaniline	Yarahmadi et al. (2023)
Cu MOF, Cu MOF-NH_2	BDC (waste PET)	Modified alkaline hydrolysis	$Cu(NO_3)_2 \cdot 3H_2O$ (Merck)	Catalytic reduction of deleterious 4-nitrophenol into the value-added 4-aminophenol	Cu MOF-NH_2 shows increased adsorption capability and reduction efficiency over Cu MOF	Farshchi et al. 2023)
Zr-MOF	BDC (waste PET)	Depolymerization using auto-clave reactor	$ZrOCl_2 \cdot 8H_2O$ (Sigma–Aldrich)	Energy (supercapacitor) application	Zr-MOF exhibited high specific surface area and mesoporosity, resulting in exceptional specific capacitance values measured by GCD and CV techniques	Al-Enizi et al. (2024)

resulting MOF. Commonly used metal ions include zinc, copper, and chromium, chosen for their ability to form stable coordination bonds with organic linkers.

A suitable metal ion must be chosen as the initial stage in the metal coordination process. The selection is contingent upon several elements, including the targeted use, the preferred characteristics of the MOF, and availability and pricing considerations. MOF structure's reactivity, porosity, and catalytic capabilities are all influenced by the distinct features that each metal ion contributes to the structure. The organic linker, which is frequently made of terephthalic acid, is produced by depolymerizing PET, and the metal ion forms a coordination bond once the metal ion has been selected. The MOF is constructed from a repeating unit that is created when coordination bonds are formed between the carboxyl groups of the organic linker and the metal ion. A major factor in MOFs' crystalline structure is their coordination bonds, which are usually metal-carboxylate linkages. The MOF has a distinct three-dimensional structure because of these strong, directed connections (Alavi and Morsali 2014). Strong linkages like this guarantee the MOF's stability in a variety of situations, which is important for uses like catalysis and gas storage. The coordination process involves more than just a chance pairing of organic linkers and metal ions. Researchers carefully regulate the stoichiometry and reaction conditions to create MOFs with the correct structure and characteristics. The overall symmetry and porosity of the MOF are influenced by the geometry of the metal coordination environment, which enables customization for particular uses (Jhung et al. 2005; Dang et al. 2020; Bailey et al. 2024).

Utilizing recycled PET as a source of acidic linkers brings the sustainability of MOF synthesis, bringing it to the forefront. Metal coordination needs carboxyl groups, which are provided by terephthalic acid, which is produced by PET depolymerization. Utilizing recycled materials to generate useful and functional materials is in line with the circular economy's tenets. Plastic waste is being repurposed. Apart from its positive impact on the environment, MOFs' capabilities take on a new dimension when recovered PET is utilized in metal coordination (Wu et al. 2015). The coordination process may be influenced by differences in the structure and chemical makeup of terephthalic acid obtained from recycled PET, which could result in MOFs with distinct characteristics from those developed using conventional procedures. The interaction between metal ions and recycled PET-derived linkers expands the application of MOFs to sustainable catalysis. The coordination environment of the metal ions determines the catalytic characteristics of MOFs. Researchers can create MOFs with increased catalytic activity and contribute to environmentally friendly and sustainable chemical processes by customizing the metal coordination process using recycled PET linkers (Table 18.2).

As the realm of MOF continues to evolve, challenges persist in the metal coordination process. Further research aims to understand the impact of recycled PET-derived linkers on MOF characteristics, optimize reaction conditions for large-scale synthesis, and achieve fine control over the coordination geometry. When moving from laboratory-scale synthesis to industrial applications, the scalability of these techniques is an important factor to take into account. Therefore, the features of these materials demonstrate how including recycled PET bottles in the metal coordination process not only promotes sustainability but also creates new opportunities for customizing MOF characteristics (Seetharaj et al. 2019). The integration of recycled PET-derived linkers and metal coordination promises to furnish MOFs with heightened performance, heralding a more environmentally friendly and sustainable future in materials science (Shanmugam et al. 2022).

18.5 Methods of Synthesis for MOF Production Process: Unlocking the Architecture of Metal–Organic Frameworks

The construction of MOFs is a multifaceted procedure that requires strategically assembling metal ions and organic linkers into a three-dimensional network. MOFs are generated by forming strong bonds between organic counterparts and inorganic (metal ions, clusters, or metal nodes) elements (Kumar et al. 2020). These compounds are synthesized in a variety of forms, each differentiated by ligand functional groups, geometry, and pore and crystal dimensions. In the synthesis process, various transition and inner transition metals serve as the inorganic constituents. Additionally, numerous organic linkers are utilized as the organic components, such as acetamido substituted isophthalic acid, substituted dicarboxylic acid, thioctic acid, and anionic form of imidazole, among others. Two primary methodologies are commonly used for MOF synthesis: Liquid-phase synthesis encompasses various methods, especially sonication, MW, SE, ionothermal, hydrothermal, electrochemical, solvothermal, and flow techniques (Fig. 18.4).

Additionally, solid-phase (mechanochemical) synthesis involves solvent-free, liquid-assisted, and ion and liquid-assisted procedures (Table 18.1). In liquid-phase synthesis, three highly pure components—solvent, organic linker, and metal salt—are necessary, and they are combined in two primary ways: (I) by mixing the organic and inorganic constituents in a solvent within a container and (II) by dissolving the organic and inorganic constituents separately in the solvent, followed by mixing the two solutions. Solid-phase synthesis methods are conducted without solvent or with minimal solvent participation (Wang et al. 2022).

18.5.1 Solvothermal Synthesis

Solvothermal synthesis is one of the most widely employed methods for MOF production. This technique involves the reaction of metal precursors and organic linkers in a solvent under high-temperature and high-pressure conditions. The controlled application of heat and pressure promotes the rapid and uniform formation of MOF crystals. Solvothermal synthesis allows for precise control over particle size,

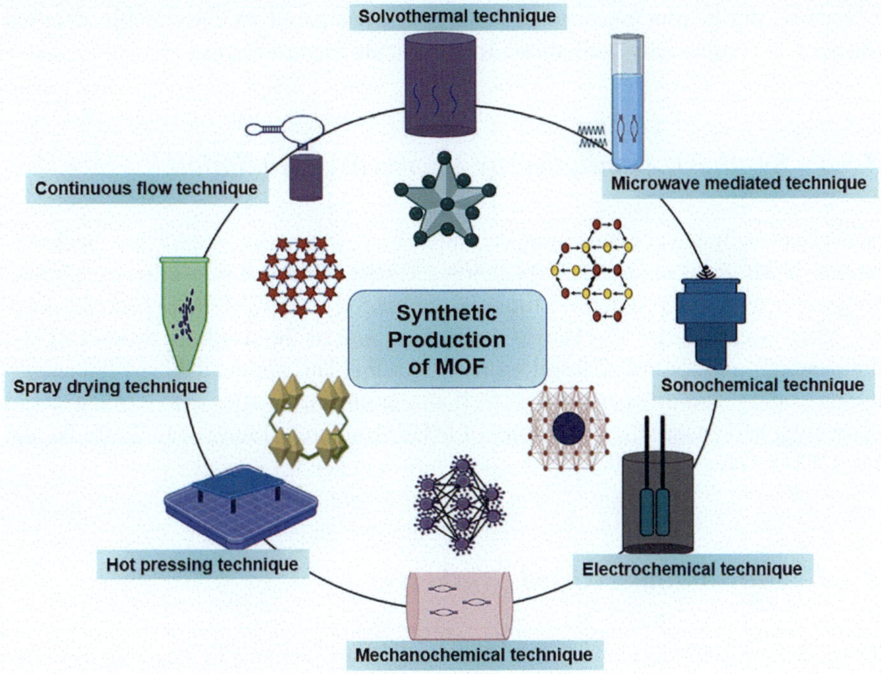

Fig. 18.4 Methods of synthesis for MOF production process

morphology, and crystallinity, influencing the overall performance of the MOF in specific applications. It is beneficial when aiming for large, well-defined MOF crystals with enhanced stability. The solvothermal technique was employed to produce MOF-5 ($Zn_4O(BDC)_3$, with BDC representing the terephthalate anion), utilizing autoclaves fabricated through 3D printing using polypropylene. Subsequently, the resultant polymer was isolated and characterized by elemental analysis and powder X-ray diffraction.

X-ray diffraction investigation verifies the crystal structures of both MOF-5 and other coordination polymers generated from zinc (II) terephthalate, identified as $\{Zn_4O(BDC)_3\}\cdot(ZnO)_{0.125}$, which was formed as a by-product (Denisov et al. 2019).

18.5.2 Hydrothermal Synthesis

Hydrothermal synthesis differs from solvothermal synthesis in that it uses water as a solvent. This approach takes advantage of the unique features of water at high temperatures and pressures to enable the controlled building of MOF structures. Hydrothermal synthesis is beneficial because it is simple and versatile, producing MOFs with a wide range of morphologies and functions (Natarajan et al. 2006).

Because of the gentler reaction circumstances compared to classic solvothermal processes, it is especially well-suited to huge-scale manufacturing.

18.5.3 Synthesis Facilitated by Microwave Irradiation

Microwave-mediated synthesis employs microwave irradiation to expedite the development of MOFs. This technology, known for its rapid reaction kinetics, significantly reduces MOF production times as compared to typical heating methodologies. Microwave energy accelerates the heating rate of the reaction, promoting the homogeneous and controlled development of MOFs. This approach has a high energy efficiency and scalability perspective. Still, thorough optimization is required to avoid overheating and ensure the uniformity of MOF structures (Jhung et al. 2005; Ni and Masel 2006; Dang et al. 2020).

18.5.4 Ultrasound-Assisted Synthesis

Ultrasound-assisted fabrication exploits ultrasonic waves to aid in the nucleation and development of crystalline MOFs. Ultrasonic waves generate acoustic cavitation, which causes confined heating and pressure oscillations, enabling effective reactant mixing and promoting mass transfer. This approach is distinguished by its high energy efficiency, quick reaction times, and capacity to generate MOFs with distinct morphologies. Ultrasound-mediated synthesis provides advantages for fine-tuning MOF characteristics and exploring novel arrangements (Esrafili et al. 2020).

18.5.5 Mechanochemical Synthesis

A more environmentally friendly method of producing MOFs can be referred to as mechanochemical synthesis, which involves milling or grinding solid reactants without the use of solvents (Pichon et al. 2006; Yuan et al. 2010). Mechanical forces drive the initiation and execution of MOF synthesis in this solvent-free technique. Mechanochemical synthesis has a number of benefits, such as simplicity, the ability to produce huge quantities of MOF, and a lower environmental impact. The omission of solvents facilitates the purifying procedure and boosts the sustainability of the overall process.

18.5.6 Electrochemical Synthesis

Electrical currents are used in electrochemical synthesis to propel the development of MOFs. Metal cations are reduced at the cathode to enable their regulated assembly into MOF structures. At the same time, organic linkers undergo oxidation at the anode. With the fine control that electrochemical synthesis offers over the reaction settings, MOFs with customized characteristics can be produced. This approach is especially useful for applications requiring precise control regarding MOF composition and architecture (Zhang et al. 2020; Pirzadeh et al. 2020).

18.5.7 Template-Assisted Synthesis

Template-assisted fabrication refers to the incorporation of pre-formed frameworks or templates to direct the formation of MOFs. These templates are able to be sacrificed or removed after synthesis, leaving only the targeted MOF architecture. Regulation over the MOF topology at the nanoscale is made possible by template-assisted techniques, which enable the construction of hierarchical structures and clearly specified morphologies. This method is useful for applications that require unique MOF topologies or architectures (Xu et al. 2023).

18.5.8 Green Synthesis

Green synthetic techniques reduce the use of toxic substances and energy-consuming procedures in order to prioritize sustainability. These methods frequently start with waste materials, renewable resources, or safe solvents. Green synthesis emphasizes resource conservation and ecological sustainability in accordance with the tenets of green chemistry. While still an evolving field, green synthesis has tremendous potential for generating MOFs with minimal adverse impact on the ecological system (Kumar et al. 2020).

To sum up, there are many different approaches to the synthesis of MOFs, and each has its advantages and limitations. The preference for synthetic strategy determines the desired MOF characteristics, the targeted application and concerns about scalability and sustainability. As this field flourishes, researchers are delving into innovative synthetic protocols to uncover the complete potential of MOFs in tackling modern world challenges across a range of scientific and industrial sectors (Boukayouht et al. 2023).

18.5.9 Unlocking the Potential: Leveraging MOFs Sourced from PET Waste Bottles to Address Environmental Issues and Beyond

MOFs are extremely porous materials with a distinctive composition made up of organic ligands and metal ions or clusters. MOFs are suitable for a variety of applications because of their outstanding porosity, flexible characteristics, and large surface areas. They also have open metal sites and ordered crystal structures. Waste PET bottles have garnered attention as a renewable resource for the synthesis of MOFs with an extensive spectrum of uses in modern times. This section delves into the versatile utilities of MOFs developed from waste PET waste that can advance technology and meet societal demands (El-Sayed and Yuan 2020; Sharifzadeh and Morsali 2022; Boukayouht et al. 2023; Kumar et al. 2020). This strategy has prospects for facilitating the development of novel materials with improved qualities in addition to providing a viable solution for handling PET waste. Because of their structural flexibility, they can be tailored for a variety of applications, including remediating the environment, storing gases, adsorption, segregation, catalytic processes, and detecting various components. The synthesis of MOFs sourced from discarded PET bottles is reviewed in the following sections, along with their multifaceted applications in a variety of industries, enlightening their potential contributions towards environmental remediation as well as sustainability (Table 18.2).

18.5.10 Adsorption and Separation

PET-derived MOFs feature large surface areas and configurable pore architectures, making them ideal for gas adsorption (hydrogen, methane, and carbon dioxide) and elimination of contaminants (heavy metals, organic compounds, and pharmaceuticals) from water sources. In this context, The research team of Onyango have innovatively employed trashed plastic bottles as a source of 1,4-benzenedicarboxylic acid (BDC) ligand for the synthesis of UiO-66(Zr) and MIL-101(Cr) MOFs (Ren et al. 2016; Dyosiba et al. 2016, 2019). The economical fabrication of MOFs prompted an investigation into hydrogen storage applications, comparing the prepared materials with commercially-derived MOFs. The study unveiled the microporous characteristics of both MOFs, showing differing BET surface areas yet comparable pore volumes. The waste-derived MIL-101(Cr) MOF displayed superior BET surface area and gravimetric hydrogen uptake compared to its commercially-derived counterpart, while UiO-66(Zr) MOF sourced from waste PET exhibited similar textural properties to commercial-derived MOFs. However, UiO-66(Zr) MOFs derived from coloured bottles showed reduced BET surface areas and hydrogen uptake capacities compared to the commercial-derived MOF. Despite these differences, the waste-derived MOFs demonstrated properties comparable to or even better than their commercially-sourced counterparts, emphasizing the economic benefits of employing waste materials for MOF construction. In yet another groundbreaking initiative, the waste PET bottles have been ingeniously repurposed as starting materials by Lo et al. for synthesizing five nanoporous trivalent MOFs—MIL-47, MIL-53(Cr), MIL-53(Al), MIL-53(Ga), and MIL-101(Cr), eliminating the need for terephthalic acid. Gas sorption measurements demonstrated the significant gas uptake capabilities of MIL-53(Cr) and MIL-101(Cr) towards CO_2 and H_2 gases (Lo et al. 2016). On the other hand, methylene blue (MB) is a dangerous cationic dye known to pose significant health risks upon prolonged exposure and has adverse effects on ecosystems, particularly in developing countries. Despite its dangers, research must prioritize detoxification practices for wastewater holding MB because of its extensive use in dyeing and coating sectors. In this scenario, Nguyen's research team fabricated Cu-containing MOF (Cu-BDC) utilizing PET waste bottles as an organic ligand and applied it as an adsorbent for aqueous MB solutions (Doan et al. 2020). The optimization of adsorption parameters revealed the high stability of CuBDC structure and long-term adsorption capability. Furthermore, the recyclability study over four cycles demonstrated the Cu-BDC's stability, affirming its long-term viability and cost-effectiveness for dye removal applications. Porous MOFs have gained recognition as prominent adsorbents for removing pharmaceutical contaminants from water sources, posing threats to public health and marine ecosystems. In this context, Kansal and co-workers constructed an economical and eco-friendly nano-cuboidal Zr-UiO(66) MOF by utilizing BDC sourced from alkaline hydrolysis of waste PET bottles. Zr-UiO(66) MOF was utilized for extended adsorption of the nonsteroidal anti-inflammatory drug ketorolac tromethamine from aqueous phases. Screening of adsorption variables, such as pH, initial KTC dose, and MOF dosage, revealed optimal conditions for efficient adsorption. Notably, the MOF demonstrated superior performance under acidic conditions, with the highest adsorption capability. Recycling tests confirmed the MOF's efficacy, maintaining maximum adsorption capability even after six cycles. This study showcases a sustainable and economically viable adsorbent for environmental remediation by swiftly removing drugs from aquatic environments (Singh et al. 2018). Jung et al. addressed the urgent issue of antibiotic overuse and its detrimental effects on human health and environmental sustainability. They focused on enhancing the cost-effectiveness and practical applicability of porous carbon composites derived from MOFs for antibiotic sequestration (Jung et al. 2020). By employing ultrasound-assisted, phase-transfer-catalyzed alkaline hydrolysis, they successfully recovered BDC from PET waste bottles under mild conditions, achieving nearly complete BDC recovery. The high-purity BDC was employed as an organic linker in the fabrication of a magnetic porous carbon composite (α-Fe/Fe3C) produced from a Fe-encompassed MOF. This composite features a mesoporous graphite framework and superparamagnetic activities, allowing for efficient adsorption and easy magnetic separation to remove tetracycline hydrochloride (TCH) from aqueous environment. The adsorption process includes several interactions, including π-π electron donor–acceptor interaction, cation-π bonding, hydrogen bonding, and pore-filling interactions. The α-Fe/Fe3C composite showed great potential for the treatment of wastewater polluted with antibiotics due to its magnetic isolation and exceptional reusability. Das and coworkers proposed a unique approach for producing an Sn(II)-based MOF that exploits trash PET bottles as a strongly binding ligand. The Sn(II)-MOF effectively removed arsenate and phosphate from aqueous solutions, with efficiencies of around 98% and 88%, respectively. Furthermore, the MOF exhibited substantial anti-interference action against contemporaneous anions and efficiently eliminated arsenate from aqueous solutions. The researchers also investigated the transformation of positively charged Sn(II)-MOF into negatively charged tin oxide nanoparticles (SnO_2 NPs) by thermal degradation. The SnO_2 NPs had a remarkable adsorption ability towards Mn^{2+} ions, with minimal interference with other ions. Their study proposes an economically effective strategy for recycling trash PET bottles into useful adsorbents, furnishing promising applications in water purification (Ghosh and Das 2021). There has been a rising curiosity in employing waste PET bottles as key reactants to produce MOFs for extraction applications. For instance, the research group of Serre established the straightforward fabrication of a zirconium-incorporated MOF with waste PET as a ligand resource. The synthesized Zr12 cluster terephthalate MOF (hcp UiO-66(Zr)) has a hexagonal prism form, good crystallinity, and a larger surface area due to ligand defects, with around three lacking linkers per Zr12 cluster. The hcp UiO-66(Zr) exhibited exceptional thermal and hydrothermal durability. Furthermore, it indicated favourable prospects for the separation of benzene and cyclohexane by competitive adsorption methods, as demonstrated by mixed-vapour analysis. On the whole, the findings demonstrate PET's applicability as a ligand resource for the direct formation of a tetravalent metal-based MOF, which facilitates the recovery and extraction of industrial chemical compounds via adsorptive selectivity (Zhou et al. 2019).

18.5.11 Sensing

The use of waste-to-value-added MOFs is a viable route, especially for fixing environmental issues across multiple sectors. MOFs derived from PET waste can be designed to have precise chemical detection capabilities. These materials were tested for sensing a variety of analytes, ranging from volatile organic compounds (VOCs) hazardous gases to explosive compounds. Their massive surface areas and customizable pore architectures facilitate accurate and selective sensing, which renders them desirable for monitoring the environment and safeguarding purposes. In an innovative investigation designed by Yang and coworkers, trashed PET bottles were effectively converted into a valuable resource for producing a lanthanide-based MOF geared towards the sensing of picric acid (TNP). BDC has been extracted from PET using customized hydrolytic conditions, serving as an important precursor for MOF synthesis. With photoluminescence characteristics designed specifically for selective sensing, the customized Tb-BDC functioned as a molecular sensor, exhibiting exceptional sensitivity and selectivity towards TNP. The imperative is to develop highly sensitive and easily deployable luminescent sensors capable of swiftly detecting low concentrations of TNP. Notably, luminescence quenching occurs solely in the presence of TNP, affirming the high selectivity of Tb-BDC for detecting picric acid. The resulting Tb-BDC MOF displayed exceptional selectivity and sensitivity towards TNP. This innovative approach not only demonstrates the effective utilization of waste PET but also holds promise for environmental protection efforts (Zhang et al. 2019). Yet another novel approach was demonstrated by Kaur et al. for the solvothermal synthesis of a copper-based MOF utilizing TPA extracted from waste soft drink bottles through base hydrolysis. Advantageously, the Cu-MOF demonstrated remarkable photocatalytic activity under UV irradiation, facilitating the detection of VOCs, such as acetone, with a detection range of approximately 20 ppm. Notably, the selectivity of Cu-MOF in sensing acetone underscores its potential to spot various organic pollutants. Their study underscores the feasibility of utilizing waste plastic as a resource for synthesising crucial sensors, positioning Cu-MOF as a favourable contender for practical sensing processes. The order of sensitivity of Cu-MOF for various solvents investigated was as follows: $CH_3OH < CH_3CN <$ DMF $< CH_3CH_2OH <$ propanol $<$ acetone (Kaur et al. 2020).

18.5.12 Catalysis

Evaluation of the PET-derived MOFs' catalytic activity in organic transformations demonstrated their effectiveness in promoting various reactions with high efficiency and selectivity, highlighting their potential for sustainable catalysis. In this regard, Rahmani et al. conducted a study focusing on recycling BDC from waste PET bottles as a linker source through a basic hydrolysis technique. The tailored Cu-BDU MOF is cost-effective and fabricated at ambient conditions without using any additives. The catalytic activity of different Cu-BDC MOFs was evaluated for sodium borohydride-mediated reductive elimination of methyl orange from aqueous medium. Among them, some of the activated Cu-BDC samples exhibited superior catalytic performance, achieving 85% dye removal in just 5 min. This performance surpassed other catalysts employing precious metals like Co, Au, Ag, and Pt nanoparticles. The mechanism of dye reduction was proposed using UV–Vis spectra and mass spectrometry, indicating the evolution of hydrogen from $NaBH_4$, followed by azo fragment reduction in methyl orange. Additionally, the catalyst's recyclability was demonstrated over six cycles without any significant loss in catalytic activity. Despite a gradual decrease in reaction rate, the catalyst maintained stable performance, highlighting its potential for practical applications (Rahmani et al. 2018).The pressing issue of climate change, driven primarily by CO_2 emissions, necessitates the development of eco-friendly catalysts for carbon capture and utilization (CCU). Utilizing CO_2 in the synthesis of valuable chemicals, such as cyclic carbonates, holds promise for mitigating environmental impact. Environmentally sustainable MOF catalysts, synthesized from recycled plastics, offer a cost-effective solution for catalyzing CO_2 conversion with epoxides, contributing to CCU efforts. In this context, the five nanoporous trivalent MOFs—MIL-47, MIL-53(Cr), MIL-53(Al), MIL-53(Ga), and MIL-101(Cr)

MOFs reported by Lo et al. were employed for the sorption of CO_2 and H_2 gases have also examined the catalytic potential of MIL-47 and MIL-101(Cr) for chemical fixation of CO_2 by converting CO_2 and epoxides into cyclic carbonates (Ren et al. 2016). MIL-47 was tested for its catalytic activity towards the coupling of propylene oxide with CO_2 to form propylene carbonate. The findings demonstrated the outstanding catalytic performance of MIL-47, achieving a maximum transformation (95%)in mild conditions. Similarly, MIL-101(Cr) was examined in the coupling of cyclohexene oxide and CO_2 to produce cyclohexene carbonate. The MIL-101(Cr)/tetra-n-butylammonium bromide catalytic system demonstrated high productivity, with over 90% CHO conversion and a cis-CHC yield of over 99%. Comparative analysis with other MOF/n-Bu_4NBr catalytic systems indicated the superior performance of MIL-101(Cr)/*tetra*-n-butylammonium bromide in smooth conditions, yielding better CHC yields compared to literature reports. Reboul et al. present a novel study showcasing the feasibility of utilizing recycled PET as an organic ligand to synthesize Ni(0)–Al_2O_3 catalysts derived from MOFs. This innovative approach aligns with the principles of the circular economy, offering a sustainable solution to plastic waste accumulation. The study demonstrates that MOFs synthesized from recycled PET exhibit physicochemical properties and synthesis yields equivalent to those from commercial BDC, ensuring no compromise in material quality. The resulting Ni(0)–Al_2O_3 catalyst, derived from PET-based MIL-53(Al), features uniform Ni(0) nanoparticles dispersed on alumina lamellae, delivering performance comparable to catalysts derived from commercial sources. Moreover, the Ni(0)–Al_2O_3 catalyst exhibits excellent catalytic performance in the reaction of dry reforming of methane (DRM), comparable to its commercial counterpart, highlighting the transformative potential of waste-derived materials in environmentally friendly catalysis and sustainable practices (Karam et al. 2021). Yarahmadi et al. present a novel approach by employing $ZnCl_2$:Urea catalysts for degrading PET-waste plastic bottles, yielding BDC for the construction of heterobimetallic MOF. The resulting Cu–Zn(BDC)-MOF showcased exceptional efficacy (>95% degradation in 6–12 min) in catalyzing the reduction of organic pollutant dyes (OPDs), with enhanced activity correlated to increased copper content. Utilizing plastic waste for sustainable catalyst creation, the high surface area of the catalyst, optimized through copper incorporation, proved effective for OPD degradation. Efficient catalyst recycling, along with stability and performance, underscores its potential for environmental remediation. The study highlights the importance of utilizing waste materials for making sustainable catalysts and offers a promising direction for environmental remediation, particularly in the context of wastewater purification contaminated with organic pollutant dyes (Yarahmadi et al. 2023). The research group of Farshchi have devised a distinctive solvent-free sustainable method for generating Cu-based MOFs employing a mechano-synthesis methodology using recycled PET-derived BDC as an organic spacer. The resultant MOFs serve as durable catalysts for the reduction of p-nitrophenol to p-aminophenol. The amino group incorporated Cu-MOF possesses higher adsorption and reduction efficiency than plain Cu-MOF CuBDC. The technique of using plastic waste-sourced MOF for the catalytic valorization of harmful substances offers a viable alternative for the long-term recycling and reusing of waste plastics. It also provides valuable insights into the environmentally friendly treatment of pollutants in aquatic environments (Farshchi et al. 2023). Furthermore, Enizi and co-workers successfully recycled PET trash bottles to create nitrogen-loaded mesoporous carbon-tailored ZnO nanocomposites employing a solvothermal process at lower temperatures. The synthesized nanocomposite has a large specific surface area and mesoporosity, enabling several active spots and effective charge transfer for H_2 and O_2 evolution processes during electrocatalysis (Ubaidullah et al. 2020).

18.5.13 Energy and Super Capacitor Applications

Waste PET-derived MOFs furnish a long-term solution to energy storage challenges. These novel materials, made from recycled plastic bottles, feature substantial specific surface areas and porosity distributions, which renders them suitable for improving energy storage capacity. PET-derived MOFs offer a substantial contribution to ecologically friendly solutions by proficiently preserving and transferring energy. Given these circumstances, economically viable ZnO and CO_3O_4-based MOFs have been developed directly from waste PET plastics via a solvothermal technique. The as-generated nanocomposites were integrated into mesoporous carbon (ZnO@MC and CO_3O_4@MC). These composites had exceptionally high specific surface surfaces. The symmetrical supercapacitors created using these composites displayed high capacitive efficacy and maximal specific capacitance. The CO_3O_4@MC composite demonstrated an impressive energy density and power density with excellent stability, displaying just ~5% loss after five thousand cycles in GCD cycles. The composites' outstanding redox activity may be attributed to aspects such as the large surface area of the CO_3O_4@MC composite, the existence of mesoporous carbon-incorporated metal-oxide nanoparticles, an advantageous Faradaic reaction in CO_3O_4@MC, and enhanced mechanical and electrical interactions. These findings highlight the composites' potential as high-performance electrode materials for symmetrical supercapacitors, providing intriguing opportunities for next-generation energy-storing devices (Al-Enizi et al. 2020a). The same research team demonstrated nickel oxide nanoparticles sustained nitrogenous porous carbon using PET-derived MOFs. Nitrogenous porous carbon improves electrical mobility, durability, and charge transfer kinetics, leading to greater supercapacitor performance. The composite has excellent specific surface regions and a high specific capacitance. Furthermore, the symmetric supercapacitor using the NiOx@NPC composite demonstrated sufficient specific capacitance with outstanding cyclic stability after five thousand cycles (Al-Enizi et al. 2020b). They also created nitrogen-doped mesoporous carbon (N-MC) with a large surface area leveraging BDC derived from trashed PET plastics via depolymerization. While evaluated in supercapacitors with a three-electrode setup and KOH electrolyte, the generated N-MC revealed outstanding electrochemical and energy-storing capabilities. Electrochemical analysis reveals enhanced specific capacitance and galvanostatic charge–discharge. Moreover, the N-MC electrode exhibits remarkable stability, suggesting its potential for environmental remediation and its suitability as a cost-effective electrode material for supercapacitor applications (Ubaidullah et al. 2021). An economical bimetallic $NiCO_2O_4$@nitrogen-doped carbon nanocomposite was constructed by Alhokbany et al. using bimetallic nickel/cobalt MOF, leveraging waste PET bottles for energy storage uses. The tailored $NiCO_2O_4$@NC exhibited good surface area, mesoporosity, and redox activity and demonstrated excellent specific capacitances. The analysis of cyclic stability revealed an outstanding retention capacity of 99% over 50 cyclic segments. The sustainable and low-cost synthesis approach, coupled with the outstanding electrochemical performance of $NiCOO_4$@NC, makes it a promising electrode material for energy storage devices (Alhokbany et al. 2020).The research team of Sun has repurposed PET waste bottles to fabricate CONi-MOF for application as Li battery anode components through solvothermal technique under lower temperatures. The synthesis process is straightforward, achieving nearly 100% conversion of the source material. Evaluation of MOFs using different Ni: Co proportions revealed that the $CO_{0.8}$Ni-MOF exhibited exceptional crystalline nature, purity, and redox activities and sustained a significant Coulombic efficiency even after 200 cycles. Therefore, owing to its environmental and economic advantages, the bimetallic PET-derived MOF emerges as a viable anode component for lithium-ion batteries (Wang et al. 2022).Waste PET bottles were successfully converted into functional MOF materials, Cu-MOF, Zr-MOF, and Ti-MOF, using a hydrothermal method. Cu-MOF demonstrated superior electrochemical performance with a capacitance of 104.8 F/g at 0.5 A/g and 78.4% diffusion contribution. Symmetrical supercapacitor devices based on these PET-derived MOFs showed promising results, particularly the Cu-MOF//Cu-MOF configuration, which exhibited excellent energy density and cyclic stability even after 10,000 runs (Dubey et al. 2023). Li et al. have established a simple and economical hydrothermal reaction to efficiently recycle discarded PET bottles into MIL-53(Al) with a 100% conversion rate, followed by the assembly of MOF-derived porous carbon into electrodes for highly efficient supercapacitors. The synthesized carbon exhibited a high specific surface area with an exclusive accordion-like architecture featuring hierarchical pores. The constructed tri-electrode supercapacitor showcased impressive specific capacitances. In addition, the formed zinc ion capacitor demonstrated exceptional capacitance, cycling stability, retaining 92% capacitance even upon 10,000 runs, and an ultra-high energy density (Li et al. 2023). The trend towards upcycling waste into valuable chemicals has become increasingly prominent, signalling a shift towards a circular economy aimed at reducing waste and conserving resources. This transition holds crucial significance in addressing global challenges related to resource depletion and waste management. In line with this, Mobin and co-workers have synthesized a Fe-based metal–organic framework material (Fe-BDC(W)) entirely from waste materials. By upcycling rust and waste PET plastic bottles to obtain the Fe salt and BDC linker, respectively, sustainable energy storage solutions are pursued, aligning with environmentally friendly and economically feasible energy storage technologies. The resulting MOF has demonstrated promising performance as an active material for supercapacitors, achieving a specific capacitance of 752 F/g at 4 A/g, comparable to MOFs produced from commercially available chemicals, such as Fe-BDC(C) (Deka et al. 2023). The research group of

Enizi has recently reported the fabrication of zirconium-metal Organic Frameworks (Zr-MOF) using PET bottle waste via the solvothermal method. The Zr-MOF exhibited a high specific surface area of 927.567 m^2/g with mesoporous characteristics. Electrochemical analysis revealed a specific capacitance of 822 F/g at a low scan rate of 2 mV/s and 890 F/g at a current density of 0.5 A/g in a 3-electrode system, with 92% capacitance retention after 5000 charge–discharge cycles. This research underscores the potential of waste PET bottles in fabricating functional Zr-MOF for advancing next-generation supercapacitors and environmental remediation efforts (Al-Enizi et al. 2024).

18.6 Conclusions and Future Prospects: Unearthing Trash PET Wastes to MOF Treasures Transformation

In recent decades, significant strides have been made towards achieving a more sustainable fabrication of MOFs by utilizing several waste materials and by-products as potential sources. The book chapter explores the feasibility of using PET waste as a source for BDC linkers, alongside methods for extracting BDC from PET and the potential for reacting various metals with in-situ produced BDC and their applications. Combining PET wastes as acid linkers offers promise for producing more ecologically sound MOFs, albeit with existing limitations. Challenges persist, requiring further investigations, product advancement, and industry collaboration for the commercialization of waste-derived MOF materials. These challenges include optimizing waste depolymerization to achieve high efficiency, ensuring scalability of waste-to-MOF synthesis methods, and surmounting limitations such as poor crystalline nature and irregular architecture often encountered in waste-derived MOFs. While techniques for MOF fabrication from various inorganic and inorganic as well as commercial/produce synthons are well-established, exploration of MOFs from waste materials remains limited. Currently, only three polymers, polyethylene terephthalate (PET), polylactic acid (PLA), and polybutylene terephthalate (PBT), have been utilized in this endeavour. Despite these challenges, significant progress has been made in using PET waste materials for MOF synthesis, marking an important transition towards a more sustainable future. Harnessing organic ligands sourced from waste has shown potential in yielding highly effective MOFs with improved activity and capability, attributed to the existence of defects and a greater number of active sites within their crystalline framework, which can augment pore dimensions and volume, thereby impacting adsorption characteristics and interactions between host and guests, paving the way for industrial-scale applications of MOF materials.

Additionally, exploring greener alternatives to toxic solvents and the commercial viability of MOFs derived from waste is imperative. As advancements in synthetic methodologies for pilot production of MOFs progress, there is a growing demand for inexpensive synthesis routes utilizing renewable and sustainable precursors in industrial applications. Emphasizing rational experimental design in waste depolymerization to achieve nearly 100% efficiency is crucial for scaling up operations. Additionally, the appropriate selection between direct utilization of polymers as linker sources or their depolymerized counterparts can enhance the quality and purity of the resulting MOFs. To conclude, concerted efforts must focus on maximizing the efficient utilization of waste materials to generate valuable products, aligning with the principles of a circular green economy for cleaner waste management and valorization.

Acknowledgements The author, Dr T. Sathiya Kamatchi, gratefully acknowledges the Department of Science and Technology (DST)-INSPIRE, Government of India, for providing financial support and for DST-INSPIRE Faculty Fellowship (DST/INSPIRE/04/2022/002322).

References

Alabi OA, Ologbonjaye KI, Awosolu O, Alalade OE (2019) Public and environmental health effects of plastic wastes disposal: a review. J Toxicol Risk Assess 5(021):1–3

Alavi MA, Morsali A (2014) Synthesis and characterization of different nanostructured copper (II) metal–organic frameworks by a ligand functionalization and modulation method. Cryst Eng Comm 16(11):2246–2250

Al-Enizi AM, Ahmed J, Ubaidullah M, Shaikh SF, Ahamad T, Naushad M, Zheng G (2020a) Utilization of waste polyethylene terephthalate bottles to develop metal-organic frameworks for energy applications: a clean and feasible approach. J Clean Prod 248:119251

Al-Enizi AM, Ubaidullah M, Ahmed J, Ahamad T, Ahmad T, Shaikh SF, Naushad M (2020b) Synthesis of NiOx@ NPC composite for high-performance supercapacitor via waste PET plastic-derived Ni-MOF. Compos B: Eng 183:107655

Al-Enizi AM, Nafady A, Alanazi NB, Abdulhameed MM, Shaikh SF (2024) Waste polyethylene terephthalate plastic derived Zr-MOF for high performance supercapacitor applications. Chemosphere 350:141080

Alhokbany N, Ahmed J, Ubaidullah M, Mutehri S, Khan MM, Ahamad T, Alshehri SM (2020) Cost-effective synthesis of $NiCO_2O_4$@ nitrogen-doped carbon nanocomposite using waste PET plastics for high-performance supercapacitor. J Mater Sci: Mater Electron 31:16701–16707

Aliakbari R, Ramakrishna S, Kowsari E, Marfavi Y, Cheshmeh ZA, Ajdari FB, Kiaei Z, Torkzaban H, Ershadi M (2022) Scalable preparation of MOFs and MOF-containing hybrid materials for use in sustainable refrigeration systems for a greener environment: a comprehensive review as well as technical and statistical analysis of patents. Res Chem Intermediat 48(7):2767–2819

Bailey T, Yang L, Humphreys E, Esat F, Douglas B, Hondow N (2024) The controlled microfluidic formation of stable mixed phase HCP/FCC-UiO-67 (Zr)-benzoic acid through modification of water concentration. J Porous Mater 31(1):267–279

Ban Y, Li Y, Liu X, Peng Y, Yang W (2013) Solvothermal synthesis of mixed-ligand metal–organic framework ZIF-78 with controllable size and morphology. Microporous Mesoporous Mater 173:29–36

Boukayouht K, Bazzi L, El Hankari S (2023) Sustainable synthesis of metal-organic frameworks and their derived materials from organic and inorganic wastes. Coord Chem Rev 478:214986

Campanelli JR, Cooper DG, Kamal MR (1994) Catalyzed hydrolysis of polyethylene terephthalate melts. J Appl Poly Sci 53(8):985–991

Dang YT, Hoang HT, Dong HC, Bui KB, Nguyen LH, Phan TB, Kawazoe Y, Doan TL (2020) Microwave-assisted synthesis of nano Hf-and Zr-based metal-organic frameworks for enhancement of curcumin adsorption. Microporous Mesoporous Mater 298:110064

Deka R, Mal DD, Mobin SM (2023) Upcycling rust and plastic waste into an Fe MOF for effective energy storage applications: transformation of trash to treasure. Dalton Trans 52(24):8204–8210

Denisov GL, Primakov PV, Korlyukov AA, Novikov VV, Nelyubina YV (2019) Solvothermal synthesis of the metal-organic framework MOF-5 in autoclaves prepared by 3D printing. Russ J Coord Chem 45:836–842

Doan VD, Do TL, Ho TMT, Le VT, Nguyen HT (2020) Utilization of waste plastic pet bottles to prepare copper-1, 4-benzenedicarboxylate metal-organic framework for methylene blue removal. Sep Sci Technol 55(3):444–455

Dubey P, Shrivastav V, Maheshwari PH, Hołdyński M, Krawczyńska A, Sundriyal S (2023) Comparative study of different metal-organic framework electrodes synthesized using waste PET bottles for supercapacitor applications. J Energy Storage 68:107828

Dutt K, Soni RK (2013) A review on synthesis of value added products from polyethylene terephthalate (PET) waste. Polym Sci Ser B 55(7):430–452

Dyosiba X, Ren J, Musyoka NM, Langmi HW, Mathe M, Onyango MS (2016) Preparation of value-added metal-organic frameworks (MOFs) using waste PET bottles as source of acid linker. Sus Mat Tech 10:10–13

Dyosiba X, Ren J, Musyoka NM, Langmi HW, Mathe M, Onyango MS (2019) Feasibility of varied polyethylene terephthalate wastes as a linker source in metal–organic framework UiO-66 (Zr) synthesis. Ind Eng Chem Res 58(36):17010–17016

El-Sayed ES, Yuan D (2020) Waste to MOFs: sustainable linker, metal, and solvent sources for value-added MOF synthesis and applications. Green Chem 22(13):4082–4104

Esrafili L, Gharib M, Morsali A, Retailleau P (2020) Rational morphology control of nano-scale amide decorated metal-organic frameworks by ultrasonic method: capability to selective and sensitive detection of nitro explosives. Ultrason Sonochem 66:105110

Farshchi ME, Bozorg NM, Ehsani A, Aghdasinia H, Chen Z, Rostamnia S, Ni BJ (2023) Green valorization of PET waste into functionalized Cu-MOF tailored to catalytic reduction of 4-nitrophenol. J Environ Manag 345:118842

Ghosh A, Das G (2021) Facile synthesis of Sn(II)-MOF using waste PET bottles as an organic precursor and its derivative SnO_2 NPs: role of surface charge reversal in adsorption of toxic ions. J Environ Chem Eng 9(4):105288

Gomes TS, Visconte LL, Pacheco EB (2019) Life cycle assessment of polyethylene terephthalate packaging: an overview. J Polym Environ 27:533–548

Jhung SH, Lee JH, Chang JS (2005) Microwave synthesis of a nanoporous hybrid material, chromium trimesate. Bull Korean Chem Soc 26(6):880–881

Jung KW, Kim JH, Choi JW (2020) Synthesis of magnetic porous carbon composite derived from metal-organic framework using recovered terephthalic acid from polyethylene terephthalate (PET) waste bottles as organic ligand and its potential as adsorbent for antibiotic tetracycline hydrochloride. Compos B: Eng 187:107867

Karam L, Miglio A, Specchia S, El Hassan N, Massiani P, Reboul J (2021) PET waste as organic linker source for the sustainable preparation of MOF-derived methane dry reforming catalysts. Mater Adv 2(8):2750–2758

Karayannidis GP, Chatziavgoustis AP, Achilias DS (2002) Poly (ethylene terephthalate) recycling and recovery of pure terephthalic acid by alkaline hydrolysis. Adv Poly Tech 21(4):250–259

Kaur R, Marwaha A, Chhabra VA, Kaushal K, Kim KH, Tripathi SK (2020) Facile synthesis of a Cu-based metal-organic framework from plastic waste and its application as a sensor for acetone. J Clean Prod 263:121492

Kumar S, Jain S, Nehra M, Dilbaghi N, Marrazza G, Kim KH (2020) Green synthesis of metal–organic frameworks: a state-of-the-art review of potential environmental and medical applications. Coord Chem Rev 420:213407

Li J, Zhang S, Hua Y, Lin Y, Wen X, Mijowska E, Tang T, Chen X, Ruoff, RS (2023) Facile synthesis of accordion-like porous carbon from waste PET bottles-based MIL-53 (Al) and its application for high-performance Zn-ion capacitor. Green Energy Environ

Lo SH, Raja DS, Chen CW, Kang YH, Chen JJ, Lin CH (2016) Waste polyethylene terephthalate (PET) materials as sustainable precursors for the synthesis of nanoporous MOFs, MIL-47, MIL-53 (Cr, Al, Ga) and MIL-101 (Cr). Dalton Trans 45(23):9565–9573

Ma X, Li Y, Huang A (2020) Synthesis of nano-sheets seeds for secondary growth of highly hydrogen permselective ZIF-95 membranes. J Membr Sci 597:117629

Ma D, Huang X, Zhang Y, Wang L, Wang B (2023) Metal-organic frameworks: synthetic methods for industrial production. Nano Res 16(5):7906–7925

Mancini SD, Zanin M (2007) Post consumer pet depolymerization by acid hydrolysis. Polym Plast Tech Eng 46(2):135–144

Mastropietro TF (2023) Metal-organic frameworks and plastic: an emerging synergic partnership. Sci Tech Adv Mat 24(1):2189890

Mishra S, Goje AS (2003) Chemical recycling, kinetics, and thermodynamics of alkaline depolymerization of waste poly (ethylene terephthalate)(PET). Polym React Eng 11(4):963–987

Mishra S, Goje AS, Zope VS (2003) Chemical recycling, kinetics, and thermodynamics of hydrolysis of poly (ethylene terephthalate)(PET) waste in sulfuric acid in presence of phosphoric acid. Polym PlastiTech Eng 42(4):581–603

Morel FL, Ranocchiari M, van Bokhoven JA (2014) Synthesis and characterization of phosphine-functionalized metal-organic frameworks based on MOF-5 and MIL-101 topologies. Ind Eng Chem Res 53(22):9120–9127

Moyses, DN, Teixeira DA, Waldow VA, Freire DM, Castro AM (2021) Fungal and enzymatic bio-depolymerization of waste post-consumer poly (ethylene terephthalate)(PET) bottles using Penicillium species. Biotech 10:435.

Nasrollahi M, Beynaghi A, Mohamady FM, Mozafari M (2020) Plastic packaging, recycling, and sustainable development. Responsible Consumption Prod 544–551

Natarajan S, Manual S, Mahata P, Rao VK, Ramaswamy P, Banerjee A, Paul AK, Ramya KV (2006) The use of hydrothermal methods in the synthesis of novel open-framework materials. J Chem Sci 118:525–536

Ni Z, Masel RI (2006) Rapid production of metal—organic frameworks via microwave-assisted solvothermal synthesis. J Am Chem Soc 128(38):12394–12395

Nyam TT, Ayeleru OO, Ramatsa IM, Olubambil PA (2024) Management strategies for plastic wastes: a roadmap toward circular economy and environmental sustainability. Plastic Waste Manage: Methods Appl 4:31–54

Onwucha CN, Ehi-Eromosele CO, Ajayi SO, Schaefer M, Indris S, Ehrenberg H (2023) Uncatalyzed neutral hydrolysis of waste PET bottles into pure terephthalic acid. Ind Eng Chem Res 62(16):6378–6385

Pichon A, Lazuen-Garay A, James SL (2006) Solvent-free synthesis of a microporous metal–organic framework. Cryst Eng Comm 8(3):211–214

Pirzadeh K, Ghoreyshi AA, Rahimnejad M, Mohammadi M (2020) Optimization of electrochemically synthesized $Cu_3(BTC)_2$ by Taguchi method for CO_2/N_2 separation and data validation through artificial neural network modeling. Front Chem Sci Eng 14:233–247

Rahmani A, Rahmani H, Zonouzi A (2018) Cu (BDC) as a catalyst for rapid reduction of methyl orange: room temperature synthesis using recycled terephthalic acid. Chem Pap 72:449–455

Ren J, Dyosiba X, Musyoka NM, Langmi HW, North BC, Mathe M, Onyango MS (2016) Green synthesis of chromium-based metal-organic framework (Cr-MOF) from waste polyethylene terephthalate (PET) bottles for hydrogen storage applications. Int J Hydrogen Energy 41(40):18141–18146

Santos JC, Pramudya Y, Krstic M, Chen DH, Neumeier BL, Feldmann C, Wenzel W, Redel E (2020) Halogenated terephthalic acid "antenna effects" in lanthanide-SURMOF thin films. Appl Mater Inter 12(46):52166–52174

Sato O, Arai K, Shirai M (2006) Hydrolysis of poly (ethylene terephthalate) and poly (ethylene 2, 6-naphthalene dicarboxylate) using water at high temperature: effect of proton on low ethylene glycol yield. Cat Today 111(3–4):297–301

Seetharaj R, Vandana PV, Arya P, Mathew S (2019) Dependence of solvents, pH, molar ratio and temperature in tuning metal organic framework architecture. Arab J Chem 12(3):295–315

Sharifzadeh Z, Morsali A (2022) Amine-functionalized metal-organic frameworks: from synthetic design to scrutiny in application. Coord Chem Rev 459:214445

Singh S, Sharma S, Umar A, Jha M, Mehta SK, Kansal SK (2018) Nanocuboidal-shaped zirconium based metal organic framework for the enhanced adsorptive removal of nonsteroidal anti-inflammatory drug, ketorolac tromethamine, from aqueous phase. New J Chem 42(3):1921–1930

Singh A, Banerjee SL, Kumari K, Kundu PP (2022) Recent innovations in chemical recycling of polyethylene terephthalate waste: a circular economy approach toward sustainability. In: Handbook of solid waste management: sustainability through circular economy. Springer Nature Singapore, Singapore, pp 1149–1176

Ubaidullah M, Al-Enizi AM, Shaikh S, Ghanem MA, Mane RS (2020) Waste PET plastic derived ZnO@ NMC nanocomposite via MOF-5 construction for hydrogen and oxygen evolution reactions. J King Saud Univ Sci 32(4):2397–2405

Ubaidullah M, Al-Enizi AM, Ahamad T, Shaikh SF, Al-Abdrabalnabi MA, Samdani MS, Kumar D, Alam AM, Khan M (2021) Fabrication of highly porous N-doped mesoporous carbon using waste polyethylene terephthalate bottle-based MOF-5 for high performance supercapacitor. J Energy Storage 33:102125

Vallés-García C, Cabrero-Antonino M, Navalon S, Álvaro M, Dhakshinamoorthy A, Garcia H (2020) Nitro functionalized chromium terephthalate metal-organic framework as multifunctional solid acid for the synthesis of benzimidazoles. J Colloid Interface Sci 560:885–893

Wang Y, Wang H, Li S, Sun S (2022) Waste PET plastic-derived CoNi-based metal–organic framework as an anode for lithium-ion batteries. ACS Omega 7(39):35180–35190

Wu Y, Xu G, Liu W, Yang J, Wei F, Li L, Zhang W, Hu Q (2015) Postsynthetic modification of copper terephthalate metal-organic frameworks and their new application in preparation of samples containing heavy metal ions. Microporous Mesoporous Mater 210:110–115

Xu X, Li J, Dymerska A, Koh JJ, Min J, Liu S, Azadmanjiri J, Mijowska E (2023) MIL-53 (Al) assisted in upcycling plastic bottle waste into nitrogen-doped hierarchical porous carbon for high-performance supercapacitors. Chemosphere 340:139865

Yarahmadi H, Salamah SK, Kheimi M (2023) Synthesis of an efficient MOF catalyst for the degradation of OPDs using TPA derived from PET waste bottles. Sci Rep 13(1):19136

Yuan W, Garay AL, Pichon A, Clowes R, Wood CD, Cooper AI, James SL (2010) Study of the mechanochemical formation and resulting properties of an archetypal MOF: Cu_3 (BTC) 2 (BTC= 1, 3, 5-benzenetricarboxylate). Cryst Eng Comm 12(12):4063–4065

Yue S, Wang P, Yu B, Zhang T, Zhao Z, Li Y, Zhan S (2023) From plastic waste to treasure: selective upcycling through catalytic technologies. Adv Energy Mat 13(41):2302008

Zhang T, Wei JZ, Sun XJ, Zhao XJ, Tang HL, Yan H, Zhang FM (2020) Rapid synthesis of UiO-66 by means of electrochemical cathode method with electrochemical detection of 2, 4, 6-TCP. Inorg Chem Commun 111:107671

Zhang F, Chen S, Nie S, Luo J, Lin S, Wang Y, Yang H (2019) Waste PET as a reactant for lanthanide MOF synthesis and application in sensing of picric acid. Polymers 11(12):2015

Zhao Y, Wu H, Emge TJ, Gong Q, Nijem N, Chabal YJ, Kong L, Langreth DC, Liu H, Zeng H, Li J (2011) Enhancing gas adsorption and separation capacity through ligand functionalization of microporous metal–organic framework structures. Chem Eur J 17(18):5101–5109

Zhou L, Wang S, Chen Y, Serre C (2019) Direct synthesis of robust hcp UiO-66 (Zr) MOF using poly (ethylene terephthalate) waste as ligand source. Microporous Mesoporous Mater 290:109674

Chapter 19
Innovative Approaches in Microbial Bioremediation: Harnessing Synthetic Biology for Environmental Pollutants Removal

Aishwarya Vetrivel, Suganya Ilango, Ramathilaga Ariyamuthu, Gopi Devarajan, and T. G. Nithya

Abstract The advancement of industry and agriculture, combined with population growth, has raised numerous concerns within the scientific community regarding the state of environmental contamination and the scarcity of natural resources. A significant pollution issue has evolved from the ongoing xenobiotic and recalcitrant compound contamination of the environment. Due to their toxicity and inability to degrade, numerous contaminants, including heavy metals, polychlorinated biphenyls, plastics, and different agrochemicals, are prevalent in the environment. Bioremediation holds a prominent role in removal of persistent pollutants from the environment. Novel bioremediation technologies are harnessed as the conventional methods of bioremediation have limitations. Recently, microbial biomass has been emphasized as an alternate approach for decontaminating environment compartments due to its distinct qualities that can be used for human advantage. Microbial bioremediation is an environmental friendly method to reduce environmental contamination brought on by various contaminants, including heavy metals, organic pollutants, and oil spills. The use of microorganisms to remove, convert, or alter environmental pollutants into harmless substances from various environmental matrices is now developing into a promising technique, attracting interest from both the scientific community and entrepreneurs. The ability to reduce environmental toxins is being made possible by the ongoing development of novel microbial bioremediation techniques. These developments have the potential to alleviate pollution issues

A. Vetrivel (✉) · R. Ariyamuthu
Department of Biotechnology, Dr. G.R. Damodaran College of Science, Coimbatore 641014, India
e-mail: aishwarya.v@grd.edu.in

S. Ilango · T. G. Nithya
Department of Biochemistry, Faculty of Science and Humanities, SRM Institute of Science and Technology, Kattankulathur, Chennai 603203, India

G. Devarajan
Department of Economics, Faculty of Science and Humanities, SRM Institute of Science and Technology, Kattankulathur, Chennai 603203, India

© The Author(s), under exclusive license to Springer Nature Switzerland AG 2025
A. Jeyaseelan and K. Murugasen (eds.), *Sustainable Environmental Remediation: Avenues in Nano and Biotechnology*, Environmental Science and Engineering,
https://doi.org/10.1007/978-3-031-78483-5_19

while fostering ecological resilience and environmental sustainability. Hence, this chapter aims to give a brief summary on the most recent approaches and developments in microbial bioremediation methods that are intended to lessen the harm that the pollutants do to ecosystems.

Keywords Bioremediation · Pollutants · Microbial biomass · Toxins · Environment · Ecofriendly

19.1 Introduction

The state of the environment as a whole is intrinsically tied to the standard of life on Earth. Waste is being generated and dumped into the environment as a result of growing urbanisation, civilization, and industrial sector advancements. An approximate of one thousand new chemicals are thought to be created each year. According to reports from the Third World Network, over 450 million kilogrammes of toxins are released into the air and water worldwide. The issues related to polluted environment are becoming more and more prominent in many countries today. The majority of contaminated lands and water bodies are the result of previous industrial operations, when people were less aware than they are now of the negative effects that hazardous materials pose on human health and the environment. Global environmental pollution is estimated to have affected a significant number of environments, and in recent years, efforts have been made on a global scale to address many of these environments, either to mitigate the risk of harmful health or environmental effects resulting from pollution or to allow for the area to be redeveloped or restored for use (Luka et al. 2018).

One of the main causes of pollution in the soil and aquatic environment is the runoff from various industrial processes. Various kinds of heavy metals are discharged as effluents during the industrial production processes. For instance, antimony, chromium, and mercury have been linked to the wastewater from dye manufacturing companies (Methneni et al. 2021). The agricultural industry releases pollutants into the environment, such as aluminium, copper, zinc, nickel, lead, and arsenic, through the use of fertilisers, pesticides, and herbicides. Similarly, untreated pollutants from agri-food industry wastewaters that are dumped into river canals and other bodies of water have negative environmental effects. Additionally, crude oil contributes significantly to environmental pollution, especially when it leaks during transportation, gets vandalised in pipelines, or spills accidentally. Certain hazardous substances that are discharged into the surrounding environment during mining include lead, arsenic, cadmium, and copper (Liu et al. 2020).

Heavy metal-containing wastes are frequently dumped inappropriately into soil and water environments. If they are dumped into water bodies, they have the potential to kill fish and other aquatic life; if not, they are biomagnified and can cause long-term illnesses in both people and animals. All of these environmental pollutants must be cleaned up using physical, chemical, or biological techniques. Although the chemical and physical methods have been used for many years, they have certain disadvantages. For example, the chemical bioremediation process requires specialised equipment and an expert, and the physical bioremediation process is costly. Nevertheless, none of them have been able to totally reduce the harmful pollutants. The primary obstacles to successful implementation of in-practice remediation methods or techniques are the production of toxic by-products/residues or secondary environmental pollutants, poor cost-effective ratio, high processing costs, and excessive usage of harsh chemicals (Mahmood et al. 2021).

The most effective, economical, and environmentally benign technology for removing pollutants is bioremediation. Within this framework, the bioremediation of hazardous environmental pollutants using microbes provides a number of benefits, including high cost-effectiveness, broad mineralization, minimal or non-existent side pollution generation, and reaction efficiency. Both plants and microorganisms can be used in biological remediation; however, microbes are preferred because they are easier to manipulate compared to plants. Recently, a variety of organisms, including plants, algae, fungus, and bacteria, have contributed to the successful design of technology for the removal of environmental hazards. Hazardous toxins can be eliminated by a range of enzymes that are commonly produced by environmental microorganisms, which use these toxins as a substrate for growth and development. Microbial enzymes have the power to break down and remove dangerous environmental pollutants, transforming them into non-toxic forms through their catalytic reaction process. The most common types of microbial enzymes that are capable of breaking down harmful environmental pollutants are laccases, oxidoreductases, hydrolases, lipases, and oxygenases (Singh and Porwal 2023).

Furthermore, in-situ treatment is a crucial component of bioremediation technology as it reduces the possibility of toxins dispersing to other areas. In addition, pollution from pesticides and heavy metals has permeated every ecosystem in today's industrialised world, making it impossible to avoid. The persistent nature, recalcitrance, high toxicity, and biological enrichment of metals and pesticides have put the ecosystem's stability and the health of living things at risk (Singh and Porwal 2023). The overuse of agricultural inputs such as pesticides, fertilisers, and mulch has contaminated soils with heavy metals, causing heavy metal pollution, a significant environmental issue that lowers crop productivity and food quality. Majority of pesticides are made of organic materials; however some also contain pure minerals, organic–inorganic compounds, or heavy metals like zinc, copper, and cadmium. Unlike organic contaminants, metals are not degradable and can therefore linger in the environment for extended period of time. When metal concentrations are high, metals can adversely impact plant metabolism. It has been suggested that a variety of microbes can effectively and affordably replace heavy metals in soil and water (Ahirwar et al. 2016).

To date, several different approaches have been developed and applied to control microbial strains with the goal of accelerating the formation of environmental dominance. To summarize, these methods involve: (1) cloning and propagating particular microbial genes that have the ability to break down the target toxic compounds; (2) expressing target enzymes with particular catalytic potentials to perform strong and eco-friendly degradation; (3) expressing degradation genes for different types of pollutants; and (4) fusing protoplasts for the purpose of degradation. While substances-degrading microorganisms are present in the environment, situations with limited growth conditions for microbes may not always yield effective contaminant degradation. As a result, its crucial to create a range of bioremediation techniques by growing microorganisms with specific catalytic potentialities based on the contaminants found in the surrounding environment (Bilal and Iqbal 2020).

The synthetic group of chlorinated aromatic hydrocarbons known as polychlorinated biphenyls, or PCBs, is widely used worldwide. However, because of their recalcitrance, potential for accumulation, widespread pollution, and ability to cause cancer, the widespread use of PCBs is posing long-term and harmful risks to human health and the environment. In such a case, genetically modified microbial cultures have the potential to greatly increase PCB removal/mitigation. The development of genetic engineering methods has made it possible to create and use genetically altered organisms to decrease the amount of hazardous substances in the environment. Thus, implementing these techniques and making them more effective will have positive effects on society and the economy by lowering the risk of disease and manage environmental pollutants, as well as achieve greater ecological stability and a greener environment (Aldhafiri et al. 2018).

19.2 Fundamentals of Microbial Bioremediation

19.2.1 Principles of Bioremediation

The process of biologically degrading organic wastes under controlled circumstances to a safe level or below concentration limits set by regulatory bodies is known as bioremediation. The goal of bioremediation is to support their activity by providing the right amounts of nutrients and other substances necessary for their metabolism in order to break down and detoxify substances that are harmful to the environment and living organisms. In order to perform bioremediation, specific microbes that utilise pollutants such as oil, solvents, and pesticides as a source of food and energy must be encouraged to grow (Saleem et al. 2022).

Toxic pollutants are remedied using two different approaches: in-situ remediation, which involves bringing the biological agent to the contaminated site to begin the decontamination process, or promotion of native organisms by creating an environment that is conducive for their proliferation. The second method is called ex-situ, and it involves moving the contaminated area to another location for processing. Although there are numerous ways for the organism to control the detoxification process, the key idea is that microorganisms utilise toxic metals as a source of nutrition (Sun et al. 2020).

The main focus of microorganism-mediated bioremediation is the reduction of environmental pollutants' toxicity through the breakdown of hazardous materials into safe by-products through the activity of microbial enzymes. Environmental microbes are a preferred methodology for remediating contaminated sites because their application is guided by factors such as pH, temperature, oxygen, soil character, moisture, appropriate level of nutrients, and bioavailability of contaminants. The identification of metal-resistant, metal-accumulating, or metal-transforming microorganisms as well as their molecular characterization have been made possible by advancements in microbiological techniques and biochemical assay methods. The specificity of how microbes interact with heavy metals have been revealed to us by molecular and genetic methods, and hence this interaction is very beneficial for successful environmental remediation efforts (Mukherjee et al. 2020).

19.2.2 Microbes Used in Bioremediation

The role of microorganisms in food chains contributes to the preservation of biological equilibrium. In the presence of hazardous substances or any waste stream, microbes can grow at temperatures as low as -196 degrees Fahrenheit and as high as 1200 degrees Fahrenheit. Because of their nature of biological systems and adaptability, microorganisms are the ideal choice for cleanup. Carbon is the most important nutrient for bacteria. Microbes from a variety of habitats, such as *Nitrosomonas, Xanthobacter, Arthrobacter, Pseudomonas, Bacillus, Mycobacterium, Corynebacterium, Flavobacterium*, and other bacteria, can be used for bioremediation (Hussain et al. 2022).

Numerous microorganisms can bioremediate different environmental contaminants when they are in an aerobic environment. Numerous complex organic materials can be broken down by aerobic bacteria such as *Bacillus, Pseudomonas, Sphingomonas, Flavobacterium, Nocardia, Rhodococcus, and Mycobacterium.* It has been shown that these microorganisms degrade insecticides, alkane hydrocarbons, and polyaromatic chemicals. Many of these microbes use these toxins as a source of carbon and energy. Oxygen limits the growth of microorganisms during the aerobic bioremediation process. Amphibian bacteria are becoming more and more popular in the bioremediation of polychlorinated biphenyls, chlorine compounds, and the chlorinated solvents trichlorethylene and chloroform. Pollutants are broken down and changed into less hazardous forms by these bacteria. In the anaerobic bioremediation process, a variety of bacteria have been used, such as sulfate-reducing bacteria, *Aeromonas*, and *Pseudomonas* (Tegene and Tenkegna 2020).

19.2.3 Microbial Metabolism and Transformation of Pollutants

Toxic substances can be changed by microorganisms into less toxic substances like water, carbon dioxide, and other substances that are then further broken down by other microbes in a process known as mineralization. Being widely distributed in nature and able to take up carbon from a variety of substrates, microbes can be found in odd places with a broad spectrum of pollutants available for their absorption. They are also more efficient because they can thrive in strange environments. The psychrophiles, for instance, flourish in cold climates, the halophiles in saline regions, and the acidophiles in acidic environments (Kour et al. 2022).

Microbes assist in the process of mineralization by converting pollutants into final products like water and carbon dioxide or other intermediate metabolic materials. Comparably, immobilisation is the process of transforming substances into a state in which they will not exist in the environment. One example would be the transformation of nitrate nitrogen into organic nitrogen. Diverse microbes from various locations and environmental conditions are being studied globally to ensure the detoxification of environmental contaminants. According to a report, the wastewaters from the olive mill contained phenol-degrading yeasts, specifically *Saccharomyces cerevisiae, P. membranifaciens, Pichia holstii,* and *Candida boidinii* (Pratush et al. 2018).

Bacillus sp., a bacterium that breaks down petroleum, was discovered by Zhang et al. (2010) in soil contaminated with oil. *Trametes versicolor*, a white rot fungus, was mentioned as a polycyclic aromatic hydrocarbon (PAH) bioremediating agent in another study (Sayara et al. 2011). Three bacterial species namely *Achromobacter insolitus*, *Bacillus cereus*, and *Sphingobacterium* sp. isolated from the field of petrochemical refineries were reported to be effective at remediating PAHs in a report by Janbandhu and Fulekar (2011).

A report by Cerqueira et al. (2012) detailed the variety of bioremediating bacterial isolates from oily sludge-contaminated soil, including *Bacillus megaterium, Bacillus cibi, Bacillus cereus, Pseudomonas aeruginosa*, and *Stenotrophomonas acidaminiphila.* It has been reported that these strains are capable of breaking down both aromatic and aliphatic compounds. The bacteria identified by Syakti et al. (2013) was found to have the potential to bioremediate the mangroves growing in soil contaminated with hydrocarbons. These bacteria includes *Bacillus aquimaris, B. megaterium*, and *B. pumilus,* as well as *Flexibacteraceae, Rhodobacteraceae*, and *Halobacillus trueperi*. In another report by Roy et al. (2014), bacteria that break down crude oil are classified into the following genera: *Achromobacter, Alcaligenes, Bacillus, Brevibacillus, Delftia, Lysinibacillus, Paenibacillus, Pseudomonas,* and *Stenotrophomonas.*

Pseudomonas sp. was isolated from the soil of a petroleum refinery during an investigation by Goudarztalejerdi et al. (2015), and the strains were found to be capable of breaking down hydrocarbons. From soil contaminated with PAHs, Godoy et al. (2016) isolated fungal species capable of bioremediating xenobiotics. The isolates of fungus isolated were identified as *Scopulariopsis brevicaulis* and *Fomes sp*. According to an investigation by Sarkar et al. (2017), a variety of bacterial strains from the genera *Bacillus, Burkholderia, Enterobacter, Kocuria, Pandoraea*, and *Pseudomonas* that degrade hydrocarbons were found in petroleum refinery waste.

There have been reports of the microbial community from the genera *Shinella, Microbacterium, Micrococcus*, and *Bacillus* bioremediating environmental pollutants containing heavy metals (cadmium, chromium, cobalt, nickel, and zinc) (Bhakat et al. 2019). *Pseudomonas nitroreducens* and *Ochrobactrum* sp. were identified as pesticide-degrading agents in another report (Góngora-Echeverría et al. 2020). *Yarrowia* sp., *Barnettozyma californica,* and *Sterigmatomyces halophilus* are the yeast cultures that Ali et al. (2021) have reported as being capable of bioremediating textile Red HE3B dye.

19.2.4 Factors Influencing Microbial Bioremediation

(A) **Environmental Factors**

A multitude of environmental conditions strongly influence microbial bioremediation. Important factors include the kind of contamination, its concentration, and the chemical and physical characteristics of the surrounding environment. Temperature, pH, moisture content, and other variables can have a big impact on microbial activity and how well bioremediation systems work. Optimizing these parameters is critical to maximize the metabolic activities and remediation capabilities of certain microbes that are adapted to particular environmental conditions.

(B) **Microbial Community Composition**

The microbial community's makeup has a significant impact on the outcome of bioremediation. Microorganisms differ in their metabolic capacities and their inclinations towards particular pollutants. The effective degradation of various contaminants requires the selection of a suitable microbial consortia. Effective bioremediation depends on knowing the microbial ecology of the contaminated location, encouraging the growth of native microorganisms, or introducing exogenous strains with appropriate metabolic pathways (Sokal et al. 2022).

(C) **Contaminant Characteristics**

The kind of contaminants has a big impact on microbial bioremediation techniques. The selection of microorganisms and bioremediation strategies are influenced by various factors, including the chemical structure, toxicity, and persistence of pollutants. While some bacteria may show broad substrate specificity, others may have evolved specialized enzyme systems to break down specific contaminants. A focused and effective remediation process is ensured by customizing bioremediation techniques to the unique properties of pollutants (Das and Dash 2019).

(D) **Nutrient Availability**

Access to vital nutrients like carbon, phosphorous, nitrogen, and trace elements is critical for microbial development and activity. During bioremediation, nutrient amendment or supplementation is frequently needed to increase microbial biomass and metabolic activities. Nutrient constraints that could reduce the effectiveness of remediation can be avoided by supplying co-substrates and balancing nutrient ratios, which will eventually encourage microbial breakdown of pollutants (Mounier-Kuhn 2012).

(E) **Oxygen Levels and Redox Conditions**

Microbial metabolism is significantly impacted by the environment's redox conditions and oxygen availability. Anaerobic bacteria are acclimated to low or no oxygen, whereas aerobic microbes flourish in oxygen-rich settings. It is essential to comprehend the redox potential of the contaminated site in order to choose the right microbial species and create bioremediation plans that are compatible with the current oxygen levels. The processes by which pollutants convert and degrade can be affected by reducing or oxidizing environments (Li et al. 2021).

19.3 Applications of Synthetic Biology in Microbial Bioremediation

A novel and exciting strategy for remediating pollution is to tailor microbial strains, which takes advantage of microorganism's special capacity to break down or alter environmental contaminants. The intentional alteration of microbial strain's genetic composition to improve particular characteristics or add new capabilities is known as customisation, and the end goal is to maximize the strain's ability to degrade pollutants. Customizing microbial strains for pollution cleanup involves genetically engineered bacteria, fungus, or algae to express enzymes that specifically target and degrade contaminants. One example is the successful engineering of bacteria by researchers to develop enzymes that can break down hydrocarbons present in oil spills. Scientists can improve the efficiency of microorganisms metabolization and removal of contaminants from the environment by adding genes that encode for these enzymes. This strategy works especially well at polluted places where more conventional cleanup techniques could be too costly or unfeasible (Ahmad et al. 2023).

Additionally, the process of customization may entail the creation of microbial consortia, which are collections of various microorganisms that collaborate to remove toxins from the environment. By providing a unique combination of enzymes or metabolic pathways, each consortium member can enhance the system's resilience and adaptability in terms of pollutant degradation. This tactic resembles the microbial communities that naturally exist in ecosystems, where different species work together to preserve ecological equilibrium. Through meticulous selection and engineering of microbial strains possessing complementary capabilities, scientists can create consortia that exhibit exceptional efficacy across a range of contamination conditions (Che and Men 2019).

To ensure the success of customized microbial strains in pollution remediation, researchers must also consider factors such as environmental conditions, nutrient availability, and microbial competition. Microorganisms may function optimally within specific temperature ranges, pH levels, and nutrient concentrations. Therefore, tailoring the strains to thrive in the target environment is crucial for their sustained activity and effectiveness. Additionally, understanding the ecological interactions between engineered strains and native microbial communities is essential to prevent unintended consequences or the displacement of indigenous species (Ahmad et al. 2023).

Another strategy that shows promise for waste management, bioenergy production, and environmental remediation involves engineering metabolic pathways for improved degradation. Cells use complex networks of biological events called metabolic pathways to change substrates into products. Researchers can optimize the breakdown of particular molecules, like waste materials or pollutants, by adjusting these pathways, which can result in more effective and environmental friendly solutions. Optimizing the movement of substrate within cells is another different approach that could be followed to enhance the bioremediation processes. Target compounds must be efficiently taken up by the cell in order to be degraded later. To improve the overall efficiency of substrate uptake, genetic changes can be made to increase the expression of transport proteins or introduce new transport systems. In the context of producing bioenergy, this phase is very important because it involves effectively using biomass as a substrate for metabolic processes (Singh et al. 2023).

Achieving accelerated degradation depends critically on the management of metabolic pathways. Enzymes and other pathway constituents can have their expression levels precisely adjusted with the use of synthetic biology technologies. This makes it possible to precisely regulate the breakdown process and avoid metabolic imbalances or undesirable byproducts. Cutting-edge methods such as CRISPR-Cas9 allow precise and targeted genome changes, making it easier to create artificial microbial strains with customized metabolic capacities. In case of targeted contaminants removal, development of tuneable microbes is a crucial component. Microorganisms with particular genetic pathways that allow them to digest and convert pollutants into innocuous byproducts can be engineered by scientists. With this method, microbial communities can be specifically designed and tuned to remove particular pollutants (Das et al. 2023).

The creation of sensor-responsive systems is another exciting direction in synthetic biology for targeted pollutant removal. Real-time pollutant detection is possible with sensors added to engineered microbes. These microbial systems detect contaminants in the environment and then trigger particular biological mechanisms for quick and focused degradation. This allows for an effective reaction to pollution on-demand. The development of designer enzymes with increased catalytic activity for the degradation of contaminants is also made possible by synthetic biology. Engineered enzymes provide a precise and effective method of bioremediation by being able to target particular contaminants and expedite their destruction. This strategy encourages ecologically friendly activities while reducing unintentional harm to non-target substances. The precision and efficiency of synthetic biology based approaches have been greatly improved by the development of gene editing technologies like CRISPR-Cas9. The genetic codes of microorganisms can now be precisely altered by researchers to maximize their capacity to remove contaminants. Microbial systems can be fine-tuned to operate better in complex environmental situations to this degree of genetic control.

Though synthetic biology has great potential for targeted pollution clearance, it presents certain difficulties and ethical issues. Careful consideration must be given to potential ecological effects, unforeseen repercussions, and the release of genetically modified organisms into the environment. Strong regulatory frameworks are necessary to enable the responsible and sustainable deployment of synthetic biology in environmental applications, as ethical concerns surround this field (Das et al. 2023).

19.4 Bioremediation Techniques

Diverse bioremediation methodologies have been devised to tackle distinct categories of contaminants and environmental circumstances. Some of them are as follows:

(A) **Microbial Bioremediation**

Microbial bioremediation uses the metabolic processes of bacteria, algae, and fungi among other microorganisms to degrade contaminants. In situ bioremediation entails direct introduction into the contaminated site using nutrient additions to promote the development and activity of the chosen microorganisms. On the other hand, ex situ bioremediation entails removing contaminated material so that it can be treated under controlled conditions. Two subcategories namely bio stimulation and bioaugmentation are concerned with importing specialized microorganisms or stimulating pre-existing microbial communities (Chandran et al. 2020).

(B) **Phytoremediation**

The process of using plants to eliminate, break down, or stabilize pollutants is known as phytoremediation. Pollutants from the soil or water can be absorbed, accumulated, and translocated by plants into their tissues. Hyperaccumulator plants are very useful for heavy metal remediation as they can accumulate large concentrations of metals. Phytoextraction entails the extraction of toxins by the plant roots whereas phytostabilization seeks to immobilize pollutants in the soil and lessen their mobility (Peer et al. 2005).

(C) **Mycoremediation**

Fumigation is a technique used in mycoremediation to clean up polluted areas. Pesticides and other complicated chemicals can be broken down by fungi, which can also absorb and accumulate heavy metals. Some species of mushrooms are capable of breaking down organic contaminants. Mycoremediation is a flexible solution that may be applied in various environmental settings and serve to be promising approach for treatment of various pollutants (Yadav et al. 2021).

(D) **Bioaugmentation**

In order to improve the capacity of the current microbial population to breakdown contaminants, bioaugmentation entails the introduction of particular microorganisms or microbial consortia into the contaminated environment. This method is frequently combined with other bioremediation strategies to target certain contaminants that may be difficult for naturally occurring microbial communities to effectively degrade (Muter 2023).

(E) **Bioventing and Biosparging**

The in situ methods of bioventing and biosparging are mostly used to clean up locations contaminated by petroleum hydrocarbons. By adding air to the soil, bioventing promotes microbial activity and accelerates the breakdown of pollutants by nature. In order to encourage aerobic microbial activity and speed up the decomposition of petroleum hydrocarbons, biosparging introduces air or oxygen into groundwater (Yadav et al. 2021).

Thus, for cleaning up the environment, bioremediation techniques provide adaptable and long-lasting solutions. Depending on the kind of contamination, the surrounding environment, and the precise objectives of remediation activities, the best technique can be chosen. The efficacy of bioremediation in resolving various pollution concerns is still attributed to ongoing research and technological improvements in the field.

19.5 Genetic Engineering in Microbial Bioremediation

The cutting-edge science of genetic engineering in microbial bioremediation uses genetic alteration of microorganisms to improve their remediation capacity from environmental contaminants. Microorganisms, including fungus and bacteria, are desirable alternatives for bioremediation since they are essential in the breakdown and detoxification of different types of pollutants. By customizing these microorganisms to particular toxins through genetic engineering, their efficacy and adaptability in environmental remediation can be enhanced. A crucial component of genetic engineering in the context of microbial bioremediation entails altering the genetic composition of bacteria to augment their capacity for degradation. To do this, genes encoding for enzymes that specifically target contaminants can be introduced. For instance, adding genes that produce specialized enzymes such as aromatic ring-cleaving enzymes to break down hydrocarbons or dehalogenases to break down chlorinated substances (Kaur and Sodhi 2022).

It is possible to create microbial strains through genetic engineering that are more suited to the surroundings of a contaminated location. This can involve changing pH preferences, strengthening resilience to high pollution levels, or strengthening tolerance to extreme temperatures. Through the refinement of these traits, engineered microorganisms can flourish in a variety of settings, guaranteeing their efficacy in a range of bioremediation situations. Genetic engineering in microbial bioremediation is aided by developments in synthetic biology. With the application of synthetic biology, biologists can create completely new biological systems, such as microbial strains with genetic circuits specifically engineered to detect and react to external stimuli. These specially designed systems have the ability to adjust microbial activity in response to shifting environmental factors or contaminant concentrations (Thai et al. 2023).

In order to produce a synergistic impact, genetic engineering makes it easier to form microbial consortia or to enhance already-existing microbial communities. This entails creating microbial communities in which each member performs a distinct role, such as producing growth-promoting agents or degrading various contaminants. The overall bioremediation efficiency can be greatly increased by carefully designing these consortia. Regulation and ethical issues are raised by the use of genetic engineering in microbial bioremediation. Safe and appropriate use of genetically modified microbes in environmental remediation requires striking a balance between innovation and potential hazards. Establishing regulations that address the possible ecological impact and unintended consequences of releasing modified microbes into the environment requires collaboration between governments, regulatory agencies, and the scientific community. Application of genetic engineering to modify microorganisms to effectively break down toxins in microbial bioremediation offers enormous promise for solving environmental problems. To maximize the benefits of genetic engineering in microbial bioremediation while reducing hazards, a thorough grasp of the ecological ramifications and the development of strong regulatory frameworks are crucial as the area develops (Kaur and Sodhi 2022).

Future developments in genetic methods for microbial remediation hold great promise for revolutionizing environmental biotechnology. The creation and improvement of synthetic biology techniques to create microbial strains with improved remediation capabilities is a crucial path. In order to enable microbes to effectively breakdown a wide range of pollutants, including pollutants like hydrocarbons, heavy metals, and industrial chemicals, researchers are concentrating on creating synthetic gene circuits and pathways. The precise alteration of microbial genomes made possible by the application of CRISPR-Cas and other cutting-edge gene-editing technologies enables the development of specially adapted strains with enhanced metabolic pathways and resilience to a variety of environmental stresses (Li et al. 2023).

Furthermore, a systems-level understanding of microbial responses to pollutants is made possible by the integration of omics technologies, including transcriptomics, proteomics, and genomics. This aids in the identification of important genetic targets for manipulation. Remediation efficiency can also be increased by investigating microbial consortia and creating microbial communities for mutually beneficial interactions. When combined with high-throughput screening methods, the use of machine learning algorithms to anticipate and optimize genetic alterations speeds up the process of developing and choosing efficient microbial strains for remediation.

With all things considered, synthetic biology, gene editing, omics technologies, and computational tools will come together to form the future of genetic techniques in microbial remediation, offering a comprehensive and long-lasting means for tackling various environmental problems (Guan et al. 2017).

19.6 Innovative Technologies in Microbial Bioremediation

19.6.1 Nanotechnology Applications

Today's environmental conditions has been deteriorated by human activities. As a result, oceans, lands, atmosphere, agriculture fields and ground water are prone to contamination by crude oils, plastic trash, emission of greenhouse gases, pesticides etc. These pollutants cause a great risk to the environment. The treatment strategies to be employed for removing these pollutants should be strictly cost effective and eco-friendly. There are now more options for cleaning up environmental contaminants because of the advancements in nanotechnology and nanoscience. Using microorganisms in the production of nanoparticles serves as an efficient method for bioremediation. Properties of nanoparticles such as high surface area, chemical features and smaller size are the main reasons for its efficiency in bioremediation (Baruah et al. 2019). Nanomaterials with biological processes accelerate and promote the removal of toxic compounds from the medium.

19.6.2 Nanotechnology in Bioremediation

Wastewater treatment consist of the process such as adsorption, filtration, and photocatalytic degradation. Nanoparticles are mostly used to adsorb harmful organic and inorganic contaminants from industrial waste water. Nanomaterials are made up of carbon, metal or metal oxides. Fluoride ions from waste water can be removed by activated carbon modified nano-magnets. Iron oxide magnetic nanoparticles are used to remove heavy metals from waste water (Palit and Hussain 2020). Waste water treatment can benefit from the environmental friendly creation of nanoparticles using plant extract, microorganisms, and their enzymes. Bio fabrication of nano materials and nanoparticles create nanotechnology as the most effective method for bioremediation. Green nanoparticles act as reductive agents and form metallic nano particles. Using *Aspergillus* sp. biofabricated iron oxide nanoparticles, over 90% of the heavy metals in wastewater were eliminated. Green synthesized bacterial organophosphorus hydrolase nanocrystals was used to degrade the pollutant p-nitrophenol and acid orange 7. Nano sponge of organoclays employed to degrade Triclopyr in soil. Nanocomposites using *Arthrobacter globiformis* are used to degrade herbicides in water (Mahanty et al. 2020).

19.6.3 Advanced Monitoring and Assessment Tools

Bioremediation processes takes place in in situ and ex situ methods using various biological entities. Assessing the effectiveness of the bioremediation process is very important. It is monitored by physico chemical analyses, pollutant evaluation and byproducts estimation. Most important method is ecotoxicological bioassay on the aquatic organisms, microorganisms and plants. It depends on analysis cost, nature of the matrix and advanced method of assessment in molecular analyses such as nucleic acid sequencing, metagenomics and mutagenicity (Mahanty et al. 2020).

19.6.4 Remote Sensing and GIS in Bioremediation

The science and art of remote sensing is the use of electromagnetic radiation (EMR) to gather data, such as spectral, spatial, and temporal information about material items, areas, or phenomena, without physical touch. Information that may be linked to a specific location is gathered, stored, analyzed, distributed, and altered using geographic information systems, or GIS. GIS are utilized in wide range of areas including transportation, local government, business, socioeconomic and environmental issues, and effective short- and long-term decision making.

In environmental mapping, agriculture, forestry, geology, disaster mitigation, and environmental management, remote sensing and geographic information systems (GIS) are crucial tools. A key technique for gathering data on practically every area of the planet is remote sensing combined with geographic information systems. It is impossible to directly detect groundwater pollution from Earth's surface. Remote sensing & GIS are useful to get satellite data where the pollution occurs and also to get information about the effectiveness of bioremediation. Remote sensing play an important role in water remediation. UV photography, TIR scanners are used to analyse oil contamination in water bodies. Satellite thermal infrared imagery is effectively used for effluent monitoring. CIR photography evaluate the eutrophication in water bodies. CIR and B&W photography are used to analyse the pesticides level in agricultural runoff (Kumar et al. 2013).

19.7 Emerging Technologies and Tools

19.7.1 CRISPR-Based Approaches in Microbial Bioremediation

For the bio remediation process, microorganisms and its enzymes are used as biological agents to restore its original environment. The effectiveness of these biological agents was reduced by the pollutants and its compounds. Development of CRISPR/

Cas (Clustered Regularly Interspaced Short Palindromic Repeats and associated proteins) is a game changing genomic tool that allows to improve the specific characteristic features of the biological agents. It is easy, efficient and show low mutation rate. It is the most convenient and efficient gene editing technology. The use of microbial resources in bioremediation to reduce hazardous chemicals greatly depends on the genetic engineering of microorganisms through the use of gene editing technology (Manisha and Solanki 2023).

19.7.2 Next-Generation Sequencing in Monitoring Engineered Strains

In genome research, next-generation sequencing is a potent instrument. It is used to analyze DNA and RNA molecules in an economical and high-throughput way. It enables the quick sequencing of millions of DNA fragments at once, and offers comprehensive insights into the structure of the genome, genetic variants, gene expression profiles, and epigenetic changes. The invention of next-generation sequencing has increased the scope of genetic study across all biotechnology fields. The goal of next-generation sequencing technology is to improve data analysis techniques, sequencing data sensitivity and accuracy, and the search for future, more scalable, cost-effective solutions (Satam et al. 2023).

19.7.3 Integration of Artificial Intelligence in Bioremediation Strategies

Using naturally occurring microbes and other components of the natural environment, bioremediation is a technique for eliminating contaminants from the environment. Bioremediation is the process of treating contaminated media by modifying the surrounding environment to encourage the growth and breakdown of the target pollutants by plants or microorganisms. The fields of wastewater treatment and bioremediation have been revitalized by the introduction of artificial intelligence technologies. It allows the computer to be programmed to perform specific tasks utilizing mathematically sophisticated and applied techniques including genetic algorithms, artificial neural networks, and fuzzy logics, which are largely employed in bioremediation. The utilization of artificial intelligence in parameter optimization, surrogate model analysis, historical data analysis, and AI-based technologies to improve decision-making is crucial in determining the best remediation procedures. With open-data rules and data integration, AI has the potential to revolutionize the process of remediating hazardous sites (Janga et al. 2023).

19.8 Economic and Environmental Sustainability

Remedial techniques including physical and chemical means do not yield economic gains and does not eliminate all contaminants. Furthermore, both approaches run the risk of increasing pollution and upsetting the ecosystem, which would be harmful to both people and other biota at the contaminated locations. Thus, these techniques are not regarded as environmentally benign. In contrast to these techniques, bioremediation permanently removes toxins from the environment through the action of several kinds of living organisms (Bhansal and Jha 2023).

19.8.1 Cost–Benefit Analysis of Microbial Bioremediation

Cost–benefit analysis is a method of evaluation. The main method for economically assessing different options for soil and water remediation is cost–benefit analysis. CBA works on the straightforward tenet that an intervention is acceptable if its costs outweigh its benefits. The bioremediation projects are broad and it should be carefully planned and monitored. Bioremediation depends upon certain factors such as type of soil, amendments deployed, type of contaminants, and its severity and bioremediation methods followed to create eco sustainability (Guerra et al. 2018).

19.8.2 Future Perspectives and Emerging Trends

New and serious risks to water, air, soil, ecosystems, natural resources, and human health will be posed by emerging pollutants. Unplanned elimination and inappropriate handling of household, commercial, and farming wastes would pose a serious risk to the environment and public health. As a common method of cleaning up contaminated soil and wet regions, bioremediation has emerged. It is the ideal substitute for cleaning up the environment as it offers to do it affordably and with a lower likelihood of contaminating other media. In order to support the elimination of newly discovered toxins, further research and studies must be conducted in the future to create sustainable bioremediation techniques. The creation of biosensors could monitor the environment in real time and create cutting-edge instruments to determine a pollutant's toxicity and forecast microbial availability and breakdown in the environment (Bhansal and Jha 2023).

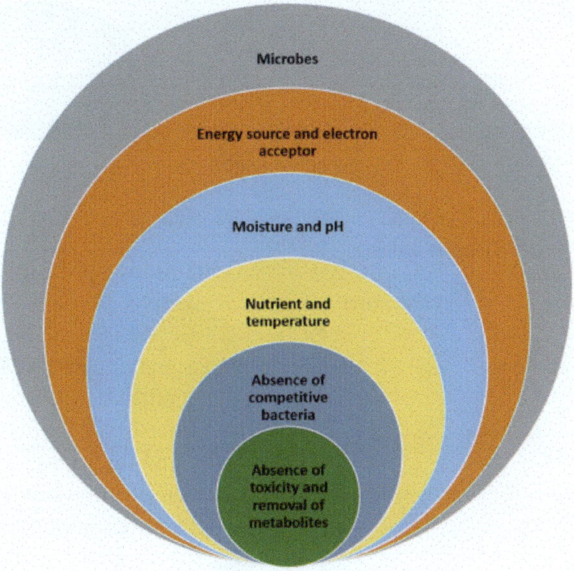

Fig. 19.1 Principles of bioremediation (Shilpi and Sharma 2012)

19.9 Conclusion

To conclude, the creative incorporation of synthetic biology into microbial bioremediation signifies a revolutionary development in the environmental science domain. The creation and design of microbial strains with improved capacities for the targeted removal of environmental pollutants is made possible by the application of synthetic biology technologies. By carefully modifying the genomes of microorganisms, scientists can enhance metabolic pathways and add new features, increasing the effectiveness and selectivity of bioremediation procedures. The combination of synthetic biology and microbial bioremediation exhibits significant potential in tackling the progressive intricate problems presented by environmental contaminants. With further study and development in this multidisciplinary field, we would be able to uncover new avenues and fully utilize synthetic biology for the efficient and sustainable elimination of environmental pollutants. Incorporating cutting-edge methods into microbial bioremediation not only demonstrates our capacity to adjust to and overcome environmental obstacles, but it also represents a proactive and revolutionary strategy to reduce the effects of human activity on the environment.

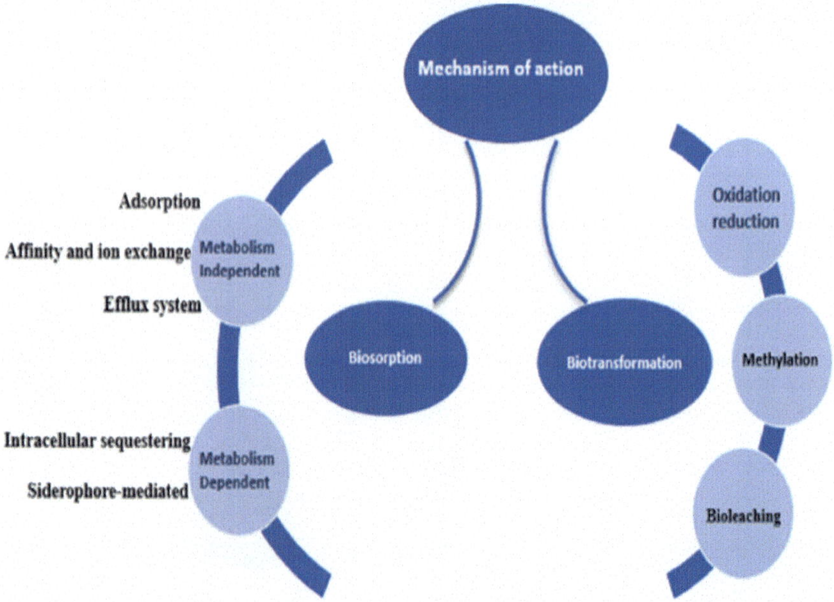

Fig. 19.2 Different mechanisms of bioremediation (Abo-Alkasem et al. 2023)

Table 19.1 Advanced monitoring tools to assess in situ bioremediation

Tool	Outline
Compound-specific isotope analysis (CSIA)	Analyze ^{13}C & ^{12} C isotopes
Quantitative polymerase chain reaction (qPCR)	Quantitative polymerase chain reaction (qPCR)
Fluorescent in situ hybridization (FISH)	Fluorescent dye is added to analyze the gene of interest
Microarrays	Identify the community base on 16S rRNA sequence
Stable-isotope probing (SIP)	Using isotope probing for byproduct analyses
Enzyme activity probes	Using enzymes to analyse the transformed product
Metagenomics	Get data on the genomes present in remediated site
Proteomics	Study of proteins or enzymes produced by microbial community

Source Kucharzyk and Rosansky (2018)

References

Abo-Alkasem MI, Hassan NH, Elsoud MMA (2023) Microbial bioremediation as a tool for the removal of heavy metals. Bull Natl Res Cent 47:31

Ahirwar NK, Gupta G, Singh R, Singh V (2016) Isolation, identification and characterization of heavy metal resistant bacteria from industrial affected soil in central India. Int J Pur Appl Biosci 4(6):88–93. https://doi.org/10.18782/2320-7051.2424

Ahmad A, Mustafa G, Rana A, Zia AR (2023) Improvements in bioremediation agents and their modified strains in mediating environmental pollution. Curr Microbiol 80(5):208

Aldhafiri S, Mahmoud H, Al-Sarawi M, Ismail WA (2018) Natural attenuation potential of polychlorinated biphenyl-polluted marine sediments. Pol J Microbiol 67(1):37–48. https://doi.org/10.5604/pjm-2018-6140

Ali SS, Al-Tohamy R, Koutra E, El-Naggar AH, Kornaros M, Sun J (2021) Valorizing lignin-like dyes and textile dyeing wastewater by a newly constructed lipid-producing and lignin modifying oleaginous yeast consortium valued for biodiesel and bioremediation. J Hazard Mater 403:123575

Bansal H, Jha S (2023) Implementation of AI techniques for bioremediation and wastewater treatment. The internet of medical things (IoMT) and telemedicine frameworks and applications. IGI Global, pp 124–134

Baruah A, Chaudhary V, Malik R, Tomer VK (2019) Nanotechnology based solutions for wastewater treatment. Nanotechnol Water Wastewater Treat 337–368. https://doi.org/10.1016/b978-0-12-813902-8.00017-4

Bhakat K, Chakraborty A, Islam E (2019) Characterization of arsenic oxidation and uranium bioremediation potential of arsenic resistant bacteria isolated from uranium ore. Environ Sci Pollut Res 26:12907–12919

Bilal M, Iqbal HMN (2020) Microbial bioremediation as a robust process to mitigate pollutants of environmental concern. Case Stud Chem Environ Eng 2:100011. https://doi.org/10.1016/j.cscee.2020.100011

Cerqueira VS, Hollenbach EB, Maboni F, Camargodo Carmo R Peralba M, Bento FM, FA (2012) Bioprospection and selection of bacteria isolated from environments contaminated with petrochemical residues for application in bioremediation. World J Microbiol Biotechnol 28:1203–1222

Chandran H, Meena M, Sharma K (2020) Microbial biodiversity and bioremediation assessment through omics approaches. Front Environ Chem 1(1):570326. https://doi.org/10.3389/fenvc.2020.570326

Che S, Men Y (2019) Synthetic microbial consortia for biosynthesis and biodegradation promises and challenges. J Ind Microbiol Biotechnol 46(9–10):1343–1358

Das S, Dash HR (2019) Microbial bioremediation: a potential tool for restoration of contaminated areas. Environ Sustain 2(4):393–401. https://doi.org/10.1007/s42398-019-00086-6

Das N, Das A, Das S, Bhatawadekar V, Pandey P, Choure K, Damare S, Pandey P (2023) Petroleum hydrocarbon catabolic pathways as targets for metabolic engineering strategies for enhanced bioremediation of crude-oil-contaminated environments. Fermentation 9(2):196

Godoy P, Reina R, Calderón A, Wittich RM, García-Romera I, Aranda E (2016) Exploring the potential of fungi isolated from PAH polluted soil as a source of xenobiotics-degrading fungi. Environ Sci Pollut Res 23:20985–20996

Góngora-Echeverría VR, García-Escalante R, Rojas-Herrera R, Giácoman-Vallejos G, Ponce-Caballero C (2020) Pesticide bioremediation in liquid media using a microbial consortium and bacteria-pure strains isolated from a biomixture used in agricultural areas. Ecotoxicol Environ Saf 200:110734

Goudarztalejerdi A, Tabatabaei M, Eskandari MH, Mowla D, Iraji A (2015) Evaluation of bioremediation potential and biopolymer production of pseudomonads isolated from petroleum hydrocarbon-contaminated areas. Int J Environ Sci Technol 12:2801–2808

Guan N, Li J, Shin HD, Du G, Chen J, Liu L (2017) Microbial response to environmental stresses: from fundamental mechanisms to practical applications. Appl Microbiol Biotechnol 101(10):3991–4008. https://doi.org/10.1007/s00253-017-8264-y

Guerra FD, Attia MF, Whitehead DC, Alexis F (2018) Nanotechnology for environmental remediation: materials and applications. Molecules 23(7):1760

Hussain A, Rehman F, Rafeeq H, Waqas M, Asghar A, Afsheen N, Rahdar A, Bilal M, Iqbal HM (2022) In-situ, ex-situ, and nano-remediation strategies to treat polluted soil, water, and air-a review. Chemosphere 289:133252

Janbandhu A, Fulekar MH (2011) Biodegradation of phenanthrene using adapted microbial consortium isolated from petrochemical contaminated environment. J Hazard Mater 187:333–340

Janga JK, Reddy KR, Kvns R (2023) Integrating artificial intelligence, machine learning, and deep learning approaches into remediation of contaminated sites: a review. Chemosphere 140476

Kaur M, Sodhi HS (2022) Genetically engineered microorganisms for bioremediation processes. In: Microbial bioremediation. Springer, Cham, pp 67–84. https://doi.org/10.1007/978-3-031-18017-0_5

Kour D, Khan SS, Kour H, Kaur T, Devi R, Rai PK et al (2022) Microbemediated bioremediation: current research and future challenges. J Appl Biol Biotechnol 10(2):6–24. https://doi.org/10.7324/JABB.2022.10s202

Kucharzyk, Rosansky (2018) Advances in the state of the practice for enhanced in situ bioremediation, Technical Report

Kumar SS, Arivazhagan S, Rangarajan N (2013) Remote sensing and GIS applications in environmental sciences–a review. J Environ Nanotechnol 2(2):92–101

Li Y, Chen Z, Chen J, Castellano MJ, Ye C, Zhang N, Miao Y, Zheng H, Li J, Ding W (2021) Oxygen availability regulates the quality of soil dissolved organic matter by mediating microbial metabolism and iron oxidation. Glob Chang Biol 28(24):1–15

Li X, Gao Y, Ning X, Li Z (2023) Research progress and hotspots on microbial remediation of heavy metal-contaminated soil: a systematic review and future perspectives. Environ Sci Pollut Res 30(10):118192–118212. https://doi.org/10.1007/s11356-023-30655-w

Liu X, Shi H, Bai Z, Zhou W, Liu K, Wang M et al (2020) Heavy metal concentrations of soils near the large opencast coal mine pits in China. Chemosphere 244:125360. https://doi.org/10.1016/j.chemosphere.2019.125360

Luka Y, Highina BK, Zubairu A (2018) Bioremediation: a solution to environmental pollution—a review. Am J Eng Res 7(2):101–109. https://doi.org/10.5281/zenodo.1208315

Mahanty S, Chatterjee S, Ghosh S, Tudu P, Gaine T, Bakshi M et al (2020) Synergistic approach towards the sustainable management of heavy metals in wastewater using mycosynthesized iron oxide nanoparticles: biofabrication, adsorptive dynamics and chemometric modeling study. J Water Proces Eng 37:101426. https://doi.org/10.1016/j.jwpe.2020.101426

Mahmood A, Bilal B, Naeem Z, Iram S (2021) Physical, chemical, and biological remediation techniques for textile effluents in context with developed and developing countries in Rhizobiont in bioremediation of hazardous waste. Springer, Singapore, pp 409–441

Manisha C, Solanki K (2023) Role of crisper-cas technique in bioremediation of pesticides. Int J Curr Microbiol App Sci 12(08):82–93

Methneni N, Morales-Gonzalez JA, Jaziri A, Mansour HB, Fernandez Serrano M (2021) Persistent organic and inorganic pollutants in the effluents from the textile dyeing industries: ecotoxicology appraisal via a battery of biotests. Environ Res 196:110956. https://doi.org/10.1016/j.envres.2021.110956

Mounier-Kuhn P (2012) Computer science in French universities: Early entrants and latecomers. Inf Cult: J Hist 47(4):414–456

Mukherjee A, Sarkar D, Das S (2020) Microbial interaction with heavy metals and its application in bioremediation. Environ Microbiol Biotechnol 45(3):123–145

Muter O (2023) Current trends in bioaugmentation tools for bioremediation: a critical review of advances and knowledge gaps. Microorganisms 11(3):710

Palit S, Hussain CM (2020) Functionalization of nanomaterials for industrial applications: recent and future perspectives. Handbook of functionalized nanomaterials for industrial applications. Elsevier, Amsterdam, pp 3–14

Peer WA, Baxter IR, Richards EL, Freeman JL, Murphy AS (2005) Phytoremediation and hyperaccumulator plants. In: Molecular biology of metal homeostasis and detoxification. Springer Berlin Heidelberg, pp 299–340

Pratush A, Kumar A, Hu Z (2018) Adverse effect of heavy metals (As, Pb, Hg, and cr) on health and their bioremediation strategies: a review. Int Microbiol 21(3):97–106. https://doi.org/10.1007/s10123-018-0012-3

Roy AS, Baruah R, Borah M, Singh AK, Deka Boruah HP, Saikia N et al (2014) Bioremediation potential of native hydrocarbon degrading bacterial strains in crude oil contaminated soil under microcosm study. Int Biodeterior Biodegrad 94:79–89

Saleem A, Naureen I, Tasleem G, Anwar R, Mairaj M, Muddassar H, Rana NJ (2022) Microbial assisted bioremediation of polluted water. Haya Saudi J Life Sci 7(4):116–127

Sarkar P, Roy A, Pal S, Mohapatra B, Kazy SK, Maiti MK et al (2017) Enrichment and characterization of hydrocarbon-degrading bacteria from petroleum refinery waste as potent bioaugmentation agent for in situ bioremediation. Bioresour Technol 242:15–27

Satam H, Joshi K, Mangrolia U, Waghoo S, Zaidi G, Rawool S, Malonia SK (2023) Next-generation sequencing technology: current trends and advancements. Biology 12(7):997

Sayara T, Borràs E, Caminal G, Sarrà M, Sánchez A (2011) Bioremediation of PAHs-contaminated soil through composting: influence of bioaugmentation and biostimulation on contaminant biodegradation. Int Biodeterior Biodegradation 65:859–865

Shilpi M, Sharma N (2012) Principles of bioremediation. Environmental Science Publishers, New Delhi

Singh AK, Pal P, Rathore SS, Sahoo UK, Sarangi PK, Prus P, Dziekański P (2023) Sustainable utilization of biowaste resources for biogas production to meet rural bioenergy requirements. Energies 16(14):5409. https://doi.org/10.3390/en16145409

Singh A, Porwal S (2023) Microbial bioremediation: A feasible method for detoxifying contaminated sites by heavy metals and pesticides. Biomed J Sci Tech Res 52

Sokal S, Palsania P, Kaushik G (2022) Bioremediation and functional metagenomics: advances, challenges, and opportunities. In: Omics insights in environmental bioremediation. Springer, pp 3–36. https://doi.org/10.1007/978-981-19-4320-1_1

Sun X, Meng J, Huo S, Zhu J, Zheng S (2020) Remediation of heavy metal pollution in soil by microbial immobilization with carbon microspheres. Int J Environ Sci Dev 11

Syakti AD, Yani M, Hidayati NV, Siregar AS, Doumenq P, Made Sudiana IM (2013) The bioremediation potential of hydrocarbonoclastic bacteria isolated from a mangrove contaminated by petroleum hydrocarbons on the cilacap Coast, Indonesia. Bioremediat J 17:11–20

Tegene BG, Tenkegna TA (2020) Mode of action, mechanism and role of microbes in bioremediation service for environmental pollution management. J Biotechnol Bioinform Res 2:1–18

Thai TD, Lim W, Na D (2023) Synthetic bacteria for the detection and bioremediation of heavy metals. Front Bioeng Biotechnol 11:1178680. https://doi.org/10.3389/fbioe.2023.1178680

Yadav P, Rai SN, Mishra V, Singh MP (2021) Mycoremediation of environmental pollutants: a review with special emphasis on mushrooms. Environ Sustain 4(3):605–618

Zhang Z, Gai L, Hou Z, Yang C, Ma C, Wang Z et al (2010) Characterization and biotechnological potential of petroleumdegrading bacteria isolated from oil-contaminated soils. Bioresour Technol 101:8452–8456

Chapter 20
Nanocomposites as a Promising Tool for Remediation of Pollutants

Hasna Abdul Salam

Abstract Nanocomposites have been gaining considerable interest as a potential tool to tackle environmental issues in the past few years. More than single or pure component structures, composite structures have been found to have more power in tackling such issues. When the strength of nanocomposites combines with the eco-friendly approach of bioremediation, the result is a promising way to reduce pollution and similar other problems. In this review, we shall revisit the existing nanocomposites used for this purpose and the possibilities of developing better means and methods for the same.

Keywords Composite · Eco-friendly · Pollution · Remediation · Nanostructures

20.1 Introduction

Ever increasing global population implies continuous demand for the availability of resources for survival which includes education, finance, food, shelter and clothing. The processes involved in the production of many of these resources impacts the environment negatively via various types of pollution. A leading global concern is water pollution which negatively affects not only human and animal lives but also agricultural and economic sectors (Sajjadi et al. 2020; Jaleh et al. 2020). Globally, heavy metals and oxyanions are the most common pollutants which come from both anthropogenic and industrial activities (Wadhawan et al. 2020; Vardhan et al. 2019; Sankhla et al. 2019). Marine ecosystem faces severe water quality degradation due to decrease in dissolved oxygen, thereby leading to environmental issues (Boeykens et al. 2017). When diffused into the groundwater and soil, contamination by hazardous heavy metals is a major risk faced by humans (Xu et al. 2018). The presence of organic and/or inorganic contaminants is a threat to ecological systems and human body owing to their toxicity, non-biodegradability and complexity. These compounds can cause

H. A. Salam (✉)
Former Science Teacher, Al Alia International Indian School, Riyadh, Saudi Arabia
e-mail: dr.hasnabiotech@gmail.com

© The Author(s), under exclusive license to Springer Nature Switzerland AG 2025
A. Jeyaseelan and K. Murugasen (eds.), *Sustainable Environmental Remediation: Avenues in Nano and Biotechnology*, Environmental Science and Engineering,
https://doi.org/10.1007/978-3-031-78483-5_20

various environmental impacts such as unhealthy air, unwanted ecological change, water scarcity and particularly water pollution (Baran 2019). Effective remediation of toxic organic pollutants from water bodies is a global concern for purification of drinking water and wastewater (Melike and Talat 2022). Constantly depending on conventional techniques is not feasible and also not enough to tackle the enormous pollution. To deal with these challenges, the reduction/degradation processes, which are vital chemical reactions in industrial and synthetic chemistry, allows the introduction of useful and/or harmless compounds for degradation of pollutants.

20.2 Remediation of Environment by Nanostructures

Various methods are used to remove toxins and other contaminants from wastewater, including physical and material precipitation, particle exchange, reverse osmosis, layer filtration, electrochemical treatment, and dissolvable extraction (Wadhawan et al. 2020; Vardhan et al. 2019). But these methods have the demerit of being expensive along with the requirement of vast quantities of chemicals, and poor selectivity in the presence of competing anions. Due to simplicity and ease of operation, adsorption has become a commonly adaptable technique for water remediation. Incorporating nanostructures in a polymer matrix is a strategic way to develop potential candidates for this purpose (Mohamed et al. 2021). The effectiveness of this technique depends on the selection of appropriate sorbent in terms of quality and cost (Wadhawan et al. 2020; Shaban et al. 2018c; Yang et al. 2019). Custom-developed nanomaterials can be chemically modified, functionalized and embedded in specific polymers and matrices to produce hybrid materials or nanocomposites with specific chemical, physical and mechanical properties, which can be used for the development of sustainable and innovative products for environmental remediation (Santhosh et al. 2016; Dontsova et al. 2019; Lu et al. 2016; Ielo et al. 2021a; AbouElmaaty et al. 2021). These hybrid organic–inorganic nanomaterials can be classified into composites, nanocomposites, hybrids and nanohybrids (Descalzo et al. 2006; Kickelbick 2007). Due to their mechanical, physical and adsorption properties, they can be utilized for the treatment of wastewater, soil and air contaminated by organic substances, heavy metals, hydrocarbons, inorganic salts and radionuclides, and also pathogens (Ucankus et al. 2018). Nanocatalyst apparently has a high potential to serve as effective catalysts for remediation of wastewater due to its facile preparation, low cost, and good catalytic activity (Melike and Talat 2022). A notable aspect of these materials concerns not only their catalytic capacity to degrade contaminants, but also the possibility of using them for bioremediation. Actually, they can act as ideal substrates for bacterial growth as well as stimulants for microorganisms to enhance decontamination of the environment (Vázquez-Núñez et al. 2020; Szczepanik 2017). There are several reports till date for the combined effect of nanotechnology and bioremediation techniques (Cappello and Mancini 2019; Scaffaro et al. 2017). Useful starting nanomaterials that can be used as functional nanofillers may be classified as metal nanomaterials, metal oxide nanomaterials, bimetallic nanomaterials, carbon-based nanomaterials, zeolite

Table 20.1 Nanostructure-based remediation of pollutants

Nanostructure used	Method of treatment	Pollutant treated	References
Silver nanoparticles PCBMA nanocomposite	Membrane filtration	Protein	Li et al. (2017)
Osmium nanoparticles	Membrane filtration	p-Nitrophenol and 10-undecylenic acid	Nechifor et al. (2021)
Au/Bi_2WO_6 nanocomposite	Photocatalysis	Benzylic alcohols and Cr(VI)	Yang et al. (2015)
Palladium-silver nanoparticles	Photocatalysis	Methylene Blue	Wali et al. (2019)
Metallic microgel	Catalysis	Nitrophenoles and cationic and anionic organic dyes	Ajmal et al. (2016)
Cellulose-AgNPs composite	Photocatalysis	Bio-fouling and methylene blue	Ali et al. (2017)
Silver nanoparticle clusters	In-situ extraction and detection	Thiabendazole, phosmet and acetamiprid	Kang et al. (2018)
AuNPs nanocomposite	In-situ detection and bioremediation	Cadmium, lead, copper, mercury	Gnanaprakasam et al. (2016)
PdNPs nanocomposite	Catalysis	Nitroarenes and organic dyes	Melike and Talat (2022)
PdNPs immobilized on bentonite	Catalysis	4-nitrophenol, Cr(VI), rhodamine B, potassium hexacyanoferrate, congo red	Sajjadi et al. (2020)

and silica-based nanomaterials, ceramic nanomaterials, polymeric nanomaterials, bio-nanomaterials, metal–organic frameworks and core–shell nanomaterials (Saleh 2020). Table 20.1 gives a handful of examples for nanostructure-based remediation of pollutants.

20.2.1 Metal and Metal Oxide Nanoparticles (MNPs and MONPs)

One of the most important characteristics of MNPs is their antibacterial properties that occur via production of reactive oxygen species, ATP depletion, damage to biomolecules, cation release, and membrane interaction (Slavin et al. 2017). They also have important optical, electrical and catalytic properties with reactivity influenced by the particle size, geometry, composition, oxidation state, and environment (Kelly et al. 2003; Doria et al. 2012; Cuenya 2010). These characteristics make these nanomaterials applicable in various sectors like nanocatalysis, sensing and biosensing, smart textiles, biomedicine, and also in the environmental remediation sector (Elmaaty et al. 2022; Ielo et al. 2021b, c; De Luca et al. 2018; Saturnino et al. 2018; Cook 2009). Bioremediation by bio surfactants facilitate the solubility of hydrophobic contaminants and remediate contaminated marine sites with a green approach. This can be bettered by using nanoparticles. Iron nanoparticles (FeNPs), for example, can stimulate the production of biosurfactants by bacteria such as *Actinobacterium nocardiopsis* (Kiran et al. 2014). Silver nanoparticles (AgNPs) have attracted great interest owing to its nontoxicity, biocompatibility, and high adsorption capacity due to large surface area to volume ratio, high antimicrobial properties, and reusability (Chondath et al. 2019; Gordienko et al. 2019). Palladium nanoparticles immobilized on bentonite has been found to effectively reduce 4-nitrophenol, Cr(VI), rhodamine B, potassium hexacyanoferrate and congo red within seconds and also exhibited antifungal and antibacterial activity (Fig. 20.1) (Sajjadi et al. 2020).

The use of MONPs in bioremediation increases the oxidation of hydrocarbons which decreases their toxic effects, along with supporting microbial growth (Kumari and Singh 2016). Tin dioxide, titanium dioxide, zirconium oxide, zinc oxide, cerium oxide, ferric oxide and their nanocomposites with natural polymers, play a key role in the immobilization of active enzymes exploited for bioremediation processes, as these enzymes detoxify hazardous and carcinogenic compounds via electron transfer reaction between the active sites on the enzyme and the substrate. Zinc oxide nanoparticles (ZnONPs) are reportedly an efficient adsorbent due to their environmentally friendly, high surface area, high catalytic efficiency, nontoxicity, biocompatibility, and chemical stability (Mustapha et al. 2020). Hence, numerous studies have been

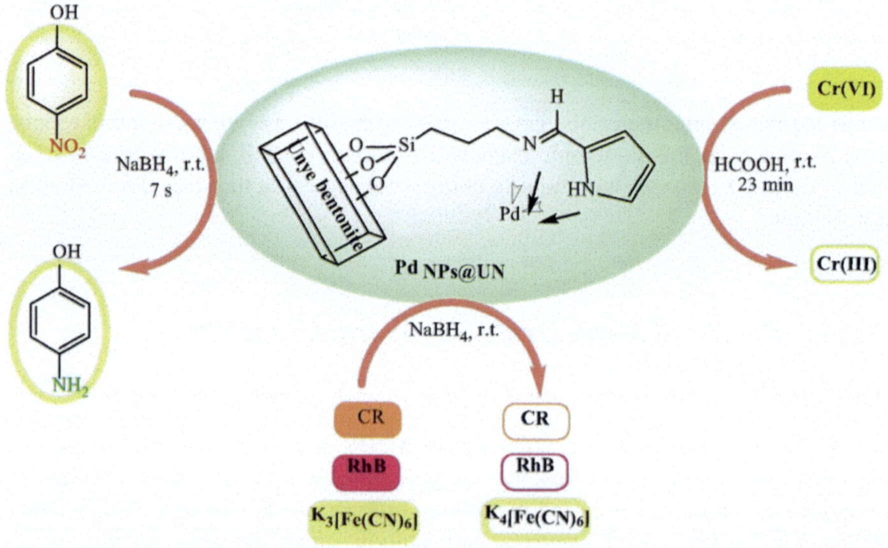

Fig. 20.1 Palladium nanoparticles immobilized on bentonite for catalytic reduction (Sajjadi et al. 2020)

carried out for the bioremediation effect of ZnO and Ag NPs for wastewater treatment (Samuel et al. 2020; Hazeem et al. 2016). Maximum photodecolorization of Blue RGB by ZnONPs has been possible under sunlight as compared to UV light while titanium dioxide nanoparticles (TiO_2-NPs) showed potential photodecolorization, with decrease in decolorization, with increasing dye concentration irrespective of the light source (Hasna and Rajeshwari 2015a, b).

20.2.2 Metal and Metal Oxide Based Nanostructures/ Nanocomposites

There are different classes of MONP composites, such as polymer-supported NPs dispersed in polymeric matrices and magnetically active polymeric particles that are reusable (Cumbal et al. 2003). The absorption of pollutants from MONP composites occurs through various mechanisms like complexation, electrostatic attraction and ion-exchange. Incorporating nanostructured metal and metal oxide in a polymer matrix is a strategic way to develop a novel candidate for water bioremediation. Polymer-inorganic hybrid nanocomposites have attracted substantial attention due to possibility of combining the properties of organic and inorganic components for applications in numerous fields. Compared with conventional preparation methods, microwave synthesis, is attracting interest in nanocomposite fabrication due to advantages, such as rapid heating, homogeneous thermal transmission, and shorter reaction time (Motshekga et al. 2013; Bogdal et al. 2003). Also, this technique stands out by its ability to regulate properties, repeatability, reproducibility, short synthesis time, low cost, purity, and environmental friendliness (Wojnarowicz et al. 2020).

Graphene oxide polyaniline (GO-PANI) nanocomposite is reportedly an effective tool for bioremediation (Asim et al. 2021). For the improvement of GO electron transportation in MFCs (microbial fuel cells), ZnO NPs have been synthetized with a green solvothermal method from lemon peel to produce a L-GO/ZnO composite. For extra mechanical strengthening in the preparation of the anode, the L-GO/ZnO electrodes are dipped into a polylactic acid (PLA) solution. The resulting anode have been tested for the production of energy and remediation of Pb^{2+} ions from wastewater through MFC, which showed excellent energy generation and remediation efficiencies and the presence of bacterial species such as *Lysinibacillus* sp., *Klebsiella* sp. and *Leucobacter* sp. detected on its surface (Yaqoob et al. 2020). Using microwave irradiation, a ZnO–Ag/polypyrrole (PPy) nanocomposite with a core/shell structure has been created by interfacial polymerization of pyrrole in the presence of ZnO nanoparticles and $AgNO_3$ as an oxidant. The antimicrobial behavior of the ZnO–Ag core, combined with the electrical properties of the conducting PPy shell created a special ZnO–Ag/PPy nanocomposite with inherent adsorption behavior and antimicrobial properties. It displayed enhanced adsorption of Cd^{2+} and PO_4^{3-} ions in the mixed solution, and antifungal activity against different fungi, along with impressive antibacterial activity against Gram-negative and Gram-positive bacteria. This efficiency can be attributed to the synergistic combination of crystallinity, coherence of the ZnO–Ag core in the PPy matrix, and the negative zeta potential. These results have significant consequences in the wastewater bioremediation field using a simple operation process (Mohamed et al. 2021; Chen et al. 2018; Husain et al. 2017). A poly(vinylidene fluoride) (PVDF) membrane grafted with silver nanoparticles, poly(carboxybetaine methacrylate) (AgNPs-PCBMA) nanocomposite, prepared by physiosorbed free-radical grafting technique, shows anti-protein and anti-bacterial fouling properties, especially against *E. coli*. In addition, water flux recovery ratio (FRR) and bovine serum albumin (BSA) rejection ratio improved by 40 and 60%, respectively, due to better hydrophilicity performances, conferred by the synergistic effect of AgNPs and zwitterionic polymer brushes of PCBMA (Li et al. 2017). Nano-sized reusable heterogenous catalyst system made by stabilizing palladium nanoparticles on composite hybrid microspheres based on chitosan and γMnO_2 has been found to catalytically reduce nitroarenes and organic dyes (Melike and Talat 2022).

PPy as a conductive polymer with metal oxide nanocomposites has brought out more fields of water treatment application. Polyaniline zinc oxide nanocomposite has been applied for removal of Cr(VI) from aqueous solutions with high adsorption capacity (Ahmad 2019). The accumulation of the ZnO nanoparticles on the bacteria surface due to the electrostatic forces could be the reason for the antibacterial effect of ZnO nanoparticles (da Silva et al. 2019). Moreover, reactive oxygen species (ROS) are generated on the surface of the particles during photocatalysis, which penetrate the cell wall and damage the DNA and enzymes of bacteria, leading to the death of the microorganism (Siddiqi et al. 2018; Shaban et al. 2018b). The release of Zn^{2+} with zeta potential -15.7 mV can disturb the metabolism of microbes via the electrostatic attraction with the negative charge of the polysaccharides of lipopolysaccharide, which predominates the amide (Shaban et al. 2018a, Diez-Pascual 2019). Some nanocomposites of Au/Bi_2WO_6, has been used as visible-light selective photocatalysts in water. They exhibited high capacity for benzylic alcohols oxidation and Cr(VI) (chromium) reduction in water under UV radiation and aerobic conditions (Yang et al. 2015). A heterostructure based on a plasmonic bimetallic photocatalyst made of Pd-AgNPs/macro porous silicon (macroPSi) has shown improve the activity of the methylene blue degradation in water under the UV light (Wali et al. 2019). A nanocomposite obtained by a simple impregnation of cellulose, obtained from citrus waste, with AgNPs, is known to have antibacterial, antioxidant and photodegradation properties. Discs made of the composite material displays more than 90% reduction of *Staphylococcus aureus* culture within two and a half hours, a moderate total antioxidant potential, minor 2,2-diphenyl 1-picrylhydrazyl (DPPH) radical scavenging activity, and a moderate photodegradation capacity under sunlight of methylene blue dye of up to 63.16% within an hour (Ali et al. 2017).

A polypropylene hollow fiber membrane functionalized with osmium nanoparticles (OsNPs) obtained in situ from a solution of osmium tetroxide in tert-butyl alcohol by reduction with molecular hydrogen, has been synthesized and used for a redox process in a membrane contactor. Due to the catalytic nature of OsNPs, the reduction of p-nitrophenol and the oxidation of 10-undecylenic acid were performed with conversions of about 90 and 80%, respectively, in t-butanol. The obtained composite membranes had advantages such as large contact reactive area, capacity to operate at low temperatures, perform in the same module oxidation and reduction processes, no contamination of working environment, thereby representing a feasible approach to recover the reaction mass by ultrafiltration (Nechifor et al. 2021).

MNPs based on copper (Cu) and cobalt (Co) have been synthetized in situ by loading them amid microgels with the aqueous metal salt solutions of Cu(II) and Co(II) ions, followed by treatment with sodium borohydride ($NaBH_4$). The obtained microgel composites (amid-p (Mac-co-AN)-M, M: Cu, Co) reported high catalytic effectiveness for the simultaneous degradation of nitrophenoles and cationic and anionic organic dyes (eosin Y, methylene blue and methyl orange), that may have existed in contaminated aquatic environments. The ability of such systems to be reused for more catalytic cycles has also been observed, owing to the strong coordinating interaction of CoNPs with amidoxime groups (Ajmal et al. 2016).

The use of MNPs composites and hybrids is not limited to environmental remediation and extraction/degradation of pollutants, but also for the production of selective sensors for some substances. An example of a detector based on MNPs is a polyurethane micelle/Ag nanoparticle (AgNPs) cluster fabricated as a surface-enhanced Raman scattering (SERS) substrate for in situ extraction and measurement of pesticide residues such as thiabendazole, phosmet, and acetamiprid (Kang et al. 2018). Another example is that of a reduced graphene oxide detector decorated with AuNPs (rGO@Au NPs), for detecting heavy metal contaminants like Cd^{2+}, Pb^{2+}, Cu^{2+} and Hg^{2+} in water bodies (Gnanaprakasam et al. 2016).

Fe–Ti–Mn (FTMO) composite oxide absorbs As(V) and As(III) by forming inner sphere surface complexes under darkness and light at the water/oxide interface. This material can also trigger the photo-oxidation and oxidation processes of As(III), due to the Ti content in the FTMO adsorbent, which under UV light produces electron–hole pairs and a series of reactive oxygen species (ROS) that oxidize the adsorbed As(III) to As(V) at the solid/liquid interface. Mn also contributes to the oxidation process, while Fe plays a key role in the absorption of the oxidated As(V) (Zhang et al. 2018). A composite based on ZnO@TiCN nanourchin has the ability to thermochemically convert adsorbed Cr(VI) to Cr(III) for the complete removal of Cr(VI) (Rasakia et al. 2018). The removal of Cr(III), Cu(II) and Ni(II) can be achieved by a dendrimer/titania nanocomposite via chelation (Barakat et al. 2013). Manganese oxide NP-impregnated graphene oxide aggregate (GO-MO) nanocomposite, due to an inner-sphere complexation mechanism, absorbs Cd(II) (cadmium) and Cu(II) with an extremely high uptake rate (>99.9%) in column operation (Wan et al. 2018). TiO_2-graphene (P25-GR) composite with a 2.5% of graphene showed the commendable for absorption, charge transportation and photocatalytic performance at high PAH concentrations in alkaline conditions (Bai et al. 2017). Titania-based composite based on zinc porphyrin covalently functionalized fullerene [C60] derivatives, prepared for the sensitization of titania as a visible-light active photocatalyst for river and wastewater remediation, showed interesting results for its potential applications in the remediation of a series of complex environmental and industrial matrices (Regulska et al. 2019).

Magnetic particles of Fe_3O_4 can be used in combination with GO, as exemplified by Fe_3O_4 NPs decorated with graphene oxide and carboxymethylcellulose (Fe_3O_4@GOCMC) core–shell structured composite beads, for the remediating phosphate and nitrate ions from an aqueous medium. A 0.1 N NaOH solution can be used to regenerate the nanocomposite material for the removal of the adsorbed anion and the regenerated Fe_3O_4@GOCMC can be dried and used for subsequent adsorption–desorption cycles (Karthikeyan and Meenakshi 2020). Magnetic TiO_2 hybridized with lignocellulosic biomass can remove different dyes (Rhodamine B, Methylene blue and Congo Red) and Cr(VI) from contaminated aquatic systems by absorption and photocatalytic processes under visible light. The lignocellulose biomass serves as a matrix for TiO_2 and Fe_3O_4 and isolate direct contact, thereby preventing the dispersion of NPs in treated water and also an unfavorable heterojunction due to increase of the electron/hole charge recombination of pure TiO_2 and Fe_3O_4, which might lower photocatalytic performances (Djellabi et al. 2019). Magnetic shell cross-linked knedel-like nanoparticles (MSCKs) have been used for complex crude oil sorption test in water (Pavia-Sanders et al. 2013).

Biochar, in combination with NPs such as Fe NPs develops nanocomposites that can be applied for remediation of heavy metals and organic compounds in the environment by absorption, reduction and advanced oxidation processes (Lyu et al. 2020). A PSB/Fe_3O_4/biochar composite is known to show that PSB-supported (*Rhodobacter capsulatus*) enhanced nutrient removal and degradation capability from wastewater; precisely, the removal efficiencies were 83.1% for chemical oxygen demand, 87.5% for NH^{4+}, and 92.1% for PO_4^{3-}. Fe_3O_4 NPs and iron oxides are known to promote the growth, metabolism and enzyme activities of bacteria and also of PSB, as they act as extracellular electron acceptors to efficiently scavenge reducing equivalents (He et al. 2017). A bionanocompound consisting of polyether amine (PEA) coated magnetite NPs immobilizes OmpA and laccase for the demulsification of oil/water emulsions, and removal/degradation of oils (Rangel-Muñoz et al. 2020).

Due to the development of tunable properties and the versatility of different nanomaterials and embedding polymers, nowadays new functional nanohybrids and nanocomposites for pollution remediation are being designed and developed. With carbon-based nanomaterials, it is possible to exploit π–π interactions, hydrogen bonds, and electrostatic interactions, beyond hydrophobic interactions, for trapping water pollutants. Adsorbent materials can thus be developed to remediate contaminants such as oil spilled on water. Nanosponges are a category of these porous materials (Moura and Lago 2009). CNTs (carbon nanotubes) are used for in situ and ex situ bioremediation processes as they increase the growth of some microbial communities in oil-contaminated freshwater sediments to promote in situ degradation of contaminants (Abbasian et al. 2016). Polyaniline/carbon nanotube (PANi/CNT) hybrid composite reportedly performs 80% COD removal from wastewater (Ali et al. 2019). A different utility of MFCs with CNT composites is represented by bare and multi-walled carbon nanotube/polypyrrole (MWCNT/Ppy) coated electrodes, used for bioelectrochemical treatment of phenol in anodic and cathodic compartments under aerobic and anaerobic conditions (Khan et al. 2020).

Composite aerogels are porous materials used for water remediation field owing to their adsorption properties. An example is graphene aerogel/Fe_3O_4/polystyrene composite. It has a crude oil absorption capacity of 40 times its own mass after 10 water–oil separation cycles, and can be easily recovered due its floating capacity. Fe_3O_4 NPs confers magnetic properties that enable collection of the exhausted aerogel with a magnet (Zhou et al. 2015).

A graphenized, sand-based composite adsorbent with large porosity and surface area has been reported by an acid and high temperature treatment of sand-coated sucrose, and tested for the absorption of Hg^{2+} from contaminated water. This is eco-friendly and economically feasible. The absorption capacity of Hg^{2+} shown by the composite material was high, due to the interaction of the metal ion with the functional groups. Since the chelated bonds between the Hg^{2+} and the composite material gets weak acidic pH, it is possible to regenerate it with a desorption process, for reuse (Bajpai et al. 2017). Green solvothermal method produces more complex hybrid nanocomposite materials for the remediation of contaminated aquatic sites, such as a nanosized nickel-benzene dicarboxylate (Ni-BDC) metal organic framework (MOF) decorated over graphene oxide (GO) or a carbon nanotube (CNT) platform. This nanocomposite (Ni-BDC@GO) shows high retention capacity of methylene blue due to π–π interaction and can be regenerated (Ahsan et al. 2020). Graphene incorporated with composite nanofibers (TZB-Gr) (titanium dioxide-zinc oxide-bismuth oxide-graphene), showed better removal of organic dyes. Under visible-light and UV-light irradiation, the TZB-Gr composite nanofibers activated organic dyes to produce $\cdot O^{2-}$ and $\cdot OH^-$ radicals which are known to be powerful oxidizing species for the degradation of most organic pollutants (Kanjwal et al. 2019).

Porous silica nanomaterials are known to have notable performance in various sectors owing to their large surface area and the presence of large pores that act as hosts for different molecules. Due to the absorption properties, they are utilized for environmental remediation for the adsorption of various contaminants from aqueous or gaseous media (Guerra et al. 2018; Singh et al. 2020). Mesoporous silica be configured as nanotubes to perform catalytic processes for water decontamination. Palladium nanoparticles (PdNPs) supported on hollow mesoporous silica nanotubes (h-mSiO2), are used for the catalytic reduction of 4-nitrophenol and hydrodechlorination of 4-chlorophenol in wastewater, due to the enhancement of the Pd active site accessibility and mass transfer of the reactants (Tian et al. 2018). Laccase and mesoporous silica composites can also be used for the bioremediation and oxidation of phytotoxic phenolic compounds present in olive mill wastewaters (Salis et al. 2009).

Zeolites are crystalline aluminosilicates (or silicates) used in the absorption and retention of environmental contaminants, due to their high selectivity for inorganic cations (zinc, lead, cadmium, copper, nickel), anions, and organic compounds such as pesticides and phenols (Boros-Lajszner et al. 2018). Other minerals based on aluminosilicates are clays, defined as nanostructured fine-grained minerals that possess strong adsorption/absorption properties due to their negatively charged surface

(Bergaya and Lagaly 2006). Magnetic graphene oxide modified zeolite (Cu-Z-GO-M) composites exhibits better removal of methylene blue than zeolite alone due to the introduction of active sites and an increased surface area (Huang et al. 2019). A bimetallic composite of a fly ash-based zeolite supported with nano zerovalent iron and nickel (nZVI/Ni@FZA) combines the absorption properties of zeolite and the catalytic properties of metal NPs. This material can also remove Cr(VI) and Cu(II) contaminants from polluted aquatic sites by reduction, adsorption, and ion exchange mechanisms. The presence of Ni in the bimetallic nanocomposite enhances the reduction properties of nano zero valent Fe (Kumar et al. 2021). TiO_2-based Mexican natural zeolite composite (T/MZ) has been analyzed for the photodegradation of metoprolol. Additionally, it has been observed that the zeolite acts not only as a support, but can also enhance the photocatalytic activity of pristine TiO_2 under near-UV light (Piedra López et al. 2021). A nanobiocatalyst of zeolite and GO is known to degrade Direct Red 23 (Mahmoodi and Saffar-Dastgerdi 2020). Biopolymer consisting of *Pseudomonas veronii* cells immobilised on xanthan gum-based biopolymer has been used as a coating on a natural zeolite to produce a nanocomposite and tested for the bioremediation of Hg^{2+} from contaminated soil (McCarthy et al. 2017). The use of clays in composite materials leads to reduction of the toxic effects of some nanomaterials, such as GO nanoplates, due to the aggregation of the clay NPs leading to a weakening of their chemical properties (Cavallaro et al. 2019). A schematic representation of photocatalysis is as given in Fig. 20.2.

Nanocomposites based on clays and metal oxide NPs show catalytic properties which can be used in combination with macrocycles such as cyclodextrin to improve their affinity and degradation for organic pollutants. An iron-clay-cyclodextrin

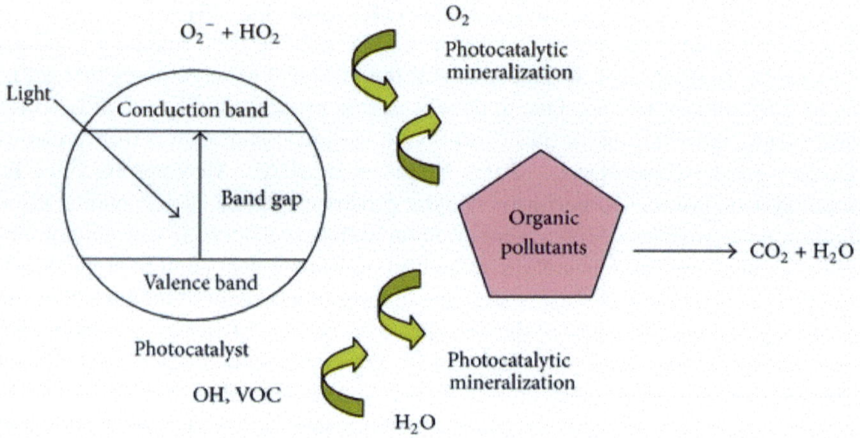

Fig. 20.2 Photocatalytic degradation of organic effluent (Gnanaprakasam et al. 2016)

composite absorbent catalyst, produced by coating montmorillonite clay with iron-oxide, and modifying the obtained composite with cyclodextrin monomers and polymers cross-linked with polyfluorinated aromatic molecules, has shown high absorption bisphenol A (BPA), carbamazepine (CBZ), and perfluorooctanoic acid (PFOA), and excellent degradation efficiency using hydrogen peroxide (Kundu et al. 2020). Chitosanclay-laccase nanobiocomposite tested for the degradation of three synthetic dyes, showed maximum decolorization percentage for anthrax quinone dye of 85%, azo dye of 82% and diazodye of 69%, and good reusability (Mehandia et al. 2020).

20.3 Conclusion

The nanostructures has unique merits such as as eco-friendly preparation methods, unlimited possibilities of functionalization., versatility in term of shape and application forms, adsorption selectivity, stability, degradation of pollutants by catalysis (photocatlysis), reusability, regeneration capacity and possibility to experiment for bioremediation utility. Also, nanohybrids and nanocomposites have lower toxicity than pristine nanomaterials. Most of the reported nanotechnological examples for (bio) are still having poor scalability for real-time applications. In certain instances, the cost/method of production does not justify the benefit achieved for their use. So, research needs to be enhanced for the development of nanostructures that can be applied real-time, can be produced in large scale and be commercially viable, without itself becoming a source of waste or secondary pollution.

References

Abbasian F, Lockington R, Palanisami T, Megharaj M, Naidu R (2016) Multiwall carbon nanotubes increase the microbial community in crude oil contaminated fresh water sediments. Sci Total Environ 539:370–380

AbouElmaaty T, Abdeldayem SA, Ramadan SM, Sayed-Ahmed K, Plutino MR (2021) Coloration and multi-functionalization of polypropylene fabrics with selenium nanoparticles. Polymers 13:2483

Ahmad R (2019) Polyaniline/ZnO nanocomposite: a novel adsorbent for the removal of Cr (VI) from aqueous solution. In: Advances in composite materials development. IntechOpen, London, UK (2019)

Ahsan MA, Jabbari V, Imam MA, Castro E, Kim H, Curry ML, Valles-Rosales DJ, Noveron JC (2020) Nanoscale nickel metal organic framework decorated over graphene oxide and carbon nanotubes for water remediation. Sci Total Environ 698:134214

Ajmal M, Demirci S, Siddiq M, Aktas N, Sahiner N (2016) Simultaneously catalytic degradation/reduction of multiple organic compounds by modifiable p (Methacrylic acid-co Acrylonitrile)-M (M:Cu, Co) microgel catalyst composites. New J Chem 40:1485–1496

Ali N, Zamanb H, Bilal M, Shah AA, Nazir MS, Iqbal HMN (2019) Environmental perspectives of interfacially active and magnetically recoverable composite materials—a review. Sci Total Environ 670:523–538

Ali A, ul Haq I, Akhtar J, Sher M, Ahmed N, Zia M (2017) Synthesis of silver nanoparticles impregnated cellulose composite material: Its possible role in wound healing and photocatalysis. IET Nanobiotechnol 11:477–484

Asim AY, Mohamad NMI, Khalid U, Showkat AB, Anish K, Abdullah MA, Mohammad RK, Mohammad A, Ahmad MA (2021) Cellulose derived graphene/polyaniline nanocomposite anode for energy generation and bioremediation of toxic metals via benthic microbial fuel cells. Polymers 13:135 (26 pages)

Bai H, Zhou J, Zhang H, Tang G (2017) Enhanced adsorbability and photocatalytic activity of TiO_2-graphene composite for polycyclic aromatic hydrocarbons removal in aqueous phase. Colloids Surf B 150:68–77

Bajpai AK, Dubey R, Bajpai J (2017) Synthesis, characterization, and adsorption properties of a graphene composite sand (GCS) and its application in remediation of Hg (II) ions. Water Air Soil Pollut 228:346

Barakat MA, Ramadan MH, Alghamdi MA, Algarny SS, Woodcock HL, Kuhn JN (2013) Remediation of Cu (II), Ni (II), and Cr (III) ions from simulated wastewater by dendrimer/titania composites. J Environ Manag 117:50–57

Baran T (2019) Bio-synthesis and structural characterization of highly stable silver nanoparticles decorated on a sustainable bio-composite for catalytic reduction of nitroarenes. J Mol St 1182:213–218

Bergaya F, Lagaly G (2006) Chapter 1 General introduction: clays, clay minerals, and clay science. Dev Clay Sci 1:2006

Boeykens SP, Piol MN, Samudio Legal L, Saralegui AB, Vázquez C (2017) Eutrophication decrease: phosphate adsorption processes in presence of nitrates. J Environ Manag 203:888–895

Bogdal D, Penczek P, Pielichowski J, Prociak A (2003) Microwave assisted synthesis, crosslinking, and processing of polymeric materials. Adv Polym Sci 163:193–263

Boros-Lajszner E, Wyszkowska J, Kucharsk J (2018) Use of zeolite to neutralise nickel in a soil environment. Environ Monit Assess 190:54

Cappello S, Mancini G (2019) Use of nanomaterials for marine bioremediation: a perspective. Ann Mater Sci Eng 4:1035

Cavallaro G, Lazzara G, Rozhina E, Konnova S, Kryuchkova M, Khaertdinov N, Fakhrullin R (2019) Organic-nanoclay composite materials as removal agents for environmental decontamination. RSC Adv 9:40553–40564

Chen J, Yu M, Wang C, Feng J, Yan W (2018) Insight into the synergistic effect on selective adsorption for heavy metal ions by a polypyrrolee/TiO_2 composite. Langmuir 34:10187–10196

Chondath SK, Poolakkandy RR, Kottayintavida R, Thekkangil A, Gopalan NK, Vasu ST, Athiyanathil S, Menamparambath MM (2019) Water-chloroform interface assisted microstructure tuning of polypyrrolee-silver sheets. ACS Appl Mater Interfaces 11:1723–1731

Cook SM (2009) Assessing the use and application of zero-valent iron nanoparticle technology for remediation at contaminated sites. Jackson State University, Jackson, MS, USA

Cuenya BR (2010) Synthesis and catalytic properties of metal nanoparticles: Size, shape, support, composition, and oxidation state effects. Thin Solid Films 518:3127–3150

Cumbal L, Greenleaf J, Leun D, SenGupta AK (2003) Polymer supported inorganic nanoparticles: characterization an environmental applications. React Funct Polym 54:167–180

da Silva BL, Caetano BL, Chiari-Andréo BG, Pietro RCLR, Chiavacci LA (2019) Increased antibacterial activity of ZnO nanoparticles: influence of size and surface modification. Colloids Surf B Biointerfaces 177:440–447

De Luca G, Bonaccorsi P, Trovato V, Mancuso A, Papalia T, Pistone A, Casaletto MP, Mezzi A, Brunetti B, Minuti L et al (2018) Tripodal tris-disulfides as capping agents for a controlled mixed functionalization of gold nanoparticles. New J Chem 42:16436–16440

Descalzo AB, Martínez-Máñez R, Sancenón F, Hoffmann K, Rurack K (2006) The supramolecular chemistry of organic-inorganic hybrid materials. Angew Chem Int Ed 45:5924–5948

Diez-Pascual AM (2019) Antibacterial nanocomposites based on thermosetting polymers derived from vegetable oils and metal oxide nanoparticles. Polymers 11:1790

Djellabi R, Yang B, Sharif HMA, Zhang J, Ali J, Zhao X (2019) Sustainable and easy recoverable magnetic TiO_2-lignocellulosic biomass@Fe_3O_4 for solar photocatalytic water remediation. J Clean Prod 233:841–847

Dontsova TA, Nahirniak SV, Astrelin IM (2019) Metaloxide nanomaterials and nanocomposites of ecological purpose. Hindawi J Nanomater 2019:1–31

Doria G, Conde J, Veigas B, Giestas L, Almeida C, Assunção M, Rosa J, Baptista PV (2012) Noble metal nanoparticles for biosensing applications. Sensors 12:1657–1687

Elmaaty TA, Raouf S, Sayed-Ahmed K, Plutino MR (2022) Multifunctional dyeing of wool fabrics using selenium nanoparticles. Polymers 14:191

Gnanaprakasam P, Jeena SE, Premnath D, Selvaraju T (2016) Simple and robust green synthesis of AuNPs on reduced graphene oxide for the simultaneous detection of toxic heavy metal ions and bioremediation using bacterium as the scavenger. Electroanalysis 28:1885–1893

Gordienko MG, Palchikova VV, Kalenov SV, Belov AA, Lyasnikova VN, Poberezhniy DY, Chibisova AV, Sorokin VV, Skladnev DA (2019) Antimicrobial activity of silver salt and silver nanoparticles in different forms against microorganisms of different taxonomic groups. J Hazard Mater 378:120754

Guerra FD, Attia MF, Whitehead DC, Alexis F (2018) Nanotechnology for environmental remediation: materials and applications. Molecules 23:1760

Hasna AS, Rajeshwari S (2015a) Photocatalytic decolorization of Blue RGB by green synthesized zinc oxide nanoparticles and its impact on cardiac cell-line (H9C2). 114:409–415

Hasna AS, Rajeshwari S (2015b) Photo-decolorization potential of green synthesized titanium dioxide nanoparticles. 11(1):39–44

Hazeem LJ, Bououdina M, Rashdan S, Brunet L, Slomianny C, Boukherroub R (2016) Cumulative effect of zinc oxide and titanium oxide nanoparticles on growth and chlorophyll content of *Picochlorum* sp. Environ Sci Pollut Res 23:2821–2830

He S, Zhong L, Duan J, Feng Y, Yang B, Yang L (2017) Bioremediation of wastewater by iron oxide-biochar nanocomposites loaded with photosynthetic bacteria. Front Microbiol 8:823

Huang T, Yan M, He K, Huang Z, Zeng G, Chen A, Peng M, Li H, Yuan L, Chen G (2019) Efficient removal of methylene blue from aqueous solutions using magnetic graphene oxide modified zeolite. J Colloid Interface Sci 543:43–51

Husain J, Kiran JS, Anjum R, Reddy N, Raghu N, Sagar J, Anjum B, Aleem M, Prasad MVNA (2017) Synthesis, characterization, and electrical conductivity of polypyrrole/zinc oxide nanocomposites. Ferroelectrics 515:118–124

Ielo I, Giacobello F, Sfameni S, Rando G, Galletta M, Trovato V, Rosace G, Plutino MR (2021a) Nanostructured surface finishing and coatings: functional properties and applications. Materials 14:2733

Ielo I, Iacopetta D, Saturnino C, Longo P, Galletta M, Drommi D, Rosace G, Sinicropi MS, Plutino MR (2021b) Gold derivatives development as prospective anticancer drugs for breast cancer treatment. Appl Sci 11:2089

Ielo I, Rando G, Giacobello F, Sfameni S, Castellano A, Galletta M, Drommi D, Rosace G, Plutino MR (2021c) Synthesis, chemical-physical characterization, and biomedical applications of functional gold nanoparticles: a review. Molecules 26:5823

Jaleh B, Karami S, Sajjadi M, Feizi Mohazzab B, Azizian S, Nasrollahzadeh M, Varma RS (2020) Laser-assisted preparation of Pd nanoparticles on carbon cloth for the degradation of environmental pollutants in aqueous medium. Chemosphere 246:125755

Kang Y, Li L, Chen W, Zhang F, Du Y, Wu T (2018) Rapid in situ SERS analysis of pesticide residues on plant surfaces based on micelle extraction of targets and stabilization of Ag nanoparticle aggregates. Food Anal Methods 11:3161–3169

Kanjwal MA, Shing Lo KK, Leung WW (2019) Graphene composite nanofibers as a high-performance photocatalyst for environmental remediation. Sep Purif Technol 215:602–611

Karthikeyan P, Meenakshi S (2020) In situ fabrication of magnetic particles decorated biopolymeric composite beads for the selective remediation of phosphate and nitrate from aqueous medium. J Environ Chem Eng 8:103530

Kelly KL, Coronado E, Zhao LL, Schatz GC (2003) The optical properties of metal nanoparticles: the influence of size, shape, and dielectric environment. J Phys Chem B 107:668–677

Khan N, Anwer AH, Ahmad A, Sabir S, Sevda S, Khan MZ (2020) Investigation of CNT/PPy-modified carbon paper electrodes under anaerobic and aerobic conditions for phenol bioremediation in microbial fuel cells. ACS Omega 5:471–480

Kickelbick G (2007) Hybrid materials: synthesis, characterization, and applications. Wiley, Hoboken, NJ, USA, ISBN 3527610480

Kiran GS, Nishanth LA, Priyadharshini S, Anitha K, Selvin J (2014) Effect of Fe nanoparticle on growth and glycolipid biosurfactant production under solid state culture by marine *Nocardiopsis* sp. MSA13A. BMC Biotechnol 14:48

Kumar G, Angaru R, Choi Y, Lingamdinne LP, Choi J, Kim D, Koduru JR, Yang J, Chang Y (2021) Facile synthesis of economical feasible fly ash-based zeolite-supported nano zerovalent iron and nickel bimetallic composite for the potential removal of heavy metals from industrial effluents. Chemosphere 267:128889

Kumari B, Singh DP (2016) A review on multifaceted application of nanoparticles in the field of bioremediation of petroleum hydrocarbons. Ecol Eng 97:98–105

Kundu S, Manor NK, Radian A (2020) Iron–montmorillonite–cyclodextrin composites as recyclable sorbent catalysts for the adsorption and surface oxidation of organic pollutants. ACS Appl Mater Interfaces 12:52873–52887

Li J, Zhang D, Ni X, Zheng H, Zhang Q (2017) Excellent hydrophilic and anti-bacterial fouling PVDF membrane based on Ag nanoparticle self-assembled PCBMA polymer brush. Chin J Polym Sci 35:809–822

Lu H, Wang J, Stoller M, Wang T, Bao Y, Hao H (2016) An overview of nanomaterials for water and wastewater treatment. Adv Mater Sci Eng 2016:1–10

Lyu H, Tang J, Cui M, Gao B, Shen B (2020) Biochar/iron (BC/Fe) composites for soil and groundwater remediation: synthesis, applications, and mechanisms. Chemosphere 246:125609

Mahmoodi NM, Saffar-Dastgerdi MH (2020) Clean Laccase immobilized nanobiocatalysts (graphene oxide-zeolite nanocomposites): from production to detailed biocatalytic degradation of organic pollutant. Appl Catal B 268:118443

McCarthy D, Edwards GC, Gustin MS, Care A, Miller MB, Sunna A (2017) An innovative approach to bioremediation of mercury contaminated soils from industrial mining operations. Chemosphere 184:694–699

Mehandia S, Sharma SC, Arya SK (2020) Immobilization of laccase on chitosan-clay composite beads to improve its catalytic efficiency to degrade industrial dyes. Mater Today Commun 25:101513

Melike C, Talat B (2022) Palladium nanoparticles embedded over chitosan/γMnO_2 composite hybrid microspheres as heterogeneous nanocatalyst for effective reduction of nitroarenes and organic dyes in water. J Organomet Chem 963:122284

Mohamed F, Allah AE, Abu Al-Ola KA, Shaban M (2021) Design and characterization of a novel ZnO–Ag/polypyrrole core–shell nanocomposite for water bioremediation. Nanomaterials 11:1688. https://doi.org/10.3390/nano11071688

Motshekga SC, Ray SS, Onyango MS, Momba MNB (2013) Microwave-assisted synthesis, characterization and antibacterial activity of Ag/ZnO nanoparticles supported bentonite clay. J Hazard Mater 262:439–446

Moura FCC, Lago RM (2009) Catalytic growth of carbon nanotubes and nanofibers on vermiculite to produce floatable hydrophobic "nanosponges" for oil spill remediation. Appl Catal B 90:436–440

Mustapha S, Ndamitso MM, Abdulkareem AS, Tijani JO, Shuaib DT, Ajala AO, Mohammed AK (2020) Application of TiO_2 and ZnO nanoparticles immobilized on clay in wastewater treatment: a review. Appl Water Sci 10:49

Nechifor G, Păncescu FM, Grosu AR, Albu PC, Oprea O, Tanczos S-K, Bungău C, Grosu V-A, Pîrţac A, Nechifor AC (2021) Osmium nanoparticles-polypropylene hollow fiber membranes applied in redox processes. Nanomaterials 11:2526

Pavia-Sanders A, Zhang S, Flores JA, Sanders JE, Raymond JE, Wooley KL (2013) Robust magnetic/polymer hybrid nanoparticles designed for crude oil entrapment and recovery in aqueous environments. ACS Nano 7:7552–7561

Piedra López JG, González Pichardo OH, Pinedo Escobar JA, de Haro del Río DA, Méndez HI, González Rodríguez LM (2021) Photocatalytic degradation of metoprolol in aqueous medium using a TiO_2/natural zeolite composite. Fuel 284:119030

Rangel-Muñoz N, González-Barrios AF, Pradilla D, Osma JF, Cruz JC (2020) Novel bionanocompounds: outer membrane protein A and laccase co-immobilized on magnetite nanoparticles for produced water treatment. Nanomaterials 10:2278

Rasakia SA, Zhang B, Liua S, Thomas T, Yang M (2018) Nanourchin ZnO@TiCN composites for Cr (VI) adsorption and thermochemical remediation. J Environ Chem Eng 6:3837–3848

Regulska E, Rivera-Nazario DM, Karpinska J, Plonska-Brzezinska ME, Echegoyen L (2019) Zinc porphyrin-functionalized fullerenes for the sensitization of titania as a visible-light active photocatalyst: river waters and wastewaters remediation. Molecules 24:1118

Sajjadi M, Yilmaz Baran N, Baran T, Nasrollahzadeh M, Tahsili MR, Shokouhimehr M (2020) Palladium nanoparticles stabilized on an novel Schiff base modified Unye bentonite: highly stable, reusable and efficient nanocatalyst for treating wastewater contaminants and inactivating pathogenic microbes. Sep Purif Technol 237:116383

Saleh TA (2020) Nanomaterials: classification, properties, and environmental toxicities. Environ Technol Innov 20:101067

Salis A, Pisano M, Monduzzi M, Solinas V, Sanjust E (2009) Laccase from Pleurotus sajor-caju on functionalised SBA-15 mesoporous silica: immobilization and use for the oxidation of phenolic compounds. J Mol Catal B Enzym 58:175–180

Samuel MS, Jose S, Selvarajan E, Mathimani T, Pugazhendhi A (2020) Biosynthesized silver nanoparticles using *Bacillus amyloliquefaciens*; application for cytotoxicity effect on A549 cell line and photocatalytic degradation of p-nitrophenol. J Photochem Photobiol B Biol 202:111642

Sankhla MS, Kumari M, Nandan M, Kumar R, Agarwal P (2019) Heavy metals contamination in water and their hazardous effect on human health—a review. SSRN Electron J 5:759–766

Santhosh C, Velmurugan V, Jacob G, Jeong SK, Grace AN, Bhatnagar A (2016) Role of nanomaterials in water treatment applications: a review. Chem Eng J 306:1116–1137

Saturnino C, Sinicropi MS, Iacopetta D, Ceramella J, Caruso A, Muià N, Longo P, Rosace G, Ielo I (2018) N-Thiocarbazolebased gold nanoparticles: synthesis, characterization and anti-proliferative activity evaluation. IOP Conf Ser Mater Sci Eng 459:012023

Scaffaro R, Lopresti F, Catania V, Santisi S, Cappello S, Botta L, Quatrini P (2017) Polycaprolactone-based scaffold for oil-selective sorption and improvement of bacteria activity for bioremediation of polluted water: porous PCL system obtained by leaching melt mixed PCL/PEG/NaCl composites: oil uptake performance and bioremediation efficiency. Eur Polym J 91:260–273

Shaban M, Mohamed F, Abdallah S (2018a) Production and characterization of superhydrophobic and antibacterial coated fabrics utilizing ZnO nanocatalyst. Sci Rep 8:3925

Shaban M, Rabia M, Fathallah W, Abd El-Mawgoud N, Hussien H, Said O (2018b) Preparation and characterization of polyaniline and Ag/polyaniline composite nanoporous particles and their antimicrobial activities. J Polym Environ 26:434–442

Shaban M, Sayed MI, Shahien MG, Abukhadra MR, Ahmed ZM (2018c) Adsorption behavior of inorganic and organic modified kaolinite for Congo red dye from water, kinetic modeling, and equilibrium studies. J Sol-Gel Sci Technol 87:427–441

Siddiqi KS, Ur Rahman A, Tajuddin Husen A (2018) Properties of zinc oxide nanoparticles and their activity against microbes. Nanoscale Res Lett 13:141

Singh B, Na J, Konarova M, Wakihara T, Yamauchi Y, Salomon C, Gawande MB (2020) Functional mesoporous silica nanomaterials for catalysis and environmental applications. Bull Chem Soc Jpn 93:1459–1496

Slavin YN, Asnis J, Häfeli UO, Bach H (2017) Metal nanoparticles: understanding the mechanisms behind antibacterial activity. J Nanobiotechnol 15:65

Szczepanik B (2017) Photocatalytic degradation of organic contaminants over clay-TiO_2 nanocomposites: a review. Appl Clay Sci 141:227–239

Tian M, Long Y, Xu D, Wei S, Dong Z (2018) Hollow mesoporous silica nanotubes modified with palladium nanoparticles for environmental catalytic applications. J Colloid Interface Sci 521:132–140

Ucankus G, Ercan M, Uzunoglu D, Culha M (2018) Methods for preparation of nanocomposites in environmental remediation. In New polymer nanocomposites for environmental remediation. Elsevier BV, Amsterdam, The Netherlands, pp 1–28

Vardhan KH, Kumar PS, Panda RC (2019) A review on heavy metal pollution, toxicity and remedial measures: current trends and future perspectives. J Mol Liq 290:111197

Vázquez-Núñez E, Molina-Guerrero CE, Peña-Castro JM, Fernández-Luqueño F, de la Rosa-Álvarez MG (2020) Use of nanotechnology for the bioremediation of contaminants: a review. Processes 8:826

Wadhawan S, Jain A, Nayyar J, Mehta SK (2020) Role of nanomaterials as adsorbents in heavy metal ion removal from waste water: a review. J Water Process Eng 33:101038

Wali LA, Alwanb AM, Dheyab AB, Hashim DA (2019) Excellent fabrication of Pd-Ag NPs/PSi photocatalyst based on bimetallic nanoparticles for improving methylene blue photocatalytic degradation. Optik 179:708–717

Wan S, Ding W, Wang Y, Wu J, Gu Y, He F (2018) Manganese oxide nanoparticles impregnated graphene oxide aggregates for cadmium and copper remediation. Chem Eng J 350:1135–1143

Wojnarowicz J, Chudoba T, Lojkowski W (2020) A review of microwave synthesis of zinc oxide nanomaterials: reactants, process parameters and morphoslogies. Nanomaterials 10:1086

Xu X, Nie S, Ding H, Hou FF (2018) Environmental pollution and kidney diseases. Nat Rev Nephrol 14:313–324

Yang J, Wang X, Chen Y, Dai J, Sun S (2015) Enhanced photocatalytic activities of visible-lightdriven green synthesis in water and environmental remediation on Au/Bi_2WO_6 hybrid nanostructures. RSC Adv 5:9771–9782

Yang J, Hou B, Wang J, Tian B, Bi J, Wang N, Li X, Huang X (2019) Nanomaterials for the removal of heavy metals from wastewater. Nanomaterials 9:424

Yaqoob AA, Ibrahim MNM, Yaakop AS, Umar K, Ahmad A (2020) Modified graphene oxide anode: a bioinspired waste material for bioremediation of Pb^{2+} with energy generation through microbial fuel cells. Chem Eng J 417:128052

Zhang W, Zhang G, Liu C, Li J, Zheng T, Ma J, Wang L, Jiang J, Zhai X (2018) Enhanced removal of arsenite and arsenate by a multifunctional Fe-Ti-Mn composite oxide: photooxidation, oxidation and adsorption. Water Res 147:264–275

Zhou S, Jiang W, Wang T, Lu Y (2015) Highly hydrophobic, compressible, and magnetic polystyrene/Fe_3O_4/graphene aerogel composite for oil-water separation. Ind Eng Chem Res 54:5460–5467

Chapter 21
Bioremediation of Aquatic Environmental System

P. Dhanalakshmi, P. Arunkumar, M. I. Farheena, A. Chithira, S. Uthirasamy, and S. Haritha

Abstract The base of a diet highly relies on quality of protein in aquatic environment. Aquaculture is growing as the demand for protein rises. Aquaculture has thus had to contend with a number of hazards including bacteria, fungus, viruses and parasites. Numerous wastes are produced by the aquaculture sector, such as nutrients, fecal matter, metabolic byproducts and beneficial and preventative materials that are crucial for maintaining water quality and preventing disease. A wide range of bioremediation tools and approach are applied to improve pond base renewal, water quality maintenance and aquatic habitat restoration. An important bioremediation approach is to use microbes for maintaining water quality. Natural antibiotics do not work on many germs, thus the government's stringent regulations for ecologically friendly therapy actions. Currently, being treated with different methods in a process called bioremediation to enhance water quality and preserve the sustainability of aquatic ecosystems. By mineralizing carbon-based materials for carbon dioxide production, nitrification, and denitrification, bioremediation can: Remove extra nitrogen from the pond; and boost primary productivity to maintain a stable and diverse pond community even in the presence of pathogens. The shape that is desired is created. In addition to heterotrophic microorganisms that break down organic materials, bioremediation

P. Dhanalakshmi
Department of Zoology, Padmavani Arts and Science College for Women (Autonomous), Opp. Periyar University, Salem, Tamil Nadu 636011, India

P. Arunkumar (✉)
Department of Zoology, Bharathiar University, Coimbatore, Tamil Nadu 641046, India
e-mail: arunkumar.ak88@gmail.com

M. I. Farheena · S. Uthirasamy
Department of Zoology, Government Arts College (Autonomous), Salem, Tamil Nadu 636007, India

A. Chithira
Department of Zoology, Thiruvalluvar Government Arts College, Namakkal, Tamil Nadu 637001, India

S. Haritha
Department of Natural Science, Rajiv Memorial College of Teacher Education, Kannur, Kerala 670702, India

© The Author(s), under exclusive license to Springer Nature Switzerland AG 2025
A. Jeyaseelan and K. Murugasen (eds.), *Sustainable Environmental Remediation: Avenues in Nano and Biotechnology*, Environmental Science and Engineering,
https://doi.org/10.1007/978-3-031-78483-5_21

also include nitrogenizing, denitrifying, and photosynthetic microbes. Controlling pond microbial communities can play a significant role in functional research aimed at enhancing global aquaculture ecology and productivity.

Keywords Bioremediation · Aquatic environment · Bacteria · Microalgae · Fungi · *Bacillus* and Hydrogen sulphide

21.1 Introduction

The demand for food resources is being driven by changes in eating habits and an increasing global population. To meet these basic demands, countries all around the world began to mostly rely on aqua farming for additional effective nutritional food scouring. Aquaculture is one of the major producers of animal protein, with human consumption of this type of protein expected to reach 58% by 2028 (Tom et al. 2021). To accommodate the rising demand for fish, a number of profitable fish farmhouses have been developed in the National of Qatar in recent decades. Environmental contaminants are without a doubt one of the major issues that provide the arena with the greatest risk. Soil, water, and air pollution are caused by a wide variety of contaminants from many sectors. Air, soil, and clean water are greatly impacted by rapid, unplanned industry and ongoing population growth. Anthropogenic sports and natural causes have introduced a variety of toxins into the environment, causing pollution (Thirumalaisamy et al. 2022). Odorous substances, whether they are xenobiotics or herbs, are essential to industry and daily life. Alkanes, polycyclic aromatic hydrocarbons, phenols, and petroleum hydrocarbons are a few types of these aromatic chemicals. It is currently employed as a precursor in a variety of processes in the production of paint, oil refineries, medicines, pesticides, explosives, and other products. However, they represent the value reason of marine pollution (Syranidou et al. 2017).

The most often used ecological and biological method to speed up natural degradation processes is bioremediation (Sattar et al. 2022). Because heavy metal (HM) pollution of soil and water is a major issue in developing countries, phytoremediation is more effective in areas with high and moderate concentrations than in areas with low concentrations (Timalsina et al. 2022). Heavy metals (HM) are dangerous to the environment and public health because they cannot be broken down by the body and can build up over time. The metabolism of residing organisms can remove heavy metals (HM) from soil and water. This process, known as bioremediation, can remove pollution by utilizing cyanobacteria, algae, bacteria, fungi, yeast, and plants. Growing industry and urbanization have exposed the environment to numerous toxins that could be harmful to living things. One of the biggest causes of pollution in the land and water is waste from various corporate processes. Industrial manufacture releases heavy metals of all kinds at some point. For example, antimony, chromium, and mercury have been linked to wastewater from paint manufacturing companies (Methneni et al. 2021).

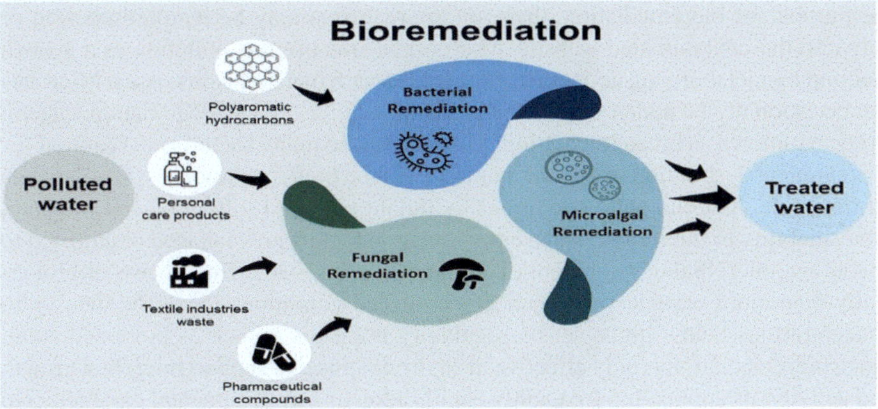

Fig. 21.1 Strategies for bioremediation (Marut Jain et al. 2022)

Thus, microbial communities expand into the biogeochemical cycles and modify their genomes to enable the absorption of new chemicals into their metabolic pathways in order to adapt to a variety of environmental challenges. As a result, it has been proposed that natural selection of mutants with substantial degradative enzymes but much less specific substrate-specificities and likely novel metabolic pathways is the only way for microbes to adapt to the metabolism of different anthropogenic compounds. But because xenobiotics have developed more quickly than bacteria, the uncontrolled exploitation of herbal resources by humans has caused serious changes in the natural environment. Consequently, the ecosystems' capacity for self-cleaning is exhausted, and pollution accumulation to hazardous levels has become the norm. Artificial plastic, which is formed of polymers generated from petroleum, is at the top of the list of pollutants that are becoming more and more prevalent. Over the past fifty years, plastic materials have been increasingly important in all aspects of human undertakings, taking the place of resources such as glass, metal, and logs in unique programs because of their low price, high toughness, and excessive electricity. Consequently, throughout the next 25 years, the global plastic manufacturing sector has tripled (Feil and Pretz 2020) (Fig. 21.1).

21.2 Aquatic Bioremediation

21.2.1 Principle of Bioremediation

Bioremediation is the process of employing living beings in combination with microorganisms (bacteria, fungus, and algae) or plants to break down and detoxify dangerous chemicals in the environment into CO_2, H_2O, microbial biomass, and metabolites (byproducts). Significantly less harmful than the discern material. For the purpose of bioremediation, these microorganisms may be remote-imported or native to the contaminated website. Microorganisms use this pollution as a growth medium by metabolizing and transforming it through metabolic processes. Microbes can occasionally be added to the contaminated site for a green degradation system, a process known as bioaugmentation. This is because many bacteria are required for the complete destruction of contaminants. Since the biodegradation process depends on favorable environmental conditions, pollutant type and solubility, and pollutant bioavailability to microorganisms, environmental factors are managed or adjusted to encourage microbial development (Tyagi and Kumar 2020). The process of biologically degrading organic wastes under controlled conditions to a safe state or to concentrations below limits set by regulatory bodies is known as bioremediation. Since bioremediation is only effective in environments that support microbial growth and activity, its application frequently entails adjusting environmental parameters to speed up microbial growth and degradation—that is, to allow the pollutants to be attacked by enzymes and transformed into harmless products. "Probiotics" refers to the use of competitive exclusion to enhance a certain environment. Probiotic therapies might be viewed as biological control techniques, or "biocontrol," which is the term for the restriction or eradication of pests by the introduction of unfavorable organisms like parasites or certain diseases.

21.2.2 Water Quality

One popular method for treating wastewater from aquaculture is the Recirculation Aquaculture System (RAS). Restoring fresh water to the aqua farming fishpond after treating the effluent is the aim (Badiola et al. 2018; Ahmad et al. 2022). In addition to promoting more effective use of water resources, RAS helps cultured fish develop healthily. But in addition to being an expensive and energy-intensive technique, RAS needs proper administration. Pilot-scale RAS systems were the subject of techno-economic analysis, which showed that these systems had significant building and upkeep costs According to Vázquez-Romero et al. (2022). Other aquaculture water treatment solutions, including aerobic filters and built wetlands, require a lot of space or energy (Brix (1999); Tom et al. 2021). Applying nitrogen to the pond in excess of its absorption capacity may cause nitrogen compounds (such as ammonia and nitrite) to build up and degrade the water quality, which could be harmful to fish and shrimp. The main uses of ammonia are in fish excretion, the mineralization of organic matter, and molecular diffusion from decreasing sediments; cyanobacterial nitrogen fixation and deposition to the environment are, however, sometimes important.

Nitrification occurs as follows:

$$NH_4^+ + 1^{1/2}O_2 - NO_{2-} + 2H^+ + H_2O$$

$$NO_{2-} + 1^{1/2}O_2 - NO_3^-$$

BFTs (Bio-Floc Technology) work best with species like carp, tilapia, and shrimp that can withstand high amounts of solids in their water. While a concentration of around 1000 mg L^{-1} is usually recommended for tilapia, 200 to 500 mg L^{-1} of solids is adequate for complete system capacity and ammonia control in favorable BFT manufacturing. Nitrate can also accumulate in extraordinarily intense cultivation under BFT conditions. The notion of non-water rejuvenation gives rise to this, as microorganisms are largely concentrated in water, which negatively impacts the growth performance of animals and increases their susceptibility to infections that might kill Nile tilapia (Hargreaves 2013; Flores-Valenzuela et al. 2021). Therefore, techniques for improving the hemato-immunological traits, overall growth performance, and water quality of aquatic organisms—like Nile tilapia produced in BFT—had been looked for globally (Silva et al. 2023). As of right now, aquaculture has demonstrated some success in using microbes as probiotics for bioremediation and biocontrol applications.

21.2.3 Excess Manufacture in Aquaculture Industry

When it comes to manufacturing inputs, aquaculture is very similar to other manufacturing businesses. Any such machine could constantly be wasting merchandise or underused inputs. This garbage is often hazardous to the environment and has very little commercial value. Aquaculture's sustainability in society has changed due to the garbage it produces. Fishing-related pollutants can take many different forms. One problem is that there is a lot of waste and uneatable food produced by the high density of fish in small cages. Dust carpets are also found around farm houses and in basements. Sea lice are another problem with pollution. The business has employed a variety of chemical treatments over time to maintain low sea lice levels on farms. Sea lice on farms are the objective of the treatment, but they can also pose a threat to nearby shrimp and other crustaceans. Furthermore, sea lice may become more resistant to toxic substances, rendering wild mussels helpless. The physiological, chemical, and organic conditions of the shrimp growing environment impact the health and productivity of the species. Stress and long-term dysfunction are caused by pollution exposure in turtles, including ammonia, carbon dioxide, and hydrogen sulfide. The waste types generated in fish farms are essentially the same. However, depending on the type of culture and cultural norms followed, there are differences in the number and quality of additions. Food waste and excrement, metabolic side effects, fertilizer residues, residues from biocides and biostats, trash produced while cleaning, and sediment algal blooms are the six categories into which industrial waste or water can be separated (Tyagi and Kumar 2020). This emphasizes the necessity of appropriate tactics to guarantee aquaculture intensification that is sustainable. Compare sustainable intensification to a manufacturing device that produces more food from the same area while reducing its impact on the environment. The generation of trash from aquaculture and management techniques in various culture systems are the main topics of this chapter.

21.2.4 *Extensive Regulating System*

Large-scale aquaculture may be particularly unique in that it's essentially a part of nature and relies on it for the provision of water, the majority of the living thing's food, and other necessities. Large aquaculture facilities thus have restricted inputs to promote fish development and survival, such as airless environments and a few primary organic fertilizers, among other things. This device typically has a modest stocking density (<500 kg/ha), and cultured organisms can be supported by the natural fuel exchange and productivity of our food system (plants and animals). Fish raised in low depth aquaculture are fed solely naturally occurring vitamins found in the pond, which results in a slight improvement in natural productivity (Soltan 2016).

21.2.5 *Intensive Water Purification System*

Higher yields are obtained with high-intensity farming than from conventional farming. 10 and up, aeration facilities must occupy at least 50% of the pool's surface. A diet rich in protein is the only prerequisite for a fish diet. Robust systems include raceways (for species of trout in temperate countries), ponds (for shrimp in tropical and subtropical settings, for example), cages (for marine fish culture in temperate seas), and tanks (for eels in Japan, for example).

In every scenario, the highest possible stocking density is contingent upon the preservation of the water quality parameters necessary for the cultivated organism. Tanks have the highest stocking density, followed by cages and highways, and ponds with the lowest density. Natural systems differ greatly from dynamic aquaculture systems. Their characteristics include: 1. really basic food chains, such feed → grown organisms; 2. high food conversion rates from specialized artificial feeds with low energy losses from feed input; 3. Total non-sustainability and no energy recycling; 4. High energy input needs (for example, for pumping, aeration, filtration, feed, and nutrients); 5. High yields relative to volume or area. High water trade prices and sometimes mechanical methods are responsible for maintaining high-quality water. Particle removal, gasoline change, and oxygen production are typically completed in indoor tanks for intensive cultivation. In intensive open structures with soil substrates and phytoplankton, there is improved gasoline change through particle sedimentation, bacterial dispersal, and mechanical aeration. Contingent on the category of system and the cultivated organisms, stocking density in dynamic systems can vary, but it is typically high (Soltan 2016).

21.3 Applications of Microbes as Disease Regulatory Mediator in Aquaculture

The primary factors contributing to the rise in pollution are the widespread production, use, and knowledge of pesticides, pharmaceuticals, personal care products (PPCPs), plastic polymers, and heavy metals. These substances are made to meet the ever-increasing demands for better health, a greater standard of living, and methods of providing food and clothing for an expanding human population. However, the Great Waste Era frequently observes their production and use. However, mitigating techniques don't always align with this. By traveling via rivers, groundwater, air, and ocean circulation, these often-resistant chemicals and compounds have impacted nearly every ecosystem on Earth (Lebreton et al. 2017; Pujari and Kapoor 2021; Roy et al. 2022). For example, deep marine sediments in the Pacific Ocean have been shown to contain organic pesticides (Ge et al. 2021). Waste water treatment systems often contain personal care goods and prescription drugs in addition to marine ecosystems (Bayen et al. 2016; Ojemaye and Petrik 2022). In remote places, such Antarctic Ocean ice (Kelly et al. 2020) and the Mariana Trench, one of the world's deepest marine areas (Peng et al. 2020), microplastics have been found. Concentrations of nanoplastics seem to be substantial (Ter Halle et al. 2017; Materić et al. 2022). Everything from organisms to the ecosystem as a whole is negatively impacted by rising pollution levels in the environment. The emergence of antimicrobial resistance genes due to antibiotic overuse and improper disposal, the bioaccumulation of methylmercury in shellfish, which can lead to human poisoning, and the eventual negative ecological effects of the widespread use of the pesticide dichlorodiphenyltrichloroethane (DDT) before it was outlawed in many countries are a few well-known examples (Kurtela and Antolović 2019). Biological methods, commonly referred to as "bioremediation," can be utilized to decompose and eliminate unwanted, dangerous contaminants from the environment to decrease their impacts and growth rates. In order to mitigate the impacts of oil spills, George M. Robinson was the first to use a bioremediation approach in the 1960s. This technique included bacteria. Most of the time, bioremediation uses plants (phytoremediation) or microorganisms (micro remediation) to break down pollutants. Notable examples of bioremediation in aquatic environments include the usage of prescription drugs (Silva et al. 2019; Chandel et al. 2022) or microalgae that absorb heavy metals (Zhang et al. 2019; Leong and Chang 2020). When working with groundwater systems, scientists and engineers typically make assumptions regarding things like protective zones for drinking water production and bioremediation on a broad scale of kilometers. For instance, based on point measurements of the aquifer's characteristics, they could determine parameters for the Darcy equation that is applied to the whole aquifer. Conversely, bioremediation methods such as pump-and-treat and reactive walls are applied on a microscale. But in this conversation, we want to see how much more we can learn and do if we include the small-scale processes in the large-scale thinking. Considering that shown the importance of micro size studies for the hyporheic zone, it is reasonable to believe that concentrating on the micro scale will be as instructive in groundwater ecosystems more broadly.

Provide an overview of the various phenomena that impact bioremediation. The smallest scale at which a phenomenon may be identified is used to label it. The engineering of bioremediation systems benefits from an understanding of the three scales: micro, meso, and macro. The term "microscale" refers to the scale at which interactions between organisms, chemicals, and microbes can be explained independently of all delivery events. The composition of microbial consortia and the kinetics and stoichiometry of transformation events are examples of microscale homes. The body scale of this event is about the size of a microbial cell. The term "mesoscale" describes the dimensions (aside from advective or mixing methods) at which machine geometry and transportation phenomena first manifest. Mesoscale phenomena include mass transfer between phases, diffusion, and absorption. Possible body scales for mesoscale processes include the size of pore channels or soil particles, the feature diffusion length, and the diameters of microbial aggregates. Advection, dispersion, and geological spatial heterogeneity are examples of macroscale phenomena. For this occurrence, a physical size of around or larger is appropriate.

21.3.1 Microscale

"Micro scale" in this context refers to sizes smaller than one millimeter. It is known that groundwater bacteria, fungi, and archaea—a kingdom of unicellular organisms that are usually smaller than 0.001 mm and lack membrane-bound organelles and a nucleus like that of bacteria—have physiological and genetic traits that distinguish them from other types of bacteria (Griebler et al. 2014). While worms, rotifers, and micro arthropods can grow as small as 1 mm, the largest groundwater arthropods are usually approximately 10 mm in size. In terms of size, protozoa, or unicellular organisms, lie between the groups of fungi, bacteria, and archaea. This text covers various techniques for microscale representative characterization, including gene probes, plate counts, degradation pathways, stochastic reactions, kinetics of reactions, chemical evaluation, electron acceptor chemistry examination, nutrient block, toxic agents, pH and water activity probes, and interactions with the aquifer matrix or soil, as well as abiotic reactions (Table 21.1).

21.3.2 Macroscale

Food webs and ecosystems are shaped by the macro scale since it determines the overall context. For instance, one of the main influences on the hydrochemical conditions is the geological background. Hardness is directly impacted by the total ion

Table 21.1 Microscale bioremediation and their representative characterization methods

S. no	Microscale	Representative characterization methods
1	Microorganisms	Plate counts, gene probes
2	Degradation pathways	Batch reaction studies
3	Reaction stoichiometry	Batch reaction studies
4	Reaction kinetics	Chemical analysis for N, P
5	Electron acceptors	Chemical analysis
6	Nutrients inhibitors toxicity	Batch reaction studies
7	Water activity, pH	Electrochemical probes
8	Reactions with soil or aquifer matrix	Abiotic reaction studies
9	Chemical equilibria	Abiotic batch sorption studies
10	Sorption (equilibrium)	Abiotic batch sorption studies

Table 21.2 Macroscale bioremediation and their representative characterization methods

S. no	Macroscale	Representative characterization methods
1	Sorption (non-equilibrium)	Abiotic batch and column sorption studies
2	Attachment/detachment (microorganisms)	Biofilm studies, attached microbe
3	Plugging/filtration	Column studies, pressure drop and flow rate
4	Interphase transport	Multiphase column studies

content and the mineral matrix's solubility. The source of water in an aquifer is determined by the recharge patterns that are determined by the stratigraphy of geological units, soil properties, and climate. If recharge and exchange are strong, there is a good chance of supplying allochthonic (foreign to the system) input and, most importantly, dissolved oxygen. However, since replenishment with dissolved oxygen, oxygenated compounds, or other resources is likely to be poor in a distant section of the aquifer, life must adapt to the limited quantity of energy that is accessible from resources there. Larger scale recharge distributions, in conjunction with dominant land use patterns, control the degree to which anthropogenic inputs (such as contamination; see Sect. 3.2) may be incorporated into aquifers. At sizes ranging from cm to km, groundwater recharge may exhibit spatial heterogeneity (Cuthbert et al. 2012). The study uses representative characterization methods, including abiotic batch and column sorption studies, attachment/detachment studies, plugging/filtration studies, and multiphase column studies (Table 21.2).

21.3.3 Mesoscale

Today, maintaining the survival of the planet's inhabitants faces several difficulties. Because they regulate pH, temperature, and marine biodiversity, protecting the

Table 21.3 Meoscale bioremediation and their representative characterization methods

S. no	Mesoscale	Representative characterization methods
1	Advection	Well elevations, pump tests, tracer studies
2	Dispersion	Conservative tracer studies
3	Spatial heterogeneity	Well logs, core permeabilities
4	Hydrologic properties and boundary conditions	As for advection and dispersion

integrity of aquatic ecosystems is one of the most important things to do (Hosseini et al. 2017). Wetlands, rivers, lakes, and coastal estuaries are the different types of aquatic habitats. Because eutrophic bacteria found in aquatic environments enable the uptake and subsequent fixation of atmospheric carbon dioxide, or CO_2, this equilibrium is particularly crucial (Prasad et al. 2021; Iglina et al. 2022). When it comes to water quality, aquatic microbes are a double-edged sword. First off, through nutrient cycling and chemical threat bioremediation, they significantly contribute to the preservation of aquatic ecosystem health and improved water quality. Microalgae and cyanobacteria, for instance, have demonstrated their effectiveness in eliminating hazardous substances such as pesticides originating from agricultural fields (Castellanos-Estupiñán et al. 2022) and heavy metals generated by industrial processes (Ankit et al. 2022). Furthermore, the decline in the diversity of aquatic plants and animals may result in less opportunities to discover novel substances with as-yet-unknown uses. Accurate interpretation and collection of these attributes are critical to maintaining the biodiversity of the current water reservoirs. Mesoscale bulk measurements yield average information about aquatic microorganisms' hidden, important details. The process of bioremediation involves using green plants, microbes, or their byproducts or enzymes to remove toxins from the environment. Organic contaminants are naturally eliminated from the environment on a mesoscale by nature. The study uses representative characterization methods, including advection, displacement, spatial heterogeneity, and hydrologic properties, to study the mesoscale behavior of a substance (Table 21.3).

21.4 Green Bioremediation Technology for Aquatic Environment

Any procedure that restores the environment to its pre-pollution state by means of bacteria, fungus, green plants, or their enzymes is known as bioremediation. In bioremediation, bacteria that can break down certain types of soil pollutants, such chlorinated hydrocarbons, are employed. Algae, fungus, microorganisms, and other substances can be employed in bioremediation. Because microbes are so prevalent in nature and because they can obtain carbon from a wide variety of substrates, they may come into contact with unusual situations where they can absorb various

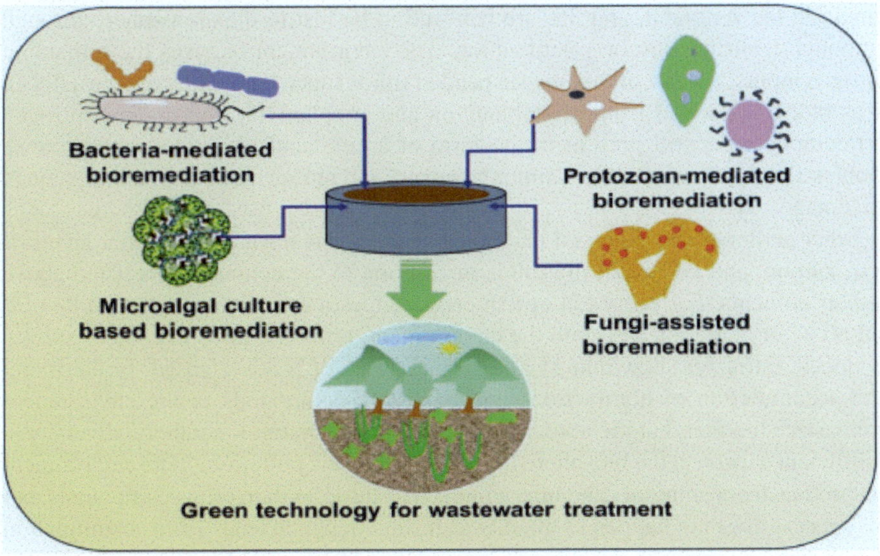

Fig. 21.2 Green bioremediations technology for aquatic environment (Singh et al. 2022)

pollutants (Kour et al. 2022). Their effectiveness is further increased by their ability to adapt to completely various situations. As an illustration, psychrophiles flourish in cold climes, acidophiles endure in acidic conditions, and halophiles endure in saline surroundings (Perera and Hemamali 2022) (Fig. 21.2).

Fish and crustaceans create ammonia, which makes up the majority of the nitrogen expelled in their bodies, by the deamination of metabolic proteins (Weihrauch and Allen 2018). For example, fish urine may include trace amounts of urea, while the gills of crustaceans and fish exchange ammonia for dissolved oxygen (Robles-Porchas et al. 2020; Rodela 2013). Ammonia production is the first step in the deterioration of water quality, necessitating the immediate dilution of these hazardous compounds or their biotransformation into less hazardous chemicals. In this sense, the aquaculture business benefits from the rapid development of bacteria, their widespread prevalence in a variety of habitats, and their nutritional needs, which are thought to be advantageous features for the biotransformation of hazardous chemicals. For example, Bender and Phillips (2004a, b) state that the "ecological success of microbial mats and their broad array of microbial activities suggest that microbial ecosystems might be useful to bioremediation of environmental pollutants." Because of their capacity to absorb heavy metals, bacteria and other microorganisms have long been employed in the bioremediation of industrial effluent. Not unexpectedly, heavy metals, petroleum pollutants, and textile colors are the main topics of the majority of scholarly writing on bioremediation employing bacteria. It is possible that most of the above-mentioned organisms are not suitable for bioremediation due to the unique toxins used in aquaculture. For example, research indicates that in shrimp farms, only 22% of the total nitrogen inputs are converted to harvested shrimp, 14%

remain in the sediment, and 3% are believed to be lost to the atmosphere through ammonia denitrification or volatilization. The environment receives the remaining nitrogen inputs. As part of organic suspended solids (unconsumed feed, feces, flocs), or as biomass derived from phytoplankton and zooplankton, the organic nitrogen (N) component is still present in the form of amino acids, peptides, and proteins (Robles-Porchas et al. 2020). Ammonia, nitrite, and nitrate make up the inorganic N percentage.

Other authors have observed that aquaculture farms release substantial levels of PO_4, carbon, and entire put off solids in addition to the contaminants listed above in their effluents. For every ton of fish produced, estimates indicate that 10 kg of P and 60 kg of N are released into the environment; however, for every ton of shrimp produced, estimates show that 21 kg of P and 56 kg of N are released; furthermore, "the accumulation of highly toxic and persistent compounds in the environment emphasizes the fact that the autochthonous microbes' natural metabolic diversity is insufficient to protect the biosphere from anthropogenic pollution." The environment can biotransform ammonia and nitrite molecules into less dangerous compounds, but this process does not happen as quickly as it should, which leads to an accumulation of poisonous chemicals. As a result, approaches like bioremediation—which uses microbes—may be regarded as long-term fixes for this issue. Despite a few isolated attempts made in previous decades, the use of microorganisms for bioremediation in aquaculture is relatively new. Through the nitrification and denitrification processes, bacteria and microalgae can act as bio-remediators for sediment and wastewater. Certain heterotrophic bacteria are also capable of decomposing organic wastes inclusively, turning them into biomass that can be utilized as a source of nutrients for farm animals or natural organisms.

Urbanization, economic growth, and population increase are the three most important and visible factors contributing to ecological degradation and illness. Solving the riddle of ecological detoxification is the aim of ecological concepts and techniques. This is why we can employ a variety of biota for the reduction of pollutants thanks to the use of cutting-edge production and technology. There are many potentially beneficial biotas that can aid in cleaning up the contaminated environment, but their application is frequently limited because of a lack of knowledge about the factors—such as pH, moisture content, temperature, redox capacity, soil type, and oxygen—that influence the growth and metabolism of microorganisms in contaminated environments. The disciplines of molecular biology, microbiology, biochemistry, analytical chemistry, chemical and environmental engineering, and molecular biology have all advanced the topic of bioremediation. Depending on the kind and extent of contamination, the bioremediation system employs either in-situ or ex-situ procedures. It has been found that, in the gift context, bioremediation is essential to restoring a contaminated atmosphere using environmentally friendly techniques.

Microbes help transform contaminants into final products including carbon dioxide, water, and other metabolic intermediates during the mineralization process. Moreover, the term "immobilization" describes the change of a substance into a form that is not found in its surroundings. For instance, nitrate nitrogen can be transformed into natural nitrogen (Pratush et al. 2018). Usually, this method is used to bioremediate heavy metals in unexpectedly contaminated environments. It is possible to immobilize both in vivo and in-situ. In order to immobilize the metal ions causing the pollution, the contaminated soil is removed from the affected area and subjected to a microbiological procedure (Ayangbenro and Babalola 2017). However, the pollution is managed by the process itself (Cao et al. 2020). Microbes like *B. cereus* and *E. asburiae* have been linked to the immobilization of heavy metals from environmental contamination, according to Fashola et al. (2020). During the bioremediation process, microbes produce efficient or hydrophobic solvent pumps to protect the cell's outer membrane from dangerous compounds (Verma and Kuila 2019).

Microbes can even use potentially harmful substances as building blocks or sources of energy. Furthermore, microbial metabolisms may be "designed" or "improved" to purposefully cause specific pollutants thanks to the resource of genetic modification. With the useful support of synthetic biology systems, microbial communities may also be designed to target certain pollution scenarios (Jaiswal et al. 2019; Borchert et al. 2021; Xiang et al. 2021). Bacillus microorganisms are particularly notable for their probiotic or bioremediation properties. Gram-poor, aerobic, and non-pathogenic Bacillus species are widely found in the environment. They might boost defensive mechanisms against infections, colonize fish gastrointestinal tracts, improve overall performance, and improve fish pond water quality (Nunes et al. 2020).

21.4.1 Cyanobacteria

Although cyanobacterial nitrogen fixation and air deposition are occasionally significant, fish excretion and sediment flow from the mineralization of organic matter and molecular diffusion from reduced sediment are the main sources of ammonia. The most feasible method of removing ammonia from closed cultivation structures is bacterial nitrification, which is often carried out in biofilters made of gravel and sand that let water to pass through. Ammonia oxidizers are found in the five genera *Nitrosomonas, Nitrosovibrio, Nitrosococcus, Nitrolobus, Nitrospira, Nitrobacter*, and *Nitrospira*. Heterotrophic nitrifiers are also known to employ organic nitrogen sources rather than ammonia or nitrite, and they are best at producing modest amounts of nitrite and nitrate. It has been demonstrated that nitrifiers in infected cultures nitrify more efficiently. By just shifting the pH into the acidic range, nitrification no longer only creates nitrate but also makes solutes available (Roveda et al. 2024).

21.4.2 Photosynthetic Bacteria

To maintain the health of the ecosystem and break down H_2 S at the bottom of ponds, aquaculture makes extensive use of photosynthetic benthic bacteria. These bacteria include chlorophyll bacteria, which, depending on the species, may perform anaerobic photosynthesis and absorb light ranging from the blue to the infrared spectrum. Purple and green sulfur bacteria are abundant in the sediment–water interface's anaerobic zone. Purple photosynthetic bacteria that do not require sulfur can decompose organic waste, H_2S, NO_2, and contaminants found in ponds. Sulfur bacteria, both purple and green, divide H_2 S to use light wavelengths that phytoplankton cannot absorb. Because purple and green sulfur bacteria get fewer electrons from H_2S at a lower energy cost than H_2O-splitting photoautotrophs, low light intensity is required for photosynthesis. The following is the general equation for this reaction:

$$CO_2 + 2H_2S(CH_2O) + H_2O + 2S$$

$$S + CO_2 + 3H_2S(CH_2O) + H_2SO_4$$

$$CO_2 + NaS_2O_3 + 3H_2O_2(CH_2O) + NaS_2O_4 + H_2SO_4$$

Homes of photosynthetic sulfur microorganisms, *Chromatiaceae* and *Chlorobiaceae*, use sunlight and sulfur dioxide to support anaerobic conditions for growth. While *Chlorobiaceae* precipitate, *Chromatiaceae* have sulfur particles within their cells. The *Rhodospirillaceae* related family no longer has an advantage in getting rid of H_2S since they primarily employ organic components, including fatty acids, as a base of hydrogen. But as it can survive in both aerobic and anaerobic environments without the need for sun energy in the dark, as a heterotrophic bacterium, it can be employed as an effective water source in subterranean mineralizers. Important photosynthetic bacteria for fisheries include: Family *Rhodospirillaceae: Rhodopseudomonas, Rhodomicrobium,* and *Rhodospirillum Thiocystis, Thiosarcina, Thiospirillum, Thiocapsa, Lamprocystis, Thiodictyon, Thiopedia, Amoebobacter,* and *Ectothiorhodospira* are all members of the family *Chromatiaceae*. The family *Chlorobiaceae* includes *Pelodictyon, Prosthecochloris, Chloropseudomonas*, and *Clathrochloris*. Probiotics for ponds can be produced by mass-culturing bacteria from the *Chromotinaceae* and *Chlorobiaceae* families for the purpose of bioremediation of H_2S poisoning (Ranjan and Bavitha 2014).

21.4.3 Autotrophic Bacteria

Anaerobic and aerobic settings both use bioremediation to treat aquaculture effluents. Under aerobic circumstances, some bacteria may do biological nitrification; a significant quantity of oxygen is required for this process. Ammonia-oxidizing bacteria and archaea that are chemolithoautotrophic use hydroxylamine to convert ammonia to nitrite. These ammonia-oxidizing bacteria belong to the β- and γ-Proteobacteria and use carbon dioxide as a carbon source, according to Koops and Pommerening-Roser (2001). The nitrite-oxidizing bacteria then change nitrite into nitrate, a metabolite that is far less hazardous to aquatic organisms. The phylum Nitrospirae, especially the α- and γ-Proteobacteria, is comprised of these bacteria.

Numerous studies focusing on biological nitrogen removal via nitrification, denitrification, and inclusively anaerobic ammonium oxidation (anammox) have investigated a broad spectrum of bacterial species. Anammox (anaerobic ammonium-oxidizing) bacteria carry out an alternative denitrification pathway; in particular, anammox accounts for 24%–67% of the nitrogen loss in marine sediments (Thamdrup and Dalsgaard 2002). Anaerobic ammonia-, nitrite-, and anaerobic ammonium oxidizing bacteria are the three major kinds of bacteria that are cultivated and employed in biofilters and bioreactors. Aquaculture effluent comes into touch with high bacterial densities, which lower ammonia burdens. Aerobic ammonia- and nitrite-oxidizing bacteria are not found in high quantities in the environment; rather, they must be cultivated and kept in bioreactors to reach high enough concentrations to perform a significant nitrification denitrification process. Anaerobic ammonium-oxidizing bacteria, on the other hand, are widely distributed in the environment, especially in areas where aquaculture farms discharge their tainted water.

Although their proliferation and excessive concentrations in the environment negatively impact the biota, anaerobic ammonium-oxidizing bacteria are essential to the denitrification of anoxic sediments derived from aquaculture. Because of this, their elevated oxygen need causes localized hypoxia or anoxia, which destroys the most delicate aerobic life forms and negatively impacts other people's physiological health. Consequently, only anaerobic bacteria should be used for bioremediation in bioreactors; in this instance, it has been predicted that employing anammox in conjunction with microbial nitrification will be more cost-effective than using other techniques. Although several anaerobic and aerobic bacterial bioreactors have been developed, the most feasible method for large-scale open farms seems to be those in which effluents are directly exposed to a series of mats inhabited by autotrophic/heterotrophic bacterial consortia. These consortia use bioaugmentation to nitrify and denitrify the nitrogenous chemicals in wastewater effluents (Audelo-Naranjo et al. 2011; Bender and Phillips 2004a, b).

Mats inoculated with microbes can be used to mimic nature, but they have a greater ability to remove pollutants. In nature, microbial mats are multilayer, autotrophic, and heterotrophic stratified ecosystems. The predominant microalgae in these communities are cyanobacteria, sulfate-reducing bacteria, anoxygenic phototrophic bacteria, and eukaryotic microalgae such as diatoms. Because both oxygenic and anoxygenic photosynthesis can be found within the same consortium, these microbial communities exhibit physiological flexibility. For example, Ebeling et al. (2006) reviewed the various pathways (photoautotrophic, autotrophic bacterial, or heterotrophic bacterial based) for ammonia removal and used half-reaction relationships to develop a set of stoichiometric balanced relationships. They also discussed the impact of these relationships on water quality. The bioremediation performance of these species may be estimated using these relationships. Several genes coding for enzymes involved in metabolic pathways to bio-transform hazardous compounds into less-harmful molecules have been investigated and discovered in additional bacterial strains used for the bioremediation of petroleum pollution. Additionally, recombinant strains of bacteria have been explored for use in bioremediation; genetically engineered bacteria have been employed for this purpose (Kumar et al. 2013). On the other hand, there hasn't been as much research done on genetically modified bacteria for aquaculture waste bioremediation.

21.4.4 Heterotrophic Bacteria

Some heterotrophic bacteria may be very beneficial for aquaculture waste bioremediation. As was covered in the last section, the primary organisms that change ammonia into nitrate are autotrophic bacteria. Heterotrophic bacteria (Cyanobacteria: *Microcoleus chthonoplastes*, *Spirulina* sp., *Oscillatoria* sp., *Schizothrix* sp., *Calothrix* sp., *Phormidium* sp., etc.) can also convert the nitrogen in ammonia into non-toxic products; in this way, ammonia is transformed into microbial biomass (Ebeling et al. 2006). In addition to turning ammonia into biomass, certain heterotrophic bacteria may break down organic wastes including uneaten food, dung, and dead animals (Martínez-Córdova et al. 2017). Heterotrophic bacteria proliferate due to the availability of nutrients in these wastes. Moreover, fish and crabs can eat biomass produced by bacteria. Thus, the productive performance of shrimp farms may be enhanced by the buildup of heterotrophic bacterial biomass in conjunction with a sufficient rate of aeration and oxygenation. An elevated is necessary to support the development of heterotrophic bacteria, according to experimental data (Ebeling et al. 2006). The carbon-based material, which is a carbon sources, will be broken down by the heterotrophic consortium, which will then absorb ammonia straight into the cellular protein. Thus, a modest food chain-link inside the similar culture component is responsible for recycling organic matter, nutrients, and inorganic nitrogenous chemicals. Heterotrophic bacteria's digestion of organic matter is particularly significant since it balances out several ecological imbalances. The purpose of using heterotrophic bacteria was not to bioremediate aquaculture effluents, but rather to promote the growth of bio-flocs, which are made up of organic matter, diatoms, algae, and heterotrophic bacteria. This is because heterotrophic bacteria produce biomass at a faster rate than autotrophic bacteria.

This biomass was used as a food source for shrimp farmed in zero water exchange systems, but there was also an added benefit from the bioremediation of pollutants (N and P compounds, organic matter). Here, nitrifying and heterotrophic bacteria are introduced, water exchange is decreased, carbon sources are added, air and oxygen are provided, and other constant adjustments to the culture conditions encourage the growth of microorganisms such as bacteria, protozoa, cyanobacteria, and microalgae that may eventually feed the farmed organisms. Furthermore, there is a decline in trash output. Lastly, by transforming organic matter, dissolved organic carbon, and nitrogenous metabolites into useable energy that is then supplied to the upper levels of the trophic chain, heterotrophic organisms enrich the chain. This creates an alternative flux of energy into higher trophic levels. The culture units in this instance can be compared to bioreactors, where nutrient cycling and a significant reduction in waste output are achieved. These systems, which go by the moniker Bio-floc Technology (BFT) systems, usually generate low alkalinity, high CO_2, and huge quantities of total suspended particles. Most of the organic matter is converted to bio-floc, however sodium bicarbonate (100–150 mg L^{-1} as $CaCO_3$) may be added to maintain the right alkalinity. According to Ebeling et al. (2006), the excess has to be eliminated and used for biofloc flour, fertilizer, etc. Other criteria of BFT include full aeration, enough mixing, extra carbon sources (cellulose, starch, molasses, etc.), and so on. Further information on the state of BFT research at this time may be found in the papers of Avnimelech (2009). Although all aquaculture systems may not be able to meet the demands of heterotrophic bacteria, they provide a viable option for the bioremediation of aquaculture pollutants. On the other hand, BFT may be performed ex situ. As a source of essential fatty acids, carbohydrates, protein, and vitamins, the biofloc is used as a feed source for cultured organisms and for other purposes. A further aquaculture bioremediation method is probiotic bacterial therapy. Even if probiotic bacteria are not used in aquaculture to eliminate or break down toxic compounds or organic debris, they can inhibit the growth and entirely eradicate the presence of some illnesses (Helmy et al. 2019).

Aquaculture farms are invaded by bacterial and viral pathogens that multiply within the culture units, multiplying and releasing more of them into the discharge effluents. Of the words of Moriarty (1999), "by adding selected bacterial species to displace deleterious normal bacteria, the microbial species composition in hatchery tanks or large aquaculture ponds can be changed." This author observed that the quantity of bright Vibrio bacteria decreased upon the addition of carefully selected probiotic strains of Bacillus spp. Reducing the quantity of antibiotics used also has the advantage of decreasing their concentration in the effluents that are released. Because aerobic bacterial consortia based only on marine strains are more successful than the majority of commercial treatments at decreasing or eradicating infections and reproducing more quickly, some shrimp farms, for instance, have started utilizing them. Farmers in the area (Northern Mexico) employ bioreactors to grow large amounts of bacteria (13 3 1012 CFU mL^{-1}) that are then transferred to culture units on a daily basis. The diseases' virulence and numbers decline when the bacterial consortium gains dominance in the ponds. Aquaculture bioremediation may include the use of bacterial probiotics. However, the effectiveness of this approach could be contingent upon a comprehensive comprehension of the dynamics of competition among certain bacterial species or strains. Furthermore, it is acknowledged that the metabolic capacity of benthic microbes—which are primarily heterotrophic—offers a cost-effective substitute for cleaning up contaminated areas. According to Ape et al. (2019), benthic prokaryotes play a role in the decomposition and transfer of sedimentary organic compounds to higher trophic levels. The microbial diversity of sediments receiving effluent discharges is significant, and these sediments may be able to handle pollutants from aquaculture operations.

A discussion of different support matrices, immobilized bacterial cultures and communities, immobilization techniques, nutrients for immobilized bacteria, contaminants treated by immobilized microbes, mass transfer considerations, factors that affect remediation using immobilized cells (e.g., salinity, pH, toxicity of pollutants), kinetics and equilibrium analyses, and design of various biofilter configurations using immobilized cells comprise the current state of knowledge regarding the use of immobilized bacteria for environmental bioremediation applications. We wrap off by quickly going over the difficulties, most recent advancements, and potential directions for the future. The majority of bacteria participating in bioremediation processes, according to Ojuederie and Babalola (2017), are members of the following genera: *Bacillus, Enterobacter, Flavobacterium, Pseudomonas, Alteromonas, Arthrobacter, Achromobacter, and Burkholderia.* The unique ability of hydrocarbonoclastic bacteria (OHCB) to degrade hydrocarbons sets them apart. Examples of Oleispira, Marinobacter, Thallassolituus, and Cycloclasticus are among them.

21.4.5 *Bacillus*

Bioremediation techniques have evolved in recent decades to successfully, cheaply, and ecologically ease polluted areas back into service. As was mentioned in the preceding section, bioremediation is a sustainable, extremely affordable, and environmentally friendly alternative to chemical and physical remediation techniques. Several bioremediation techniques can be used to clean up contaminated settings. Fungi, yeast, and algae are examples of plant-selective systems that can include bioremediation systems, which are specifically dominated by bacterial and archaeal flowers (Patel et al. 2022). In particular, a small number of research deal with bioaugmentation and/or bio stimulation techniques, which can be used to other non-bacterial microorganisms, such as yeast, fungi, and plants, in order to promote bioremediation. The bioremediation procedure has a number of drawbacks. There are two primary limits to bioremediation technique in some cases: Microbes have difficulty degrading highly chlorinated pollutants and HMW PAHs; and Microbial treatment or degradation of certain pollutants might result in products that are more hazardous and unstable than the original pollutants. Vinyl chloride, a hazardous substance with the potential to cause cancer, is produced when TCE undergoes reductive dehalogenation, for instance. Waste from moulting, leftover food and excrement, metabolic byproducts, leftovers from biocides and biostats, wastes originating from fertilizer, and collapsing algal blooms. Applying microorganisms and/or enzymes to aquaculture ponds is a chapter, environmentally friendly technique called "bioremediation" that improves water quality. Bio-remediators, also known as bioremediating agents, are microorganisms and/or their byproducts that are added to water to improve its quality.

Numerous techniques are available for managing probiotics in aquaculture, such as injection, direct addition to water, and use as a feed additive for pellet or stay feed. Certain strategies can increase probiotic effectiveness in farming. Therefore, it has been proposed that the best way to use probiotics in aquaculture is to add them to the water immediately away. This method works incredibly well for fish of any age because the fish always consumes water (Jahangiri and Esteban 2018). Therefore, the best strategy to improve the flavor of water is to add probiotics directly to it; however, probiotic supplements have shown promise as a nutritional aid. Numerous factors frequently impact the effectiveness of probiotics. Probiotic activity is influenced by a number of characteristics of high-quality water, such as dissolved oxygen, hardness, pH, temperature, mechanical friction, stress biotype, and osmotic stress. For instance, genetics, diet, and environmental factors are blamed for the contradictory outcomes between Bacillus subtilis L10 and G1 (Zokaeifar et al. 2014) and *B. subtilis* E20 in boosting water first-class.

The ways that probiotic Bacillus affects water quality are summed up in Fig. 21.3. The acronyms for chemical oxygen demand (COD), dissolved oxygen (DO), and biological oxygen demand (BOD) are respectively. The same color indicates the same mechanism lane, rounded rectangles indicate the third stage of the mechanisms, pointed pentagons indicate the fourth stage, hexagons indicate the fifth stage, and teardrops indicate the intersection of two processes, according to Hlordzi et al. (2020). Rectangular compartments indicate the first stage of the mechanisms. Prebiotics Many people are interested in the idea of biological disease control, especially in the context of employing microbiological modulators to avoid illness. Probiotics are bacterial supplements made from a single or mixed culture of particular non-pathogenic bacterial strains.

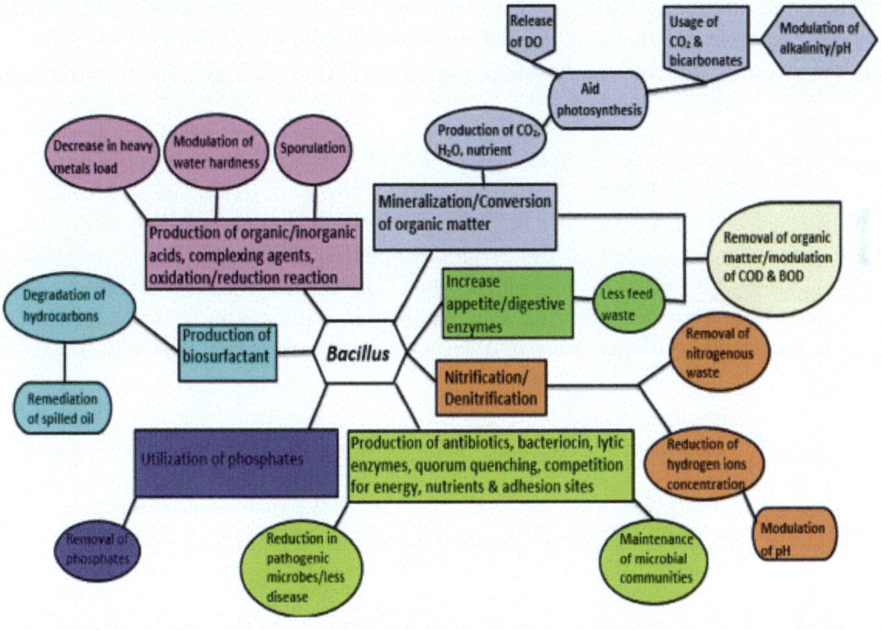

Fig. 21.3 Bacillus influences the quality of water in aquatic ecosystem

21.4.6 Fungi

According to Durairaj et al. (2015), fungi, in addition to bacteria, are known to produce certain enzymes including laccases, peroxidases, and catalases that have the ability to either break down organic contaminants or immobilize inorganic ones. The following genera of fungus have been discovered to be capable of breaking down aromatic hydrocarbons: *Aspergillus*, *Curvularia*, *Drechslera*, *Fusarium*, *Lasiodiplodia*, *Mucor*, *Penicillium*, *Rhizopus*, and *Trichoderma* (Chang et al. 2016). The culinary, pharmaceutical, textile, paper, and construction sectors are a few that employ fungal species. Since then, it has been shown that one efficient biotechnology strategy for getting rid of a lot of environmental contaminants is fungal bioremediation. Numerous studies have demonstrated the significance of fungi and their enzymes in eliminating pharmaceuticals and other persistent pollutants from a range of aquatic environments (Ferrando-Climent et al. 2015; Narayanan et al. 2022). Their inherent capacity to secrete a range of enzymes, such as laccases, peroxidases, and cytochrome P450 mixed function oxidases, has been linked to their efficacy in this area. These enzymes modify pharmaceutically active compounds through reduction, oxidation, hydroxylation, dehalogenation, dehydrogenation, deamination, formylation, and other processes (Ferrando-Climent et al. 2015; Narayanan et al. 2022) (Fig. 21.4).

Mycoremediation is a fungal-based technique that uses fungus to break down and remove contaminants. It's also known as myco-deterioration or myco-degradation. It is one of the most popular, economical, and environmentally responsible techniques that has recently attracted a lot of attention. There has been a proposal recently that

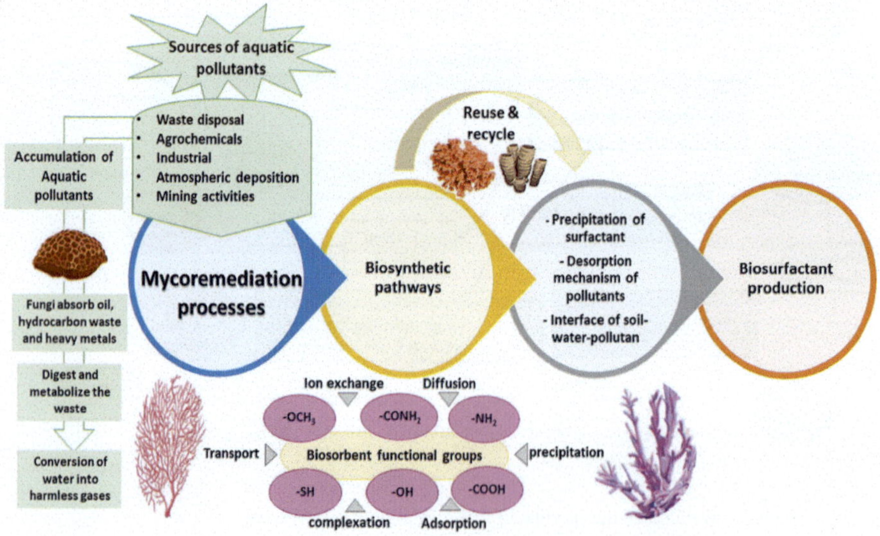

Fig. 21.4 Occurrence of fungi in aquatic environment (AbuQamar et al. 2023)

suggests using biotechnology to eliminate toxins by manipulating naturally existing bacteria. Using biological processes to degrade, convert, and remove hazardous substances from the environment is known as bioremediation, and it is an effort to reduce the frequency and severity of environmental pollution. Yeasts and molds are examples of fungi, or eukaryotes, that can contain one or more cells. They have the ability to convert complex chemicals into simpler ones through the use of digestive enzymes. Owing to their capacity to adapt to a broad variety of environmental circumstances, fungi are effective bioremediation agents. While fungi have been the subject of much study on land, less is known about them in aquatic settings, particularly those that are marine. Using naturally occurring processes, mycoremediation is often considered a safe and effective way for fungi to break down or sequester environmental pollutants. Several fungal species have the ability to efficiently absorb and eliminate metals and other contaminants from residual and overflow water. Regardless of whether they are alive or in the form of dried biomass, fungi are capable of decomposing xenobiotic chemicals and biosorbing metallic metals such as copper, iron, zinc, and manganese. According to AbuQamar et al. (2023) mycofiltration may eliminate microbiological pathogens from rainfall, cleanse industrial brewery effluent, repair drinking water sources polluted with heavy metals (HM), and completely remove nitrogen (N) and phosphorus (P) from dairy lagoon wastes.

21.4.7 Microalgae

Microalgae, primarily green algae from the genera Scenedemus, Selenastrum, or Chlorella, have been shown to be useful in breaking down polycyclic aromatic hydrocarbons including pyrene, naphthalene, and phenanthrene as well as immobilizing metals (DeLlasera et al. 2016). Through mechanisms primarily mediated by the production of exopolysaccharides, which can mediate the uptake of contaminants on the cell surface and/or their complexation into less bioavailable forms, microalgae remove toxic compounds by reducing their bioavailability and toxicity (Casillo et al. 2018). Depending on the microalgal taxa, the contaminant adhering to the membrane or cell wall exopolysaccharides may be absorbed and chelated by molecules of the phytochelatin classes, or it may stay adherent (Khullar and Reddy 2019). Using microalgal strains in conjunction with bacteria and fungus may be a feasible bioremediation technique for cleaning up petroleum hydrocarbon-contaminated marine habitats, according to current understanding.

Algae are simple aquatic plants lacking stalks, roots, or leaves. They are identified as photosynthetic eukaryotes with a very varied morphology. One kind of carbohydrate found in the cell walls of many algae is called cellulose. Numerous them are microscopic single-celled algae that are only visible under a microscope. The majority of algae are known to be unicellular, floating organisms that go by the name phytoplankton. There are between one and ten million algae in total, with a large percentage of microalgae. Microalgae are a very diverse group of unicellular organisms that develop quickly, don't require a lot of land for production, and have more high-lipid materials than food crops. Algae are classified as either freshwater or marine plants and are known to have extremely high carbon collecting and photosynthetic efficiency when compared to terrestrial plants. They also have a short growth phase. The ability of microorganisms, particularly microalgae, to bioaccumulate and absorb heavy metals has recently come to light. The ability of microalgae to differentiate between other types of metals and the critical ones required by the organisms makes them special. And this has been explained by the molecular mechanism that distinguishes them from other environmental monitors.

Microalgae have been used in many different biotechnological applications up to this point. They are regarded as useful tools and biological resources that are used in the evaluation and observation of pollutants or environmental poisons, including heavy metals. Additionally, the biological remediation of polluted water makes use of microalgae. This process is well-known for producing useful biological materials that aid in the manufacture of bioactive chemicals and biofuel. It also enables the efficient and speedy recycling of nutrients. Using algae to clean up the environment is believed to be more sustainable and beneficial to the ecology because they don't produce new pollutants. Moreover, microalgae may be effectively and possibly used to produce a broad range of high-value chemicals for biotechnological applications, including polyunsaturated fatty acids (PUFA), carotenoids, phytobiliproteins, and polysaccharides. Microalgae are also useful resources in the synthesis of biodiesel with high lipids production efficiency, according to Halim et al. (2012). Certain microalgae, like spirulina, have been utilized as food or health food due to their high nutritional content.

According to Priyadarshani et al. (2011), microalgae are tiny, unicellular, photosynthetic aquatic plants that may be found in freshwater and marine habitats. They are really different from one another. They were also characterized as minute and minuscule producers of metabolites and photosynthetic enzymes. This specific collection of phytoplankton has shown outstanding affinity, efficiency, sequestration of heavy metals, and maintained good performance when compared to other forms of bioremediators. Moreover, they have been recognized for their noteworthy contribution to controlling the levels of heavy metals in lakes and oceans.

21.4.7.1 Microalgae for Freshwater, Marine Water and Brackish Water Aquatic Environment Bioremediation

Algae are diverse organic compounds found in fresh, brackish, and marine waters. Microalgae including Spirulina sp., *Haematococcus pluvialis*, and *Dunaliella salina* are also utilized as a natural supply of colors for salmonid fish, prawn, and ornamental fish production. Three main types of algae have been effectively employed to create high amounts of important substances, including lipids, proteins, and pigments: Dunaliella sp., Chlorella sp., and Spirulina sp. (Abd El-Baky et al. 2002). The primary market criteria for attractive fish are color and look, hence Chlorella and Spirulina spp. are frequently added to fish meals (Sergejevová and Masojídek 2011). Certain microalgae are organisms that inhabit complex habitats that fluctuate in salinity, temperature, nutrients, and UVVis irradiation. As a result, in order to survive, these organisms must quickly adapt to their new surroundings and produce a wide range of secondary (biologically active) metabolites that are unique to them (Priyadarshani and Rath 2012). Microalgae pigments that have been transmitted to zooplankton have to potentially enhance nutritional value. “Green” microalgae (Tetraselmis spp.) are popular sources of the pigment lutein, which may be utilized to enhance Artemia’s nutritional content. When aquatic crustaceans were utilized as food, the lutein from halibut larvae could be transformed into vitamin A. Dunaliella salina is cultivated as a source of beta-carotene, a pigment used in photosynthetic processes. Beta-carotene is taken as a vitamin C supplement and as an orange color. High-value bioproducts that are isolated from microalgae include the following: biotin and vitamin E—*Euglena gracilis*; polyunsaturated fatty acids—Chlorella sp., Schizochytrium sp.; astaxanthin and leutin—*Haematococcus pluvialis*; and phycocyanin—*Spirulina platensis.* Micro-algae have a limited capacity to be used as food coloring sources. Food coloring made from algae is not photo stable, and cooking tends to fade the color.

Marine ecosystems are aquatic environments, such those found in and around seas, that have significant amounts of dissolved salt. the special configuration of biotic (life) and abiotic (nonliving) components that makes the oceanic environment special. Each maritime ecosystem has its own distinct features and habitats, including the deep sea, the polar regions, beaches, coastlines, coral reefs, and tides. Marine pollution is the term used to describe substances such as chemicals, plastics, oil, industrial and agricultural wastes, and home trash that end up in the ocean. This harms the nation’s economic structures, public health, and environmental quality. Fertilizer runoff into rivers and oceans is a result of chemical contamination, also referred to as nutrient pollution. Man-made objects (such plastic and bottles) that end up in the ocean are referred to as “marine trash”. The maritime ecology may also be negatively impacted by storm gusts, littering, and inefficient waste management. According to Muñoz-Palazon et al. (2023), there are two ways to address marine pollution: cleanup and prevention. The many persistent pollutants that naturally affect marine ecosystems make cleaning efforts more challenging. Thus, a primary goal for researchers is to develop a thorough remediation process for restoring damaged marine habitats that may be marketed after the required laboratory testing.

21.5 Screening of Bioremediatory Microbes

Several researchers assessed the effectiveness of both Gram-positive and Gram-negative bacteria as bio-remedial agents by culture; Table 21.4 shows their findings. Bacillus is employed the most frequently, followed by Aeromonas and Pseudomonas. A variety of enzymes produced by beneficial bacteria break down proteins and carbohydrates into smaller molecules so that other organisms can utilize them as fuel. One method to look into the prospect of bioremediation in eutrophicated settings is to find local bacteria that can break down a range of naturally occurring organic molecules,

Table 21.4 Organisms used as bio-remediators

Bio-remediator	Source	Used on	Method of application
Gram-positive bacteria			
Bacillus sp. 48	Common snook	*Centropomus undecimalis*	Added to water; reduced salinity
Bacillus sp	Commercial product	Penaeids	Water
Bacillus sp	Commercial product	Channel catfish	Spread in pond water
Mixed culture, mostly *Bacillus* sp.	Commercial product	*Brachionus plicatilis*	Mixed with water
Gram-negative bacteria			
Aeromonas media	Unknown	*Crassostrea gigas*	Mixed with water
Aeromonas CA2	Unknown	*Crassostrea gigas*	Mixed with water
Photorhodobacterium sp.	Unknown	*Penaeus chinensis*	Mixed with water
Pseudomonas fluorescence	*Onchorhynchus mykiss*	*Onchorhynchus mykiss*	Mixed with water to 10^5 or 10^6 cells ml^{-1}
Pseudomonas sp.	*Onchorhynchus mykiss*	*Onchorhynchus mykiss*	Mixed in water
Roseobacter sp. BS 107	Unknown	Scallop larvae	Mixed in water

such as proteins, carbohydrates, and lipids. One advantage of microbial bioremediation operations is the extensive number of routes that bacteria offer to metabolize or co-metabolize a large range of chemicals. Furthermore, because of the high costs of evolutionary adaptation and the germs' often short technology time, microbes are very sensitive to new chemicals. Bioremediation procedures can be used either ex-situ—where the contaminated matrix is removed and the treatment is carried out somewhere else—or in-situ—where pollutants are handled at the scene of the occurrence. One of the most popular applications of microbial bioremediation is the cleanup of industrial spills, such as oil spills (Biswas et al. 2022).

21.6 Bioremediation of Nitrogenous Mixes

The rise in popularity and expansion of intensive fish farming can be attributed to the growing worldwide need for animal protein. This is the result of two technologically sophisticated systems, recirculation aquaculture systems (RAS) and biofloc generation (BFT) (Zimmermann et al. 2023). By stocking fish in proportion to cubic meters of water, these structures allow for higher production than traditional clay pond fish farming (Owatari et al. 2018; Khanjani et al. 2022). Fish aquaculture is steadily increasing the number of fish available for the arena. Oreochromis niloticus, often known as the Nile tilapia, is the third most widely grown species globally, making for 9% of all finfish produced in inland aquaculture. Three mechanisms exist for nitrogenous compound organic stabilization in the Bio-Floc era: ammonia elimination by algae, ammonia conversion to nitrate by autotrophic bacteria, and nitrogenous compound immobilization in water by heterotrophic bacteria that also break down carbohydrates to produce protein. In extremely intense BFT systems, bacteria have the ability to ingest and attach nitrogenous substances, which lowers toxicity and maintains the stable lifestyle. However, solids still have a tendency to accumulate and become an issue when they reach unfavorable levels. This occurs as a result of the water's increasing percentage of chemical and biological oxygen requirement, which draws the attention of general solids. Since most fish ponds lack sediment filters, nitrates build up in them. To assist convert nitrates to nitrogen, a denitrifying filter is used. It converts nitrate to nitrogen gas by fostering the growth of anaerobic bacteria in an anaerobic environment. After being produced by nitrification, nitrate can proceed via a number of metabolic processes (Manduca et al. 2020).

21.7 Bioremediation of Hydrogen Sulphide

Aquaculture is interested in sulfur because of its importance in anoxic sediments. Natural sulfur breaks down into sulfide in aerobic conditions, which is subsequently oxidized to sulfate. In water, sulfate is clearly soluble, whereas in sediments, it dissolves gradually. Microorganisms in sediments facilitate sulfide oxidation, while chemical processes can also be used to initiate it completely. Sulfate can be substituted for oxygen in anaerobic conditions for microbial metabolism. Hydrogen sulfide gas is produced during this process. A sequence of reductions mediated by microbes results in the production of H_2 S.

$$SO_4^{2-} + 4H_2 + 2H^+ H_2S + 4H_2O$$

Natural loading has the potential to decrease benthic fauna range and H_2S generation. Since H_2S dissolves in water, it has been suggested that it causes fish illnesses such as gill rot. Fish are extremely poisoned by H_2S compounds at quantities that are expected to occur in aquaculture and natural waters. According to bioassays conducted on a variety of fish species, detectable levels of H_2S should be regarded as harmful to fish production (Ranjan and Bavitha 2014).

21.8 Conclusion

One method that shows promise for reviving aquatic ecosystems is aquatic bioremediation. In order to eliminate or lessen the effects of environmental pollutants in water bodies, microorganisms are used. The use of bioremediation is an economical and sustainable method. Microorganisms have many benefits when used in bioremediation, two of which are their easy manipulation and quick development. Diverse bacterial, fungal, and algal communities have been utilized for the purpose of eliminating diverse environmental contaminants. However, a number of variables, including as the kind and quantity of microorganisms, the chemical makeup of the contaminants, and the accessibility of vital nutrients like phosphorus and nitrogen, affect how well bioremediation works. Different metabolic pathways found in different types of microorganisms can be employed for bioremediation. A further strategy to enhance the efficiency of these microorganisms as aquaculture wastewater bio-remediators appears to be genetic alteration of bacteria. Microorganisms by themselves are unable to fully bioremediate aquaculture-generated effluents due to their unique properties. The different kinds of bioremediations and their applicability for different circumstances and settings are listed in this chapter. The chapter also includes information on the microorganisms that are important, potential influences on the bioremediation process, and how the use of newly developed technologies can increase the efficiency of bioremediation procedures. It also summarizes some of the mechanisms by which various bioremediation procedures operate. Many commercial products can be used in aquaculture, especially for in-depth aquaculture, to improve fish health, maintain a pleasant pond bottom, and ease the bottom. Because of the growing number of antibiotic-resistant bacterial strains, stricter environmental remediation laws, and cost effectiveness, the role of beneficial microorganisms in pathogen manipulation will be especially important in agriculture. Given the negative effects of aquaculture, research on the control of pond microbial ecology has the potential to increase the productivity and environmental "friendliness" of the global fishing sector. Employing bioremediation merchants can help the aquaculture industry as it expands gradually. The availability of microbial bio-remediators may be a factor in aquaculture's future success, given that they are properly deployed and confirmed via extensive experimental research.

CRediT Author Statement

P. Dhanalakshmi—Conceptualization, methodology, data curation, formal analysis, supervision, writing—original draft preparation, Writing—Review & Editing

P. Arunkumar—Conceptualization, methodology, data curation, formal analysis, visualization, investigation, supervision, writing—original draft preparation, Writing—Review & Editing

M.I. Farheena—methodology, data curation, formal analysis, writing—original draft preparation

A. Chithira—Conceptualization, formal analysis, visualization, investigation, Writing—original draft preparation, Writing—Review & Editing

S. Uthirasamy—methodology, data curation, investigation, writing—original draft preparation

S. Haritha—methodology, data curation, visualization, writing—original draft preparation

Acknowledgements Authors would like to thank Aquatic Biotechnology Unit, Department of Zoology, Bharathiyar University, Coimbatore (District), Tamil Nadu, India for their technical support.

Funding This research did not receive any specific grant from funding agencies in the public, commercial, or not-for-profit sectors.

Conflict of Interest All authors declare that they have no conflict of interest.

Data Availability Statement The data that support the findings of this study are available from the corresponding author, upon reasonable request.

References

Abd El-Baky HH, Moawd A, El-Behairy AN, ElBaroty GS (2002) Chemoprevention of benzo[a]pyreneinduced carcinogen and lipid peroxidation in mice by lipophilic algae extracts (phycotene). J Med Sci 2:185–193

AbuQamar SF, Abd HI, El-Fattah MM, Nader RA, Zaghloul TA, El-Mageed A, Selim S, Omar BA, Mosa WF, Saad AM, El-Tarabily KA, El-Saadony MT (2023) Exploiting fungi in bioremediation for cleaning-up emerging pollutants in aquatic ecosystems. Mar Environ Res 190:106068

Ahmad AL, Chin JY, Mohd Harun MHZ, Low SC (2022) Environmental impacts and imperative technologies towards sustainable treatment of aquaculture wastewater: a review. J Water Process Eng 46:102553. https://doi.org/10.1016/j.jwpe.2021.102553

Ankit, Bauddh K, Korstad J (2022) Phycoremediation: use of algae to sequester heavy metals. Hydrobiology 1:288–303. https://doi.org/10.3390/hydrobiology1030021

Ape F, Manini E, Quero GM, Luna GM, Sará G, Vecchio P, Mirto S (2019) Biostimulation of in situ microbial degradation processes in organically-enriched sediments mitigates the impact of aquaculture. Chemosphere 226:715–725

Audelo-Naranjo JM, Martínez-Córdova LR, Voltolina D, Gómez-Jiménez S (2011) Water quality, production parameters and nutritional condition of *Litopenaeus vannamei* (Boone, 1931) grown intensively in zero water exchange mesocosms with artificial substrates. Aquacult Res 42(9):1371–1377

Avnimelech Y (2009) Biofloc technology: a practical guide book. The World Aquaculture Society, Bato Rouge, LA

Ayangbenro AS, Babalola OO (2017) A new strategy for heavy metal polluted environments: a review of microbial biosorbents. Int J Environ Res Public Health 14(1):94

Badiola M, Basurko OC, Piedrahita R, Hundley P, Mendiola D (2018) Energy use in recirculating aquaculture systems (RAS): a review. Aquac Eng 81:57–70. https://doi.org/10.1016/j.aquaeng.2018.03.003

Bayen S, Estrada ES, Juhel G, Kit LW, Kelly BC (2016) Pharmaceutically active compounds and endocrine disrupting chemicals in water, sediments and mollusks in mangrove ecosystems from Singapore. Mar Pollut Bull 109:716–722

Bender J, Phillips P (2004a) Microbial mats for multiple applications in aquaculture and bioremediation. Bioresour Technol 94(3):229–238

Bender J, Phillips P (2004b) Microbial mats for multiple applications in aquaculture and bioremediation. Biores Technol 94(3):229–238

Biswas JK, Banerjee A, Biswas S (2022) Chapter 22—microbes and marine oil spills: oil-eating bugs can cure oily sea sickness. In: Das P, Manna S, Pandey JK (eds) Advances in oil-water separation. Elsevier, pp 393–422

Borchert E, Hammerschmidt K, Hentschel U, Deines P (2021) Enhancing microbial pollutant degradation by integrating eco-evolutionary principles with environmental biotechnology. Trends Microbiol 29:908–918

Brix H (1999) How 'green' are aquaculture, constructed wetlands and conventional waste water treatment systems? Water Sci Technol 40:45–50. https://doi.org/10.1016/S02731223(99)00418-7

Cao H-L, Liu C, Cai F-Y, Qiao X-X, Dichiara AB, Tian C (2020) In situ immobilization of ultra-fine Ag NPs onto magnetic Ag@RF@Fe_3O_4 core-satellite nanocomposites for the rapid catalytic reduction of nitrophenols. Water Res 179:115882

Casillo A, Lanzetta R, Parrilli M, Corsaro MM (2018) Exopolysaccharides from marine and marine extremophilic bacteria: structures, properties, ecological roles and applications. Mar Drugs 16:69

Castellanos-Estupiñán MA, Carrillo-Botello AM, Rozo-Granados LS, Becerra-Moreno D, García-Martínez JB, Suarez NAU, López-Barrera GL, Barajas-Solano AF, Bryan SJ, Zuorro A (2022) Removal of nutrients and pesticides from agricultural runoff using microalgae and cyanobacteria. Water 14(4):558. https://doi.org/10.3390/w14040558

Chandel N, Ahuja V, Gurav R, Kumar V, Tyagi VK, Pugazhendhi A (2022) Progress in microalgal mediated bioremediation systems for the removal of antibiotics and pharmaceuticals from wastewater. Sci Total Environ 825:153895

Chang YT, Lee JF, Liu KH, Liao YF, Yang V (2016) Immobilization of fungal laccase onto a nonionic surfactant-modified clay material: application to PAH degradation. Environ Sci Pollut Res 23:4024–4035

Cuthbert M, Riley, MS, Sidhu SH, Renshaw JC, Tobler DJ, Phoenix VR, Mackay R (2012) Controls on the rate of ureolysis and the morphology of carbonate precipitated by *S. Pasteurii* biofilms and limits due to bacterial encapsulation. Ecol Eng 41:32–40

DeLlasera MPG, Olmos-Espejel JJ, Díaz-Flores G, Montaño-Montiel A (2016) Biodegradation of benzo(a)pyrene by two freshwater microalgae *Selenastrum capricornutum* and *Scenedesmus acutus*: a comparative study useful for bioremediation. Environ Sci Pollut Res 23:3365–3375

Durairaj P, Malla S, Nadarajan SP, Lee PG, Jung E, Park HH, Kim BG, Yun H (2015) Fungal cytochrome P450 monooxygenases of Fusarium oxysporum for the synthesis of-hydroxy fatty acids in engineered Saccharomyces cerevisiae. Microb Cell Factories 14:1–16

Ebeling JM, Timmons MB, Bisogni J (2006) Engineering analysis of the stoichiometry of photoautotrophic, autotrophic, and heterotrophic removal of ammonia nitrogen in aquaculture systems. Aquaculture 257(1–4):346–358

Fashola MO, Ngole-Jeme VM, Babalola OO (2020) Heavy metal immobilization potential of indigenous bacteria isolated from gold mine tailings. Int J Environ Res 14(1):71–86

Feil A, Pretz T (2020) Mechanical recycling of packaging waste. In: Plastic waste and recycling, pp 283–319

Ferrando-Climent L, Cruz-Morató C, Marco-Urrea E, Vicent T, Sarrà M, Rodriguez-Mozaz S (2015) Non-conventional biological treatment based on Trametes versicolor for the elimination of recalcitrant anticancer drugs in hospital wastewater. Chemosphere 136:9–19. https://doi.org/10.1016/j.chemosphere.03.051

Ge M, Wang X, Yang G, Wang Z, Li Z, Zhang X (2021) Persistent organic pollutants (POPs) in deep-sea sediments of the tropical western Pacific Ocean. Chemosphere 277:130267

Griebler C, Malard F, Lefébure T (2014) Current developments in groundwater ecology—from biodiversity to ecosystem function and services. Curr Opin Biotechnol 27:159–167

Halim R, Danquah MK, Paul A (2012) Webley. Extraction of oil from microalgae for biodiesel production: a review. Biotechnol Adv 30:709–732

Hargreaves JA (2013) Biofloc production systems for aquaculture. Southern Regional Aquaculture Center, Stoneville, MS, pp 1–11

Helmy Q, Kardena E, Gustiani S (2019) Probiotics and bioremediation microorganisms. IntechOpen

Hlordzi V, Kuebutornye FKA, Gyamfua A, Abarike ED (2020) The use of Bacillus species in maintenance of water quality in aquaculture: a review. Aquacult Rep 18:100503

Hosseini N, Johnston J, Lindenschmidt K-E (2017) Impacts of climate change on the water quality of a regulated prairie river. Water 9(3):199. https://doi.org/10.3390/w9030199

Iglina T, Iglin P, Pashchenko D (2022) Industrial CO_2 capture by algae: a review and recent advances. Sustainability 14(7):3801. https://doi.org/10.3390/su14073801

Jahangiri L, Esteban MA (2018) Administration of probiotics in the water in finfish aquaculture systems: a review. Fishes 3:30–33

Jain M, Khan SA, Sharma K, Jadhao PR, Pant KK, Ziora ZM, Blaskovich MAT (2022) Current perspective of innovative strategies for bioremediation of organic pollutants from wastewater. Bioresourc Technol 344(Part B):126305

Jaiswal S, Singh DK, Shukla P (2019). Gene editing and systems biology tools for pesticide bioremediation: a review. Front Microbiol 10

Kelly A, Lannuzel D, Rodemann T, Meiners KM, Auman HJ (2020) Microplastic contamination in east Antarctic Sea ice. Mar Pollut Bull 154:111130

Khanjani MH, Sharifinia M, Hajirezaee S (2022) Recent progress towards the application of biofloc technology for tilapia farming. Aquaculture 552:738021

Khullar S, Reddy MS (2019) Cadmium and arsenic responses in the ectomycorrhizal fungus Laccaria bicolor: glutathione metabolism and its role in metal(loid) homeostasis. Environ Microbiol Rep 11:53–61

Koops H-P, Pommerening-Röser A (2001) Distribution and ecophysiology of the nitrifying bacteria emphasizing cultured species. FEMS Microbiol Ecol 37(1):1–9

Kour D, Khan SS, Kour H, Kaur T, Devi R, Rai PK (2022) Microbe mediated bioremediation: current research and future challenges. J Appl Biol Biotechnol 10(2):6–24

Kumar S, Dagar VK, Khasa YP, Kuhad RC (2013) Genetically modified microorganisms (GMOs) for bioremediation biotechnology for environmental management and resource recovery. Springer, pp 191–218

Kurtela A, Antolović N (2019) The problem of plastic waste and microplastics in the seas and oceans: impact on marine organisms. Croatian J Fish 77:51–56

Lebreton LCM, van der Zwet J, Damsteeg J-W, Slat B, Andrady A, Reisser J (2017) River plastic emissions to the world's oceans. Nat Commun 8:15611

Leong YK, Chang J-S (2020) Bioremediation of heavy metals using microalgae: recent advances and mechanisms. Bioresource Technol 303:122886

Manduca LG, da Silva MA, de Alvarenga ER, de Oliveira Alves GF, de Araújo Fernandes AF, Assumpção AF, Cardoso CC, de Sales SCM, de Alencar Teixeira E, de Almeida e Silva M, Turra EM (2020) Effects of a zero exchange biofloc system on the growth performance and health of Nile tilapia at different stocking densities. Aquaculture 521:735064

Martínez-Córdova LR, Martínez-Porchas M, Emerenciano MGC, Miranda-Baeza A, Gollas-Galván T (2017) From microbes to fish the next revolution in food production. Crit Rev Biotechnol 37(3):287–295

Materić D, Holzinger R, Niemann H (2022) Nanoplastics and ultrafine microplastic in the Dutch wadden Sea—the hidden plastics debris? Sci Total Environ 846:157371

Methneni N, Morales-Gonzalez JA, Jaziri A, Mansour HB, Fernandez-Serrano M (2021) Persistent organic and inorganic pollutants in the effluents from the textile dyeing industries: ecotoxicology appraisal via a battery of biotests. Environ Res 196:110956

Moriarty DJ (1999) Disease control in shrimp aquaculture with probiotic bacteria. In: Paper presented at the proceedings of the eighth international symposium on microbial ecology

Narayanan M, El-Sheekh M, Ma Y, Pugazhendhi A, Natarajan D, Kandasamy G (2022) Current status of microbes involved in the degradation of pharmaceutical and personal care products (PPCPs) pollutants in the aquatic ecosystem. Environ Pollut 300:118922–118912. https://doi.org/10.1016/j.envpol.2022.118922

Nunes AJP, Castro LF, Sabry-Neto H (2010) Microbial flocs spare protein in white shrimp diets. Glob Advocate 10:28–30

Ojemaye CY, Petrik L (2022) Pharmaceuticals and personal care products in the marine environment around false bay, cape town, south Africa: occurrence and risk-assessment study. Environ Toxicol Chem 41:614–634

Ojuederie OB, Babalola OO (2017) Microbial and plant-assisted bioremediation of heavy metal polluted environments: a review. Int J Environ Res Public Health 14:1504

Owatari MS, Jesus GFA, de Melo Filho MES, Lapa MR, Martins ML, Mouriño JLP (2018) Synthetic fibre as biological support in freshwater recirculating aquaculture systems (RAS). Aquacult Eng 82:56–62

Palazón BM, Martínez MH, Rosa-Masegosa A, Susanna (2023) GorrasiSimultaneous removal of nitrate and pesticides from contaminated groundwater using aerobic granular biomass technology. J Environ Chem Eng 11(2)

Patel AK, Singhania RR, Albarico FPJB, Pandey A, Chen C, Dong C (2022) Organic wastes bioremediation and its changing prospects. Sci Total Environ 824:153889

Peng G, Bellerby R, Zhang F, Sun X, Li D (2020) The ocean's ultimate trashcan: Hadal trenches as major depositories for plastic pollution. Water Res 168:115121

Perera IC, Hemamali EH (2022) Genetically modified organisms for bioremediation: current research and advancements. Bioremediat Environ Pollut: Emerg Trends Strat 1:163–186

Prasad R, Gupta SK, Shabnam N, Oliveira CYB, Nema AK, Ansari FA, Bux F (2021) Role of microalgae in global CO_2 sequestration: Physiological mechanism, recent development, challenges, and future prospective. Sustainability 13(23):13061. https://doi.org/10.3390/su132313061

Pratush A, Kumar A, Hu Z (2018) Adverse effect of heavy metals (As, Pb, Hg, and cr) on health and their bioremediation strategies: a review. Int Microbiol 21(3):97–106

Priyadarshani I, Rath B (2012) Commercial and industrial applications of micro algae—a review. J Algal Biomass Utln 3(4):89–100

Priyadarshani I, Sahu D, Rath B (2011) Microalgal bioremediation: current practices and perspectives. J Biochem Technol 3(3):299–304

Pujari M, Kapoor D (2021) 1—heavy metals in the ecosystem: sources and their effects. In: Kumar V, Sharma A, Cerdà A (eds) Heavy metals in the environment. Elsevier

Ranjan R, Bavitha MS (2014) Bioremediation—a potential tool for management of aquatic pollution. Int J Multidiscip Res Develop 1(7):335–340

Robles-Porchas GR, Gollas-Galván T, Martínez-Porchas M, Martínez-Cordova LR, Miranda-Baeza A, Vargas-Albores F (2020) The nitrification process for nitrogen removal in biofloc system aquaculture. Rev Aquaculture 12(4):2228–2249

Rodela T (2013) The humble beginnings of the study of nitrogen excretion in fish. J Exp Biol 216(2):162–163

Roveda M, de Menezes CCA, Bolívar-Ramírez NC, Owatari MS, Jatobá A (2024) Acidifying remediation and microbial bioremediation decrease ammoniacal nitrogen, orthophosphates, and total suspended solids levels in intensive Nile tilapia farming under biofloc conditions. Aquaculture 580(Part 1)

Roy P, Mohanty AK, Misra M (2022) Microplastics in ecosystems: their implications and mitigation pathways. Environ Sci Adv 1

Sattar S, Hussain R, Shah SM, Bibi S, Ahmad SR, Shahzad A, Zamir A, Rauf Z, Noshad A, Ahmad L (2022) Composition, impacts, and removal of liquid petroleum waste through bioremediation as an alternative clean-up technology: a review. Heliyon 8:e11101

Sergejevová M, Masojídek J (2011) Chlorella biomass as feed supplement for freshwater fish: Sterlet, *Acipenser ruthenus*. Aquaculture Res

Silva A, Delerue-Matos C, Figueiredo SA, Freitas OM (2019) The use of algae and fungi for removal of pharmaceuticals by bioremediation and biosorption processes: a review. Water 11:1555

Silva VF, Pereira SA, Martins MA, Rezende PC, Owatari MS, Martins ML, Mouriño JLP, do N. Vieira F (2023) Hemato-immunological parameters can be influenced by microalgae addition and fish feed supplementation in the integrated rearing of Pacific whitshrimp and juvenile Nile tilapia using biofloc technology. Aquaculture 574:739622

Singh D, Goswami RK, Agrawal K, Chaturvedi V, Verma P (2022) Bio-inspired remediation of wastewater: a contemporary approach for environmental clean-up. Curr Res Green Sustain Chem 5:100261

Soltan MA (2016) Aquaculture systems. In: Aquaculture. Chapter

Syranidou E, Karkanorachaki K, Amorotti F, Franchini M, Repouskou E, Kaliva M, Vamvakaki M, Kolvenbach B, Fava F, Corvini PFX, Kalogerakis N (2017) Biodegradation of weathered polystyrene films in seawater microcosms. Sci Reports 7:17991

Ter Halle A, Jeanneau L, Martignac M, Jardé E, Pedrono B, Brach L (2017) Nanoplastic in the north Atlantic subtropical gyre. Environ Sci Technol 51:13689–13697

Thamdrup B, Dalsgaard T (2002) Production of N_2 through anaerobic ammonium oxidation coupled to nitrate reduction in marine sediments. Appl Environ Microbiol 68(3):1312–1318

Thirumalaisamy R, Suriyaprabha R, Prabhu M, Sakthi Thesai A (2022) Role of nanomaterials in environmental remediation: recent advances—a review. Strat Tools Pollutant Mitig 51–68

Timalsina H, Gyawali T, Ghimire S, Paudel SR (2022) Potential application of enhanced phytoremediation for heavy metals treatment in Nepal. Chemosphere 306:135581

Tom AP, Jayakumar JS, Biju M, Somarajan J, Ibrahim MA (2021) Aquaculture waste water treatment technologies and their sustainability: a review. Energy Nexus 4:100022. https://doi.org/10.1016/j.nexus.2021.100022

Tyagi B, Kumar N (2020) Bioremediation: principles and application in environmental management, toxicity, mechanisms of contaminants degradation, detoxification and challenges. In: Bioremediation: principles and application in environmental management. Publisher: Elsevier

Valenzuela EF, Miranda AM, Baeza, Vega MER, Arizmendi VM, Beltrán-Ramírez O, Emerenciano MGC (2021) Water quality and productive response of *Litopenaeus vannamei* reared in biofloc with addition of commercial strains of nitrifying bacteria and *Lactobacillus rhamnosus*. Aquaculture 542:736869

Vázquez-Romero B, Perales JA, de Vree JH, Böpple H, Steinrücken P, Barbosa MJ, Kleinegris DMM, Ruiz J (2022) Techno-economic analysis of microalgae production for aquafeed in Norway. Algal Res 64:102679. https://doi.org/10.1016/j.algal.2022.102679

Verma S, Kuila A (2019) Bioremediation of heavy metals by microbial process. Environ Technol Innov 14:100369

Weihrauch D, Allen GJ (2018) Ammonia excretion in aquatic invertebrates: new insights and questions. J Exp Biol 221:2

Xiang L, Li G, Wen L, Su C, Liu Y, Tang H (2021) Biodegradation of aromatic pollutants meets synthetic biology. Synthetic Syst Biotechnol 6:153–162

Zhang Z, Yan K, Zhang L, Wang Q, Guo R, Yan Z (2019) A novel cadmium-containing wastewater treatment method: bio-immobilization by microalgae cell and their mechanism. J Hazard Mater 374:420–427

Zimmermann HH, Stoof-Leichsenring KR, Dinkel V, Harms L, Schulte L, Hütt M, Nürnberg D, Tiedemann R, Herzschuh U (2023) Marine ecosystem shifts with deglacial sea-ice loss inferred from ancient DNA shotgun sequencing. Nat Commun 14:1650

Zokaeifar H, Babaei N, Saad CR, Kamarudin MS, Sijam K, Balcazar JL (2014) Administration of Bacillus subtilis strains in the rearing water enhances the water quality, growth performance, immune response, and resistance against *Vibrio harveyi* infection in juvenile white shrimp, *Litopenaeus vannamei*. Fish Shellfish Immunol 36:68–74

Chapter 22
Review on Microplastics as an Emerging Pollutant in Ecosystem and Its Impact: Integrated Approaches of Microplastic Biodegradation

Suganya Ilango, Aishwarya Vetrivel, Gopi Devarajan, and T. G. Nithya

Abstract Microplastics are tiny plastic particles (<5 mm), which are prevalent and slowly degrading pollutant in soils and waterways, having long shelf-life period, extremely stable, high disintegration potential and capable of adsorbing other pollutants. Microplastics are easily acquired by plankton and invertebrates and are present in many marine species. At that point, microplastics pass on with food chains, causing physical damages and reduction in the nutritional content of the diet, which causes various effects in living organisms. Scientists remain unknown that, microplastics consumed by living organisms are threatening to the health, and if that was the case, what specific risks could they face. And many countries are making efforts to lessen the amount of microplastics in the environment. Microplastics are also more likely to carry various contaminants due to their adsorption properties. In fact, there is a need of modern techniques to enhance degradation of microplastics in terrestrial, aquatic and other ecosystems. Bioremediation is an emerging technique for eliminating the toxic waste, from contaminants. In the bioremediation process, contaminated regions are cleaned up using a range of microorganisms, including aerobes and anaerobes. Microorganisms are crucial for the bioremediation process because it removes, breaks down, eliminates, and inhibits toxic wastes and contaminants. Therefore, this review aims to focus on understanding the evolution, fate and impact of microplastics in environment and discuss in detail about the modern biotechnological methods to enhance microplastic bioremediation and to explore the manner in

S. Ilango (✉) · T. G. Nithya
Department of Biochemistry, Faculty of Science and Humanities, SRM Institute of Science and Technology, Kattankulathur, Chennai, Tamil Nadu 603203, India
e-mail: suganyai@srmist.edu.in

A. Vetrivel
Department of Biotechnology, Dr. G. R. Damodaran College of Science, Coimbatore, Tamil Nadu 641014, India

G. Devarajan
Department of Economics, Faculty of Science and Humanities, SRM Institute of Science and Technology, Kattankulathur, Chennai, Tamil Nadu 603203, India

© The Author(s), under exclusive license to Springer Nature Switzerland AG 2025
A. Jeyaseelan and K. Murugasen (eds.), *Sustainable Environmental Remediation: Avenues in Nano and Biotechnology*, Environmental Science and Engineering,
https://doi.org/10.1007/978-3-031-78483-5_22

which bacteria, fungi, and algae contributes to biodegradation and proposed biotechnological techniques to speed up the process, such as bioinformatics and gene editing tools.

Keywords Bioremediation · Microplastics · Ecosystem · Pollutants · Emerging technology

22.1 Introduction

Materials made of plastic are becoming an essential part of our daily existence. These pieces of polymers with long chains are entering the ecosystem for the sake of industrialized human activities. Plastics' flexibility to a broad range of conditions has led to significant worldwide growth in both product production and consumption. Stability, light weight, affordability, and adaptability were among the qualities that allowed it to be used in a variety of applications. Over 350 million tons of plastic are manufactured annually, according to recent studies. Polyethylene, acrylonitrile butadiene styrene, nylon, polyvinyl alcohol, polycarbonate, polyvinyl chloride, polyamide, polypropylene, polystyrene and other types are among the various forms of plastics that can be produced (Jaiswal et al. 2022). The huge plastic pieces that are exposed to various environmental factors such as UV radiation, impact abrasion, hydrolysis, soil erosion, or biological processes, eventually generate solid fragments known as microplastics that have a particle size of less than 5 mm (Ya et al. 2021). Because they have detrimental impacts on living things, microplastics have acquired a lot of attentiveness in the realm of science. Technically speaking, microplastics are tiny particles (less than 5 mm) made from unprocessed plastic components. Examples of these include the tiny spheres found in toothpaste, soap, and cosmetics (Ricardo et al. 2021).

Plastics in the environment, which are frequently the result of consumer products that have been improperly disposed off, slowly deteriorate due to photo- and thermooxidative processes, as well as biodegradation to a lesser extent. This weakens the material integrity of the plastic, which causes it to fragment into pieces fewer than 5 mm, which are known as secondary microplastics. Primary microplastics are plastic particles of this size that are purposefully created for use in products (like toothpaste or exfoliants) or by businesses (like air blasting) (Prata et al. 2020).

Numerous aquatic ecosystems worldwide have been shown to be contaminated by microplastics, which has a detrimental effect on a wide range of creatures. Microplastic pollution of the environment is currently regarded as a developing danger to the functioning of ecosystems and biodiversity (Boots et al. 2019).

Plastics can linger in the waters of the ocean for a number of years after they get there. In the interim, they are broken down by mechanical and photochemical reactions and mechanisms to produce microplastics (less than 5 mm) (Issac and Kandasubramanian 2021).

Plastic debris is found in all major ocean basins worldwide as a result of plastic waste generated on land entering the marine environment through dumping of plastic waste into the sea, leaching from open dumpsites or sewage effluents, riverine runoffs, spilling during transport or accidents, and washing off beaches. Because of their small size and larger surface area, microplastics and nanoplastics have the ability to penetrate tissues and cells, and react with other molecules in the surrounding environment. There is growing evidence that microplastic pollution occurs not only in the marine environment but also in the terrestrial environment due to the physical or chemical breakdown of stronger plastic items. Microplastic particles can be produced by the aging, degradation, and fragmentation of plastic goods that are widely used in daily life. The accumulation of microplastics in soil can be attributed to two main processes: the fragmentation of discarded plastic items or the leaching of buried waste materials from landfills (Thakur et al. 2022).

Microplastic contamination of soil can affect its qualities in both positive and negative ways. Furthermore, it has been demonstrated that microplastics interact with a broad range of contaminants, such as antibiotics, heavy metals, pesticides, and persistent organic pollutants. They may also serve as a vector for the transmission of toxins in terrestrial circumstances. Microplastics and the chemicals they are linked to have the ability to migrate through food webs and accumulate at different trophic levels, which could have a negative impact on the health of both people and other living things (Dissanayake et al. 2022).

Rising environmental contamination brought on by human activity is a major source of concern. In addition to microplastics, from throwaway products, there are other harmful contaminants include air pollutants like CO_2, CH_4, and SO_2 from industrial and agricultural processes, and aromatic compounds, halogens, pesticides, dyes, and heavy metals from factories. These harmful substances may result in malignancies as well as cardiovascular and respiratory conditions (Tran et al. 2021).

Nowadays, bioremediation is very popular because of its financial benefits, protection, and longevity. Chemical and physical mechanisms are used in the operation of conventional remediation techniques. After being introduced into another medium—such as air or water—and gathered in a concentrated state, contaminants are either eliminated or broken down—for example, by oxidation—into a smaller, safer form. Chemical techniques can be expensive and have limited efficiency. In addition, the chemicals used in decomposition reactions can result in hazardous intermediates that must be removed. Living things, like bacteria, use pollutants and harmful substances as growth nutrients to carry out bioremediation. Through a series of physiological reactions known as metabolism, pollutants are either entirely eliminated from living things or converted into non-hazardous compounds. Compared to other cleansing techniques, bioremediation has numerous benefits. Because bioremediation only uses the organic metabolic processes found in living systems, it creates few byproducts and causes the least amount of environmental harm. Furthermore, bioremediation eliminates the need for large, fossil fuel-powered machinery that releases toxins into the air since living things are miniature factories with all the components needed to break down hazardous substances. As a result, compared to other techniques, bioremediation is less expensive and cleaner (Tran et al. 2021).

A biotechnological procedure known as "bioremediation" modifies pollutants using biological means in order to manage them through mineralization and detoxification while also repairing the recalcinated environment. Microbial enzyme remediation of persistent organic pollutants (POPs) is regarded as a novel, cost-effective, environmentally benign, and promising approach. The bioremediation method is not without its limits, though. Only a few kinds of bacteria that can produce particular enzymes have shown their strong ability to digest contaminants thus far, and it is a slow process. Because genetically modified microbes produce a significant number of required enzymes under optimal conditions, they are preferred for bioremediation (Bhandari et al. 2021).

Microplastics' ecotoxicity to living things poses a continuous risk to the ecosystem. Since the majority of their surfaces are negatively charged, they can attract positively charged contaminants. Furthermore, animals or microbes may potentially absorb them. They can also set off aberrant immunological responses that can harm health, as well as aberrant metabolic processes in the bodies of living things (Zhou et al. 2022).

22.2 Evolution and Fate of Microplastics

The main sources of plastic particles are the transportation lines, sewage treatment plants, air blasting, fishing industries, apparel and cosmetics industries, plastic production plants, automobile and truck tires, and shipping companies (Fig. 22.1). Plastic particles can travel through a number of channels, including stormwater drainage, sewage treatment plant emissions, wind, and domestic wastewater discharge. The primary sources and routes for the release of microplastics, including various synthetic fibers and micro-exfoliates, into the environment are the wastewater discharged from various household activities (Dey et al. 2021).

Plastic wastes are commonplace and accumulate through rainwater runoff in the terrestrial system, including cultivable soil, roadways, and landfills, as a result of ineffective waste management. Road drainage is one way that microplastics from tires and wear and tear on the road can enter the environment. In fact, 42% of plastic particles that migrate from rivers into marine ecosystems do so via this route. While freshwater and terrestrial ecosystems are thought to be the main sources of microplastics, oceans are thought to be the primary sinks for plastic particles. Microplastics may break up even more during migration and end up in the sediments along the shore and in bodies of water (Dey et al. 2021).

It is well accepted that microplastics eventually find their way into the ocean after they enter a body of water. According to estimates, around 70% of marine debris ends up as sediment at the ocean's bottom. Of the remaining 30%, 15% float on the surface of the sea, and the remaining 15% are found in coastal areas (Wu et al., 2019a, b).

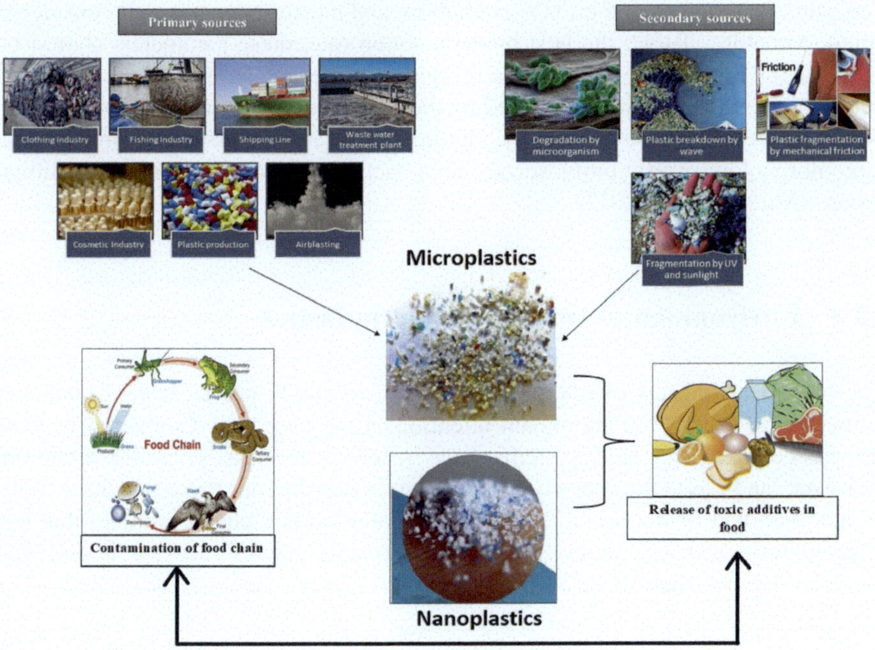

Fig. 22.1 Sources of plastics and its effects

22.3 Mechanisms of Microbial Degradation of Microplastics

By means of their enzymatic machinery, the applied microorganisms transform contaminants into CO_2, H_2O, and inorganic substances. As an example, the oxidation of ethylene groups is carried out by oxidative enzymes, such as peroxidase, monooxygenase, manganese peroxidase, oxidase, and dehydrogenase (Miloloža et al. 2022).

In general, biodegradation occurs in two main phases. The first is the polymer's molecular weight decrease, which permits molecules to pass past cell membranes and target enzymes within the cells. The second stage involves the metabolization of the carboxylic acids through the Krebs cycle and β-oxidation once they have been transformed from hydrocarbons (Miloloža et al. 2022).

Both biotic and abiotic variables can influence the biodegradation of microplastics. The kind and quantity of living cells, the existence of a single or mixed culture, the hydrophobicity of cells, the variety of enzymatic systems, the release of extracellular enzymes, and the potential for the synthesis of biosurfactants are all regarded as biotic factors. Environmental parameters including the media's pH and temperature, salinity, oxygen content, UV irradiation, and moisture level are all considered abiotic variables. To get the best biodegradation rate, these parameters should be changed for each of the applied microorganisms separately. The mechanism and rate of biodegradation are determined by the properties of polymers, which include flexibility, hydrophobicity, elasticity, melting temperature, glass transition temperature, morphology, crystallinity, surface area, tacticity, molecular weight, functional groups (Miloloža et al. 2022).

22.4 Environmental Impact of Microplastics

The biological impacts of microplastics, an increasingly prevalent environmental pollutant, have garnered significant attention in the past year. Compared to their aquatic counterparts, land systems have received far less scientific attention. However, land-based microplastic contamination may be four to twenty-three times greater than that of the ocean. Microplastic pollution is a global problem that has affected freshwater systems, coastal waters, marine waters, soil ecosystems, and even the polar regions (Sana et al. 2020).

22.4.1 *Soil*

Numerous significant ecosystem services, including food production, gas exchange and carbon sequestration, soil formation and structure maintenance, insect control, organic matter decomposition, and cycling, are provided by soil ecosystems. Plastic mulches, sewage sludge land use, plastic-coated fertilizer and compost, soil supplements, wastewater-irrigation, road runoff, and air deposition are some of the sources of microplastics entering the soil environment. Of which compost and mulching films used in agriculture were thought to be the leading origin of microplastics in the soil environment. The soil environment inherently influences the growth of terrestrial plants, which are an essential part of the soil ecosystem. In addition to decreasing soil microbial activity and biomass and functional diversity, microplastics in the soil can also disrupt the plant nutrition cycle, which in turn affects the germination of plant seeds and the growth of seedlings (Du and Wang 2021).

22.4.2 *Air*

Urban dust, abrasion of rubber tires, furniture, synthetic textile apparel, and construction materials are the main sources of microplastics in the air. Furthermore, the airborne microplastics that wind up in aquatic or terrestrial habitats are a result of resuspended soil from automobiles that is suspended in the atmosphere and gets transported away. By wind erosion, airborne microplastics spread throughout terrestrial habitats, posing a danger of exposure from dust-borne particles that can be inhaled (Sridharan et al. 2021).

22.4.3 Water

These contaminants are found in large quantities in coastal areas, oceanic surface waters, the water column, and sediments, and the oceans are emerging as important sinks for them. According to studies, microplastics pose a major risk to both humans and the marine ecosystem. Both live things and non-living environmental elements have previously shown the detrimental consequences of marine microplastic pollution. Microplastics are known to be harmful to fish, higher marine species, phytoplankton, and zooplankton, despite the fact that their exact mechanisms are yet unknown (Sharma et al. 2023).

22.5 Impact of Microplastics on Human Health

We now know very well that there are a number of ways that microplastics can enter our bodies, including through eating, skin contact, and inhalation. The most frequent route for humans to be exposed to microplastics among these is by consuming food. Microplastics can be found in plastic food packaging and utensils, as well as in tainted seafood and drinking water. Microplastics have the ability to build up in the digestive system and other organs after consumption. Another method that people might be exposed to microplastics is by inhalation, especially if they work in industries where plastic is produced or handled. Because they can escape during production, shipping, and disposal, microplastics may be found in the atmosphere. However, using personal care products that include microbeads, such as body washes and face scrubs, might result in cutaneous exposure to microplastics. Synthetic textiles and apparel, which can shed microfibers after washing and wear, can also contain microplastics. Although human health effects of microplastic exposure are still being investigated, there is proof that microplastic contact has a number of detrimental effects on human health. Numerous studies demonstrate that microplastics may cause genotoxicity, cellular damage, and inflammation in human cells (Table 22.1). To safeguard both human health and ecosystems, it is crucial to limit exposure to microplastics and their discharge into the environment (Ghosh et al. 2023).

The consumption of contaminated food is one of the main ways that nano- and microplastics enter the human body. According to a 2019 study by Cox et al., there were 0.44 microplastics/g of microplastics in sugar, 0.11 microplastics/g in salt, 0.03 microplastics/g in alcohol, and 0.09 microplastics/g in pure water. Additionally, plants (fruits and vegetables) that absorb microplastics from contaminated soil may

Table 22.1 Effects of microplastics and nanoplastics

Biological effects	Material	Size	Inference	References
Cell membranes	Polystyrene	100 μm	Synthetic non-ionic surfactant bilayers' hydrophobic core was penetrated by polystyrene chains, which altered the bilayers' elastic and structural characteristics	Radlinska et al. (1995)
	Polystyrene	5 μm	Polystyrene microplastics treatment, inhibits the plasma membrane ABC transporter activity	Wu et al. (2019a, b)
		0.1 μm	Showed stronger inhibitory effect than 5 μm	
	Polystyrene	50 and 100 μm	Cell membrane breakdown	Bhattacharjee et al. (2014) and Ruenraroengsak et al. (2012)
Oxidative stress	Polystyrene	20 μm	Induced ROS production in Hs27, T98G, HeLa and EAhy926 cell lines	Meindl et al. (2015); Poma et al. (2019); Schirinzi et al. (2017)
	Polyethylene (PE) and polystyrene	3–16 μm and 10 μm	Induced ROS production in T98G and HeLa cell lines	Schirinzi et al. (2017)
	Polystyrene	100 nm	The activation of the NLRP3 inflammasome is induced, which in turn causes human macrophages to produce IL-1β	Lunov et al. (2011)
	Polystyrene	200 nm	Polystyrene particles cause oxidative stress	Meindl et al. (2015)
	Polystyrene	20 nm	Induced ROS production in HepG2 cells	He et al. (2020)
Inflammatory effects	Polystyrene	50 nm and 100 nm	Modulating the inflammatory mediators in cell lines, including A549, AGS, and human alveolar epithelial type 1	Ruenraroengsak et al. (2012); Wu et al. (2019a, b); Xu et al. (2019)
	Polystyrene	50 nm	Increased IL-6 and IL-8 secretion	Ruenraroengsak et al. (2012)

(continued)

Table 22.1 (continued)

Biological effects	Material	Size	Inference	References
Genotoxicity	Polystyrene	50 nm	THP-1 macrophages and Calu-3 cells experienced induced genotoxicity	Paget et al. (2015)
	Polystyrene	30–500 nm	Effects of size-dependent DNA damage on murine macrophage cell line cells RAW 264.7	Zhang et al. (2011)
	Polystyrene	100 nm	Induced DNA damage	Poma et al. (2019)
	Polystyrene	20 nm	Induced DNA damage	Ferraro et al. (2016)
Apoptosis	Polystyrene	20 and 40 nm	Apoptosis induction in the Caco-2 cells	Thubagere and Reinhard (2010)
	Polystyrene	25 and 70 nm	Induce apoptosis in A549 cells	Xu et al. (2019)
	Polystyrene	50 and 100 nm	Apoptosis induced in alveolar macrophages	Ruenraroengsak and Tetley (2015)
	Polystyrene	56–64 nm	Apoptosis triggered in $CaCO_2$, HT-29, and LS174T cell lines	Inkielewicz-Stepniak et al. (2018)
Autophagy	Polystyrene	59–94 nm	Autophagic activation and disruption of energy metabolism	Lim et al. (2019)
	Polystyrene	60 nm	Induce autophagic cell death in RAW 264.7 and BEAS-2B cell lines	Chiu et al. (2015)
	Polystyrene	120 nm	Induce autophagy in leukemia cells	Loos et al. (2014)
	Polyethylene and nylon fibres	0.1–12 mm	Lysosomal dysfunctions, which ultimately result in cell death and lysosomal blocking of autophagic flow	Wang et al. (2018)
	Polystyrene	50 nm	Activation of autophagy in HeLa cell lines	Song et al. (2015)
Disruption of the energy homeostasis and metabolism	Polystyrene	12.4–54 nm	Metabolic alterations linked to autophagy and ER stress, including an increase in amino acids and intermediate metabolites of the tricarboxylic acid cycle (TCA)	Lim et al. (2019)

(continued)

Table 22.1 (continued)

Biological effects	Material	Size	Inference	References
	Polystyrene	100 nm	Significant deposition of lipids and an upregulation of mdt-15 and sbp-1 expression	Yang et al. (2020)
	Polystyrene	0.5 μm, 15 μm	Markedly decreased fat and protein levels, which can be a sign of a deterioration in the nutritional value and energy reserve of marine demersal fish (*Sebastes schlegelii*)	Yin et al. (2019)
	Polystyrene	5 μm	At the transcriptomic, physiological, and biochemical levels, prevent development and cause malfunction in the liver's glycolipid metabolism in adult zebrafish	Zhao et al. (2020)

provide an estimated 80 g of microplastics to humans each day (Campanale et al. 2020).

22.6 Modern Methods for Microplastic Bioremediation

22.6.1 Microorganisms for Bioremediation

The cycle starts with the choice of a suitable chassis and the enzymes needed to break down pollutants. Genetic engineering approaches are then used to realize the designed systems. The created microbial scavengers are assessed and looked into for potential improvement (Fig. 22.2) (Tran et al. 2021).

In both aerobic and anaerobic conditions, microorganisms break down plastics, releasing methane, CO_2, and water into the atmosphere (Table 22.2). They obtain their energy from carbon. The biodegradation of plastic involves several processes, including assimilation, depolymerization, and bio-deterioration. In biodeterioration, microorganisms break down polymers into minute fragments through the combined action of abiotic stimuli. Several microorganisms release catalytic enzymes with free radicals after biodeterioration, which breaks down the polymer into dimers and monomers. And depolymerization is the term for this process. Following this,

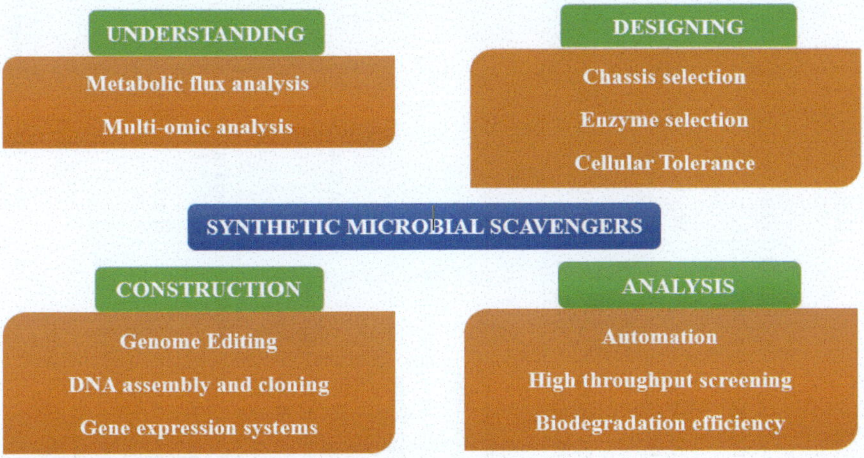

Fig. 22.2 Synthetic microbial scavenger construction

the microbial community on the plastic surface created a biofilm. Microorganisms release Extracellular Polymeric Substances (EPS) that break down the plastic surface. Nucleic acids, proteins, and polysaccharides are found in EPS. While in the cytoplasm, these smaller monomers and dimers are absorbed by the microbial cell membrane and used as an energy source to produce energy, metabolites, biomass, and storage vesicles. Experts term this process as assimilation. Following assimilation, mineralization takes place, releasing CO_2, N_2, CH_4, and H_2O into the atmosphere (Fig. 22.3) (Kumar et al. 2022).

22.6.2 Gene Editing Tools for Plastic Waste Degradation

Molecular scissors, or designed nucleases, are used in gene editing to change DNA sequences. The three primary gene editing technologies are CRISPR-Cas, ZFN, and TALENs. The aforementioned tools function by breaking down double strands in the gene target sequence, which are then repaired by the mechanisms of homology-directed repair (HRD) or nonhomologous end joining (NHEJ) (Thakur et al. 2022).

Additionally, genome editing facilitates the manipulation of a target gene, enabling the performance of gain- and loss-of-function studies that modify the expression of various genes. Genes encoding enzymes linked to the breakdown of microplastics, including laccase, polyethylene terephthalate hydrolase, esterase, depolymerase, and dehalogenase, can be effectively included via this method (Anand et al. 2023).

Table 22.2 Role of microorganisms in Biodegradation

S. no	Microorganism	Microplastic	Size	Inference	References
Fungus					
1	*Aspergillus flavus*	Polyethylene	125 and 200 μm	*A. flavus* strain was efficient polyethylene degrading strain	Zhang et al. (2020)
2	*Aspergillus* spp., *Fusarium oxysporum* and *Penicillium granulatum*	Polyethylene (PE) and polypropylene (PP)	5 mm–1 μm	Potential for PP degradation and PE degradation	Wróbel et al. (2023)
3	*Zalerion maritimum*	Polyethylene	250–1000 μm	Biodegradation of Polyethylene	Paço et al. (2017)
Bacteria					
4	*Bacillus* and *Rhodococcus* species	Polypropylene	250,000 of molecular weight	Polypropylene biodegradation	Auta et al. (2018)
5	C. testosteroni F4 and Alkali-resistant testosterone F6	Polyethylene terephthalate	10 μm	Polyethylene terephthalate biodegradation	Gong et al. (2018)
6	*Bacillus gottheilii* and *Bacillus cereus*	Polypropylene, polystyrene, polyethylene terephthalate, and Polyethylene	75 μm	Biodegradation of microplastics by the bacteria's B. *cereus* and B. *gottheilii*	Auta et al. (2017)
7	*Agromyces mediolanus* and *Bacillus cereus*	Polyethylene terephthalate (PET)	300–425 μm	Polyethylene terephthalate (PET) degrading activity	Torena and Reza (2020)
8	*Pseudomonas* and *Rhodococcus* species	Polypropylene	1 mm	*Pseudomonas* sp. ADL15 and *Rhodococcus* sp. ADL36 possess a promising polypropylene biodegradation activity	Habib et al. (2020)

(continued)

Table 22.2 (continued)

S. no	Microorganism	Microplastic	Size	Inference	References
9	*Bacillus* species, *Bacillus cereus*, and *Bacillus paramycoides*	Polyethylene (PE) and polypropylene (PP)	250 nm	Biodegradation of *Bacillus cereus*, *Bacillus sp.* (BS-2), and *Bacillus paramycoides* (BP)	Devi et al. (2021)
10	*Klebsiella pneumoniae*, *Pseudomonas fluorescens*, *Enterobacter* ludwigii, *Chryseobacterium sp.*	Polyethylene (PE) and polypropylene (PP)	5 mm–1 μm	Potential for PP degradation and PE degradation	Wróbel et al. (2023)
11	*Anoxybacillus flavithermus* ST3 *and Anoxybacillus* sp. ST6	Polyethylene	50 μm	Microplastics can biodegrade in the presence of bacteria	Özdemir et al. (2022)
12	*Anoxybacillus flavithermus* Gecek4s and *Bacillus firmus* ST5	Polyethylene	50 and 150 μm	Degradation of microplastics	Akarsu et al. (2022)
13	*Pseudomonas, Pandoraea and Dyella* species	Polyvinyl chloride, Polyhydroxyalkanoates, Polystyrene, Polyamide, Polyhydroxy butyrate, Polypropylene, Polybutylene succinate, Polylactic acid, Polycaprolactone, Polyvinyl alcohol and Polyethylene	13 μm	Potential for the degradation of microplastics	Ren and Ni (2023)

(continued)

Table 22.2 (continued)

S. no	Microorganism	Microplastic	Size	Inference	References
14	*Rhodococcus ruber, Pseudomonas stutzeri, Bacillus cereus, and Brevibacillus borstelensis*	Poly (ethylene terephthalate) (PET) and polyethylene (PE)	5 mm	Degradation of microplastics	Hooda and Mondal (2023)
Microalgae					
15	*Spirulina* sp.	Polyethylene Terephthalate and Polypropylene	1–2 μm	*Spirulina* sp. plays an important role in the degradation of Polyethylene Terephthalate and Polypropylene	Khoironi et al. (2019)

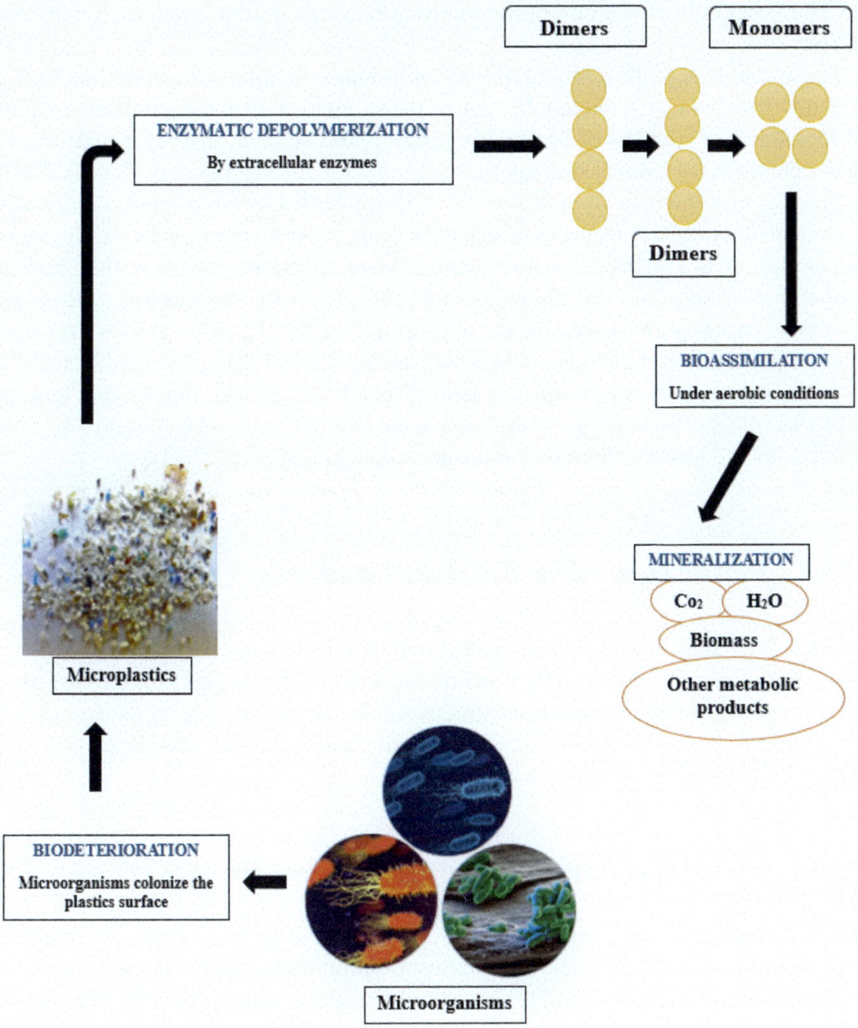

Fig. 22.3 Microbial degradation of plastics

22.6.3 Bioinformatics

Bioinformatics is another helpful method for enhancing the biodegradation of plastic waste, especially microplastic particles (Purohit et al. 2020). Numerous databases that provide information on the metabolic pathways, microbial enzymes, and genes involved in the process have been produced to evaluate the biodegradation process. These databases include The Environmental Contaminant Biotransformation Pathway Resource, MetaCyc database pertaining to biodegradation pathways, and The University of Minnesota Biocatalysis/Biodegradation Database (Caspi et al. 2020).

These databases and computational techniques facilitate the identification of enzymes involved in a metabolic pathway of interest and the prediction of the biodegradation pathways of hazardous compounds, thereby offering a platform for the development of a novel strategy for the biodegradation of plastic (Ali et al. 2021).

The biggest drawback of bioinformatics, despite all of its advantages, is the dearth of research data and the accuracy of that data, which is crucial for further study. Moreover, minimal is known about the many classes of microorganisms that degrade synthetic polymers and the enzymes that catalyse them. As a result, a thorough analysis is necessary to determine the appropriate metabolic pathways for the breakdown of polymers and the related enzymes. In the not-too-distant future, a variety of methods including bioinformatics, genetics, metabolic engineering, system biology, and molecular and technologies may assist us in identifying a workable and long-term solution for the biodegradation of microplastics (Anand et al. 2023).

22.6.4 Plastic Degrading Enzymes Databases

Microbiological enzymes for the breakdown of plastic have been sequenced since plastic pollution is a major problem in our day and age. Databases provide records of these enzymes, pertinent genes, and microbes. In the future, this breakthrough will be helpful for analysing, modifying, and cloning genes that degrade plastic (Ali et al. 2021).

22.6.4.1 Plastic Microbial Biodegradation Database (PMBD)

Being a naturally occurring process, microbes' biodegradation of materials is commonly accepted in eco-conscious communities. The Plastics Microbial Biodegradation Database (PMBD) is created to support research on the biodegradation of plastics. The Plastic Microbial Biodegradation Database is used to gather and disseminate information on this subject (PMBD). This database (PMDB) contains 949 connections and 79 genes related to microorganisms and plastics that have been constructed by hand and validated by literature research (Ali et al. 2021).

22.6.4.2 PlasticDB

Data about microbes and proteins connected to plastic biodegradation in scholarly literature can be found in PlasticDB. As of right now, it contains information from 421 scholarly works, which correspond to 562 different microbial species. Since a single plastic and a single reference account for a separate record, these species include

1462 records (i.e., one microbial taxon may be associated with the breakdown of numerous plastic kinds) (Gambarini et al. 2022).

22.6.4.3 Plastics-Active Enzymes Database (PAZy)

Data about published activities were gathered and combined into an extensive open-access database. The Plastics-Active Enzymes Database (PAZy) is a list of enzymes that have been identified and validated through experimentation and that function on synthetic polymers that are sourced from fossil fuels. About 3000 homologs of active enzymes for polyethylene terephthalate (PET) were discovered using profile hidden markov models. More than 2000 homologs of the active enzyme polyurethane (PUR) were discovered using BLAST. PET- and PUR-active enzyme sequence motifs were created, and conservation analysis based on multiple sequence alignments was used to identify the most conserved amino acids (Buchholz et al. 2022).

22.6.5 Metagenomic Analysis

Different microbial communities can be identified, and it's possible that they aid in the degradation of microplastics by producing enzymes. Not every enzyme that the microbial population produces is effective enough to take part in the breakdown process. A metagenomic investigation of the microbial population involved in plastic biodegradation is suggested as a solution to this challenge in order to comprehend the nature of the microbial community and predict its potential to participate in *in-situ* biodegradation. Because of developments in sequencing techniques and bioinformatics, it is now feasible to do an in-depth metagenomic study of microbiomes and to mine related genes and enzymes involved in biodegradation. Consequently, metagenomics may be a useful technique to find efficient genes and enzymes in uncultivable microbial communities that break down plastic (Reddy et al. 2023).

22.7 Conclusion

The current research clarifies the known sources of microplastics as well as how they end up and get into the environment. Furthermore, the aforementioned consider defining the potential impacts of microplastics in the different environmental niches. Nevertheless, compared to primary microplastics, secondary microplastics contribute more heavily to harmful environmental effects. Although these two forms of microplastics travel in different ways, their fates and impacts are essentially the same. In addition, these minuscule particles exhibit remarkable persistence in the ecosystem. Should plastics be stopped entering the environment, the marine litter currently in place will serve as a reservoir for further microplastics, which could have detrimental effects on the aquatic and terrestrial environments. Thus, in addition to determining the origins, destiny, and impacts of microplastics, we must consider removal strategies and long-term corrective action.

References

Ali SS, Elsamahy T, Koutra E, Kornaros M, El-Sheekh M, Abdelkarim EA, Sun J (2021) Degradation of conventional plastic wastes in the environment: a review on current status of knowledge and future perspectives of disposal. Sci Total Environ 771:144719. https://doi.org/10.1016/j.scitotenv.2020.144719

Anand U, Dey S, Bontempi E, Ducoli S, Vethaak AD, Dey A, Federici S (2023) Biotechnological methods to remove microplastics: a review. Environ Chem Lett 1–24

Auta HS, Emenike CU, Fauziah SH (2017) Screening of Bacillus strains isolated from mangrove ecosystems in Peninsular Malaysia for microplastic degradation. Environ Pollut 231:1552–1559

Auta HS, Emenike CU, Jayanthi B, Fauziah SH (2018) Growth kinetics and biodeterioration of polypropylene microplastics by Bacillus sp. and Rhodococcus sp. isolated from mangrove sediment. Mar Pollut Bull 127:15–21

Bhandari S, Poudel DK, Marahatha R, Dawadi S, Khadayat K, Phuyal S, Parajuli N (2021) Microbial enzymes used in bioremediation. J Chem 2021:1–17

Bhattacharjee S, Ershov D, Islam MA, Kämpfer AM, Maslowska KA, van der Gucht J, Rietjens IM (2014) Role of membrane disturbance and oxidative stress in the mode of action underlying the toxicity of differently charged polystyrene nanoparticles. RSC Adv 4(37):19321–19330. https://doi.org/10.1039/C3RA46869K

Boots B, Russell CW, Green DS (2019) Effects of microplastics in soil ecosystems: above and below ground. Environ Sci Technol 53(19):11496–11506. https://doi.org/10.1021/acs.est.9b03304

Buchholz PC, Feuerriegel G, Zhang H, Perez-Garcia P, Nover LL, Chow J, Pleiss J (2022) Plastics degradation by hydrolytic enzymes: the plastics-active enzymes database—PAZy. Proteins: Struct Function Bioinform 90(7):1443–1456

Campanale C, Massarelli C, Savino I, Locaputo V, Uricchio VF (2020) A detailed review study on potential effects of microplastics and additives of concern on human health. Int J Environ Res Public Health 17(4):1212

Caspi R, Billington R, Keseler IM, Kothari A, Krummenacker M, Midford PE, Karp PD (2020) The MetaCyc database of metabolic pathways and enzymes-a 2019 update. Nucl Acids Res 48(D1):D445–D453. https://doi.org/10.1093/nar/gkz862

Chiu HW, Xia T, Lee YH, Chen CW, Tsai JC, Wang YJ (2015) Cationic polystyrene nanospheres induce autophagic cell death through the induction of endoplasmic reticulum stress. Nanoscale 7(2):736–746. https://doi.org/10.1039/C4NR05509H

Cox KD, Covernton GA, Davies HL, Dower JF, Juanes F, Dudas SE (2019) Human consumption of microplastics. Environ Sci Technol 53(12):7068–7074. https://doi.org/10.1021/acs.est.9b01517

Dey TK, Uddin ME, Jamal M (2021) Detection and removal of microplastics in wastewater: evolution and impact. Environ Sci Pollut Res 28:16925–16947

Dissanayake PD, Kim S, Sarkar B, Oleszczuk P, Sang MK, Haque MN, Ok YS (2022) Effects of microplastics on the terrestrial environment: a critical review. Environ Res 209:112734

Du H, Wang J (2021) Characterization and environmental impacts of microplastics. Gondwana Res 98:63–75. https://doi.org/10.1016/j.gr.2021.05.023

Ferraro D, Anselmi-Tamburini U, Tredici IG, Ricci V, Sommi P (2016) Overestimation of nanoparticles-induced DNA damage determined by the comet assay. Nanotoxicology 10(7):861–870. https://doi.org/10.3109/17435390.2015.1130274

Gambarini V, Pantos O, Kingsbury JM, Weaver L, Handley KM, Lear G (2022) PlasticDB: a database of microorganisms and proteins linked to plastic biodegradation. Database 2022:baac008

Ghosh S, Sinha JK, Ghosh S, Vashisth K, Han S, Bhaskar R (2023) Microplastics as an emerging threat to the global environment and human health. Sustainability 15(14):10821

Gong J, Kong T, Li Y, Li Q, Li Z, Zhang J (2018) Biodegradation of microplastic derived from poly (ethylene terephthalate) with bacterial whole-cell biocatalysts. Polymers 10(12):1326

He Y, Li J, Chen J, Miao X, Li G, He Q, Wei Y (2020) Cytotoxic effects of polystyrene nanoplastics with different surface functionalization on human HepG2 cells. Sci Total Environ 723:138180. https://doi.org/10.1016/j.scitotenv.2020.138180

Hooda S, Mondal P (2023) Insights into the degradation of high-density polyethylene microplastics using microbial strains: effect of process parameters, degradation kinetics and modeling. Waste Manage 164:143–153

Inkielewicz-Stepniak I, Tajber L, Behan G, Zhang H, Radomski MW, Medina C, Santos-Martinez MJ (2018) The role of mucin in the toxicological impact of polystyrene nanoparticles. Materials 11(5):724. https://doi.org/10.3390/ma11050724

Issac MN, Kandasubramanian B (2021) Effect of microplastics in water and aquatic systems. Environ Sci Pollut Res 28:19544–19562. https://doi.org/10.1007/s11356-021-13184-2

Jaiswal KK, Dutta S, Banerjee I, Pohrmen CB, Singh RK, Das HT, Kumar V (2022) Impact of aquatic microplastics and nanoplastics pollution on ecological systems and sustainable remediation strategies of biodegradation and photodegradation. Sci Total Environ 806:151358

Khoironi A, Anggoro S, Sudarno (2019). Evaluation of the interaction among microalgae Spirulina sp, plastics polyethylene terephthalate and polypropylene in freshwater environment. J Ecol Eng 20(6):161–173

Kumar A, Dalal J, Poonia A, Kumari A, Lata P, Sharma R, Verma N (2022) Cost effective and sustainable plastics bio-degradation using microbiota. Shodhsamhita 9:80–95

Lim SL, Ng CT, Zou L, Lu Y, Chen J, Bay BH, Ong CN (2019) Targeted metabolomics reveals differential biological effects of nanoplastics and nanoZnO in human lung cells. Nanotoxicology 13(8):1117–1132. https://doi.org/10.1080/17435390.2019.1640913

Loos C, Syrovets T, Musyanovych A, Mailänder V, Landfester K, Simmet T (2014) Amino-functionalized nanoparticles as inhibitors of mTOR and inducers of cell cycle arrest in leukemia cells. Biomaterials 35(6):1944–1953. https://doi.org/10.1016/j.biomaterials.2013.11.056

Lunov O, Syrovets T, Loos C, Nienhaus GU, Mailänder V, Landfester K, Simmet T (2011) Amino-functionalized polystyrene nanoparticles activate the NLRP3 inflammasome in human macrophages. ACS Nano 5(12):9648–9657. https://doi.org/10.1021/nn203596e

Meindl C, Kueznik T, Bösch M, Roblegg E, Fröhlich E (2015) Intracellular calcium levels as screening tool for nanoparticle toxicity. J Appl Toxicol 35(10):1150–1159. https://doi.org/10.1002/jat.3160

Miloloža M, Cvetnić M, Kučić Grgić D, Ocelić Bulatović V, Ukić Š, Rogošić M, Bolanča T (2022) Biotreatment strategies for the removal of microplastics from freshwater systems. A review. Environ Chem Lett 20(2):1377–1402. https://doi.org/10.1007/s10311-021-01370-0

Nanthini Devi K, Raju P, Santhanam P, Dinesh Kumar S, Krishnaveni N, Roopavathy J, Perumal P (2021) Biodegradation of low-density polyethylene and polypropylene by microbes isolated from Vaigai River, Madurai, India. Arch Microbiol 203:6253–6265. https://doi.org/10.1007/s00203-021-02592-0

Özdemir S, Akarsu C, Acer Ö, Fouillaud M, Dufossé L, Dizge N (2022) Isolation of thermophilic bacteria and investigation of their microplastic degradation ability using polyethylene polymers. Microorganisms 10(12):2441

Paço A, Duarte K, da Costa JP, Santos PS, Pereira R, Pereira ME, Rocha-Santos TA (2017) Biodegradation of polyethylene microplastics by the marine fungus *Zalerion maritimum*. Sci Total Environ 586:10–15

Paget V, Dekali S, Kortulewski T, Grall R, Gamez C, Blazy K, Lacroix G (2015) Specific uptake and genotoxicity induced by polystyrene nanobeads with distinct surface chemistry on human lung epithelial cells and macrophages. PLoS ONE 10(4):e0123297. https://doi.org/10.1371/journal.pone.0123297

Poma A, Vecchiotti G, Colafarina S, Zarivi O, Aloisi M, Arrizza L, Di Carlo P (2019) In vitro genotoxicity of polystyrene nanoparticles on the human fibroblast Hs27 cell line. Nanomaterials 9(9):1299. https://doi.org/10.3390/nano9091299

Prata JC, da Costa JP, Lopes I, Duarte AC, Rocha-Santos T (2020) Environmental exposure to microplastics: an overview on possible human health effects. Sci Total Environ 702:134455. https://doi.org/10.1016/j.scitotenv.2019.1344

Purohit J, Chattopadhyay A, Teli B (2020) Metagenomic exploration of plastic degrading microbes for biotechnological application. Curr Genomics 21(4):253–270. https://doi.org/10.2174/1389202921999200525155711

Radlinska EZ, Gulik-Krzywicki T, Lafuma F, Langevin D, Urbach W, Williams CE, Ober R (1995) Polymer confinement in surfactant bilayers of a lyotropic lamellar phase. Phys Rev Lett 74(21):4237. https://doi.org/10.1103/PhysRevLett.74.4237

Reddy CN, Kallem P, Mounika KVSSN, Muqeet A, Raj JC J, Aishwarya CVS, Banat F (2023) Review of microplastic degradation: Understanding metagenomic approaches for microplastic degrading organisms. Polym Testing 108223

Ricardo IA, Alberto EA, Júnior AHS, Macuvele DLP, Padoin N, Soares C, Trovo AG (2021) A critical review on microplastics, interaction with organic and inorganic pollutants, impacts and effectiveness of advanced oxidation processes applied for their removal from aqueous matrices. Chem Eng J 424:130282. https://doi.org/10.1016/j.cej.2021.130282

Ruenraroengsak P, Tetley TD (2015) Differential bioreactivity of neutral, cationic and anionic polystyrene nanoparticles with cells from the human alveolar compartment: robust response of alveolar type 1 epithelial cells. Part Fibre Toxicol 12(1):1–20. https://doi.org/10.1186/s12989-015-0091-7

Ruenraroengsak P, Novak P, Berhanu D, Thorley AJ, Valsami-Jones E, Gorelik J, Tetley TD (2012) Respiratory epithelial cytotoxicity and membrane damage (holes) caused by amine-modified nanoparticles. Nanotoxicology 6(1):94–108. https://doi.org/10.3109/17435390.2011.558643

Sana SS, Dogiparthi LK, Gangadhar L, Chakravorty A, Abhishek N (2020) Effects of microplastics and nanoplastics on marine environment and human health. Environ Sci Pollut Res 27:44743–44756. https://doi.org/10.1007/s11356-020-10573-x

Schirinzi GF, Pérez-Pomeda I, Sanchís J, Rossini C, Farré M, Barceló D (2017) Cytotoxic effects of commonly used nanomaterials and microplastics on cerebral and epithelial human cells. Environ Res 159:579–587. https://doi.org/10.1016/j.envres.2017.08.043

Sharma S, Bhardwaj A, Thakur M, Saini A (2023) Understanding microplastic pollution of marine ecosystem: a review. Environ Sci Pollut Res. https://doi.org/10.1007/s11356-023-28314-1

Song W, Popp L, Yang J, Kumar A, Gangoli VS, Segatori L (2015) The autophagic response to polystyrene nanoparticles is mediated by transcription factor EB and depends on surface charge. J Nanobiotechnol 13(1):1–12. https://doi.org/10.1186/s12951-015-0149-6

Thakur S, Mathur S, Patel S, Paital B (2022) Microplastic accumulation and degradation in environment via biotechnological approaches. Water 14(24):4053

Thubagere A, Reinhard BM (2010) Nanoparticle-induced apoptosis propagates through hydrogen-peroxide-mediated bystander killing: insights from a human intestinal epithelium in vitro model. ACS Nano 4(7):3611–3622. https://doi.org/10.1021/nn100389a

Tran KM, Lee HM, Thai TD, Shen J, Eyun SI, Na D (2021) Synthetically engineered microbial scavengers for enhanced bioremediation. J Hazard Mater 419:126516. https://doi.org/10.1016/j.jhazmat.2021.126516

Wang Z, Chen M, Zhang L, Wang K, Yu X, Zheng Z, Zheng R (2018) Sorption behaviors of phenanthrene on the microplastics identified in a mariculture farm in Xiangshan Bay, southeastern China. Sci Total Environ 628:1617–1626. https://doi.org/10.1016/j.scitotenv.2018.02.146

Wróbel M, Szymańska S, Kowalkowski T, Hrynkiewicz K (2023) Selection of microorganisms capable of polyethylene (PE) and polypropylene (PP) degradation. Microbiol Res 267:127251

Wu P, Huang J, Zheng Y, Yang Y, Zhang Y, He F, Gao B (2019a) Environmental occurrences, fate, and impacts of microplastics. Ecotoxicol Environ Saf 184:109612. https://doi.org/10.1016/j.ecoenv.2019.109612

Wu B, Wu X, Liu S, Wang Z, Chen L (2019b) Size-dependent effects of polystyrene microplastics on cytotoxicity and efflux pump inhibition in human Caco-2 cells. Chemosphere 221:333–341. https://doi.org/10.1016/j.chemosphere.2019.01.056
Xu M, Halimu G, Zhang Q, Song Y, Fu X, Li Y, Zhang H (2019) Internalization and toxicity: a preliminary study of effects of nanoplastic particles on human lung epithelial cell. Sci Total Environ 694:133794. https://doi.org/10.1016/j.scitotenv.2019.133794
Ya H, Jiang B, Xing Y, Zhang T, Lv M, Wang X (2021) Recent advances on ecological effects of microplastics on soil environment. Sci Total Environ 798:149338
Yang Y, Shao H, Wu Q, Wang D (2020) Lipid metabolic response to polystyrene particles in nematode *Caenorhabditis elegans*. Environ Pollut 256:113439. https://doi.org/10.1016/j.envpol.2019.113439
Yin L, Liu H, Cui H, Chen B, Li L, Wu F (2019) Impacts of polystyrene microplastics on the behavior and metabolism in a marine demersal teleost, black rockfish (*Sebastes schlegelii*). J Hazard Mater 380:120861. https://doi.org/10.1016/j.jhazmat.2019.120861
Zhang M, Li J, Xing G, He R, Li W, Song Y, Guo H (2011) Variation in the internalization of differently sized nanoparticles induces different DNA-damaging effects on a macrophage cell line. Arch Toxicol 85:1575–1588. https://doi.org/10.1007/s00204-011-0725-y
Zhang J, Gao D, Li Q, Zhao Y, Li L, Lin H, Zhao Y (2020) Biodegradation of polyethylene microplastic particles by the fungus *Aspergillus flavus* from the guts of wax moth *Galleria mellonella*. Sci Total Environ 704:135931
Zhao Y, Bao Z, Wan Z, Fu Z, Jin Y (2020) Polystyrene microplastic exposure disturbs hepatic glycolipid metabolism at the physiological, biochemical, and transcriptomic levels in adult zebrafish. Sci Total Environ 710:136279. https://doi.org/10.1016/j.scitotenv.2019.136279
Zhou Y, Kumar M, Sarsaiya S, Sirohi R, Awasthi SK, Sindhu R, Awasthi MK (2022) Challenges and opportunities in bioremediation of micro-nano plastics: a review. Sci Total Environ 802:149823. https://doi.org/10.1016/j.scitotenv.2021.1498

Chapter 23
Systematic Review on Role of Microorganisms in Bioremediation

Gopi Devarajan, Suganya Ilango, Aishwarya Vetrivel, and T. G. Nithya

Abstract Worldwide, environmental contamination and its remediation are major problems. Heavy metals, dyes, hydrocarbons, pesticides, and other pollutants are some of the main causes of environmental deterioration. Eliminating residual contaminants is also one of the toughest challenges. Since bioremediation restores the damaged site to its previous state, it is one of the most successful methods for lowering environmental toxins. Through the action of bacteria, fungus, and plants, it degrades, removes, alters, immobilizes, or detoxifies numerous chemicals and physical wastes from the environment. Microorganisms' natural biodegradation of complex organic compounds implies that they can be reduced to more basic inorganic forms. Bioremediation, is a technology developed by humans that employs microorganisms to reduce pollutants, methods such as bioaugmentation, biostimulation, bioventing, and bio attenuation are some of the types of bioremediation used to increase the capacity of microorganisms to degrade pollutants. These organisms will use the pollutants as a source of energy. Using symbiotic microorganisms, genetically engineered microorganisms have been used to improve pollutant cleanup and ensure safe biodegradation. Hence, this review focuses on the most recent developments in bioremediation techniques, the mechanisms by which microorganisms degrade various pollutants, and the possible uses of bioremediation as a worldwide pollution mitigation strategy.

Keywords Bioremediation · Microorganisms · Emerging technology · Pollutants · Bioremediation techniques

---

G. Devarajan (✉)
Department of Economics, Faculty of Science and Humanities, SRM Institute of Science and Technology, Kattankulathur, Chennai, Tamil Nadu 603203, India
e-mail: gopid1@srmist.edu.in

S. Ilango · T. G. Nithya
Department of Biochemistry, Faculty of Science and Humanities, SRM Institute of Science and Technology, Kattankulathur, Chennai, Tamil Nadu 603203, India

A. Vetrivel
Department of Biotechnology, Dr. G.R. Damodaran College of Science, Coimbatore, Tamil Nadu 641014, India

© The Author(s), under exclusive license to Springer Nature Switzerland AG 2025

A. Jeyaseelan and K. Murugasen (eds.), *Sustainable Environmental Remediation: Avenues in Nano and Biotechnology*, Environmental Science and Engineering,
https://doi.org/10.1007/978-3-031-78483-5_23

23.1 Introduction

Environmental contamination has recently emerged as one of the world's most sensitive topics. The rapid growth of the human population, urbanization, hazardous farming practices, and industrialization all had a part in the vast environmental pollution that resulted from the discharge of various waste products. Pollutants found in sewage, industrial effluent, municipal and household solid waste include heavy metals Cd, Ld, Mercury, Cu, and Pb; organic chemicals and volatile organic compounds like Benzene, Toluene, Xylenes, Dichloromethane; and inorganic chemicals like NH_3, N_2, P, NO_3, and NO_2. Chlorine-containing compounds, dyes such as anthraquinone, vat dye, triphenylmethane, Azodye and greenhouse gas emissions, polymers, nuclear waste, petroleum hydrocarbons, paints, disinfectants, polychlorinated biphenyls, and other biphenyls, lubricants and detergents, contaminate agriculture soil, ground water, rivers and air in general terrestrial and aquatic ecosystem (Tegene 2020).

Natural biodegradation by microorganisms suggests that complex organic materials can be broken down into simpler inorganic forms. While bioremediation is a human-engineered method that uses microorganisms to lessen pollution. Natural attenuation, biostimulation, bioaugmentation are the methods used to boost the ability of microorganisms to fight against pollutants. Pollutants will be used as a source of energy by these microorganisms. Bacteria, fungus, algae, and protozoa are some of the potential microorganisms that can breakdown environmental pollutants (Fouad et al. 2022). There are two methods for carrying out the bioremediation process: ex situ and in situ bioremediation (Tyagi and Kumar 2021).

To treat and manage these wastes, many processes—physical, chemical, and biological methods or in combination—are available. Some uneconomical physicochemical approaches, are gradually losing momentum due to intrinsic issues with secondary contamination and nonsustainable contaminant control (Brar et al. 2006).

Although they have been used for a long time, both chemical and physical approaches have disadvantages. For example, the chemical method requires specialized knowledge and equipment, and the physical method is expensive. This has increased demand for biological remediation (bioremediation), a superior substitute. The most economical, environmentally friendly, and efficient method of changing pollutants is bioremediation. Both microbes and plants can be employed in biological remediation; however, microbes are preferred since they are easier to control and require less time to grow than plants (Ayilara and Babalola 2023).

Environmental contaminant remediation from a microbiological standpoint is both eco-friendly and superior to conventional procedures. Pollutants are eliminated by microbes by enzymatic or metabolic activities such as elimination, immobilization, oxidation, or detoxification. Various biotic or abiotic elements may influence the degradation pathway, however various in-situ or ex-situ bioremediation strategies are employed worldwide to breakdown pollutants (Tak et al. 2022).

The goal of this review is to represent the current trend of microorganism application and role in bioremediation and to give important background for efficient evaluation of bioremediation processes.

23.2 Types of Bioremediation

There are various treatment technologies or procedures used in bioremediation processes (Fig. 23.1).

23.2.1 In-Situ Bioremediation

This involves relocating the contaminants to a different area for treatment once they have been removed or excavated. The polluted site's location, treatment costs, kind of pollution, pollution depth, and pollution intensity are the main determinants of "ex-situ" bioremediation processes. "Ex-situ" bioremediation methods can treat a

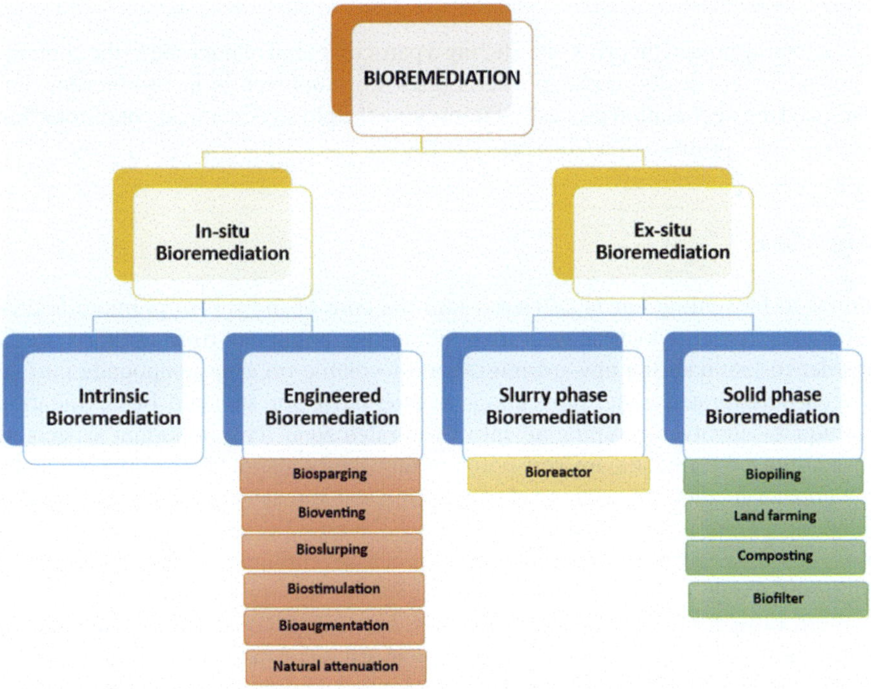

Fig. 23.1 Types of bioremediation

wider range of toxins and soils and are simpler to administer. Because the material is regularly blended, well-aerated, and abundant in nutrients, the breakdown of contaminants is significantly faster than in "in-situ" techniques (Paul et al. 2021). When assessing ex-situ bioremediation processes, the following variables are typically taken into account: the cost of treatment, the kind and depth of pollutants, the level of pollution, the location of the polluted site, and its geology (Azubuike et al. 2020).

23.2.1.1 Intrinsic Bioremediation

An in-situ bioremediation method known as natural reduction or intrinsic bioremediation involves passively cleaning contaminated areas without the need for force or human intervention. Native or naturally occurring microbial communities can be used in this technique. The method uses both anaerobic and aerobic microbial processes to biodegrade contaminated elements, including resistant ones. The process appears to be less expensive than previous in-situ procedures because there isn't an external force involved (Sharma 2019).

23.2.1.2 Engineered In-Situ Bioremediation

The second approach involves introducing a particular kind of microbe to the contaminated site. Genetically modified microorganisms employed in in-situ bioremediation speed the degradation process by improving the physicochemical conditions that facilitate their proliferation (Sharma 2019).

Biosparging

Similar to bioventing, air is delivered into the core of the soil to promote microbiological activity, which subsequently eliminates pollutants from polluted areas. In order to facilitate the upward migration of volatile organic compounds into an adjacent unsaturated zone, bioventing—a alternative for standard biodegradation techniques—involves pumping air into a saturated zone. Two important aspects of biosparging performance are soil porosity and the biodegradability of pollutants. One popular method for getting rid of gasoline and kerosene from water sources is called biosparging. During enhanced bioremediation, microorganisms are exposed to oxygen to speed up the biodegradation processes. organic toxins BTEX can be eliminated using a variety of techniques, including adsorption, microbial decomposition, biosparging, permeability reactive barriers (PRBs), and the application of synthetic or modified zeolites. Still, not much study has been done on easily accessible, moderately cost materials for BTEX adsorption, such as natural zeolite.

Bioventing

One of the most popular techniques for cleaning up soil tainted with petroleum hydrocarbons is bioventing. Bio-inventing supplies air to an unsaturated soil zone by using blowers and pumps to generate a vacuum in the desired area and continuously inject small amounts of air (Sharma 2019).

Bioslurping

Through vacuum-enhanced pumping and bioventing, the Bioslurping approach indirectly provides oxygen and promotes the biodegradation of pollutants, thereby remediating soil and groundwater. This technique mimics the fun sip-and-slurp action kids love with their favourite drinks to extract liquids from your carpet. A well is connected to a movable "slurp tube" to form the basis of the Bioslurping system. The well's vacuum-induced negative pressure zone draws in trapped LNAPL from minuscule pore gaps above the water table and promotes LNAPL flow there. This process for decontaminating soils containing organic chemicals that are volatile and semi-volatile (Awasthi et al. 2019).

Biostimulation

The most common method of biostimulation involves infusing particular nutrients into the contaminated areas. One of the primary benefits of using native microorganisms in the biostimulation process is their good appropriateness and spatial dispersion in the subsurface, which helps to prevent potential issues with their incompatibility with the surroundings (Dehnavi and Ebrahimipour 2022).

Bioaugmentation

In order to expand the volume of a NAPL degrader, bioaugmentation entails the cultivation of microorganisms that introduce believable microbial communities that fill up on their own under particular conditions. Moreover, the plants accelerate the elimination of NAPL by facilitating microbial re-foundation in contaminated soils and water. This is achieved by regular oxygen delivery through root zone air circulation and the provision of nutrients for microbial advancement through fixation and exudation. Moreover, the processes that produce wetlands also serve as antidotes against contamination of the water and soil. Many factors are considered while selecting any kind of bioremediation technique, such as the scenario type, the site, and the level, kind, and amount of pollution (Muthukumaran 2022).

Natural Attenuation

One sort of remediation approach called enhanced natural attenuation (RENA) turns the soil to encourage microbial activity, aeration, sustenance, hydration, and breakdown in order to lessen soil contamination. RENA mainly regulates pollution caused by crude oil by using a range of microorganisms. By immobilizing the contaminants or using them as carbon sources, these microorganisms lessen the number of environmental pollutants that plants may absorb. The ability of the microorganisms utilized in RENA is enhanced by the application of various organic and inorganic substrates. Numerous enzymes produced by microbes, including lipase, hydrogenase, laccase, and others, aid in the digestion of a variety of pollutants. This is a crucial process of pollution that has to be thoroughly investigated, particularly when many contaminants are present in the environment. Numerous variables, such as the pH, oxygen content, temperature, and nutrients in the surrounding environment, might affect how effective RENA bioremediation is. Since these conditions directly influence the microorganisms, the bacteria intended to carry out the bioremediation process perish when these conditions are adverse (Ayilara et al. 2023).

23.2.2 Ex-Situ Bioremediation

23.2.2.1 Slurry Phase Bioremediation

The process of slurry-phase bioremediation proceeds a little quicker than that of other treatment methods. In a bioreactor, contaminated soil is mixed with oxygen, nutrients, and water to create the ideal environment for microorganisms to decompose the contaminants. In this stage of the procedure, the contaminated soil has to have the stones and debris removed. The amount of water supplied is influenced by soil physicochemical properties, pollutant content, and biodegradation rate. After that, the dirt is extracted and dried using centrifuges, vacuum filters, and pressure filters. The soil must then be disposed of, and the resultant fluids must be treated beforehand (Sharma 2019).

Bioreactor

A bioreactor can be either closed or open, but its purpose is to maximize interaction between contaminants and microorganisms. A thorough understanding of the biological system, including its growth dynamics, microbial metabolism, and genetic composition, is necessary to design an appropriate bioreactor for a given treatment method. Several kinds of bioreactors have been created over the last 20 years. Sequential or combination bioreactors and hybrid bioreactors are examples of contemporary technologies being added to many other bioreactors based on modifications of conventional technology. And others are stirred-tank bioreactors, airlift bioreactors, fluidized bed bioreactors, wave bioreactors. Bioreactors, whether sequential or mixed, are engineered to enable the simultaneous execution of physical, chemical, and biological processes (Vishwakarma et al. 2020).

23.2.2.2 Solid Phase Bioremediation

An ex-situ technique called solid-phase bioremediation involves digging out and stacking the contaminated soil. Together with home, industrial, and municipal garbage, it also includes organic wastes like leaves, animal dung, and agricultural wastes. Pipes are positioned all over the heaps to transport bacterial growth. Air needs to draw itself through the pipes in order for microbes to breathe and ventilate. Compared to slurry-phase operations, solid-phase procedures need more room and take longer to clean up. Additional techniques for treating solid phases include land farming, composting, biopiles, windrows, and others (Kulshreshtha et al. 2014).

Biopiling

Biopiles are a cross between composting and land gardening. Numerous aerobic and anaerobic microorganisms can thrive in the rich conditions that biopiles provide. The aqueous reactors, which are pumped up from a designated area, remediate a contaminated environment ex-situ. It involves using specifically created technologies to bioremediate an environment that has been contaminated. Engineered cells are produced with the intention of regulating the physical losses of surface contaminants through leaching and volatilization. Biopiling is said to be an efficient and cost-effective way to treat contaminated soils (Sayqal and Ahmed 2021).

Land Farming

Land farming is a simple process that involves excavating contaminated soil over a prepared area and tilling the soil periodically until the pollution is eliminated by microorganisms. The method is limited to treating a small portion of the soil. The technique is easy to use and quite effective, particularly on petroleum-contaminated soil. Unfortunately, the method can only be used to treat a tiny area of the upper soil (10–35 cm) (Sivakumar et al. 2014).

Composting

A composting method requires combining polluted soil with non-hazardous organic agricultural wastes to permit the growth of high microbial number with higher temperatures 40–65 °C. The method is used to a variety of soils, including disturbed soils, biosolids, wood chips, and plant and animal wastes contaminated with organic contaminants such as pesticides and petroleum hydrocarbons (Vidali 2001).

Biofilters

It has been discovered that biofilters are an appropriate, environmentally friendly, and simple technique for eliminating a variety of pollutants from the aquatic environment. Biofilters are an important new technology that function by using live biological organisms like viruses, bacteria, plants, algae, yeast, and mixed microorganisms as catalysts. Because of their flexibility, the necessary designs for these biological filters were built based on available space and financial resources. The advantages of an environmentally friendly biofilter approach over traditional water treatment methods led to its selection. The fact that the pollutants were transformed into biodegradable wastes and new pollutants did not emerge after a predefined amount of time was a major benefit of biofiltration techniques (Pachaiappan et al. 2022).

23.3 Microorganisms Used in Bioremediation

Microorganisms contribute to food chains, which helps to maintain biological equilibrium. Bioremediation is the process of eliminating contaminated materials from the environment by means of bacteria, algae, fungi, and yeast. Microbes can survive in temperatures as low as $-196°$ Fahrenheit and as high as $1200°$ Fahrenheit when hazardous pollutants or any type of waste stream are present. Microorganisms' biological systems and adaptability make them a great alternative to traditional cleaning products. For bacteria, carbon is the most vital nutrient. The process of bioremediation was carried out by using microorganisms from a variety of environments (Enerijiofi 2021).

23.3.1 Aerobic

In an aerobic environment, a wide variety of microorganisms can bioremediate several sorts of environmental pollutants. Numerous complex organic molecules can be broken down by aerobic bacteria, including *Sphingomonas*, *Rhodococcus*, *Bacillus*, *Flavobacterium*, *Mycobacterium*, *Nocardia,* and *Pseudomonas*. It has been demonstrated that these bacteria break down polyaromatic chemicals, alkane hydrocarbons, and pesticides. Many of these microorganisms get their energy and carbon from these contaminants. Microorganisms' ability to develop is restricted by oxygen in the aerobic bioremediation process.

23.3.2 Anaerobic

Researchers are utilizing amphibious bacteria for the bioremediation of chlorinated solvents like trichlorethylene and chloroform, as well as polychlorinated biphenyls and chlorine compounds. These microbes degrade contaminants and transform them into less harmful forms. The anaerobic bioremediation technique has made use of a number of bacteria, including sulfate-reducing bacteria, *Aeromonas*, and *pseudomonas*. Microbial degradation of azo dyes under varied environmental conditions. Utilizing reduction techniques that make use of electrons generated by the oxidation of the organic substrate (or substrates), azo dyes can be broken down anaerobically. The efficacy of colour removal would be significantly impacted by the electrochemical characteristics of the microorganisms due to these regulated dye decolorization events. For commercial purposes, time-varying decolorized metabolites (DMs) were produced using anaerobic dye decolorization. To create a more accurate system, however, external augmentation of DMs collected under specific circumstances was done for a more in-depth study (Tegene and Tenkegna 2020) (Table 23.1).

23.4 Role of Genetically Engineered Microorganisms in Bioremediation

The usage of bioremediation and the development of genetically modified microorganisms (GEMs) were prompted by the increasing environmental degradation. Utilizing engineered microorganisms is less expensive and safer than other methods. Bacteria are changed genetically to produce a more robust protein that enhances a desired characteristic; this process produces gene-engineered microorganisms, or GEMs. Since biotechnologically produced, bacteria may quickly adjust to new pollutants they come into contact with or co-metabolize, they are more potent than naturally occurring ones and may be able to break down toxins faster. The field of genetic engineering has great promise for improving human health and the environment in the long run (Rafeeq et al. 2023).

23.5 Recent Advancement in Bioremediation

23.5.1 Bioremediation Tools Based on Omics

Understanding genomes, transcriptomics, metabolomics, proteomics, and metabolomics can be useful for bioremediation research. This technique links DNA sequences to the number of proteins, metabolites, and mRNA, making it easier to assess the *in-situ* bioremediation process (Villegas-Plazas et al. 2019) (Fig. 23.2).

Table 23.1 Role of microorganisms in bioremediation

S. No.	Contaminants	Species	Inference	References
Bacteria				
1	Naphthalene-polycyclic aromatic hydrocarbon (PAH) and Heavy metals	*Bacillus licheniformis* and *Bacillus sonorensis* strains	Bioremediation of PAH and heavy metals	Rabani et al. (2022)
2	Oil spills (Hydrocarbon)	*Azospirillum* species, *pseudoalcaligenes*, *Acinetobacter* species, *Alcaligenes* species, *Pseudomonas* species, *Pseudomonas stutzeri*, and *Paenibacillus* species	Remediation of oil spills	Machado et al. (2019)
3	Benzene, toluene, ethylbenzene, xylene isomers	*Pseudomonas putida*, *Pseudomonas synxantha*, and *Pseudomonas azotoformans*	Degradation of xylene isomers	Miri et al. (2022)
4	Hydrocarbons and polycyclic aromatic hydrocarbons	*Pseudoalteromonas* and *Agarivorans* species	Strains could be favoured by hydrocarbon contamination and may actively contribute to the degradation of aliphatic hydrocarbons and polycyclic aromatic hydrocarbons	Dell'Anno et al. (2021)
5	Hexavalent chromium	*Escherichia coli*	Degradation of chromium	Mohamed et al. (2020)
6	Aflatoxin is a highly toxic mycotoxin	*P. aeruginosa*	Promising candidate for aflatoxin biodegradation	Song et al. (2019)
7	Arsenic	*Microbacterium*, *Micrococcus*, *Shinella*, and *Bacillus*	Metal bioremediation	Bhakat et al. (2019)
8	Copper, cadmium and lead	*Lysinibacillus* species	Reduced the heavy metal concentrations with 88% in copper, 41% in cadmium, and 71% in lead	Kharangate-Lad and D'Souza (2021)

(continued)

Table 23.1 (continued)

S. No.	Contaminants	Species	Inference	References
9	Crude oil	*P. aeruginosa*	Bioremediation of crude oil	Mukjang et al. (2022)
10	Heavy metals	*Saccharomyces cerevisiae, Cunninghamella elegans, Rhodomicrobium, Beggiatoa thiothrix*	Bioremediation of heavy metals	Shah and Jain. (2020)
11	Chloroethenes	*Dehalococcoides* species	Bioremediation of chloroethenes	Mohamed et al. (2020)
12	Polycyclic aromatic hydrocarbons (PAHs)	*Novosphingobium, Pseudomonas, Comamonas, Burkholderia, Bacillus*, Pseudoxanthomonas, *Paenibacillus* and *Actinobacteria*	Degradation of polycyclic aromatic hydrocarbons	Mohapatra and Phale (2021)
13	Polybrominated diphenyl ethers (BDE)	*P. chrysosporium*	*P. chrysosporium* might be a type of ideal microorganisms in the removal of BDE pollution	Liu et al. (2018)
14	Cadmium	*Cladosporium* species, *A. fumigatus, Paecilomyces* species, *Paecilomyces* species, *Terichoderma* species, *Cladosporium* species, and *A. versicolor*	Potential biodegradation of cadmium by microorganisms	Unuofin et al. (2021)
15	Oils	*Corynebacterium propinquum, P. aeruginosa, Alcaligenes odorans* and *Fusarium* species	Biodegradation of oils	Pande et al. (2020)
16	Chromium	*Ulothrix tenuissima, Spirulina* species and *C. reinhardtii*	Biodegradation of chromium	Aregbesola et al. (2020)
17	Aromatic hydrocarbons	*Microbacterium* species, *Pseudomonas* species, *Acinetobacter* species, and *Ralstonia* species	Biodegradation of Aromatic hydrocarbons	Basu et al. (2018)
18	Cadmium, lead and chromium	*Aerococcus* species, and *Rhodopseudomonas palustris*	Biodegradation of Cadmium, lead and chromium	Sravya and Sangeetha (2022)
19	Phenol	*B. Subtilis, P. aeruginosa, Alcaligenes odorans* and *Corynebacterium propinquum*	Biodegradation of Phenol	Gaur et al. (2018)

(continued)

Table 23.1 (continued)

S. No.	Contaminants	Species	Inference	References
20	Vat dyes	*K. oxytoca*, *B. firmus*, *B. macerans*, and *Staphylococcus aureus*	Biodegradation of Vat dyes	Sangkharak et al. (2020)
21	Lead, nickel and dichromate	*Chlorella* species and *Spirulina* species	Biodegradation of dichromate, nickel and Lead	Geetha et al. (2021)
22	Heavy metals and mercury	*Saccharomyces cerevisiae* and *Cunninghamella elegans*	Biodegradation of heavy metal components	Duc et al. (2021)
23	Dyes	*Bacillus licheniformis*	Biodegradation of dyes	Mousavi et al. (2021)
24	Iron and zinc	*Pseudomonas fluorescence* and *Bacillus subtilis*	Biodegradation of iron and zinc	Li et al. (2018)
25	Chromium	*Bacillus*, *Escherichia*, *Shewanella*, *Enterobacter*, *Thermus* and *Pseudomonas* species	Biodegradation of chromium	Mousavi et al. (2021)
Fungi				
26	Xylene, benzene, toluene, organochlorines and N-heterocyclic explosives	*Phanerochaete chrysosporium*	Biodegradation of organic compounds	Singh et al. (2020)
27	4,4 Dibromodiphenyl ether	*Phanerochaete chrysosporium*	Biodegradation of 4,4 dibromodiphenyl ether	Sen et al. (2019)
28	Polyaromatic hydrocarbon (PAH) and total petroleum hydrocarbons (TPH)	*Aspergillus niger*, *Aspergillus flavus*, *Varicosporium elodeae* and *Cochliobolus lunatus*	Degradation of petroleum hydrocarbons	Gaur et al. (2022)
29	Arsenic	*Saccharomyces cerevisiae*	Biodegradation of Arsenic	Verma and Kuila (2019)
30	Arsenic	*Aspergillus* species	Biodegradation of Arsenic	Mohd et al. (2019)

(continued)

Table 23.1 (continued)

S. No.	Contaminants	Species	Inference	References
31	4-Hydroxy-3,5-dichlorobiphenyl	*Coprinus comatus*	Biodegradation of dichlorobiphenyl	Li et al. (2018)
32	Aliphatic hydrocarbons, Polycyclic aromatic hydrocarbons and chlorophenols	*Aspergillus* species and *Penicillium* species	Biodegradation of Aliphatic hydrocarbons, Polycyclic aromatic hydrocarbons and chlorophenols	Li et al. (2020)
33	N-hexadecane	*Aspergillus* species	Biodegradation of N-hexadecane	Al-Hawash et al. (2018)
34	Phenanthrene	*Phomopsis liquidambari*	Biodegradation of Phenanthrene	Fu et al. (2018)
35	Pyrene	*Ganoderma lucidum*	Biodegradation of Pyrene	Agrawal et al. (2018)
36	Cobalt and copper	*Trichoderma* species, *Penicillium* species and *Aspergillus* species	Biodegradation of Cobalt and copper	Dusenge et al. (2020)
Algae				
37	Arsenic	*Microcystis aeruginosa*	Biodegradation of Arsenic	Wang et al. (2018)
38	Arsenic	*Chlamydomonas reinhardtii* and *S. almeriensis*	Biodegradation of Arsenic	Saavedra et al. (2018)
39	Zinc	*Fucus vesiculosus*	Biodegradation of Zinc	Brinza et al. (2020)
40	Iron	*Chlorococcum humicola*	Biodegradation of Zinc	Chugh et al. (2022)
41	Phenol	*Isochrysis galbana, Phaeodactylum tricornutum* and *Chlorella species*	Biodegradation of Phenol	Wu et al. (2022)
42	Chromium, nickel, cadmium and lead	*F. vesiculosus*	Biodegradation of heavy metals	Moreira et al. (2019)

(continued)

Table 23.1 (continued)

S. No.	Contaminants	Species	Inference	References
43	Nickel and cadmium	*Cystoseria indicant*	Biodegradation of heavy metals	Moreira et al. (2019)
44	Chromium and cadmium	*Chlamydomonas reinhardtii*	Biodegradation of heavy metals	Nowicka et al. (2020)
45	Cadmium	*Microcystis aeruginosa*	Biodegradation of heavy metals	Deng et al. (2020)
46	Cadmium	*Cyclotella species, Scenedesmus protuberans* and *Scenedesmus accuminatus*	Biodegradation of heavy metals	Vo et al. (2020)
47	Cobalt	*Chlorococcum humicola*	Biodegradation of heavy metals	Chugh et al. (2022)

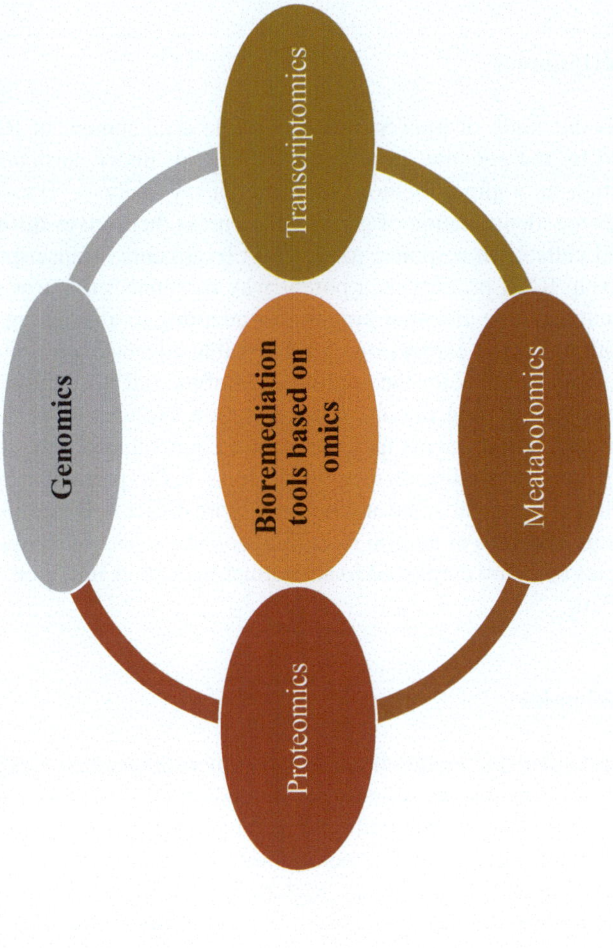

Fig. 23.2 Bioremediation tools based on omics

23.5.1.1 Genomics

Understanding the physiology of the microorganisms participating in bioremediation is made easier with the use of the genomics method. The whole or almost whole genome sequences of the majority of the microorganisms used in bioremediation are accessible. Complete genome sequences that are available can be identified using whole genome DNA microarrays. Studying gene expression in the presence of pollutants can benefit further from this.

23.5.1.2 Transcriptomics

Transcriptomics is the study of transcripts—the whole arrangement of RNA transcripts determined by the genome, in a particular cell, or under particular conditions—by using high-throughput methods like microarray analysis. Transcriptome correlation enables the identification of genes that express themselves differently in different microbial cells or in response to different treatments. Transcriptomics, a method for analysing RNA profiles, is a potent way to connect the proteome, the genome, and phenotype. One essential process for adapting to a changing environment is the regulation of gene expression. A remarkable advancement in transcriptomics, DNA microarrays allow for the confirmation of the mRNA expression level of any required gene in a living organism. Information explanation is the testing problem that microarray experiments face the most. In general, different genes may be up- or down-regulated depending on the type of stress. In this particular case, a few factual issues exhibit systemic mistake and become more difficult. By examining the mRNA transcriptional profiles of natural microbial communities, transcriptomics or meta transcriptomics tools are employed to add valuable insights into their activities (Kashyap et al. 2020).

23.5.1.3 Metabolomics

During their physiological and metabolic processes, microorganisms release certain metabolites into their immediate surroundings. Metabolomics is the process of identifying and quantifying these compounds. By producing distinctive peaks, NMR and MS spectroscopy are used to detect metabolites. The peak pattern aids in both metabolite identification and measurement. Metabolite detection and quantification have also been accomplished through the use of techniques like liquid and gas chromatography. The following areas are where metabolomics can be useful: 1. Being aware that the host environment's equilibrium is compromised by the way contaminants change the environment. 2. Being aware of how signalling occurs during quorum sensing in bacteria and how variations in microbial density affect the degree of gene expression. 3. It is crucial for pathway analysis, drug discovery, and pharmacogenomics. 4. It also serves as a significant environmental health indicator (Pulido et al. 2016).

23.5.1.4 Proteomics

A proteomics study looks at every protein found in a living thing to give a comprehensive picture of how the proteins function within a particular context. When used in conjunction with other omics technologies, the proteomics approach offers a thorough understanding of the protein complement of biological systems. This helps us uncover the mechanisms underlying these cellular processes, which in turn advances the development of environmental biotechnologies. The proteome changes over time and from cell to cell, but an organism's DNA remains relatively stable, making proteomics a more sophisticated and valuable tool than genomics. The study of protein sequences generated by bacterial cultures and environmental samples, known as proteomics and metaproteomics, can provide insights into differences in the synthesis and composition of proteins as well as reveal a multitude of proteins that are critical to the physiological response of microbes to pollutants.

Moreover, metaproteomics has been applied to determine the protein expression profiles of microbial communities in ambient samples without the need to cultivate any bacteria in the samples. With this approach, the actual functional activities of microorganisms in a particular ecosystem are captured. With the advent of two-dimensional gel electrophoresis, protein sequencing, mass spectrometry, and improvements in protein structure databases, this approach is becoming more and more useful.

23.6 Conclusion

Microbiologists face a difficult problem when trying to develop a good bioremediation approach since it requires a thorough understanding of degradative microbial communities. As a result, proteomics, metabolomics, and genome analysis have become crucial tools for discovering all of the microbial communities that have not yet been identified but are capable of decomposing pollutants and for identifying the many metabolites that organisms produce in reaction to stress. Bioremediation techniques have been found to be significantly more successful and efficient than conventional remediation approaches, and they also help to maintain the biological balance of the ecosystem. It will be necessary to evaluate in situ sediment bioremediation on a large scale by increasing laboratory or small-scale research and figuring out the financial and environmental implications in order to reach a conclusion. They are therefore crucial components for creating eco-friendly and long-lasting bioremediation treatments for contaminated locations.

References

Agrawal N, Verma P, Shahi SK (2018) Degradation of polycyclic aromatic hydrocarbons (phenanthrene and pyrene) by the ligninolytic fungi Ganoderma lucidum isolated from the hardwood stump. Bioresour Bioprocess 5(1):1–9. https://doi.org/10.1186/s40643-018-0197-5

Aguiar-Pulido V, Huang W, Suarez-Ulloa V, Cickovski T, Mathee K, Narasimhan G (2016) Metagenomics, metatranscriptomics, and metabolomics approaches for microbiome analysis: supplementary issue: bioinformatics methods and applications for big metagenomics data. Evol Bioinform 12:EBO-S36436

Al-Hawash AB, Zhang J, Li S, Liu J, Ghalib HB, Zhang X et al (2018) Biodegradation of n-hexadecane by Aspergillus sp. RFC-1 and its mechanism. Ecotoxicol Environ Saf 164:398–408. https://doi.org/10.1016/j.ecoenv.2018.08.049

Aregbesola OA, Kumar A, Mokoena MP, Olaniran AO (2020) Cloning, overexpression, purification, characterization and structural modelling of a metabolically active Fe2+ dependent 2, 6-dichloro-p-hydroquinone 1, 2-dioxygenase (CpsA) from bacillus cereus strain AOA-CPS_1. Int J Biol Macromol 161:247–257. https://doi.org/10.1016/j.ijbiomac.2020.05.268

Awasthi A, Gupta A, Bansal M (2019) Bioremediation: a sustainable approach to restore contaminated soil. In: Xenobiotics interaction: consequences and remediation, p 204

Ayilara MS, Babalola OO (2023) Bioremediation of environmental wastes: the role of microorganisms. Front Agron 5:1183691

Ayilara MS, Adeleke BS, Adebajo MT, Akinola SA, Fayose CA, Adeyemi UT et al (2023) Remediation by enhanced natural attenuation; an environment-friendly remediation approach. Front Environ Sci

Azubuike CC, Chikere CB, Okpokwasili GC (2020) Bioremediation: an eco-friendly sustainable technology for environmental management. In: Bioremediation of industrial waste for environmental safety: volume I: industrial waste and its management, pp 19–39

Basu S, Rabara RC, Negi S, Shukla P (2018) Engineering PGPMOs through gene editing and systems biology: a solution for phytoremediation? Trends Biotechnol 36(5):499–510

Bhakat K, Chakraborty A, Islam E (2019) Characterization of arsenic oxidation and uranium bioremediation potential of arsenic resistant bacteria isolated from uranium ore. Environ Sci Pollut Res 26(13):12907–12919. https://doi.org/10.1007/s11356-019-04827-6

Brar SK, Verma M, Surampalli RY, Misra K, Tyagi RD, Meunier N, Blais JF (2006) Bioremediation of hazardous wastes—a review. Pract Period Hazard Toxic Radioact Waste Manag 10(2):59–72

Brinza L, Geraki K, Cojocaru C, Holdt SL, Neamtu M (2020) Baltic Fucus vesiculosus as potential bio-sorbent for zn removal: mechanism insight. Chemosphere 238:124652. https://doi.org/10.1016/j.chemosphere.2019.124652

Chugh M, Kumar L, Shah MP, Bharadvaja N (2022) Algal bioremediation of heavy metals: an insight into removal mechanisms, recovery of by-products, challenges, and future opportunities. Energy Nexus 7:100129. https://doi.org/10.1016/j.nexus.2022.100129

Dehnavi SM, Ebrahimipour G (2022) Comparative remediation rate of biostimulation, bioaugmentation, and phytoremediation in hydrocarbon contaminants. Int J Environ Sci Technol 19:11561–11586. https://doi.org/10.1007/s13762-022-04343-0

Dell'Anno F, Rastelli E, Tangherlini M, Corinaldesi C, Sansone C, Brunet C et al (2021) Highly contaminated marine sediments can host rare bacterial taxa potentially useful for bioremediation. Front Microbiol 12:584850. https://doi.org/10.3389/fmicb.2021.584850

Deng J, Fu D, Hu W, Lu X, Wu Y, Bryan H (2020) Physiological responses and accumulation ability of microcystis aeruginosa to zinc and cadmium: implications for bioremediation of heavy metal pollution. Bioresour Technol 303:122963. https://doi.org/10.1016/j.biortech.2020.122963

Duc H, Hung N, Oanh N (2021) Anaerobic degradation of endosulfans by a mixed culture of Pseudomonas sp. and Staphylococcus sp. Appl Biochem Microbiol 57(3):327–334. https://doi.org/10.1134/S0003683821030030

Dusengemungu L, Kasali G, Gwanama C, Ouma KO (2020) Recent advances in biosorption of copper and cobalt by filamentous fungi. Front Microbiol 11:3285. https://doi.org/10.3389/fmicb.2020.582016

Enerijiofi KE (2021) Bioremediation of environmental contaminants: a sustainable alternative to environmental management. In: Bioremediation for environmental sustainability. Elsevier, Amsterdam, The Netherlands, pp 461–480

Fouad FA, Youssef DG, Shahat FM, Abd El-Ghany MN (2022) Role of microorganisms in biodegradation of pollutants. Handbook of biodegradable materials. Springer International Publishing, Cham, pp 1–40

Fu W, Xu M, Sun K, Hu L, Cao W, Dai C et al (2018) Biodegradation of phenanthrene by endophytic fungus Phomopsis liquidambari in vitro and in vivo. Chemosphere 203:160–169. https://doi.org/10.1016/j.chemosphere.2018.03.164

Gaur N, Narasimhulu K, PydiSetty Y (2018) Recent advances in the bio-remediation of persistent organic pollutants and its effect on environment. J Clean Prod 198:1602–1631. https://doi.org/10.1016/j.jclepro.2018.07.076

Gaur VK, Tripathi V, Manickam N (2022) Bacterial-and fungal-mediated biodegradation of petroleum hydrocarbons in soil. In: Development in wastewater treatment research and processes. Elsevier, Amsterdam, pp 407–427

Geetha N, Bhavya G, Abhijith P, Shekhar R, Dayananda K, Jogaiah S (2021) Insights into nanomycoremediation: secretomics and mycogenic biopolymer nanocomposites for heavy metal detoxification. J Hazard Mater 409:124541. https://doi.org/10.1016/j.jhazmat.2020.124541

Kashyap BK, Solanki MK, Kamboj DV, Pandey AK (eds) (2020) Waste to energy: prospects and applications. Springer Singapore, Imprint

Kharangate-Lad A, D'Souza NC (2021) Current approaches in bioremediation of toxic contaminants by application of microbial cells; biosurfactants and bioemulsifiers of microbial origin. Rhizobiont Bioremediat Hazard Waste 217–263. https://doi.org/10.1007/978-981-16-0602-1_11

Kulshreshtha A, Agrawal R, Barar M, Saxena S (2014) A review on bioremediation of heavy metals in contaminated water. IOSR J Environ Sci Toxicol Food Technol (IOSR-JESTFT) 8(7):44–50

Li N, Xia Q, Niu M, Ping Q, Xiao H (2018) Immobilizing laccase on different species wood biochar to remove the chlorinated biphenyl in wastewater. Sci Rep 8(1):1–9. https://doi.org/10.1038/s41598-018-32013-0

Li Q, Liu J, Gadd GM (2020) Fungal bioremediation of soil co-contaminated with petroleum hydrocarbons and toxic metals. Appl Microbiol Biotechnol 104(21):8999–9008. https://doi.org/10.1007/s00253-020-10854-y

Liu Z, Shao B, Zeng G, Chen M, Li Z, Liu Y et al (2018) Effects of rhamnolipids on the removal of 2, 4, 2, 4-tetrabrominated biphenyl ether (BDE-47) by Phanerochaete chrysosporium analyzed with a combined approach of experiments and molecular docking. Chemosphere 210:922–930. https://doi.org/10.1016/j.chemosphere.2018.07.114

Machado LF, de Assis LDC, da Costa RCTC, Paes JE, Martins EF, Peixoto RS et al (2019) Tracking mangrove oil bioremediation approaches and bacterial diversity at different depths in an in situ mesocosms system. Front Microbiol 10:2107. https://doi.org/10.3389/fmicb.2019.02107

Miri S, Rasooli A, Brar SK, Rouissi T, Martel R (2022) Biodegradation of p-xylene—a comparison of three psychrophilic pseudomonas strains through the lens of gene expression. Environ Sci Pollut Res 29(15):21465–21479. https://doi.org/10.1007/s11356-021-17387-5

Mohamed MS, El-Arabi NI, El-Hussein A et al (2020) Reduction of chromium-VI by chromium-resistant Escherichia coli FACU: a prospective bacterium for bioremediation. Folia Microbiol 65:687–696. https://doi.org/10.1007/s12223-020-00771-y

Mohapatra B, Phale PS (2021) Microbial degradation of naphthalene and substituted naphthalenes: metabolic diversity and genomic insight for bioremediation. Front Bioeng Biotechnol 9:602445. https://doi.org/10.3389/fbioe.2021.602445

Mohd S, Kushwaha AS, Shukla J, Mandrah K, Shankar J, Arjaria N et al (2019) Fungal mediated biotransformation reduces toxicity of arsenic to soil dwelling microorganism and plant. Ecotoxicol Environ Saf 176:108–118. https://doi.org/10.1016/j.ecoenv.2019.03.053

Moreira V, Lebron Y, Lange L, Santos L (2019) Simultaneous biosorption of cd (II), Ni (II) and Pb (II) onto a brown macroalgae fucus vesiculosus: mono- and multi-component isotherms, kinetics and thermodynamics. J Environ Manage 251:109587. https://doi.org/10.1016/j.jenvman.2019.109587

Mousavi SM, Hashemi SA, Iman Moezzi SM, Ravan N, Gholami A, Lai CW et al (2021) Recent advances in enzymes for the bioremediation of pollutants. Biochem Res Int 2021:1–12

Mukjang N, Chitov T, Mhuantong W, Champreda V, Pathom-Aree W, Sattayawat P et al (2022) Bacterial communities associated with crude oil bioremediation through composting approaches with indigenous bacterial isolate. Life 12(11):1712. https://doi.org/10.3390/life121117123

Muthukumaran M (2022) Advances in bioremediation of nonaqueous phase liquid pollution in soil and water. In: Biological approaches to controlling pollutants. Woodhead Publishing, pp 191–231

Nowicka B, Fesenko T, Walczak J, Kruk J (2020) The inhibitor-evoked shortage of tocopherol and plastoquinol is compensated by other antioxidant mechanisms in Chlamydomonas reinhardtii exposed to toxic concentrations of cadmium and chromium ions. Ecotoxicol Environ Saf 191:110241. https://doi.org/10.1016/j.ecoenv.2020.110241

Pachaiappan R, Cornejo-Ponce L, Rajendran R, Manavalan K, Femilaa Rajan V, Awad F (2022) A review on biofiltration techniques: recent advancements in the removal of volatile organic compounds and heavy metals in the treatment of polluted water. Bioengineered 13(4):8432–8477

Pande V, Pandey SC, Sati D, Pande V, Samant M (2020) Bioremediation: an emerging effective approach towards environment restoration. Environ Sustain 3(1):91–103. https://doi.org/10.1007/s42398-020-00099-w

Paul O, Jasu A, Lahiri D, Nag M, Ray RR (2021) In situ and ex situ bioremediation of heavy metals: the present scenario. J Environ Eng Landsc Manag 29(4):454–469

Rabani MS, Sharma R, Singh R, Gupta M (2022) Characterization and identification of naphthalene degrading bacteria isolated from petroleum contaminated sites and their possible use in bioremediation. Polycyclic Aromat Compd 42(3):978–989. https://doi.org/10.1080/10406638.2020.1759663

Rafeeq H, Afsheen N, Rafique S, Arshad A, Intisar M, Hussain A, Iqbal HM (2023) Genetically engineered microorganisms for environmental remediation. Chemosphere 310:136751

Saavedra R, Muñoz R, Taboada ME, Vega M, Bolado S (2018) Comparative uptake study of arsenic, boron, copper, manganese and zinc from water by different green microalgae. Bioresour Technol 263:49–57. https://doi.org/10.1016/j.biortech.2018.04.101

Sangkharak K, Choonut A, Rakkan T, Prasertsan P (2020) The degradation of phenanthrene, pyrene, and fluoranthene and its conversion into medium-chain-length polyhydroxyalkanoate by novel polycyclic aromatic hydrocarbon-degrading bacteria. Curr Microbiol 77(6):897–909. https://doi.org/10.1007/s00284-020-01883-x

Sayqal A, Ahmed OB (2021) Advances in heavy metal bioremediation: an overview. Appl Bionics Biomech

Sen SK, Patra P, Das CR, Raut S, Raut S (2019) Pilot-scale evaluation of bio-decolorization and biodegradation of reactive textile wastewater: an impact on its use in irrigation of wheat crop. Water Resour Ind 21:100106. https://doi.org/10.1016/j.wri.2019.100106

Shah H, Jain S (2020) Bioremediation: an approach for environmental pollutants detoxification. In: Waste to energy: prospects and applications. Springer, New York, pp 121–142

Sharma J (2019) Advantages and limitations of in situ methods of bioremediation. Recent Adv Biol Med 5(2019):10941

Sharma I (2020) Bioremediation techniques for polluted environment: concept, advantages, limitations, and prospects. In: Trace metals in the environment-new approaches and recent advances. IntechOpen

Singh RK, Tripathi R, Ranjan A, Srivastava AK (2020) Fungi as potential candidates for bioremediation. In: Abatement of environmental pollutants. Elsevier, Amsterdam, pp 177–191
Sivakumar D, Kandaswamy A, Gomathi V, Rajeshwaran R, Murugan N (2014) Bioremediation studies on reduction of heavy metals toxicity. Pollut Res 33(3):553–558
Song J, Zhang S, Xie Y, Li Q (2019) Purification and characteristics of an aflatoxin B1 degradation enzyme isolated from pseudomonas aeruginosa. FEMS Microbiol Lett 366(5):fnz034. https://doi.org/10.1093/femsle/fnz034
Sravya K, Sangeetha S (2022) Feasibility study on bioremediation techniques to contaminated soils. Mater Today Proc 51:2556–2560. https://doi.org/10.1016/j.matpr.2021.12.364
Tak Y, Kaur M, Tilgam J, Kaur H, Kumar R, Gautam C (2022) Microbes assisted bioremediation: a green technology to remediate pollutants. In: Suyal DC, Soni R (eds) Bioremediation of environmental pollutants. Springer, Cham. https://doi.org/10.1007/978-3-030-86169-8_2
Tegene BG (2020) Mode of action, mechanism and role of microbes in bioremediation service for environmental pollution management. J Biotechnol Bioinform Res 2(3):1–18
Tegene BG, Tenkegna TA (2020) Mode of action, mechanism and role of microbes in bioremediation service for environmental pollution management. J Biotechnol Bioinform Res 2:1–18
Tyagi B, Kumar N (2021) Bioremediation: principles and applications in environmental management. In: Bioremediation for environmental sustainability. Elsevier, pp 3–28
Unuofin JO, Falade AO, Aladekoyi OJ (2021) Applications of microbial laccases in bioremediation of environmental pollutants: potential issues, challenges, and prospects. Bioremed Environ Sustain 519–540
Verma S, Kuila A (2019) Bioremediation of heavy metals by microbial process. Environ Technol Innov 14:100369. https://doi.org/10.1016/j.eti.2019.100369
Vidali M (2001) Bioremediation. An overview. Pure Appl Chem 73(7):1163–1172
Villegas-Plazas M, Sanabria J, Junca H (2019) A composite taxonomical and functional framework of microbiomes under acid mine drainage bioremediation systems. J Environ Manag 251:109581
Vishwakarma GS, Bhattacharjee G, Gohil N, Singh V (2020) Current status, challenges and future of bioremediation. In: Bioremediation of pollutants. Elsevier, pp 403–415
Vo M-T, Nguyen V-T, Dao T-S (2020) Responses of green algae and diatom upon exposure to chromium and cadmium. Vietnam J Sci Technol Eng 62(1):69–73. https://doi.org/10.31276/VJSTE.62(1).69-73
Wang Z, Luo Z, Yan C, Rosenfeldt RR, Seitz F, Gui H (2018) Biokinetics of arsenate accumulation and release in microcystis aeruginosa regulated by common environmental factors: practical implications for enhanced bioremediation. J Clean Prod 199:112–120. https://doi.org/10.1016/j.jclepro.2018.07.131
Wu P, Zhang Z, Luo Y, Bai Y, Fan J (2022) Bioremediation of phenolic pollutants by algae-current status and challenges. Bioresour Technol 350:126930. https://doi.org/10.1016/j.biortech.2022.126930

Chapter 24
Microalgae Based Dye Remediation: A Sustainable Approach to Dye Abatement

V. L. Vibash Kalyaan, S. Gopinath, and M. Kamaraj

Abstract One potential answer to the urgent environmental problem of effluents tainted with dyes is wastewater treatment using microalgae. This article delves deep into the possibilities of microalgae for dye remediation, highlighting their capacity to extract pollutants and produce beneficial biomass. Controlling factors including temperature, light intensity, pH, and dissolved oxygen levels are crucial for maximizing growth and pollutant uptake. It is critical to choose microalgae species that can adapt to different types of wastewater and tolerate dye components. Photobioreactors and harvesting technology are examples of equipment that can affect the feasibility of such systems. Also, microalgal biomass quality varies, which impacts its potential for downstream uses. Because of its complicated chemistry and possible toxicity, industrial dye effluent poses significant issues. There are a number of different harvesting procedures to choose from, and each has its own set of advantages and challenges that make things more complicated. Optimizing growth conditions for microalgae is still a challenging process, but there are possible opportunities for improvement through genetic engineering and biotechnological approaches. A crucial part of long-term environmental management plans might be microalgae-based wastewater treatment, which could be achieved through persistent research and development, and creative solutions.

Keywords Microalgae · Wastewater · Biomass · Dye remediation · Adsorption · Uptake

V. L. Vibash Kalyaan · S. Gopinath · M. Kamaraj (✉)
Department of Biotechnology, Faculty of Science and Humanities, SRM Institute of Science and Technology, Ramapuram, Chennai, Tamil Nadu 600089, India
e-mail: drkamarajm@gmail.com

© The Author(s), under exclusive license to Springer Nature Switzerland AG 2025
A. Jeyaseelan and K. Murugasen (eds.), *Sustainable Environmental Remediation: Avenues in Nano and Biotechnology*, Environmental Science and Engineering,
https://doi.org/10.1007/978-3-031-78483-5_24

24.1 Introduction

Commodity dyes listed in the US Color Index have proliferated recently, indicating a tremendous boom in the dye sector. Surprisingly, azo dyes make up an astounding 80% of the 60,000 tons of dyes released into the environment every year (Shah 2018). China, a major player in the chemical, textile, and light industrial sectors, produces the vast majority of the azo dyes in the world (between 70–80%). The release of an unprecedented 900 million tons of effluent each year is just one of the many far-reaching effects of this dye inflow (Nanjani and Keharia 2021; Song et al. 2023). Due to the presence of biologically hazardous organics with carcinogenic, teratogenic, and mutagenic qualities, standard treatment technologies face major challenges when dealing with this complicated wastewater that contains a high concentration of azo dyes (Ismail et al. 2019). Consequently, novel treatment methods are required immediately to ensure thorough remediation of dye wastewater and conform to ever-tightening discharge criteria. When it comes to dye wastewater, particularly that contains azo dyes, the utilization of Carbon–Water–Oxygen (CWO) technology shows tremendous potential in resolving these issues. Algal treatment is a tempting choice for the treatment of complex wastewater streams that offers both environmental relevance and practical application (Pilli et al. 2020; Ahmad et al. 2022). The enormous chemical use and water consumption of the textile industry are the main causes of its substantial impact on the environment. Dyeing and finishing procedures are responsible for 17–20% of the water pollution caused by textile factories, which use an average of 1.6 million liters of water daily to produce fabric. However, out of 72 harmful compounds found in textile effluents, 30 have shown resistance to traditional treatment procedures, making environmental deterioration even worse (Raja et al. 2018; Toprak and Anis 2017).

Popular because of their affordability and stability, synthetic dyes are extremely harmful to the environment because of their longevity of accumulation in the water and their ease of dissolution. The textile business relies on azo dyes, which are highly complex to work with and have negative impacts on ecosystems (Pavlović et al. 2022). Not only does their release into the environment harm aquatic life, but it also blocks sunlight, which inhibits the surrounding ecosystems. High chemical oxygen demand (COD) and biological oxygen demand (BOD) levels indicate that inorganic and organic nutrients are being discharged into the environment, which contributes to eutrophication and other environmental issues, especially through excessive phosphorus (P) and nitrogen (N) loads (Dhakad et al. 2022). Issues like the production of solid waste and harmful pollutants into the air are brought about by this phenomenon. Several microalgae species have proven to be highly tolerant to wastewater (WW) contaminants and to efficiently thrive in WW environments (Abdelfattah et al. 2023). These species include *Scenedesmus*, *Chlorella*, *Euglena*, *Oscillatoria*, *Chlamydomonas*, and *Ankistrodesmus*. Consequently, there has been a lot of interest in microalgae-based approaches for cleaning wastewater from many sources, including municipalities, industries, agro-industrial sites, and livestock.

Microalgae are a versatile biological treatment option because they may remove P and N components, which can reduce eutrophication (Al-Jabri et al. 2021). They have the ability to transform organic and inorganic waste into useful biomass. Also, microalgae from certain treatment ponds can be used as a feedstock for different goods and as a source of food. They have several uses in different sectors. Improving the effectiveness of wastewater treatment (WWT) by co-cultivating microalgae with other bacteria is another potential strategy. Symbiotic relationships between microalgae and heterotrophic microbes such as bacteria, fungi, and yeast allow microalgae to increase their biomass yield and enhance bioremediation (Das et al. 2022). The production of biofuels and biomass from microalgae has also been made easier by this co-culturing technique. Collaborative cultivation methods use various microbial interactions to cultivate species with complementing enzymatic activities, further optimizing biofuel and biomass output. These methods seek to solve particular performance-enhancing requirements by forming artificial consortia based on mutualistic interactions (Mahesh et al. 2021). A potential solution to one of the biggest social and economic problems of the twenty-first century—the impending worldwide water scarcity crisis—may lie in WWT derived from microalgae (Posadas et al. 2017). This article explores the potential of microalgae, with their symbiotic interactions with other bacteria and their ability to perform several functions, to provide a viable and efficient method for treating wastewater, all while helping to produce biomass and biofuels. An overview of various topics discussed in this article is depicted in Fig. 24.1.

24.2 Classification and Chemistry of Dyes

24.2.1 Azo Dyes

More than 70% of dyes worldwide, and 60–70% of textile dyes are azo dyes, which contain one or more azo groups (–N=N–). Popularly adaptable, their rapidity, stability, and brilliant coloration make them useful in a wide range of industries. The most common azo dye colors are yellow, orange, and red, although they can cover a broad variety of hues. Nevertheless, aromatic compounds that may be carcinogenic can be obtained by cleaving their azo group. Using diazonium salts and coupling components, oxidation/reduction or diazotization/coupling processes are employed in their synthesis process (Benkhaya et al. 2017).

24.2.2 Reactive Dyes

Although reactive dyes may not be homogeneous, they have better-wet strength. Covalently bonding with substrates, they produce a variety of colors on cotton while

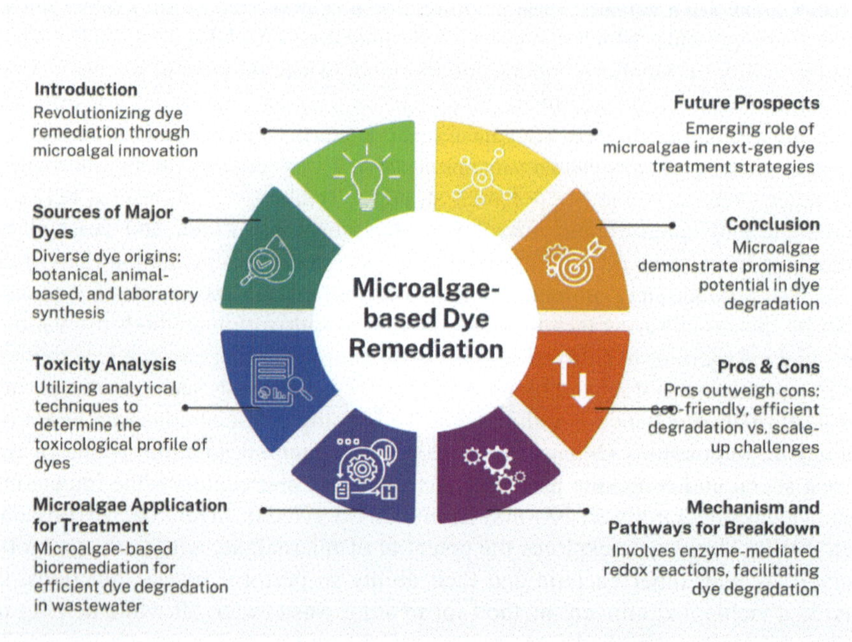

Fig. 24.1 An overview of the areas focused in this review on Microalgae-based dye remediation

maintaining exceptional light and wash fastness. Persisting trouble in fixing things has been an ongoing problem. Because they are insoluble in water, these dyes offer high resistance to light, bleach, and washing when applied to cellulose fibers like cotton. Indigoid and anthraquinone are two examples of alternative vat dye classes that include indigo derivatives, some of which have halogenated versions (Benkhaya et al. 2017; Crowle 2017).

24.2.3 Sulfur Dyes

Sulfur dyes were produced worldwide at 110,000–120,000 tons per annum in the year 1996, with a share of 9.1% in the overall US dye production and 15.8% in cellulosic fiber dyes. At 8,500 tons in 2007, sulfur dyes were the third most produced dye in China. Half of the cellulosic fiber dyes, especially black sulfur dyes, were sulfur-based (Nguyen and Juang 2013). The Color Index assigns constitution numbers to dyes that are most commonly used for dying fabrics; these dyes have a narrow color spectrum but are very light- and wash-fast (Mahdi et al. 2021).

24.2.4 Acid Dyes

Acid dyes can be used on silk, wool, or nylon as long as the pH is between 3.1 and 7.0. Their lightfastness usually falls anywhere between 5.0 and 6.0 on the blue scale, and their wet-fastness is moderate to good. The acidic conditions of formic or acetic acid are used to apply these dyes; they have varied brightness and mild washing fastness. Sulphonated groups give them their characteristic water solubility, and interactions between the sulphonate groups and the ammonium groups on the wool, in addition to Van der Waals forces, assist their bonding to the fiber (Omotosho and Ameuru 2019; Pifferi et al. 1993).

24.2.5 Basic Dyes

Acrylic, paper, and nylon are the usual substrates for basic dyes, while modified polyester has seen considerable use as well. They work well with retarders, particularly at high temperatures, because of their low migratory characteristics. Substrates with a negative charge are electrostatically attracted to these water-soluble dyes because they produce colored cations in solution. Triarylmethane, xanthene, and acridine are three examples of basic dyes that display either dispersed charges across the dye cation or localized positive charges on ammonium groups (Manikandan et al. 2022).

24.2.6 Direct Dyes

Some factors, such as chromophore, fastness qualities, and application features, are used to classify direct dyes. Formazan, anthraquinone, quinolone, and thiazole are lesser classes that are part of this larger family, which includes chromophoric types including azo, stilbene, phthalocyanine, and dioxazine (Benkhaya et al. 2022). Their moderate wash-fastness belies their large color range and ease of application. Because of their better wet and washing fastness on cellulosic surfaces, reactive dyes have partially replaced direct dyes.

24.3 Sources and Toxicity

The following is a list of synthetic dyes along with their origins, toxicity, and effects on the environment.

24.3.1 Methylene Blue

There are many industries that rely on methylene blue (MB), but if not handled properly, it can cause serious harm to people and the planet. Research has shown that MB is harmful to embryos and can cause teratogenic consequences in animals, including rats and angelfish. Textile, paint, pharmaceutical, and cosmetic industries all generate trash-containing dyes into the environment, which contributes to MB pollution. Approximately 120 cubic meters of industrial wastewater is discharged per ton of fiber generated by the textile industry, which is responsible for almost 67% of the dyestuff market. Because of the high toxicity of MB—which is recognized as carcinogenic and non-biodegradable—this careless discharge, which frequently occurs in natural water sources, poses a threat to aquatic life, beneficial bacteria, and human health (Oladoye et al. 2022). Microalgal species like *Chlorella vulgaris* show potential in the abatement of methylene blue from wastewater. The results of the microalgae-based removal of methylene blue, along with some other common dyes are shown in Table 24.1.

In addition to genitourinary and dermatological issues, MB exposure can cause problems with the gastrointestinal tract, respiratory system, cardiovascular system,

Table 24.1 Treatment of a few common dyes with microalgae

Microalgal species	Dye in subject	Efficiency of dye removal (%)	Concentration of the dye (mg/L)	References
Spirulina platensis	Acid black 210	98.55	125	Unit (2017)
Spirulina platensis	Acid blue 7	97.05	125	Unit (2017)
Chlorella vulgaris	Aniline blue	58	25	Carbajo Arteaga et al. (2018)
Nostoc sp.	Malachite green	97.13	100	Gajare (2012)
Chlorella vulgaris	Methylene blue	83.04	100	Chin et al. (2020)
Chlorella vulgaris	Congo red	100	50	Noor and Jasim (2019)
Oscillatoria sp.	Safranin	52	5	Gezahegn Gelebo et al. (2020)
Coelastrella sp.	Rhodamine B	80	100	Baldev et al. (2013)
Chlorella vulgaris	Direct blue 71	78	200	Ishchi and Sibi (2019)
Chlorella vulgaris	Disperse red 1	84	300	Ishchi and Sibi (2019)
Chlorella vulgaris	Reactive black 5	80	200	Ishchi and Sibi (2019)

and central nervous system. In addition to its high molar absorption coefficient, which hinders sunlight transmission and affects chemical and biological oxygen demand levels, MB's unique tint can aesthetically deteriorate aquatic bodies. Thus, the whole aquatic environment is adversely affected by MB pollution. These risks highlight the critical need to synthesize existing studies on MB-laden wastewater remediation systems (Krishna Moorthy et al. 2021).

24.3.2 *Azo Dyes*

The contamination of rivers and groundwater by azo dyes and nitrated polycyclic aromatic hydrocarbons is a common problem in areas around dyeing companies. The effects on human health, particularly in the food business, continue to be a source of worry, notwithstanding laws in several nations. The vast majority of the colorants approved by the US Food and Drug Administration for usage in food, medicine, and personal care products were azo dyes, which accounted for more than 3,000 tons in 1991 (Chung 2016). Cancers of the bladder, sarcomas, hepatocarcinomas, and chromosomal abnormalities have been linked to some azo dyes in both laboratory animals and dye workers. Mammalian hepatic and bacterial azoreductases both convert azo compounds to amines, but bacterial azoreductases are more effective in transforming some azo dyes into amines that might cause cancer and mutation (Mahmood et al. 2016). The dyestuff in effluent reduces growth by lowering food consumption and decreasing protein, carbohydrate, and fat contents in various tissues. Fish exposed to sublethal quantities of effluent also experience changes in feeding, absorption, and conversion rates. There are fatal amounts of naphthalene in fish organs and brain, and sublethal concentrations cause neurosensory impairment and metabolic stress (Solangi and Overstreet 1982). Cellular metabolic depression and a transition to anaerobic metabolic pathways to fulfill elevated energy needs in stressful situations are indicated by the inhibition of succinic dehydrogenase activity in many tissues. Anaerobic reduction of azo dye bonds may produce aromatic amines, which are poisonous and carcinogenic. In the process of degradation, an estimated 130 of 3,200 azo dyes can produce aromatic amines that can cause cancer (Ismail et al. 2019; Ikram et al. 2023).

24.3.3 *Anthraquinone Dyes*

A possible genotoxic route for cancer induction has been demonstrated by the carcinogenic effects of Disperse Blue 1, an anthraquinone derivative, which includes bladder tumors in rats and is mutagenic in bacteria (Mathur et al. 2012; Haws et al. 1994). Less dependence on experimental animals and quick, inexpensive evaluations have contributed to the rise in popularity of in vitro toxicity methods using microbes. *Selenastrum capricornutum* has demonstrated susceptibility to a range of substances, including herbicides and metals. There are a variety of aquatic ciliates, like *Tetrahymena pyriformis*, that can be used to assess toxicity. In addition to cytotoxicity tests, supplementary physiological and morphological tests have been used on *T. pyriformis*. When looking for evidence of genotoxicity in different kinds of environmental samples, the Ames test—which uses *Salmonella typhimurium*—is often used (Dias and Lima 2002; Novotný et al. 2006).

24.3.4 Other Major Dyes

Among the possible exposure concerns highlighted by the Ecological and Toxicological Association of Dyes and Organic Pigments Manufacturers (ETAD) are reactive dyes, which are known to cause skin sensitivity and respiratory difficulties in occupational contexts (Starovoytova and Odido 2014). To lessen the likelihood of this happening, it is advised to use liquid dyes or low-dusting formulas, wear protective gear, and stay away from dye dust. Due to the elimination of the reactive group and the great wash fastness of the fabric, reactive dyes no longer have the same toxicological qualities after fixing. This helps to reduce allergy reactions among consumers. Even though they don't show much sweat fastness, dispersed dyes can nevertheless cause allergic reactions. Because of its excellent sweat fastness, polyester dyed with dispersed dyes usually does not cause any problems (Al-Etaibi and El-Apasery 2019). Nevertheless, disperse-dyed rayon made of polyamide or acetate has the potential to cause skin responses due to its low sweat fastness, which allows the colors to travel to the skin. An elevated heart rate, nausea, vomiting, cyanosis, jaundice, quadriplegia, Heinz body development, and tissue necrosis are some of the adverse consequences linked to cationic dyes in humans (Soniya and Muthuraman 2015). Because of their acidic and water-soluble characteristics, anionic dyes (such as acid, azo, direct, and reactive dyes) are difficult to remove. Covalent bonds are formed between fibers and reactive dyes such as vinyl sulphone, chlorotriazine, trichloropyrimidine, and difluorochloropyrimidine when the dyeing process is underway (Taylor 2000).

24.4 Microalgae as Potential Dye Treaters

Microalgae come in all shapes and sizes, but some of them are particularly good at removing dyes like methylene blue and malachite green through biosorption. Biosorption experiments using ineffective algal biomass in artificial dye solutions were the subject of numerous investigations. *Microspora* sp. is one example. After the lipids were extracted, the algal biomass showed incredible biosorbent efficacy for methylene blue. In just 24 h with 150 rpm agitation, the dye was completely removed (Maurya et al. 2014). Also tested for biosorption of methylene blue was defatted biomass of the oleaginous microalga *Scenedesmus dimorphus*. Results showed that defatted biomass had a maximum adsorption capacity that was on par with both untreated and acid-pretreated biomass (Sarat Chandra et al. 2015). Biosorbents for dye removal have been developed from waste products of the algae biodiesel industry. As an example, a biosorbent for methylene blue was found to be charcoal made from *Spirulina platensis* after the oil was extracted. Biochar, which is made from the leftover biomass of *Ulothrix zonata* after pigment extraction, was also found to be an inexpensive biosorbent for Congo red, crystal violet, and malachite green. *Spirulina platensis* dried biomass also outperformed *Gracilaria edulis* seaweed in terms of Reactive Blue 19 dye biosorption capability (Venkataraghavan et al. 2020). In addition, a relatively inexpensive and effective biosorbent for malachite green was found to be *Chlorella*-based biomass, which is a byproduct of algae production.

Pretreatment of algal biomass improves dye biosorption, according to multiple studies. Case in point: *Pithophora* sp. heat activation. Compared to untreated algae, algae treated at 300 °C had an improved biomass sorption ability for eliminating malachite green (Vasanth Kumar et al. 2005). Studies on the removal of Synazol reactive dye by *Spirogyra* sp. have shown that autoclaving is an efficient way to boost color removal capacity. The use of living microalgae and cyanobacteria has opened the door to the treatment of effluents containing dyes, as they successfully remove their colours. *Anabaena flos-aquae* UTCC64, *Phormidium autumnale* UTEX1580, and *Synechococcus* sp. are cyanobacteria strains that were mentioned. *Synechococcus* sp. Strain PCC7942 was tested for its ability to break down textile dyes and effluents containing dyes, thereby decreasing the toxicity and colouring caused by the dyes. Specifically, indigo dye was entirely broken down after 19 days by *Phormidium autumnale* UTEX1580 (Dellamatrice et al. 2017). *Chlorella vulgaris*, when cultured in High-Rate Algal Ponds (HRAP), decreased ammonia and phosphate pollutants and eliminated as much as half of the dye from textile effluent, according to another study. Additionally, immobilized microalgae have shown effective in decolorizing wastewater and textile dyes. For example, compared to suspension cultures, *Chlorella vulgaris* immobilized in 2% alginate was able to remove 44% of the color from Lanaset Red 2GA textile dye from textile wastewater (Chu and Phang 2019).

When it comes to biodegrading organic contaminants, microalgae use a multi-stage enzymatic mechanism. Figure 24.2 shows various mechanisms adopted by microalgae for dye abatement. In order to increase the pollutant's hydrophilicity through oxidation–reduction processes or hydrolysis, cytochrome P450 enzymes such as hydroxylase, carboxylase, decarboxylase, and monooxygenase first aid in the addition or exposure of hydroxyl groups. Then, enzymes such as glucosyltransferases and glutathione-S-transferases help to protect cells from oxidative damage by conjugating glutathione with different molecules that include electrophilic centers (such as COOH, epoxide ring, or $CONH_2$) (Xiong et al. 2019). Stage III of detoxification involves a wide variety of enzymes that break down the molecules into less or non-toxic forms, including dehydrogenase, glutamyl-tRNA reductase, carboxylase, mono(di)oxygenase, laccases, transferase, hydrolases, pyrophosphatase, and dehydratase (Abdelfattah et al. 2023; Ding et al. 2017).

Chemical, physicochemical, and biological techniques are used in conventional wastewater treatment (WWT) methods to address dye contamination. Among these methods are adsorption, ozonation, flocculation, ultrafiltration, and coagulation.

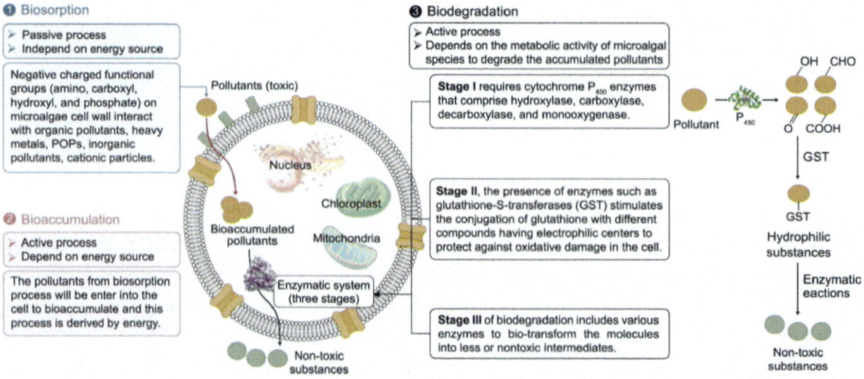

Fig. 24.2 Various mechanisms of dye treatment by microalgae (Reused from Abdelfattah et al. (2023) with permissions obtained from Copyright © 2023 Published by Elsevier Ltd)

But they come with certain problems, like making sludge and toxic byproducts and not being very good at removing dye. This means that new approaches are required. Phycoremediation, a form of bioremediation that makes use of microalgae, is becoming more popular because it is both cost-effective and safe. Phycoremediation is a process that uses plants to absorb carbon dioxide (CO_2) from waste air and remove or change contaminants like heavy metals and colors from wastewater. The use of microalgae for dye absorption has numerous benefits, such as being eco-friendly, highly efficient, easy to implement, and inexpensive (Sharma et al. 2022). Uses for this approach include biosensing dangerous chemicals, degrading compounds, treating wastewater, and removing nutrients.

24.5 Methods of Microalgal Treatment

24.5.1 Biosorption

The physicochemical process known as biosorption involves the sorption of chemicals in or on a biological surface, and it employs both living and dead biomass to remove pollutants. The potential of several biomasses to degrade synthetic dyes and other contaminants has been investigated. Figure 24.3 provides insight into the process of biosorption of dyes by microalgal cells. These biomasses include plants, bacteria, fungi, and algae. *Spirulina platensis* is a popular blue-green microalga that is grown all over the world for a cheap price, and recent studies have shown that it is an effective, reliable, and cost-effective algal biosorbent. This microalga has functional groups like as carboxyl, hydroxyl, and phosphate, which allow it to bind dyes more effectively. It is rich in biomolecules and proteins. *S. platensis* has been the subject of different investigations, able to efficiently extract heavy metals and dyes from water solutions. Furthermore, growing *S. platensis* produces beneficial pigment compounds such phycocyanin, β-carotene, and chlorophyll-a. Phycocyanin is a naturally occurring blue pigment that has several uses owing to its luminous nature and medicinal properties. These include food coloring, cosmetics, and biochemical tests. Nevertheless, waste biomass is produced during phycocyanin manufacturing, therefore finding a sustainable way to use it is essential. In a study, the possibilities of phycocyanin-free biomass of *S. platensis* have been found as an inexpensive biosorbent for the C.I. (Color Index) Basic Red 46 (C.I. BR 46). Industrial environments often make use of the azo dye Basic Red 46 (C.I. BR 46). Results of kinetic and equilibrium analysis showed that the biosorbent was successful in removing BR 46. The waste biomass was analyzed using Fourier transform infrared (FT-IR) spectroscopy, which confirmed the presence of many functional groups that may bind dyes. The Langmuir model was shown to be a good match, matching the maximal biosorption capacity of virgin algae at 23.06 mg g^{-1}. The results of this research highlight the promise of recycling *S. platensis* biomass as a cost-effective bioremediation strategy for polluted water, which provides an eco-friendly option for dye removal (Sharma et al. 2022).

Scenedesmus quadricauda, a green microalga, was immobilised in alginate beads in order to remove Reactive Blue 19 (RBBR) in an additional investigation. At 30 °C

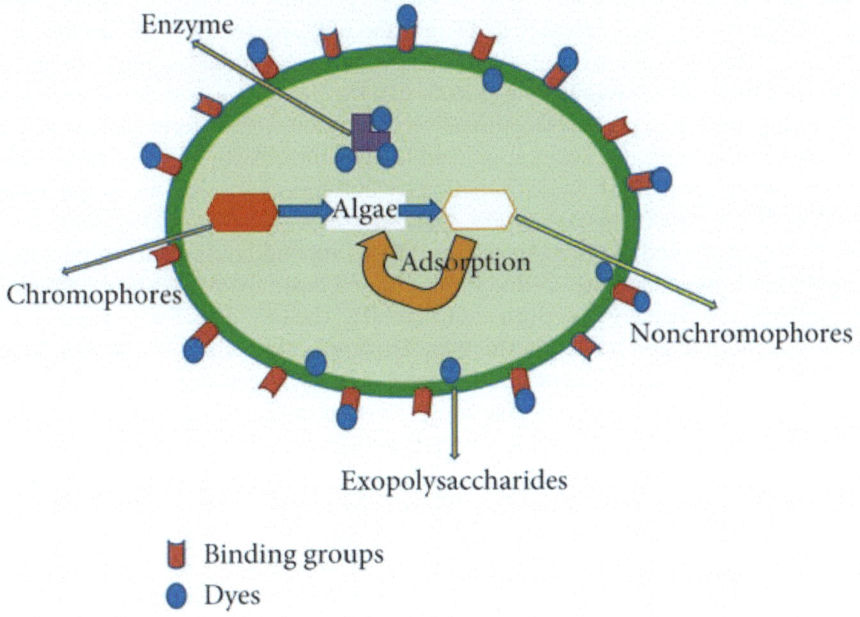

Fig. 24.3 Schematic diagram showing the biosorption of dye compounds by microalgae. (Image reused from Ayele et al. (2021) under the Creative Commons CC-BY-NC-ND license)

and pH 2.0, the dye uptake of both live and heat-inactivated biomass was at its peak. Live biomass had a maximum adsorption capacity of 45.7 mg/g, while dead biomass had 48.3 mg/g. With monolayer capacities of 68 mg/g for live biomass and 95.2 mg/g for dead biomass, respectively, the Langmuir, Freundlich, and Dubinin-Radushkevich models provided the most satisfactory fits to the biosorption data. According to the results of the kinetic analysis, pseudo-second-order kinetics is a good fit (Ergene et al. 2009).

24.5.2 Bioaccumulation

Three species of freshwater microalgae—*Chlorella*, *Cosmarium*, and *Euglena*—were studied for their potential to biologically decolorize triphenylmethane dye (malachite green, MG). The starting dye concentration, reaction duration, pH, temperature, and algal concentration were among the development factors that were fine-tuned. The *Chlorella*, *Cosmarium*, and *Euglena* species achieved maximum color removal rates of 92, 91, and 87%, respectively, under ideal conditions of 45 °C, 180 min of contact duration, pH value of 9.0, inoculum concentration of 9×10^6 cells/mL, and initial dye concentration of 10 mg/L (Kumar et al. 2020). Researchers looked into whether *Chlorella vulgaris* might remove the monoazo dye Tectilon Yellow 2G. At 50, 200, and 400 mg/L of starting Tectilon Yellow 2G concentration, removal efficiencies were found to be 69, 66, and 63%, respectively. As acclimated, *C. vulgaris* led to an increase of 88, 87, and 88% in those places, respectively. A study was conducted to examine the efficacy of a freshwater alga species *Cosmarium* in removing malachite green. Decolorization was tested with different values of pH, algal concentration, dye concentration, and temperature. A study was conducted to examine the decolorization and elimination capabilities of various microorganisms, including *Nostoc lincki*, *Volvox aureus*, *Chlorella vulgaris*, *Oscillatoria rubescens*, *Elkatothrix viridis*, and *Lyngbya lagerlerimi*. The compounds tested included Orange II, methyl red, basic cationic, G-Red (FN-3G), and basic fuchsin. When it came to removing colors, each algae performed differently. The best dyes for decolorization were basic cationic and basic fuchsin dyes. The removal efficiencies for the orange II and G-Red dyes were 43.7 and 59.12% for *C. vulgaris*, respectively, and 5.02 and 3.25% for *V. aureus*, respectively. The most effective initial pH level for decolorization was 92.4% under ideal circumstances. Another study investigated which algae species could decolorize specific colors. The removal efficiencies for the Orange II and G-Red dyes were 43.7 and 59.12%, respectively, for *Chlorella vulgaris* and *Volvox aureus*, respectively (El-Sheekh et al. 2009).

24.5.3 *Decolorization*

The presence of algae in both freshwater and saltwater habitats has led to their recognition as potential biosorbents. Their high binding affinity and huge surface area give them excellent biosorption capacity. It is worth mentioning that the greatest algal concentrations were able to nearly completely remove malachite green from water-based solutions in just 7 h with *Dunaliella salina* and 3 h with *Chlorella vulgaris*. *Chlorella vulgaris* has a cell wall, which is missing in *Dunaliella salina*, which probably helps it remove dye faster (Pathak et al. 2014). A key component of biosorption is the characteristic of algal cell walls that allows for electrostatic attraction and complexation. Algal cell surfaces contain functional groups such as hydroxyl, carboxylate, amino, and phosphate, which are thought to trap pollutants in wastewater. The effectiveness of dye removal is enhanced when dye ions accumulate on algal biopolymers and subsequently diffuse onto the solid phase of the biopolymer. Both *Dunaliella salina* and *Chlorella vulgaris* showed a progressive increase in the percentage of malachite green removed as the pH in the media rose. On the other hand, at pH 7, both types of algae removed the most methylene blue by percentage. Noteworthy, *Dunaliella salina* cannot thrive in environments with pH values higher than 9.

Previous research has shown that changing the pH level has a major impact on the pace of dye decolorization. For example, the decolorization rate of Malachite Green for microalga *Cosmarium* sp. increased thrice at pH 6.0 compared to pH 4.0 and reached a maximum of 92.4% at pH 9.0. In a similar vein, when the pH went from 3.0 to 8.0, the uptake of Basic Yellow dye by *Caulerpa scalpelliformis* went up from 17 to 27 mg/g. The efficacy of color removal is reduced at lower pH levels due to effective competition between H^+ ions and dye cations, while increased electrostatic attraction with positively charged dye cations is observed at higher pH levels due to negatively charged biomass surfaces (Sarkar and Dey 2021; Mohamed Saad Abd-El-Kareem HMT 2012).

24.6 Comparative Studies of Microalgal Treatment

24.6.1 *Comparison Between Spirulina platensis and Commercial Activated Carbon*

One investigation looked at *S. platensis* FTIR spectra in order to determine which functional groups are responsible for dye adsorption, it examined *S. platensis* and commercial activated carbon (AC). In the case of *S. platensis*, the stretching of O–H bonds and NH_2 groups was suggested by the strong peaks at 3370 cm^{-1}, while the stretching of CH_2 groups was indicated by bands at 2920 and 2859 cm^{-1}. At 1659 and 1535 cm^{-1}, the NH_2 groups were seen to be bent like scissors. The amide or amine groups were proposed to be stretched C–N at peaks at 1224, 1149, and

$1021\ cm^{-1}$. The aromatic C–H stretching vibrations, as well as the P–O and S–O absorption bands, were located between 750 and 900 cm^{-1}. O–H stretching, CH_2 stretching, aromatic C–C ring stretching, and carbonyl groups of carboxylic acid were all detected in the absorption bands of AC. Minimal changes in the FTIR spectra following RR-120 dye adsorption were indicative of low-energy interactions (Gally et al. 2014).

Using Energy-dispersive X-ray Spectroscopy (EDS), it was found that in *S. platensis*, the carbon, nitrogen, oxygen, phosphorus, and sulfur atoms are accompanied by carboxyl, hydroxyl, sulphate, phosphate, aldehyde, and ketone functional groups. AC included aromatic groups, COOH, and OH groups. The interaction between the RR-120 dye and the adsorbents in aqueous solutions is facilitated by these groups. The RR-120 dye may be effectively removed by the adsorbents *S. platensis* and AC. By conducting FTIR and SEM characterizations as well as analyzing N_2 adsorption/desorption curves, the ideal pH and contact time were ascertained. According to equilibrium isotherms, the adsorption process was best described by the general order kinetic model and by the Liu model. Both AC and *S. platensis* had maximum adsorption capabilities of 267.2 and 482.2 mg/g, respectively. It was suggested by the thermodynamic parameters ($\Delta H°$, $\Delta S°$, and ΔG) that the RR-120 dye and both adsorbents interacted physically. Even in extremely salty conditions, *S. platensis* and AC were able to obtain excellent removal efficiencies of 97.1 and 96.5%, respectively, in simulated industrial textile effluent treatment (Cardoso et al. 2012).

24.6.2 Comparison Within Algal Species

Senedesmus sp., *Chlorella* sp., *Chlamydomonas*, *Chara* sp., *Oscillatoria* sp., and *Dunaliella* sp. are just a few of the microalgae that have been found to be good bioaccumulators that may remove color from textile dyes and effluents. Using immobilized *Scenedesmus quadricauda* to treat indigo blue dye resulted in 100% decolorization, according to a study (Chia et al. 2014). Using doses of 0.01, 0.05, and 0.1 mg/L of *Chlorella vulgaris*, another study was able to remove up to 75% of the malachite green in batch culture studies (Raymond and Omo 2019). The effect of dye concentration on decolorization was studied, and it was found that the best results were achieved with dye solutions and algae at a concentration of 1 mg/L. There was no discernible effect of dye colour on decolorization, indicating that microalgae had comparable absorption and light spectrum characteristics. There was a noticeable tendency of decolorization increasing from 1 mg/L to 5 mg/L to 20 mg/L to 10 mg/L. The toxicity of the dye was a contributing factor, as the decolorization was highest at 20 mg/L compared to 10 mg/L. Biological decolorization capacity is typically diminished at high dye concentrations. Direct observation revealed more discoloration at a dye concentration of 1 mg/L (Raymond and Omo 2019).

24.7 Advantages and Challenges in Microalgal Dye Treatment

Conventional wastewater treatment procedures can present issues, such as the production of enormous volumes of sludge and the release of considerable CO_2 emissions. Phycoremediation provides a viable solution to these limits by cleaning wastewater, producing biomass, and sequestering CO_2. (Ansari et al. 2021; Souza et al. 2019). This method has been used effectively in tertiary and secondary wastewater treatment procedures, as well as in initial treatment as a biosorbent for contaminants and dyes. By combining pollutant and dye removal with algal biomass, a successful biological textile wastewater treatment may be constructed, while simultaneously producing useful algal biomass. This biomass may be used to manufacture high-value products like lipids, pigments, and fertilizers, therefore offsetting treatment costs and supporting environmentally acceptable dye and wastewater treatment techniques. Despite its potential, phycoremediation has obstacles in terms of practical use. The majority of investigations are carried out in laboratory settings using synthetic wastewater, leaving their effectiveness in environmental environments questionable. Field tests are required to evaluate their performance under varying pH, temperature, and effluent properties. Furthermore, sunlight availability is critical for microalgae development and pollution cleanup, necessitating enough exposure in large-scale applications. This presents issues in terms of acreage needs, particularly for high biomass productivity and efficient pollution removal (Carvalho et al. 2011). To solve these issues, alternative technologies such as high-efficiency photobioreactors and floating devices to effectively exploit water bodies should be investigated (Zhu et al. 2019; Pavlik et al. 2017). Other key issues include discovering strong algal species, scaling up treatment procedures, and developing cost-effective systems. Furthermore, while phycoremediation focuses on pollution removal, there is rising interest in using the created biomass, notably for natural pigment creation to replace synthetic colors used in textiles. Several algae species have been investigated for their ability to create natural dyes, providing a sustainable option for textile uses (Nair et al. 2023).

24.8 Conclusion and Future Prospects

There are advantages and disadvantages to using microalgae to detoxify colored wastewater. While microalgae have great potential as a biomass producer and pollutant removers, there are several considerations to think about before using them in dye treatment. Maintaining a stable growth environment is the main obstacle to use the microalgae in open pond systems to treat wastewater. Temperatures between 18 and 24 °C and light intensities between 80 and 200 mol/m^2/s are usually ideal for microalgae growth. Nitrogen and phosphorus can be successfully removed by treatments based on microalgae, but heavy metals and other contaminants may still be present. The selection of microalgae species, growth modes, and equipment expenses all impact overall viability, which adds to the issue of economic feasibility. For example, photobioreactors are expensive since they are enclosed, need to be sterilized, and use artificial light. Important factors to consider when selecting microalgae species include their resilience and adaptability to different environments. Additionally, the presence of heavy metal ions and low nutrient levels in industrial wastewater are two problems that hinder the efficiency of broad-scale treatment. There are several different harvesting procedures, and the overall efficiency can be affected by the pros and cons of each. Optimizing microalgae growth conditions is still a challenging undertaking, requiring a balance between controlled and uncontrollable elements, despite the increasing research into biotechnological techniques and genetic engineering. Taking on these obstacles and making the most of bioproduct yields is essential for a smooth transition to large-scale operations. There are many methods for extracting microalgae from dye wastewater, each with its own set of advantages and disadvantages. Again, optimization becomes a hard task, even with continuing research into genetic engineering and biotechnological techniques. It involves balancing variables like ambient light and temperature, which are uncontrollable, with variables like wastewater composition, which are controllable. In conclusion, microalgae show potential for treating dye wastewater, however, there are a lot of obstacles to their broad use. Creative problem-solving and an all-encompassing strategy taking environmental, economic, and technological considerations into account are essential for meeting these difficulties. Dye treatment systems based on microalgae may one day play a crucial role in environmentally responsible wastewater management initiatives, provided that this area of study is sufficiently advanced.

References

Abdelfattah A, Ali SS, Ramadan H, El-Aswar EI, Eltawab R, Ho SH, Elsamahy T, Li S, El-Sheekh MM, Schagerl M, Kornaros M, Sun J (2023) Microalgae-based wastewater treatment: mechanisms, challenges, recent advances, and future prospects. Environ Sci Ecotechnol 13

Abd-El-Kareem MS, Tahab HM (2012) Decolorization of malachite green and methylene blue by two microalgal species. Int J Chem Environ Eng 3

Ahmad A, Banat F, Alsafar H, Hasan SW (2022) Algae biotechnology for industrial wastewater treatment, bioenergy production, and high-value bioproducts. Sci Total Environ 806

Al-Etaibi AM, El-Apasery MA (2019) Dyeing performance of disperse dyes on polyester fabrics using eco-friendly carrier and their antioxidant and anticancer activities. Int J Environ Res Public Health 16. https://doi.org/10.3390/ijerph16234603

Al-Jabri H, Das P, Khan S, Thaher M, Abdulquadir M (2021) Treatment of wastewaters by microalgae and the potential applications of the produced biomass—a review. Water (Switzerland) 13

Ansari FA, Nasr M, Rawat I, Bux F (2021) Artificial neural network and techno-economic estimation with algae-based tertiary wastewater treatment. J Water Process Eng 40. https://doi.org/10.1016/j.jwpe.2020.101761

Ayele A, Getachew D, Kamaraj M, Suresh A (2021) Phycoremediation of synthetic dyes: an effective and eco-friendly algal technology for the dye abatement. J Chem 2021

Baldev E, MubarakAli D, Ilavarasi A, Pandiaraj D, Ishack KASS, Thajuddin N (2013) Degradation of synthetic dye, Rhodamine B to environmentally non-toxic products using microalgae. Colloids Surf B Biointerfaces 105. https://doi.org/10.1016/j.colsurfb.2013.01.008

Benkhaya S, El Harfi S, El Harfi A (2017) Classifications, properties and applications of textile dyes: a review

Benkhaya S, M'rabet S, Lgaz H, El Bachiri A, El Harfi A (2022) Dyes: classification, pollution, and environmental effects

Carbajo Arteaga L, Ponce Zavaleta M, Moreno Eustaquio W, Mendoza Bobadilla J (2018) Removal of aniline blue dye using live microalgae Chlorella vulgaris. J Energy Environ Sci 2. https://doi.org/10.32829/eesj.v2i1.51

Cardoso NF, Lima EC, Royer B, Bach MV, Dotto GL, Pinto LAA, Calvete T (2012) Comparison of Spirulina platensis microalgae and commercial activated carbon as adsorbents for the removal of Reactive Red 120 dye from aqueous effluents. J Hazard Mater 241–242. https://doi.org/10.1016/j.jhazmat.2012.09.026

Carvalho AP, Silva SO, Baptista JM, Malcata FX (2011) Light requirements in microalgal photobioreactors: an overview of biophotonic aspects. Appl Microbiol Biotechnol 89

Chia MA, Odoh OA, Ladan Z (2014) The indigo blue dye decolorization potential of immobilized Scenedesmus quadricauda. Water Air Soil Pollut 225. https://doi.org/10.1007/s11270-014-1920-2

Chin JY, Chng LM, Leong SS, Yeap SP, Yasin NHM, Toh PY (2020) Removal of synthetic dye by Chlorella vulgaris microalgae as natural adsorbent. Arab J Sci Eng 45. https://doi.org/10.1007/s13369-020-04557-9

Chu WL, Phang SM (2019) Biosorption of heavy metals and dyes from industrial effluents by microalgae. In: Microalgae biotechnology for development of biofuel and wastewater treatment

Chung KT (2016) Azo dyes and human health: a review. J Environ Sci Health C Environ Carcinog Ecotoxicol Rev 34. https://doi.org/10.1080/10590501.2016.1236602

Crowle G (2017) Effects of cotton structure and surface interactions on dye uptake

Das PK, Rani J, Rawat S, Kumar S (2022) Microalgal co-cultivation for biofuel production and bioremediation: current status and benefits. Bioenergy Res 15

Dellamatrice et al. 2017Dellamatrice PM, Silva-Stenico ME, de Moraes LAB, Fiore MF, Monteiro RTR (2017) Degradation of textile dyes by cyanobacteria. Braz J Microbiol 48. https://doi.org/10.1016/j.bjm.2016.09.012

Dhakad S, Chowdhury P, Arora S (2022) Efficacy of algae in the bioremediation of pollutants during wastewater treatment: future prospects and challenges. In: Biotechnological innovations for environmental bioremediation

Dias N, Lima N (2002) A comparative study using a fluorescence-based and a direct-count assay to determine cytotoxicity in Tetrahymena pyriformis. Res Microbiol 153. https://doi.org/10.1016/S0923-2508(02)01326-8

Ding T, Yang M, Zhang J, Yang B, Lin K, Li J, Gan J (2017) Toxicity, degradation and metabolic fate of ibuprofen on freshwater diatom Navicula sp. J Hazard Mater 330. https://doi.org/10.1016/j.jhazmat.2017.02.004

El-Sheekh MM, Gharieb MM, Abou-El-Souod GW (2009) Biodegradation of dyes by some green algae and cyanobacteria. Int Biodeterior Biodegrad 63. https://doi.org/10.1016/j.ibiod.2009.04.010

Ergene A, Ada K, Tan S, Katircioğlu H (2009) Removal of Remazol Brilliant Blue R dye from aqueous solutions by adsorption onto immobilized Scenedesmus quadricauda: equilibrium and kinetic modeling studies. Desalination 249. https://doi.org/10.1016/j.desal.2009.06.027

Gajare SM, Menghani S (2012) Biosorption of malachite green by naturally grown algal biomass from Girna river, Jalgaon District, Maharashtra. Research Article J Algal Biomass Utln 2012:2229–6905

Gally CR, Reis GS, Lima EC, Puchana MJ, Adebayo MA, Prola LDT, Saucier C, Dotto GL (2014) Kinectics studies of adsorption of RR-120 dye using spirulina platensis microalgae and commercial activated carbon as adsorbents from aqueous effluents

Gezahegn Gelebo G, Tessema LH, Kehshin KT, Gebremariam HH, Gebremikal ET, Motuma MT, Ayele A, Getachew D, Benor S, Suresh A (2020) Phycoremediation of synthetic dyes in an aqueous solution using an indigenous Oscillatoria sp., from Ethiopia. Ethiopian J Sci Sustain Dev 7

Haws LC, Jackson BA, Turnbull D, Dressler WE (1994) Two approaches for assessing human safety of disperse blue 1. Regul Toxicol Pharmacol 19. https://doi.org/10.1006/rtph.1994.1007

Ikram M, Zahoor M, Naeem M, Islam NU, Shah AB, Shahzad B (2023) Bacterial oxidoreductive enzymes as molecular weapons for the degradation and metabolism of the toxic azo dyes in wastewater: a review. Zeitschrift fur Physikalische Chemie 237

Ishchi T, Sibi G (2019) Azo dye degradation by chlorella vulgaris: optimization and kinetics. Int J Biol Chem 14. https://doi.org/10.3923/ijbc.2020.1.7

Ismail M, Akhtar K, Khan MI, Kamal T, Khan MA, Asiri MA, Seo J, Khan SB (2019) Pollution, toxicity and carcinogenicity of organic dyes and their catalytic bio-remediation. Curr Pharm Des 25. https://doi.org/10.2174/1381612825666191021142026

Krishna Moorthy A, Govindarajan Rathi B, Shukla SP, Kumar K, Shree Bharti V (2021) Acute toxicity of textile dye Methylene blue on growth and metabolism of selected freshwater microalgae. Environ Toxicol Pharmacol 82. https://doi.org/10.1016/j.etap.2020.103552

Kumar S, Mandotra SK, Sharma C, Ahluwalia AS, Charaya MU (2020) Role of algae in the decolourization of wastewater. In: Algae and sustainable technologies

Mahdi AB, Aljeboree AM, Alkaim AF (2021) Adsorption and removal of pollutants (dyes) from wastewater using different types of low-cost adsorbents: a review. J Chem Health Risks 11. https://doi.org/10.22034/jchr.2021.682032

Mahesh R, Panda SK, Das M, Yashavanth PR, Dhull S, Negi BB, Jakhwal P, Maiti SK (2021) Advances in biotechnological tools for bioremediation of wastewater using bacterial-algal symbiotic system. In: Wastewater treatment: cutting-edge molecular tools, techniques and applied aspects

Mahmood S, Khalid A, Arshad M, Mahmood T, Crowley DE (2016) Detoxification of azo dyes by bacterial oxidoreductase enzymes. Crit Rev Biotechnol 36

Manikandan A, Thanrasu K, Dinesh A, Kanmani Raja K, Durka M, Almessiere MA, Slimani Y, Baykal A (2022) Photocatalytic applications of magnetic hybrid nanoalloys and their nanocomposites. In: Handbook of magnetic hybrid nanoalloys and their nanocomposites

Mathur N, Bhatnagar P, Sharma P (2012) Review of the mutagenicity of textile dye products. Universal Journal of Environ Rese Technol 2

Maurya R, Ghosh T, Paliwal C, Shrivastav A, Chokshi K, Pancha I, Ghosh A, Mishra S (2014) Biosorption of methylene blue by de-oiled algal biomass: equilibrium, kinetics and artificial neural network modelling. PLoS One 9. https://doi.org/10.1371/journal.pone.0109545

Nair VK, Selvaraju K, Samuchiwal S, Naaz F, Malik A, Ghosh P (2023) Phycoremediation of synthetic dyes laden textile wastewater and recovery of bio-based pigments from residual biomass: an approach towards sustainable wastewater management. Processes 11

Nanjani S, Keharia HK (2021) Alterations in microbial community structure and function in response to azo dyes. In: Microbiome-host interactions

Nguyen TA, Juang RS (2013) Treatment of waters and wastewaters containing sulfur dyes: a review. Chem Eng J 219

Noor SN, Jasim MS (2019) Effect of temperature variation on the efficacy of Chlorella vulgaris in decolorization of congo red from aqueous solutions. Biochem Cell Arch 19. https://doi.org/10.35124/bca.2019.19.2.4169

Novotný Č, Dias N, Kapanen A, Malachová K, Vándrovcová M, Itävaara M, Lima N (2006) Comparative use of bacterial, algal and protozoan tests to study toxicity of azo- and anthraquinone dyes. Chemosphere 63.https://doi.org/10.1016/j.chemosphere.2005.10.002

Oladoye PO, Ajiboye TO, Omotola EO, Oyewola OJ (2022) Methylene blue dye: toxicity and potential elimination technology from wastewater. Results Eng 16

Omotosho OO, Ameuru US (2019) Synthesis and dyeing properties of acid dyes derived from 1-amino-2-naphthol-4-sulphonic acid. World J Appl Chem 4. https://doi.org/10.11648/j.wjac.20190404.14

Pathak VV, Singh DP, Kothari R, Chopra AK (2014) Phycoremediation of textile wastewater by unicellular microalga Chlorella pyrenoidosa. Cell Mol Biol 60. https://doi.org/10.14715/cmb/2014.60.5.7

Pavlik D, Zhong Y, Daiek C, Liao W, Morgan R, Clary W, Liu Y (2017) Microalgae cultivation for carbon dioxide sequestration and protein production using a high-efficiency photobioreactor system. Algal Res 25. https://doi.org/10.1016/j.algal.2017.06.003

Pavlović J, Farkas Z, Kraková L, Pangallo D (2022) Color stains on paper: fungal pigments, synthetic dyes and their hypothetical removal by enzymatic approaches. Appl Sci (Switz) 12

Pifferi PG, Bonora V, Spagna G, Tramontini M (1993) Immobilization of catalase on macromolecular supports activated with acid dyes. Process Biochem 28. https://doi.org/10.1016/0032-9592(94)80033-2

Pilli S, Sellamuthu B, Pandey AK, Tyagi RD (2020) Treatment of wastewater containing pharmaceuticals: biological treatment. In: Current developments in biotechnology and bioengineering: environmental and health impact of hospital wastewater

Posadas E, Alcántara C, García-Encina PA, Gouveia L, Guieysse B, Norvill Z, Acién FG, Markou G, Congestri R, Koreiviene J, Muñoz R (2017) Microalgae cultivation in wastewater. In: Microalgae-based biofuels and bioproducts: from feedstock cultivation to end-products

Raja ASM, Arputharaj A, Saxena S, Patil PG (2018) Water requirement and sustainability of textile processing industries. In: Water in textiles and fashion: consumption, footprint, and life cycle assessment

Raymond ES, Omo KM (2019) Biodecolourization of yellow dye using Chlorella vulgaris and Sphaerocystics schroeteri. Themat J Geogr 8. https://doi.org/10.26643/tjg.v8i8.8102

Sarat Chandra T, Mudliar SN, Vidyashankar S, Mukherji S, Sarada R, Krishnamurthi K, Chauhan VS (2015) Defatted algal biomass as a non-conventional low-cost adsorbent: surface characterization and methylene blue adsorption characteristics. Bioresour Technol 184. https://doi.org/10.1016/j.biortech.2014.10.018

Sarkar P, Dey A (2021) Phycoremediation—an emerging technique for dye abatement: an overview. Process Saf Environ Prot 147

Shah MP (2018) Azo dye removal technologies. In: Bioremediation: advances in research and applications

Sharma R, Mishra A, Pant D, Malaviya P (2022) Recent advances in microalgae-based remediation of industrial and non-industrial wastewaters with simultaneous recovery of value-added products. Bioresour Technol 344

Solangi MA, Overstreet RM (1982) Histopathological changes in two estuarine fishes, Menidia beryllina (Cope) and Trinectes maculatus (Bloch and Schneider), exposed to crude oil and its water-soluble fractions. J Fish Dis 5. https://doi.org/10.1111/j.1365-2761.1982.tb00453.x

Song Y, Wang L, Qiang X, Gu W, Ma Z, Wang G (2023) An overview of biological mechanisms and strategies for treating wastewater from printing and dyeing processes. J Water Process Eng 55

Soniya M, Muthuraman G (2015) Recovery of methylene blue from aqueous solution by liquid–liquid extraction. Desalin Water Treat 53. https://doi.org/10.1080/19443994.2013.866055

de Souza Celente G, Colares GS, Machado ÊL, Lobo EA (2019) Algae turf scrubber and vertical constructed wetlands combined system for decentralized secondary wastewater treatment. Environ Sci Pollut Res 26. https://doi.org/10.1007/s11356-019-04425-6

Starovoytova D, Odido D (2014) Assessment of toxicity of textile dyes and chemicals via materials safety data sheets

Taylor JA (2000) Recent developments in reactive dyes. Rev Prog Color Relat Topics 30. https://doi.org/10.1111/j.1478-4408.2000.tb03785.x

Toprak T, Anis P (2017) Textile industry's environmental effects and approaching cleaner production and sustainability: an overview. J Text Eng Fash Technol 2. https://doi.org/10.15406/jteft.2017.02.00066

Unit B (2017) Adsorption of azo dyes from textile wastewater by Spirulina platensis. Eurasian J Environ Res 1

Vasanth Kumar K, Sivanesan S, Ramamurthi V (2005) Adsorption of malachite green onto Pithophora sp., a fresh water algae: equilibrium and kinetic modelling. Process Biochem 40. https://doi.org/10.1016/j.procbio.2005.01.007

Venkataraghavan R, Thiruchelvi R, Sharmila D (2020) Statistical optimization of textile dye effluent adsorption by Gracilaria edulis using Plackett-Burman design and response surface methodology. Heliyon 6. https://doi.org/10.1016/j.heliyon.2020.e05219

Xiong JQ, Kim SJ, Kurade MB, Govindwar S, Abou-Shanab RAI, Kim JR, Roh HS, Khan MA, Jeon BH (2019) Combined effects of sulfamethazine and sulfamethoxazole on a freshwater microalga, Scenedesmus obliquus: toxicity, biodegradation, and metabolic fate. J Hazard Mater. https://doi.org/10.1016/j.jhazmat.2018.07.049

Zhu C, Zhai X, Xi Y, Wang J, Kong F, Zhao Y, Chi Z (2019) Progress on the development of floating photobioreactor for microalgae cultivation and its application potential. World J Microbiol Biotechnol 35

Chapter 25
Mathematical Modelling to Control the Spread of Water Contamination of Uppanar River in Cuddalore District

H. A. Bhavithra, N. Kiruthika, and S. Sindu Devi

Abstract Water pollution stands out as a significant environmental challenge confronting developing nations worldwide. The formulation of a mathematical model addressing both soluble and insoluble water pollutants revolves around ordinary differential equations. Through sensitivity analysis, an exploration of parameter values linked to the reproduction number is conducted, aiming to mitigate water pollution. The determination of the basic reproduction number for the water pollution model and an investigation into the model's steady state are carried out. Numerical analysis underscores the pivotal role of water treatment in securing high-quality water for human consumption.

Keywords Sensitivity analysis · Modeling · Reproduction number · Water pollution · Equilibrium points

25.1 Introduction

The art of mathematical modeling involves translating challenges from practical domains into manageable mathematical symbols, equations, and formulas. Theoretical and numerical analyses of these models provide insights, solutions, and guidelines for the original real-world applications. Across numerous disciplines such as medicine, agricultural administration, and the social sciences, mathematical models play a pivotal role. Whether linear or non-linear, stochastic or deterministic, these models offer versatile approaches. In the health sector, mathematical models

H. A. Bhavithra · S. Sindu Devi (✉)
Department of Mathematics, Faculty of Science and Humanities, SRM Institute of Science and Technology, Ramapuram, Chennai, Tamil Nadu 600089, India
e-mail: sindudes@srmist.edu.in

N. Kiruthika
Department of Microbiology, Mahalashmi Women's College for Arts and Science, Avadi, Chennai, Tamil Nadu 600071, India

© The Author(s), under exclusive license to Springer Nature Switzerland AG 2025
A. Jeyaseelan and K. Murugasen (eds.), *Sustainable Environmental Remediation: Avenues in Nano and Biotechnology*, Environmental Science and Engineering,
https://doi.org/10.1007/978-3-031-78483-5_25

prove valuable for predicting and managing epidemic outbreaks. They contribute significantly to efforts aimed at the prevention and treatment of diseases.

In recent times, there has been a continuous utilization of mathematical models to analyze the dynamics of infectious diseases within both individual hosts and populations. Containment, quarantine, vaccination, and medication represent essential strategies for managing most infectious diseases, as indicated by mathematical models (Yavuz and Necati Özdemir 2020). The current global concern is the Coronavirus Disease 2019 (COVID-19) pandemic, where the daily increase in new infections, variants, and death cases is alarming. Numerous researchers have employed epidemiological models in their studies, aiming to address the transmission challenges posed by COVID-19 (Dourado-Neto et al. 1998; El-Sheikh and El-Marouf 2004). Before COVID-19, epidemic diseases such as HIV, HBV, Ebola, H1N1, cancer, malaria, etc., were under scrutiny, and controlling their spread among the population presented a significant challenge.

In recent times, mathematical models have been employed to explore the dynamics of water pollution. Water is acknowledged as an essential life requirement, constituting a fundamental part of the body and sustaining our existence. The presence of water is crucial for human survival, and its absence would jeopardize both human and other living organisms, leading to extinction (Momoh et al. 2013). Water plays a vital role in our daily activities, utilized for domestic purposes like cooking, drinking, bathing, as well as for industrial and irrigation needs. However, human actions have led to the contamination of these precious natural resources. The discharge of various wastes, ranging from domestic to industrial, has led to the contamination of water bodies. Pollution is explained by Atangana (Bonyah and Atangana 2017) as "it involves the introduction of energy or substances into the environment. These substances can pose threats to human health, damage structures, harm living resources, disrupt ecological systems, or interfere with the legitimate use of the environment". When human and natural activities contaminate water bodies such as oceans, lakes, rivers, and groundwater, it is termed water pollution. Pollutants, including liquid wastes, oil, pesticides, insecticides, and other harmful substances discharged from industries, schools, and health centers, contribute to the formation of pollutants in water bodies (Nazir et al. 2019).

Access to clean and pure water is essential for both humans and animals (Farcas and Lesnic 2006), but due to human activities and natural causes, water bodies become infected with pollutants, leading to water pollution. The use of contaminated water for domestic purposes like drinking and cooking can result in waterborne diseases. Serious infectious diseases such as dysentery, malaria, typhoid, cholera, amoebiasis, hookworm, giardiasis, and others may be contracted due to the use of polluted water bodies for drinking and other domestic purposes (Farcas and Lesnic 2006; Mohtashemi et al. 2006). Waterborne diseases like earache, pink eye, hepatitis, gastroenteritis, rashes, respiratory infections, and others can be contracted from bathing in seawater contaminated with pollutants (Ozcaglar et al. 2012).

Various pollutants, including domestic waste, untreated sewage discharge, pesticides, insecticides, agricultural chemicals, oil spillage, heavy metals, and industrial waste, contribute to water source contamination (Farcas and Lesnic 2006;

Mohtashemi et al. 2006). These pollutants are classified into two categories: soluble and insoluble water pollutants (Nazir et al. 2019). Soluble water pollutants dissolve in water without treatment, such as nitrates, chlorides, and soaps. Insoluble water pollutants, including plastics, oil, chemicals from industries, and polythene bags, do not dissolve in water without treatment.

In contemporary times, human and animal activities have led to the pollution of most water bodies in Ghana. Activities like farming along river banks, mining (galamsey operation), refuse dumping, and liquid waste discharge into water bodies have resulted in the depletion of these resources. Rivers such as Aboabo, Korle Lagoon, Pra, Daboase, Ankobra, and others in Ghana have been polluted by human activities (Ozcaglar et al. 2012). Lagoons like Chemu and Korle Lagoon, serving as sources of fishing for tilapia and crabs, are polluted due to the discharge of untreated liquid waste into them (Zhou et al. 2014).

Water pollution, including nutrient pollution (nitrogen, urea, and phosphates), oil pollution, and chemical contamination like hydrocarbons, mercury, and pesticides, can lead to cancer and hormonal problems, liver and kidney damage, or slower growth and development. Some pollutants adversely affect marine organism development and result in damage to the nervous system, reduced reproduction, and death.

Most recently, Shah et al. (Farcas and Lesnic 2006) introduced the Water contaminant—soluble water contaminants—insoluble water contaminants (WcSI) epidemiological model to explore the dynamics of water contamination in India. The objective of their study was to identify the most effective optimal management approach to mitigate India's water body contamination. Their mathematical framework was grounded in the solubility and insolubility of water contaminants. Their stability assessment revealed that the basic reproduction number R0 is less than one, indicating that implementing control measures for insoluble water contaminants will be beneficial in reducing water contaminants. This document will utilize the water transmission model suggested by Shah et al. (Farcas and Lesnic 2006), incorporating a new compartment. The fuzzy basic reproduction number was studied by Bhavithra et al. (2024) on the other hand several numerical experiments related to one- and two-dimensional heat sources were studied by Siraj-ul-Islam et al. (2018). Ahsan et al. (Meshless 2020; Ahsan et al. 2021, 2022, 2023a, b, 2024) found a simple linearization procedure is used to convert the non-linear problem into a linear form. Moaddy et al. (2018) have investigated numerical simulation for fractional derivatives. Zaman et al. (2017) have discussed about optimal control strategy for SIR epidemic model. De Camino-Beck et al. (Camino-Beck et al. 2008) gave a graph theoretical approach for basic reproduction in epidemiology. De Barros et al. (2003) discussed fuzzy transmission analysis for SI epidemiological model. Diekmann et al. (1990) discussed about basic reproduction number ratio for epidemiological models. Vanden Driessche and Watmough discussed epidemiological models with various compartments (Vanden Driessche and Watmough 2002). The mathematical model for diabetics was studied by Kwach et al. (2011). Saddle node bifurcation analysis was studied by Gao et al. (2007). The Spread of COVID 19 and its control strategy by sensitive analysis was discussed by Mishra et al. (2021). Sensitive analysis for dengue was studied by Phaijoo et al. (2018). Odibat (2006) studied the approximation of fractional derivatives for mathematical models. Dynamical behaviour and control strategy for measles was studied by Zhou et al. (2019).

25.2 Preliminaries

Definition: 2.1 Let X be a nonempty crisp set. A fuzzy subset S of X is denoted by $\widetilde{S}$ and is defined as

$$\tilde{S} = \left\{\left(x, \mu_{S(x)}\right) : x \in X\right\}$$

where μ_S: X → [0,1] is a membership function associated with a fuzzy set $\widetilde{S}$ which describes the degree of belongingness of x with X. Here we use the membership function μ(x) to indicate the fuzzy subsets $\tilde{S}$. Also μ(x) is called fuzzy number if X is the set of real numbers.

Definition: 2.2 A Fuzzy set is called a Triangular fuzzy number if the membership value can be represented by a Triangular Function. This function by a three parameters F(x: a, b, c) such as:

$$\mathrm{F(x : a, b, c)} = \begin{cases} 0 & x < a \\ \frac{x-a}{b-a}, & a \le x \le b \\ \frac{c-x}{c-b}, & b < x \le c \\ 1 & x > c \end{cases}$$

Definition: 2.3 Let Ω be a nonempty set and P(Ω) denote the set of all subsets of Ω. Then μ: Ω → [0,1] is a fuzzy measure (Ahsan et al. 2023a). If.

(1) μ(ϕ) = 0 and μ(Ω) = 1
(2) for A,B ϵ P(Ω), μ(A) ≤ μ(B) if A ⊂ B

Let μ: Ω → [0,1] be an uncertain variable, i.e.) μ is a fuzzy subset and μ a fuzzy measure on Ω. Then fuzzy expected value (FEV) of μ is the real number, defined by the sugeno measure (Ahsan et al. 2023a).

$$FEV(\mu) = \int \mu d\mu = \sup\{\min(\alpha, k(\alpha))\}, 0 \le \alpha \le 1$$

where $k(\alpha) = \mu\{\omega \in \Omega : \mu(\omega) \ge \alpha\}$

25.3 Fuzzy Mathematical Model:

A modified version of Shah et al. (Ahsan et al. 2021) model is the one that is suggested for the transmission of water pollution. The model is split into four sections. W is a measure of the total number of contaminated water sources. S stands for water sources that could get contaminated. T stands for water sources that have been contaminated by toxins. T_r is the quantity of water sources that have undergone treatment and have been extracted from the insoluble class.

$$\frac{dW}{dt} = a - b(\beta)WS - c(\beta)WT + jc(\beta)T - eW$$

$$\frac{dS}{dt} = b(\beta)WS + fT - g(\beta)S - eS$$

$$\frac{dT}{dt} = c(\beta)WT - jc(\beta)T - fT - h(\beta)T - eT$$

$$\frac{dT_r}{dt} = g(\beta)S + h(\beta)T - eT_r$$

Mathematical models for water contamination are crucial for understanding, predicting, and managing the quality of water in various environments. However, they come with several limitations:

Complexity of Natural Systems: Water bodies are incredibly complex, involving numerous physical, chemical, and biological interactions. Mathematical models often simplify these interactions, which can lead to inaccuracies in predictions.

Data Availability: The accuracy of a mathematical model significantly depends on the quality and quantity of the input data. In many cases, consistent and comprehensive data on water quality, pollution sources, and environmental factors are hard to obtain.

Parameter Uncertainty: Many models require the estimation of parameters that are difficult to measure directly or that vary over time and space. This uncertainty can lead to errors in model predictions.

Simplification of Biological Processes: Biological processes, such as the growth and die-off of bacteria, are often simplified in models due to their complexity. This can lead to inaccuracies, especially in models dealing with biological treatment processes or the impacts of pollution on aquatic life.

Changing Environmental Conditions: Climate change and other altering environmental conditions can impact the assumptions underlying water pollution models, making predictions less reliable over time.

Human Behaviour and Policy Changes: Models often have difficulty incorporating unpredictable elements such as changes in human behaviour, policy changes, or emergency situations (e.g., industrial accidents), which can significantly affect water quality.

Despite these limitations, mathematical models are essential tools in environmental management. They provide valuable insights into the potential impacts of different pollution control strategies and help in making informed decisions. Advances in technology, data collection, and modeling techniques continue to reduce these limitations, improving the reliability and accuracy of water pollution models.

The assumptions made in water pollution mathematical models are critical for simplifying complex environmental systems into manageable equations and simulations. These assumptions can vary depending on the type of model, its scale, and its intended application. Common assumptions include:

Steady-State Conditions: Many models assume steady-state conditions, where the inputs (e.g., pollutant loads) and environmental conditions are constant over time. This simplification can make models less applicable to real-world situations where conditions change seasonally or due to specific events.

Homogeneity: It's often assumed that the water body is homogeneous in terms of its physical, chemical, and biological characteristics. In reality, water bodies can have significant spatial variations due to temperature gradients, flow patterns, and the distribution of pollutants.

Constant Flow Rates: For models that involve the movement of water, such as rivers or wastewater treatment plants, it's common to assume a constant flow rate. Natural water systems, however, experience significant variations in flow rates due to seasonal changes, weather events, and human activities.

Uniform Sediment Characteristics: In models that consider sediment–water interactions, the sediment is often assumed to be uniform in composition and reactivity. However, sediments can vary significantly within a water body, affecting pollutant binding, release, and transformation.

No Human or Ecological Feedback: Models usually do not account for the potential feedback mechanisms between human activities (such as increased water treatment in response to pollution) or ecological responses (such as algal blooms due to nutrient enrichment) and water quality.

By making these assumptions, models can be made sufficiently simple to be analytically or computationally solvable (Table 25.1).

25.4 Analysis of Fuzzy Systems

Analysis of fuzzy system in triangular manner is given below

Table 25.1 Descriptions of model variables

Parameter	Descriptions of model variables
a	The rate of water contaminants
b	The rate at which soluble water pollutants are transmitted
c	The rate at which insoluble water pollutants are transmitted
j	The rate at which water contaminants that are insoluble turn into water pollution
e	Each compartment's detachable water pollution rate
f	How quickly water contaminants that are insoluble become soluble after being treated
g	The rate at which the volume of soluble water contaminants is cleaned in the compartment
h	The rate at which the volume of insoluble water contaminants is cleaned in the compartment
β	Spreadable dosage

$$\rho(\alpha) = \begin{cases} 0, \ if \ \alpha < \overline{\alpha} - x, \\ \frac{\alpha - \overline{\alpha} + x}{x}, \ if \ \overline{\alpha} - x \leq \alpha \leq \overline{\alpha}, \\ \frac{-(\alpha - \overline{\alpha} - x)}{x}, \ if \ \overline{\alpha} < \alpha \leq \overline{\alpha} + x, \\ 1, \ if \ \alpha > \overline{\alpha} + x. \end{cases}$$

where $\overline{\alpha}$ is the central value, and x is the dispersion of each one of fuzzy sets assumed by ν. Figure 25.1 represents ρ(α) graphically

The rate at which soluble and insoluble water pollutants are transmitted is treated as a fuzzy parameter and its membership function is given by

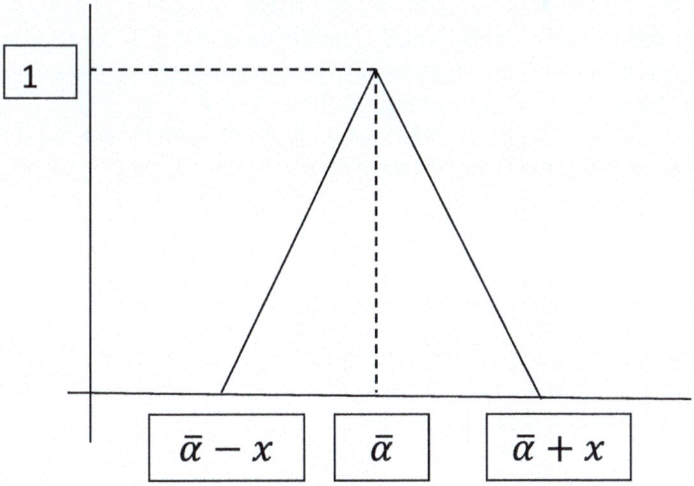

Fig. 25.1 Triangular fuzzy number

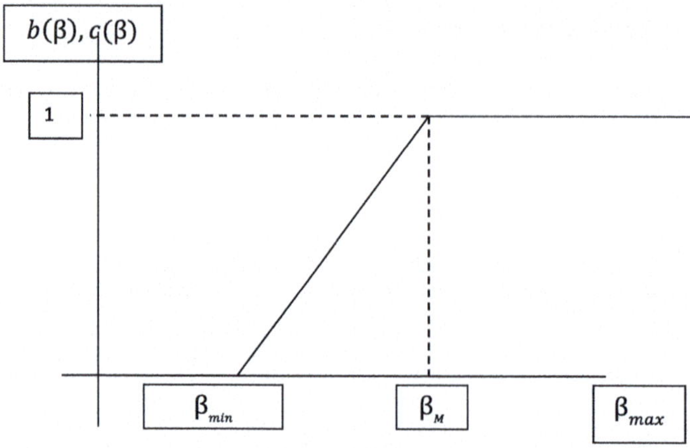

Fig. 25.2 Membership function of soluble and insoluble water pollutants are transmitted

$$b(\beta) = \begin{cases} 0, \ if \ \beta < \beta_{\min}, \\ \frac{\beta - \beta_{\min}}{\beta_M - \beta_{\min}}, \ if \ \beta_{\min} \leq \beta \leq \beta_M, \\ 1, \ if \ \beta_M \leq \beta \leq \beta_{\max} \end{cases}$$

and

$$c(\beta) = \begin{cases} 0, \ if \ < \beta_{\min}, \\ \frac{\beta - \beta_{\min}}{\beta_M - \beta_{\min}}, \ if \ \beta_{\min} \leq \beta \leq \beta_M, \\ 1, \ if \ \beta_M \leq \beta \leq \beta_{\max} \end{cases}$$

"Where $\beta_{\min}$ is the least quantity of spreadable dosage of pollutants, β_M is a moderate quantity of spreadable dosage of pollutants and $\beta_{\max}$ is the highest possible quantity of spreadable dosage of pollutants. The diagrammatic representation of the membership function is as follows (Fig. 25.2).

Let $g(\beta), h(\beta)$ be the rate at which the volume of soluble and insoluble water contaminants is cleaned in the compartment

$$g(\beta) = \frac{(\varepsilon_{\min} - 1)}{\beta_{\max}}\beta + 1, \ \text{if } 0 < \beta < \beta_{\max} \ \text{ and}$$

$$h(\beta) = \frac{(\varepsilon_{\min} - 1)}{\beta_{\max}}\beta + 1, \ \text{if } 0 < \beta < \beta_{\max}$$

The membership function of soluble and insoluble water contaminants is cleaned in the compartment is represented in Fig. 25.3

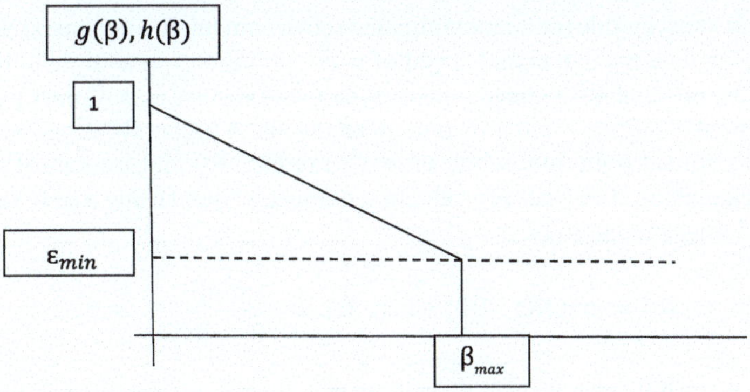

Fig. 25.3 The membership function of soluble and insoluble water contaminants

25.5 Contaminant Free Equilibria

In the context of water pollution mathematical models, contaminant-free equilibrium points refer to specific conditions or states of the system where the concentrations of pollutants are stabilized at levels that are considered non-detrimental or harmless to the environment and human health. Essentially, these are theoretical states where the dynamics of the system encompassing natural processes like dilution, degradation, and absorption balance out the inputs of pollutants, resulting in a stable state with minimal or no pollution. In mathematical terms, finding a pollution-free equilibrium point involves setting the rate of change of pollutant concentrations to zero and solving the resulting equations for the conditions that satisfy this criterion. These conditions can include specific levels of pollutant inputs, rates of natural degradation processes, and other factors influencing the system's dynamics. Achieving such a balance in real-world systems is challenging but serves as a critical reference for efforts aimed at reducing water pollution and protecting ecosystems.

The probability of contaminated water sources is given by contaminant-free equilibria $E_0 = \left(\frac{a}{e}, 0, 0, 0\right)$ which is obtained by equating $\frac{dW}{dt}, \frac{dS}{dt}, \frac{dT}{dt} and \frac{dT_r}{dt}$ to zero and assuming that water sources that could get contaminated, water sources that have been contaminated by toxins, and water sources that have undergone treatment and been extracted from the insoluble class are zero.

25.6 Basic Reproduction Number

The basic reproduction number, often denoted as R0, is a concept commonly used in epidemiology to estimate the potential for disease spread. In the context of water pollution mathematical modeling, a similar concept could be applied to assess the potential for pollutant dissemination. It represents the average number of secondary cases (or pollution incidents) that one primary case (or initial pollutant release) generates in a susceptible (or uncontaminated) environment. In this context, R0 would help to understand the extent and speed of water pollution spreading within a given system. The value of R0 depends on various factors, such as pollutant properties, environmental conditions, and transport mechanisms. A higher R0 suggests a greater risk of widespread pollution, while efforts to reduce R0 aim to control and limit pollutant dispersion. The basic reproduction number is found using a next-generation matrix and it was found to be

$$\mathrm{R0} = \max\{\mathrm{R1},\ \mathrm{R2}\} \text{ where } \mathrm{R1} = \frac{ab(\beta)}{e(e+g(\beta))} \text{ and}$$

$$\mathrm{R2} = \frac{ac(\beta)}{e(c(\beta)j + f + h(\beta) + e)}$$

25.7 Fuzzy Basic Reproduction Number

The adaptation of a fuzzy basic reproduction number in the context of water pollution mathematical modeling can be understood as a method to address uncertainties and complexities in environmental systems. Fuzzy logic provides a way to reason about uncertainty by allowing variables to have degrees of truth, rather than binary true or false values. Here's why incorporating fuzzy logic into a threshold parameter like a basic reproduction number or its equivalent in water pollution models is beneficial:

Dealing with Uncertainties: Water quality models deal with various uncertainties, including parameter values (like pollutant decay rates, which can vary depending on environmental conditions), inputs (such as irregular pollutant discharges), and processes (like unexpected changes in water flow). Fuzzy logic helps to manage these uncertainties by considering them in the model's calculations.

Reflecting Variability in Natural Processes: Natural processes and human activities affecting water bodies are not always consistent or predictable. Fuzzy logic can accommodate the natural variability and imprecision in these processes, providing more flexible and potentially more accurate models.

Improving Decision-Making: By incorporating fuzzy logic, models can provide outputs that reflect a range of possible outcomes based on varying levels of certain inputs or parameters. This can help policymakers and environmental managers make better-informed decisions under conditions of uncertainty.

The basic reproduction number which is found by spectral radius method $R_0 = \max\left\{\frac{ab(\beta)}{e(e+g(\beta))}, \frac{ac(\beta)}{e(c(\beta)j+f+h(\beta)+e)}\right\}$ cannot be found to be a fuzzy set as its So have of $R_0(\phi)$ to be less than 1. Thus $\varepsilon_{\min}\ R_0(\phi) \le 1$, Where by $\varepsilon_{\min} R_0(\phi)$ is a fuzzy set and FEV[$\varepsilon_{\min} R_0(\phi)$] is well defined. We before find the FEV value of $R_0(\phi)$, $\beta p(\phi) < 1$ and $\gamma p(\phi) < 1$. so we obtain $\max\{\left\{\frac{ab(\beta)}{e(e+g(\beta))}, \frac{ac(\beta)}{e(c(\beta)j+f+h(\beta)+e)}\right\} < 1$. In this view,

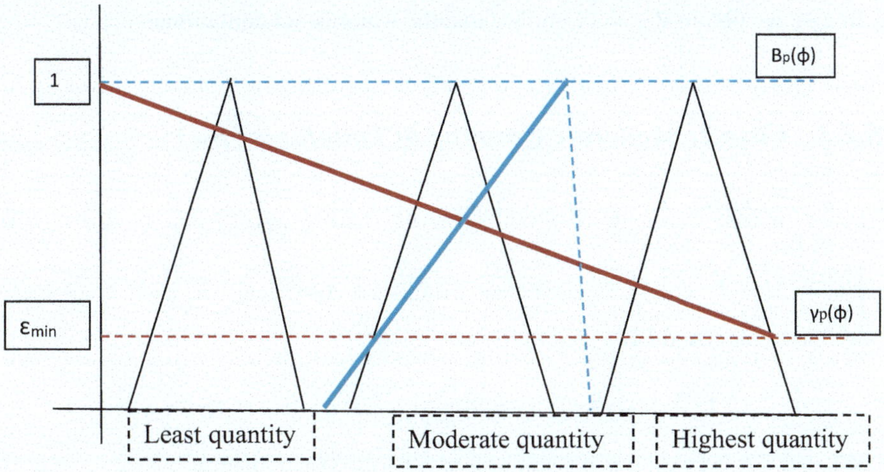

Fig. 25.4 Broad classification of contamination by fuzzy membership function

we introduce the fuzzy basic reproduction number (Mohtashemi et al. 2006), which is given by $R_{OF} = \frac{1}{\varepsilon_{min}}$ FEV $[\varepsilon_{min} R_0(\phi)]$ To obtain FEV we need to define a fuzzy measure of ϕ where *FEV* $(\varepsilon_{min}\ R_0(\phi)) = \sup\{\inf(\alpha, k(\alpha))\}$, $0 \le \alpha \le 1$, $k(\alpha) = \mu\{v : \varepsilon_{min} R_0(\phi) \ge \alpha\} = \mu(X)$, which is a fuzzy measure. we find FEV, by using the fuzzy measure. For this purpose, the possibility measure is given by Moaddy et al. (2018)

$\mu(X) = \sup \rho(\phi), \forall \phi \in X, X \subset R$

From *FEV* $(\varepsilon_{min} < R_0(\phi))$, it is clear that $R_0(\phi)$ is not decreasing with ϕ, where the set,

$X = [\overline{\beta}, \beta_{max}]$, and $\overline{\beta}$ is the solution of the following equation

$$\max\left\{\frac{ab(\beta)}{e(e+g(\beta))}, \frac{ac(\beta)}{e(c(\beta)j+f+h(\beta)+e)}\right\} = \alpha$$

Thus, $k(\alpha) = \mu\ [\beta', \beta_{max}] = \sup \rho(\beta)$ with $\beta' \le \beta \le \beta_{max}$, where $k(0) = 1$ and $k(1) = \rho(\beta_{max})$ (Fig. 25.4).

25.7.1 *Case A: Least Quantity of Pollutant* (β_{min})

Here the spreadable dosage of pollutants is very low. (i.e.) when $\overline{\beta} + x \le \beta_{min}$,

Here $b(\beta) = 0$, $c(\beta) = 0$, $g(\beta) = \frac{(\varepsilon_{min}-1)}{\beta_{max}}\beta + 1$ and $h(\beta) = \frac{(\varepsilon_{min}-1)}{\beta_{max}}\beta + 1$. We have calculated FEV $[\varepsilon_{min} R_0(\phi)]$.

$$\begin{aligned} \text{FEV}\ [\varepsilon_{min} R_0(w)] &= \sup\{\min(\alpha, w(w : 0 \ge \alpha))\}, 0 \le \alpha \le 1 \\ &= 0 \end{aligned}$$

In case we obtain $R_{0F} = 0$, implies that the water is not contaminated.

25.7.2 Case 2: Moderate Quantity of Pollutants (β_M)

Here the spreadable dosage of pollutants is moderate (i.e.) when $\overline{\beta} - x \geq \beta_{\min}$ and $\overline{\beta} + x \leq \beta_M$.

Here $b(\beta) = \frac{\beta - \beta_{\min}}{\beta_M - \beta_{\min}}$, $c(\beta) = \frac{\beta - \beta_{\min}}{\beta_M - \beta_{\min}}$ and $g(\beta) = \frac{(\varepsilon_{\min} - 1)}{\beta_{\max}}\beta + 1$ and $h(\beta) = \frac{(\varepsilon_{\min} - 1)}{\beta_{\max}}\beta + 1$.

we have calculated FEV $[\varepsilon_{\min} R_0(\beta)]$.

$$\text{FEV } [\varepsilon_{\min} R_0()] = \sup\{\min(\alpha, k(\alpha))\}, 0 \leq \alpha \leq 1$$

When $0 < \alpha < 1$with α is the solution of the following equation.

$$\max\left\{\frac{ab(\beta)}{e(e + g(\beta))}, \frac{ac(\beta)}{e(c(\beta)j + f + h(\beta) + e)}\right\} = \alpha$$

For $0 < \alpha < 1$ we divide into three Parts. $k(\alpha)$ for $0 \leq \alpha \leq 1$

$$k(\alpha) = \begin{cases} 1, \text{ if } 0 < \alpha \leq x_{\min} R_0(\overline{\beta}) \\ \rho(\beta), \text{ if } x_{\min} R_0(\overline{\beta}) < \alpha \leq x_{\min} R_0(\overline{\beta} + x) \\ 0, \text{ if } x_{\min} R_0(\overline{\beta} + x) < \alpha \leq 1 \end{cases}$$

so, if $\delta > 0$, $k(\alpha)$ is continuous and decreasing function with $k(0) = 1$ and $k(1) = 0$. Hence, $FEV(\gamma_0\ R_0(\beta))$ is the fixed point of k and R_{0F}

$$x_{\min} R_0(\overline{\beta}) \leq FEV(x_{\min} R_0(\beta)) \leq x_{\min} R_0(\overline{\beta} + x),$$

$$R_0(\beta) \leq R_{0F} \leq R_0(\beta + x)$$

As the function $R_0(\beta)$ is increasing and continuous function then by the intermediate value theorem there exists β with $\overline{\beta} < \beta < \overline{\beta} + x$ such that

$$R_{0F} R_0(\beta) > R_0(\overline{\beta})$$

So here we conclude that water is polluted for some extend.

25.7.3 *Case 3: Highest Quantity of Pollutant* ($\beta_{\max}$)

Here the spreadable dosage of pollutants is high. (i.e.) when $\overline{\beta}+x \leq \beta_M$ and $\overline{\beta}+x \leq \beta_{\max}$, Here $b(\beta) = 1$, $c(\beta) = 1$, $g(\beta) = \frac{(\varepsilon_{\min}-1)}{\beta_{\max}}\beta + 1$ and $h(\beta) = \frac{(\varepsilon_{\min}-1)}{\beta_{\max}}\beta + 1$, we have calculate FEV $[\varepsilon_{\min}R_0(\phi)]$.

FEV $[\varepsilon_{\min}R_0(\phi)] = \sup\{\min(\alpha, k(\alpha))\}, 0 \leq \alpha \leq 1$.

When $0 < \alpha < 1$ with α is the solution of the following equation.

$$\max\left\{\frac{ab(\beta)}{e(e+g(\beta))}, \frac{ac(\beta)}{e(c(\beta)j+f+h(\beta)+e)}\right\} = \alpha$$

For $0 < \alpha < 1$ we divide into three Parts. $k(\alpha)$ for $0 \leq \alpha \leq 1$

$$k(\alpha) = \begin{cases} 1, \textit{if } 0 < \alpha \leq x_{\min}R_0\left(\overline{\beta}\right) \\ \rho(\beta), \textit{if } x_{\min}R_0\left(\overline{\beta}\right) < \alpha \leq x_{\min}R_0\left(\overline{\beta}+x\right) \\ 0, \textit{if } x_{\min}R_0\left(\overline{\beta}+x\right) < \alpha \leq 1 \end{cases}$$

$$x_{\min}R_0\left(\overline{\beta}\right) \leq FEV(x_{\min}R_0(\beta)) \leq x_{\min}R_0\left(\overline{\beta}+x\right),$$

$$R_0(\beta) \leq R_{0F} \leq R_0(\beta+x)$$

Thus, $R_{0F} > 1$; in this case, water is highly polluted. It cannot be used by human beings.

25.8 Sensitivity Analysis to Control Water Contamination

In this section, we describe the methodology used by Ozcaglar et al. (2012) to identify parameter values that significantly affect the basic reproduction number. Conducting sensitivity analysis allows us to understand the significance of each parameter in disease transmission. This analysis is crucial not only for designing exploratory research but also for data assimilation and mitigating complex non-linear models (Zhou et al. 2014). Sensitivity analysis is primarily used to assess the resilience of model predictions to various parameter values. This will establish a link between the initial transmission of water pollutants and basic reproduction numbers, as well as the direct correlation between water pollutant prevalence and the endemic equilibrium point. By doing so, we gain insights into the importance of each parameter in water pollutant transmission and prevalence. It's essential to approach the evaluation of highly sensitive parameters with care, as even a slight deviation can lead to significant quantitative changes, influencing the qualitative results. Conversely, assessing the sensitivity index of an insensitive parameter requires less rigorous estimation, as minor deviations in such a parameter are unlikely to result in substantial changes to the areas of interest (Nazir et al. 2019). Identifying and focusing on these parameters

Table 25.2 Parameter value of sensitive parameter

Parameters	Values	Sensitive index
a	0.8	1
b	0.18	1
g	0.2	−0.33333
e	0.4	−1.6667

is vital for developing strategies to control water pollution and minimize the spread of pollutants in water sources. The determination of sensitivity indices is carried out to understand the proportional alteration in a state variable when a parameter undergoes a change. This can be achieved through a standardized forward method, beneficial for ascertaining the sensitivity index of R0 concerning the dependent parameter of R0. The standardized forward sensitivity index of a variable in relation to a parameter is articulated as the proportion of the relative change in the variable to the relative change in the parameter. Alternatively, the sensitivity index can be expressed using partial derivatives when the variable is a function that is differentiable with respect to the parameter (Table 25.2).

25.8.1 Positive Sensitivity Index

Positive Impact: A positive sensitivity index indicates that an increase in the value of the parameter leads to an increase in the model output. This suggests a direct, positive relationship between the parameter and the output variable. A higher positive sensitivity index implies that the parameter has a greater influence on the model output. Therefore, it is crucial to accurately estimate and understand this parameter to make reliable predictions and inform management decisions. In our model, the rate of water contaminants and the rate at which soluble water pollutants are transmitted have positive index. So these parameters should be minimized to reduce the pollution in water.

25.8.2 Negative Sensitivity Index

Inverse Impact: A negative sensitivity index suggests that an increase in the value of the parameter leads to a decrease in the model output. This indicates an inverse relationship between the parameter and the output variable. Similarly, a higher negative sensitivity index indicates that the parameter has a stronger influence on the model output, but in the opposite direction. Understanding the dynamics of this relationship is essential for effective management and intervention strategies. In our model the rate at which the volume of soluble water contaminants is cleaned in the compartment

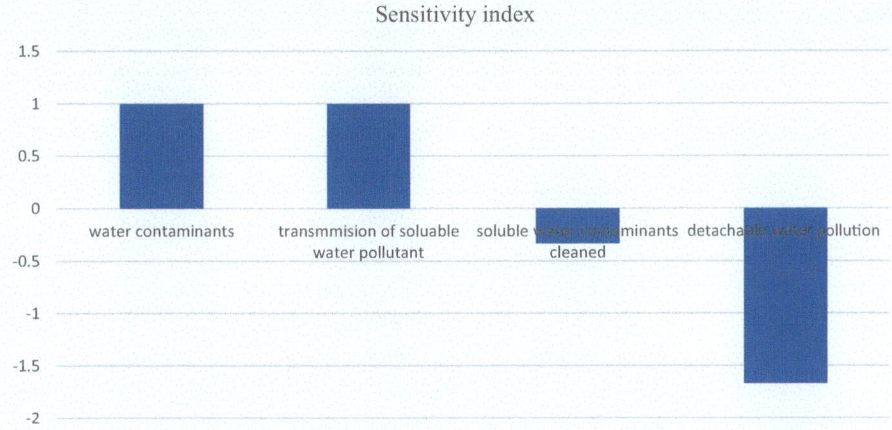

Fig. 25.5 Sensitive index of parameter

and each compartment's detachable water pollution rate have negative index. Hence these parameters should be maximized to reduce the pollution in water (Fig. 25.5).

25.9 Case Study

"Central Pollution Control Board (CPCB) monitors water Quality of aquatic resources at 4484 locations in 28 States and 7 Union Territories spread over the country including 2108 locations on Rivers, 713 on stagnant water bodies (Lakes, Ponds and Tanks), 64 on Creeks/marine, 1235 on Wells and 364 on other water bodies (drains, canals, WTPs/STPs) in association with State Pollution Control Boards/ Committees.

The current evaluation of water quality concerning nutrients and trace elements took place in the Uppanar river estuary, Southern India. Twenty-five sampling points were observed in the study area, primarily affected by human-driven activities like agriculture, settlements, and a fishing harbour in the downstream section. Conversely, the upstream and midstream sections are primarily characterized by barren terrain and an industrial centre, respectively. Concentrations of nutrients and trace elements such as nitrate, phosphate, iron, and lead surpass the acceptable standards set by WHO for drinking water. The water quality index, based on WHO acceptable limits, shows that most of the samples are classified as ranging from fair to good. Noticeably, the water quality ranges from marginal to fair in the midstream to downstream areas, where human activities are prevalent. The Uppanar River is known as one of the most contaminated rivers along the southeast coast of India due to industrial activities. Located on the banks of the Uppanar River in Cuddalore is SIPCOT (Small Industries Promotion Corporation of Tamil Nadu), which encompasses around 520 acres and

Table 25.3 Parameter values and reference

Parameter	Values	References
a	0.0008	Ahsan et al. (2021)
b	0.00018	
c	0.00002	
j	0.00025	
e	0.0004	
f	0.0003	
g	0.2	
h	0.0005	
W	500	
S	400	
T	100	
T_r	0	

includes 52 industries. The initial conditions and parameter values are as follows (Table 25.3).

We used Euler method to get the following values for 50 days (Table 25.4 and Fig. 25.6).

25.10 Observation and Findings for Real-World Problems:

The Uppanar River in Cuddalore has a contaminated water source. After continuous treatment for 50 days, it has been found that the quality of water has improved from 0 to 60%. While treating the water sources, the susceptible water source gradually decreased and reached the lowest percentage, around 0.03, after 50 days. On the other hand, water contaminated by toxins also gradually increased its percentage, varying from 10 to 12%. However, after treatment, the quality of water changed to 60%. Incorporating water pollution mathematical models into real-world managerial practices offers several practical implications and benefits. Understanding these can help stakeholders from environmental managers and policymakers to industry leaders realize the value of these models beyond theoretical scenarios. Here are key managerial implications:

25.10.1 Strategic Planning and Decision Making

Resource Allocation: This model can help identify pollution hotspots and critical sources of contaminants, guiding the efficient allocation of resources for remediation and pollution control efforts.

Table 25.4 Comparison of water sources

Days	Contaminated water sources	Susceptible water sources	Water sources contaminated by toxins	Water sources under goes treatment
0	500	400	100	0
1	462.8000005	355.87	100.8799995	80.05
2	432.0357412	314.2293105	101.6926883	151.24242
3	406.5477372	275.7247572	102.4493541	214.0786315
4	385.3749578	240.6774004	103.1594254	269.1891762
5	367.7305194	209.1717849	103.8307347	317.2685603
6	352.9744003	181.1303013	104.4697719	359.0279252
7	340.5875232	156.3713146	105.0819108	395.1626092
8	330.1490356	134.6524891	105.6716037	426.331348
9	321.3172589	115.7018018	106.2425448	453.1441491
10	313.8141238	99.23889095	106.7978045	476.1563731
11	307.4126437	84.98911836	107.3399398	495.8670876
12	301.9269148	72.69231234	107.8710844	512.7202344
13	297.204163	62.10773208	108.3930223	527.1075443
14	293.118423	53.01642225	108.9072473	539.3724442
15	289.5655055	45.22181962	109.4150124	549.8144333
16	286.4589814	38.5492337	109.9173701	558.693579
17	283.7269626	32.84464184	110.4152051	566.234907
18	281.3095131	27.97310406	110.9092617	572.632549
19	279.1565552	23.81700484	111.4001666	578.0535714
20	277.2261702	20.27425826	111.8884476	582.641451
21	275.4832123	17.25656334	112.374549	586.5191903
22	273.8981727	14.68776124	112.8588451	589.7920826
23	272.4462462	12.50232271	113.3416505	592.5501474
24	271.1065602	10.6439777	113.8232301	594.8702627
25	269.8615363	9.064488928	114.3038061	596.8180218
26	268.69636	7.722564732	114.783565	598.4493443
27	267.598539	6.582902335	115.2626627	599.8118693
28	266.5575342	5.615351015	115.7412293	600.9461563
29	265.5644513	4.794183582	116.2193732	601.8867187
30	264.6117818	4.097464656	116.6971841	602.6629104
31	263.6931859	3.50650463	117.1747359	603.2996867
32	262.8033102	3.005388971	117.6520892	603.8182552
33	261.9376336	2.580573359	118.1292932	604.2366317
34	261.0923381	2.220536116	118.6063877	604.5701164
35	260.2641997	1.915480288	119.0834038	604.8316987

(continued)

Table 25.4 (continued)

Days	Contaminated water sources	Susceptible water sources	Water sources contaminated by toxins	Water sources under goes treatment
36	259.4504961	1.65707863	119.560366	605.0324038
37	258.6489292	1.438255559	120.0372929	605.1815868
38	257.8575593	1.25300092	120.5141979	605.2871839
39	257.0747497	1.096211031	120.9910902	605.3559263
40	256.2991198	0.96355314	121.4679754	605.3935217
41	255.5295057	0.851349891	121.9448559	605.4048089
42	254.7649262	0.756480933	122.4217316	605.3938894
43	254.0045551	0.676299139	122.8986002	605.3642389
44	253.2476969	0.608559323	123.3754574	605.3188023
45	252.4937665	0.551357596	123.8522972	605.2600744
46	251.7422723	0.503079807	124.3291125	605.190168
47	250.9928018	0.462357709	124.8058948	605.1108725
48	250.245009	0.428031715	125.2826347	605.0237026
49	249.4986042	0.399119254	125.759322	604.9299408
50	248.7533455	0.374787897	126.2359457	604.8306723

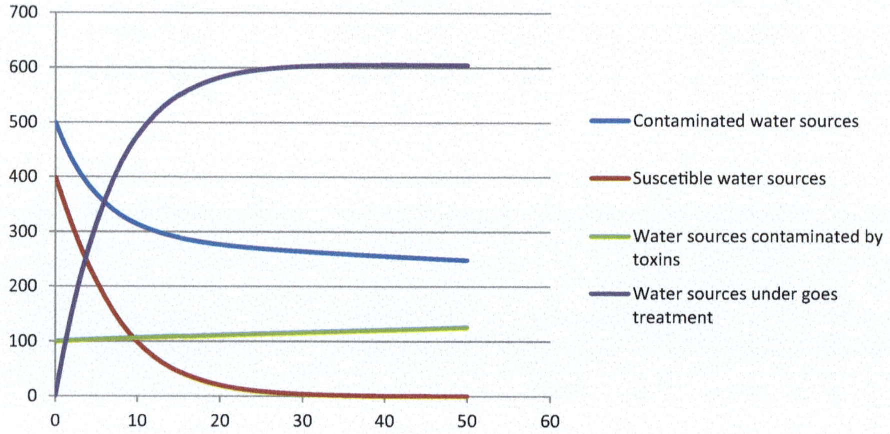

Fig. 25.6 Comparison of water contamination and its treatment

Infrastructure Development: Insights from models can inform the design and placement of wastewater treatment plants, stormwater management systems, and other infrastructure projects to maximize pollution reduction.

25.10.2 Regulatory Compliance and Policy Formulation

Regulatory Impact Analysis: This model can simulate the effects of existing or proposed regulations (e.g., discharge limits, land-use changes) on water quality, aiding in the development of more effective policies.

25.10.3 Risk Management

Predictive Analysis: By forecasting the impacts of various pollution scenarios, including accidental spills or increased industrial activity, this models enable proactive risk management and emergency planning.

Climate Change Adaptation: Modelling can help predict how climate change might alter water quality dynamics, guiding adaptations in water management practices to mitigate negative impacts.

Scenario Testing: Managers can use models to test the outcomes of various management strategies under different scenarios, helping to plan for uncertainty and change.

To maximize these benefits, it's crucial for managers to understand the assumptions, limitations, and contexts in which these models operate. Collaboration between modelers, environmental scientists, and stakeholders is essential to ensure models are accurately applied and interpreted in the context of real-world challenges. Additionally, ongoing monitoring and data collection efforts are vital for updating models to reflect changing environmental conditions and management practices, thereby improving their utility and relevance in practical applications.

25.11 Conclusion

In this study, we explored the factors contributing to water contamination. The primary elements leading to water pollution were recognized as soluble and insoluble water contaminants originating from chemical sectors, healthcare facilities, and improper disposal of waste into aquatic environments. We developed a mathematical model for the propagation of water pollution. We scrutinized parameters influencing the dispersion of water pollutants. Utilizing mat lab software, we conducted stability and numerical analyses of water pollution transmission to complement analytical findings. According to simulations, an escalation in the transmission rates of soluble and insoluble water leads to an increase in the basic reproduction number, causing the spread of water pollutants into the surroundings. This indicates a swift contamination of water sources with harmful pathogens, potentially resulting in societal disease outbreaks. The results of water contamination treatment signify a reduction in the hazards of pollutants spreading to vulnerable water populations. Consequently, encouraging the treatment of water pollutants before discharge into water bodies is imperative to minimize pollution. This suggests that implementing a water treatment strategy ensures prompt water purification, underscoring the importance of effectively managing our water resources for pollutant-free supplies.

References

Ahsan M, Lin S, Ahmad M (2021) A Haar wavelet-based scheme for finding the control parameter in nonlinear inverse heat conduction equation. Open Phys 19(1):722–734. https://doi.org/10.1515/phys-2021-0080

Ahsan M, Hussain I, Ahmad M (2022) A finite-difference and Haar wavelets hybrid collocation technique for non-linear inverse Cauchy problems. Appl Math Sci Eng 30(1):121–140. https://doi.org/10.1080/17415977.2022.2026350

Ahsan M, Hussain I, Ahmad M (2023a) A high-order reliable and efficient Haar wavelet collocation method for nonlinear problems with two point-integral boundary conditions. Alexandria Eng J 71:185–200. https://doi.org/10.1016/j.aej.2023.03.011

Ahsan M, Hussain I, Ahmad M (2023b) A higher-order collocation method based on Haar wavelets for integro-differential equations with two-point integral condition. Phys Scr 99(1):015211. https://doi.org/10.1016/j.aej.2023.11.066

Ahsan M, Hussain I, Ahmad M (2024) A higher-order collocation technique based on Haar wavelets for fourth-order nonlinear differential equations having non local integral boundary conditions. Alexandria Eng J 86:230–242. https://doi.org/10.1016/j.aej.2023.11.066

De Barros LC, Leite MBF, Bassanez RC (2003) The SI epidemiological models with a fuzzy transmission parameter. Comput Math Appl 45:1619–1628. https://doi.org/10.1016/s0898-1221(03)00141-x

Bhavithra HA, Sindu Devi S (2024) Feasibility and stability analysis for basic measles model using fuzzy parameter. Contemp Math 5(1):897–912. https://doi.org/10.37256/cm.5120242428

Bonyah E, Atangana A (2017) Modeling the spread of computer virus via Caputo fractional derivative and the beta-derivative. Asia Pac J Comput Eng 4(1):1–15. https://doi.org/10.1186/s40540-016-0019-1

De Camino-Beck T, Lewis MA, vanden Driessche P (2008) A graph-theoretic method for the basic reproduction number in continuous time epidemiological models. J Math Biol 59:503–516. https://doi.org/10.1007/s00285-008-0240-9

Diekmann O, Heesterbeek JAP, Metz JAJ (1990) On the definition and the computation of the basic reproduction ratio R0 in models for infectious diseases in heterogeneous populations. Math Biol 28:365–382. https://doi.org/10.1007/bf00178324

Dourado-Neto D, Teruel DA, Reichardt K, Nielsen DR, Frizzone JA, Bacchi OOS (1998) Principles of crop modeling and simulation: I. Uses of mathematical models in agricultural science. Sci Agric (Piracicaba, Braz.) 55:46–50. https://doi.org/10.1590/S0103-90161998000500008

El-Sheikh MMA, El-Marouf SAA (2004) On stability and bifurcation of solutions of an SEIR epidemic model with vertical transmission. Int J Math Math Sci 56:2971–2987. https://doi.org/10.1155/S0161171204310380

Farcas A, Lesnic D (2006) The boundary-element method for the determination of a heat source dependent on one variable. J Eng Math 54(4):375–388. https://doi.org/10.1007/s10665-005-9023-0

Gao S, Teng Z, Nieto JJ, Torres A (2007) Analysis of an SIR epidemic model with pulse vaccination and distributed time delay. J Biomed Biotechnol 1–10. https://doi.org/10.1016/j.vaccine.2006.05.018

Kwach B, Ongati O, Simwa R (2011) Mathematical model for detecting diabetes in the blood. J Math Biol 5(6):279–286. https://doi.org/10.1038/s41598-020-69629-0

Zaheer-ud-Din Meshless (2020) Analysis of non-local boundary value problems in an isotropic and in homogeneous media. Appl Sci 8(11):2045. https://doi.org/10.3390/math8112045

Mishra BK, Keshri AK, Saini DK, Ayesha S, Mishra BK, Rao YS (2021) Mathematical model, forecast and analysis on the spread of COVID-19. Chaos Solitons Fractals 147:110995. https://doi.org/10.1016/j.chaos.2021.110995

Moaddy K, Freihat A, Al-Smadi M, Abuteen E, Hashim I (2018) Numerical investigation for handling fractional-order Rabinovich-Fabrikant model using the multistep approach. Soft Comput 22:773–782. https://doi.org/10.1007/s00500-016-2378-5

Mohtashemi M, Szolovits P, Dunyak J, Mandl KD (2006) A susceptible infected model of early detection of respiratory infection out breaks on a background of influenza. J Theor Biol 241:954–963. https://doi.org/10.1016/j.jtbi.2006.01.031

Momoh AA, Ibrahim MO, Uwqanta IJ, Manga SB (2013) Mathematical model for control of measles epidemiology. Int J Pure Appl Math 87:707–718. https://doi.org/10.12732/IJPAM.V8715.4

Nazir S, Shahzad S, Wirza R, Amin R, Ahsan M, Mukhtar N (2019) Birthmark based identification of software piracy using Haar wavelet. Math Comp Simul 166:144–154. https://doi.org/10.1016/j.matcom.2019.04.010

Odibat Z (2006) Approximations of fractional integrals and Caputo fractional derivatives. Appl Math Comput 178:527–533. https://doi.org/10.1016/j.amc.2005.11.072

Ozcaglar C, Shabbeer A, Vandenberg SL, Yener B, Bennett KP (2012) Epidemiological models of mycobacterium tuberculosis complex infections. Math Biosci 236:77–96. https://doi.org/10.1016/mbs.2012.02.003

Phaijoo GR, Gurung DB (2018) Sensitivity analysis of SEIR-SEI model of dengue disease. Bio Med Res Int 6:41–50. https://doi.org/10.1155/2020/1508613

Siraj-ul-Islam (2018) A multi-resolution collocation procedure for time-dependent inverse heat problems. Quot Int J Therm Sci 128:160–174. https://doi.org/10.1016/j.ijthermalsci.2018.01.001

Vanden Driessche P, Watmough PJ (2002) Reproduction numbers and sub-threshold endemic equilibria for compartmental models of disease transmission. Math Biosci 180:29–48. https://doi.org/10.1016/s0025-5564(02)00108-6

Yavuz M, Necati Özdemir N(2020). Analysis of an epidemic spreading model with exponential decay law. Math. Sci. Appl. E-Notes 8(1):142–154. https://doi.org/10.36753/mathenot.691638

Zaman G, Kang YH, Cho G, Jung IH (2017) Optimal strategy of vaccination & treatment in an SIR epidemic model. Math Comput Simul 136:63–77. https://doi.org/10.1016/j.atccom.2016.11.010

Zhou T, Zhang W, Lu Q (2014) Bifurcation analysis of an SIS epidemic model with saturated incidence rate and saturated treatment function. Appl Math Comput 226(11):288–305. https://doi.org/10.1016/j.amc.2013.10.020

Zhou L, Wang Y, Xiao Y, Li MY(2019) Global dynamics of a discrete age-structured SIR epidemic model with applications to measles vaccination strategies. Math Biosci 308:27–37. https://doi.org/10.1038/s41563-021-01093-1

Chapter 26
A Review on AI Technologies for Air Toxic Waste and Human Health Prediction

V. Gowthami and H. A. Bhavithra

Abstract Since air toxic waste affects both the atmosphere and people, it is a serious problem on a worldwide scale. This effort grants a thorough conversation of the AI approaches and ML algorithms used in ecological toxic waste predicting and early-cautionary schemes. Moreover, the current effort highlights additional on AI methods used for predicting numerous main toxins in part. Further attention is specified to both AI and ML methods in foreseeing long-lasting air network infections and the estimate of weather variations and temperature surfs. The bases and effects of toxins on the atmosphere and human health were enclosed in aspect, as well as the current research status on ecological toxic waste predicting methods. Related to particular AI models, the composite model makes improved and proposals developed prediction and cautionary system precision. This study absorbed on routine estimate fault guides such as R2, RMSE, MAE, and MAPE, assessing the efficiency of several AI models.

Keywords RMSE · MAE · MAPE · MAUAT · EIBoost

26.1 Introduction

For all life on Earth to exist and flourish, air is essential. Air toxic waste has developed a worldwide concern due to the dissolute growth of modern engineering and conveyance, as well as the epidemic rise in development and populace (Zhou et al. 2019). In addition to having an effect on people's lives and careers, poor

V. Gowthami (✉)
Department of Computer Science (UG), Kristu Jayanti College (Autonomous), Bengaluru, Karnataka 560077, India
e-mail: gowthami.v@kristujayanti.com

H. A. Bhavithra
Department of Mathematics, Faculty of Science and Humanities, SRM Institute of Science and Technology, Ramapuram, Chennai, Tamil Nadu 600089, India

© The Author(s), under exclusive license to Springer Nature Switzerland AG 2025

A. Jeyaseelan and K. Murugasen (eds.), *Sustainable Environmental Remediation: Avenues in Nano and Biotechnology*, Environmental Science and Engineering,
https://doi.org/10.1007/978-3-031-78483-5_26

airborne excellence also hinders economic growth and contributes to weather variation (Kampa and Castanas 2008). Affording to a statement by the (WHO) World Health Organization (WHO), the threat posed by small particles in outdoor air toxic waste is greater than previously thought (Aayush et al. 2020). The WHO estimates that air toxic waste claims the lives of almost 7 million people annually. In together municipal and countryside ranges, outdoor air toxic waste resulted in 5.1 squillion precipitate expiries worldwide in 2016. South and East Asia was determined to have the highest percentage of deaths (15%) from air toxic waste and related dangers. Even if the number of deaths linked to air toxic waste decreased between 1990 and 2017, developing countries like India saw a 14% increase in mortality (Mannucci and Franchini 2017). The effects of anthropogenic toxic waste on the atmosphere and people are depicted in Fig. 26.1. In addition to causing air toxic waste, man-made toxic waste sources including the burning of fossil fuels, deforestation, and vehicle releases also give to worldwide weather variation.

Ecological health factors include physical, biological, psychological, and social aspects of the atmosphere that affect human health and life quality (Asha et al. 2022). Weather patterns and temperatures variation as a outcome of weather variation. In spite of the fact that these variations are ordinary, subsequently the 19th period, people consume the key basis of weather variation, mostly owing to the scorching of remains coals, which releases CO_2 (Carbon-dioxide). In the significant Libyan city of Misurata, Ahmed et al. measured the regional and periodic variation of CO_2 and more generic once every three days in the eleventh month and second month (Elsunousi et al. 2021). The findings remained extremely helpful in understanding the seasonal and geographical variations in air toxic waste. Cetin et al. (2016) analysed the particulate matter and CO_2 concentrations in different parts of Bursa city and assessed the temporal and geographical variations in air toxic waste. The study's discoveries exposed that while CO_2 was not numerically significant according to the period, particulate matter had a 99% assurance level according to the period (Cetin et al. 2019). In one case study looked at how the indoor CO_2 concentration varied in the exam room and found that, within 10 min of the exam starting, indoor CO_2 levels were higher than 1500 ppm (Cetin 2016). It is evident that air toxic waste poses a serious threat to the ecology and ecosystem since researchers in another fascinating study collected soil samples from 50 different places in order to investigate the Pb and Cr contamination. The rate of worldwide warming is higher right now than it has ever been. Increased temperatures are causing weather patterns to variation and upsetting the natural balance of the atmosphere. Hurricanes, overflows, emotionless snaps, and temperature surfs are predicted to have a bigger economical effect due to weather variation. It is anticipated that heat waves (HWs) would become more frequent and powerful due to human-caused weather variation. As an outcome of anthropogenic weather variation, HW are unquestionably significant occurrences that have the potential to quickly alter patterns of biodiversity as well as the structure and function of ecosystems (Hobday et al. 2016). Predicting air toxic waste, weather, and weather variation is critical to addressing the aforementioned problems and putting an early-cautionary scheme in habitation.

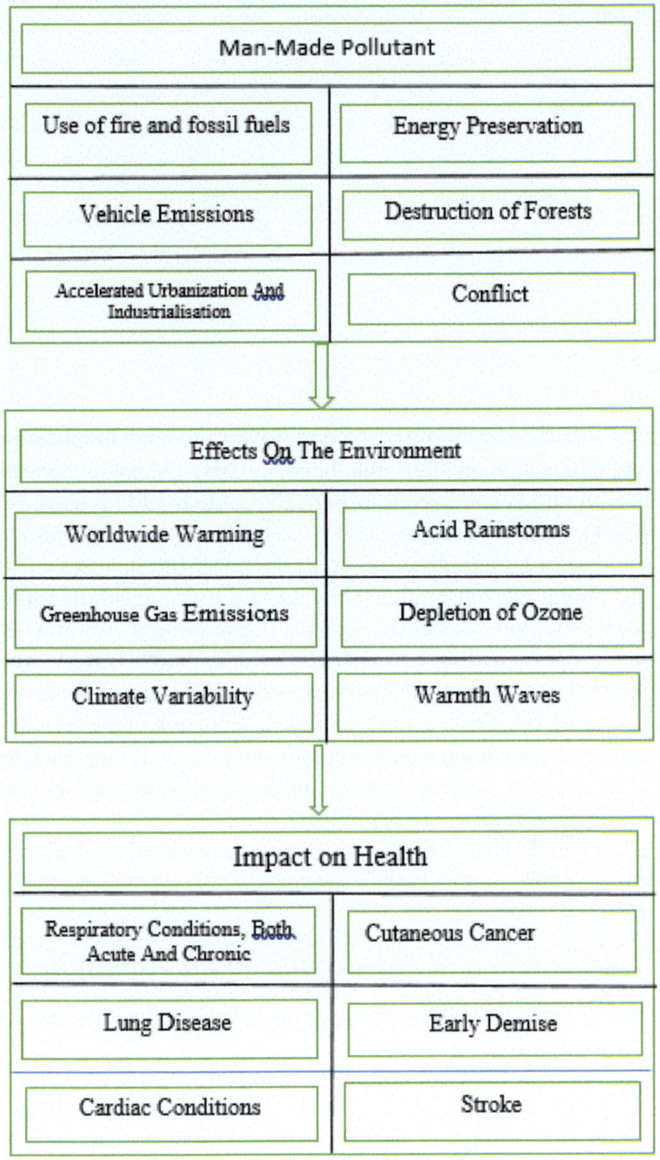

Fig. 26.1 Architecture for possible connection between ecological effects, health effects, and man-made toxins

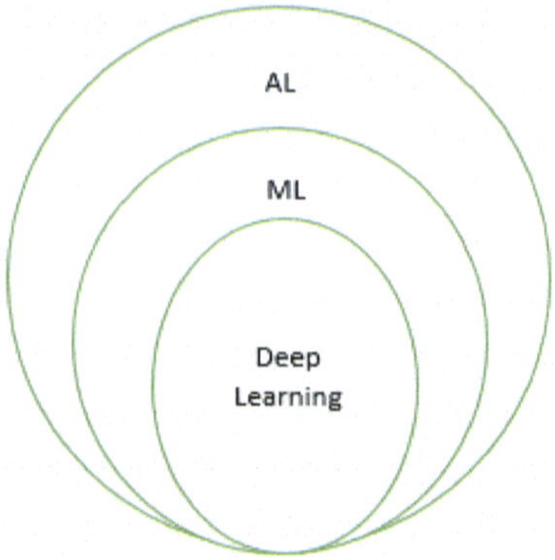

Fig. 26.2 AI, ML, and deep learning relationship

AI (Artificial Intelligence) mimics the cognitive processes involved in problem-solving and decision-making. The link between ML (Machine Learning), deep learning, and AI methods is depicted in Fig. 26.2. AI should be used as a key tool for ecological toxic waste predicting and to assistance policymakers create better ecological protection regulations in order to decrease the dangers of toxic waste exposure to the public (Jerrett et al. 2005). AI is a more accurate approach to air toxic waste prediction and predicting because it can handle complex and indirect exchanges among topological aspects (Fernández and Vico 2013). AI has an exceptional ability to track the current state of toxic waste and shows a rapid and accurate method for classifying toxic waste hotspot ranges. A portion of the AI technique used to anticipate and predict extreme weather conditions is called deep learning. AI's key competence is its capacity to grip great quantities of information and complicated real-world situations with ease. It also improves weather predicting and modelling accuracy, which is beneficial to representatives and decision-makers (Lee et al. 1894).

26.1.1 Toxin Sources and Possessions on the Atmosphere and Human Health

Because air toxic waste directly affects both human and ecological health, it is considered a larger calamity in several portions of the ecosphere. This is because air toxic waste causes numerous ecological problems, most notably health crises like cardiovascular disease, respiratory illnesses, early death, and stroke (Ong et al. 2016). Numerous crises are caused by the damaging and toxic combination of elements and vapors that are released into the air as a result of human and natural activity (Kemp et al. 2011). Toxic toxins such SO_2 (Sulphur-dioxide), CO_2, NO_2 (Nitrogen monoxide), CO, and particulate matter were released by natural calamity like abrupt forest fires and volcanic eruptions (Bai et al. 2018). These days, the air we breathe is more contaminated due to increased gas emissions from automobiles and industry,

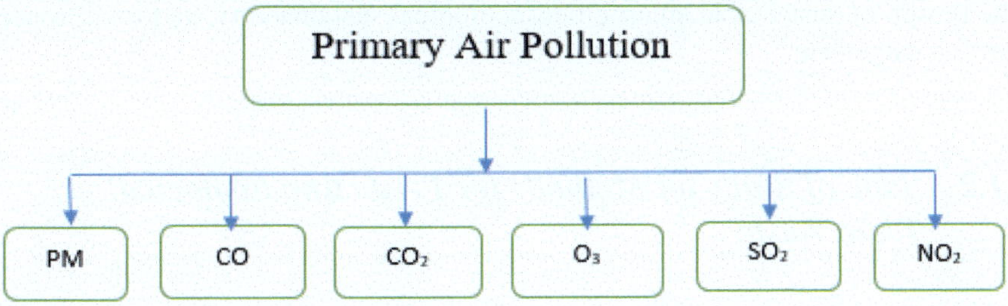

Fig. 26.3 Main air toxins, together with their causes and impacts

making the atmosphere more toxic. The primary air toxins that harm both people and the ecology are particulate matter, CO_2, NO_2, SO_2, and CO_2. The main air toxins and the effects they have on the atmosphere and people are shown in Fig. 26.3. A number of serious air toxins enter the atmosphere and contribute to the conservatory consequence, ozone depletion, and photochemical pollution, which worsen the atmosphere and cause major ecological calamity (Najjar 2011; Han et al. 2017). These calamities are also to blame for a number of serious health problems, including chronic disease. The main sources of sulphur dioxide are industrial expansion and volcanic eruptions. Acid rain, which damages the ecology and exacerbates respiratory disorders in humans, may also be caused by SO_2 emissions in the atmosphere (Zhu et al. 2019a).

The main cradle of nitrogen-dioxide emissions is the burning of fuels in transportation and industry (Aayush et al. 2020; Han et al. 2017). Exposure to NO_2 has negative consequences on both human beings and ecosystems, which clues to a rise in the prevalence of lung diseases. Ozone gas at ground level is highly oxidizing, which has detrimental effects on human health (Aayush et al. 2020). Our ability to breathe deeply is severely compromised by prolonged exposure to air toxic waste. A reduction in lung function can lead to lung cancer, asthma, and long-lasting disruptive respiratory illness (Mannucci and Franchini 2017; Han et al. 2017). Studies show that while particles a lesser amount of five micro-meter could enter the deep-region of the bronchi, scrap larger than ten micro-meter could travel along the respiratory tract (Zhu et al. 2019a). Consequently, decreasing air toxic waste can be a useful strategy for both helping society and fighting the diseases mentioned above (Aayush et al. 2020; Masood and Ahmad 2021). The substance for effective toxic waste resistor procedures and the creation of ecological rules to progress airborne excellence is laid by airborne excellence predicting, observing, and early cautionary systems. These preventive measures also performance a foremost protagonist in the design of bearable keen cities, ecological bear ability, and toxic waste resistor administration. For effective toxic waste resistor procedures and the creation of ecological laws to improve airborne excellence, airborne excellence predicting, early cautionary systems, and observing serve as preventive measures. They are also highly beneficial in the design of sustainable smart cities, ecological sustainability, and toxic waste control management.

26.1.2 State of Study on Methods for Predicting Ecological Toxic Waste

It is often recognized that the impacts of ecological toxic waste are further plain than those of dust and H_2O (Water) toxic waste (Masood and Ahmad 2021). Air toxic waste is an inevitable story of being. Governments everywhere are passing ecological legislation in an effort to enhance the quality of the air. Lacking a hesitation, the popular of developed nations wish to incorporate AI methods and technologies into their current ecological strategies. The three kinds of predicting models are bodily, numerical, and AI. The bodily predicting model focuses on studying weather phenomena and climatological conditions. The goal of Models of Chemical Transfer (MCT) is to build a calculated ideal that clarifies climatological and compound procedures in the air, with a specific importance on the attentions of air toxins and their release, transfer, and involvement (Liu et al. 2021). One of the key margins of bodily models is the huge quantity of calculations they request, and the efficiency of these inevitable approaches be contingent on a great size of information from bases of toxic waste (Feng et al. 2015). The bodily prototypical is simpler to practice and analyze than the numerical model (Liu and Yang 2021). The categories of several air toxic waste predicting models are exposed. Both the depressing model (DM) (Zhou et al. 2010) and the autoregressive integrated moving average (MAUAT) (Liu et al. 2021) are examples of conventional numerical models for air toxic waste predicting. Predicting precision is challenging to attain, and numerical copies designate the relationship among variable quantity created on numerical average and probability.

The Cohesive Affecting Typical with Auto-Regressive Function (MAUAT) model was previously used to forecast $PM_{2.5}$ well (Liu and Yang 2021; Zhang et al. 2018). Numerical models with complex, chaotic, and highly nonlinear regional characteristics can have nonlinear fitting as one of its disadvantages (Liu and Yang 2021; Chaloulakou et al. 2003). AI models perform better than numerical models and physical predicting methods, according numerous academic research (Chen et al. 2020). AI mimics human vision, knowledge, and intellectual to resolve multifaceted complications. One subcategory of AI is ML (Machine Learning). As of its in height accurateness and forcefulness, ML is measured any of the optimum methods for toxic waste predicting. It accomplishes healthy in arrangement and reversion sequences. An ANN, which pretends the humanoid intelligence and anxious system, is a frequently castoff forecaster that predicts indirect sequence (Aayush et al. 2020). AI methods like Neural Network Backpropagation Effort (NNBE), Abrupt Learning Device (ALD), Extended Short-Term Memory (ESTM), Neural Networks using Generalized Regression (NNGR), Access Controlled Recurrent Unit (ACRU), Wavelet-Based Neural Network Performance (WNNP), and additional copies like

Ambiguous Sense Prototypical and Route Appliance Support (RAS) are commonly used for ecological toxic waste prediction. Additionally, predicting algorithm-based deep learning models can execute multi-layered functions. Much effort has recently been put into examining and studying the appearances of dissimilar kinds of intellectual prediction copies in the arena of airborne excellence. Due to these limitations, it was widely believed that no single predictor could be proficient in every area of modelling and that no single clever method could be applied to every unique challenge. Composite models, in general, use the benefits of each component—in addition to combining distinct methods or methods—to produce higher performance (Gu et al. 2022). Table 26.1 displays the numerous composite AI models for toxin predicting.

Recently, AI has developed the greatest general technical gadget for guiding and sinking the harmful effects of different air toxins, generating a excessive deal of awareness in the arenas of atmospheric and wellbeing discipline (Mo et al. 2019). AI methods have been widely used by researchers as an improved healthcare decision support tool to identify, treat, and prevent diseases brought on by air toxic waste (Masood and Ahmad 2021). One such tool is AI, which has the enormous potential to accelerate weather variation mitigation and alteration strategies in industries like drive, property routine, and calamity reaction. Nevertheless, around are currently a sum of difficulties and bottlenecks that keep AI as of realizing its complete possible in this field (Stein 2020).

Table 26.1 Different composite AI methods for predicting ecological degradation

Composite AI model	Predicting of toxins	References
EMCED-MLLS	$PM_{2.5}$	Bai et al. (2019a)
PCA-CH-MLLS	$PM_{2.5}$	Sun and Sun (2017)
WPD-PSO-NNB-Adaboost	$PM_{2.5}$	Liu et al. (2019)
PSO-ALD	CO_2	Sun and Huang (2022)
GA-RF-NNBE	PM_{10}	Dotse et al. (2018)
CMCED-OPSGSA-RSV-NNGR	$PM_{2.5}$	Zhu et al. (2018)
WPD-CMCED-MLLS-AOACPSGSA-GSA	$PM_{2.5}$	Gan et al. (2018)
AM-EE-MLLS	AQ1	Wu and Lin (2019)
CMCED-CH-AGWO-RAS	NO_2 & SO	Zhu et al. (2019b)
RAA II	$PM_{2.5}$	Liu et al. (2020)

EMCED—Empirical mode for complementary ensembles Decomposition; MLLS—Machine learning with least squares; PCA—Principal component analysis; CH—Cuckoo hunt; WPD—Wavelet-packet dissection; PSO—Optimization of particle swarm; NNB—neural network in back-propagation; ALD—Abrupt Learning Device; GA—Genetic Algorithm; RF—Random Forest; NNGR—Neural Networks using Generalized Regression; OPSGSA—Optimization of particle swarms and gravitational search algorithm; RSV—Relying on Support Vectors; AOACPSGSA—Adaptive optimization approach for chaotic particle swarm and gravity search algorithm; GSA—Gravitational Search Algorithm; AM—Alternating mode; EE—Example entropy; AGWO—Algorithm of Grey Wolf Optimizer; RAS—Route Appliance Support; RAA—Route Appliance Assistance

This narrative review study provides information on the current state of AI application in airborne excellence forecasts and observing schemes as of this year (2015–2022). The core detached of the daily is to examine the several AI skills that are applied to airborne excellence prediction, initial cautionary schemes, and observing; this learning concentrations specifically on the furthermost widely used AI and ML approaches in air toxic waste predicting. In addition, a lot of composite models are existence highlighted due to their better performance compared to single predictors. Diagnostics for chronic obstructive pulmonary disease and digital healthcare may also make use of AI. The current study also focused on different contaminants and the methods used to forecast them. Furthermore, covered is the use of AI and ML in the diagnosis of breathing ailments and COVID-19 brought on by air toxic waste. The current body of literature is analysed in the context of numerous study questions and AI created knowledges. Furthermore, the application of AI methods to predict heat waves and weather variation is also covered.

26.2 Methods

The essay is structured in the manner of a traditional narrative review. While the examine consequences and orientations were revealed to enable the reader to make a learnt choice, the identification of literature sources, search strategies, and study selection are all taken into consideration even if this is not a comprehensive systematic review. The writers of this narrative review incorporated abstracts into the conversation because this is still a relatively new issue. Academic and scientific databases were used in a literature search to find papers from reputable research journals and publications. AI in air toxic waste predicting, ML, ambiguous sense, air toxic waste, and AI were the key terms used for these literary studies. AI in the observing of airborne excellence, AI and human health, ML in Numerical Fitness The atmosphere, airborne excellence observing, air toxic waste prediction, weather variation, worldwide warming, assistive vector machine ecological well-being, synthetic neural net-effort, Deep learning, EIBoost, decision trees, gradient boosting machines, random forests, and generalized additive models anticipating heat waves, the reasons behind worldwide warming and weather variation, AI composite models for predicting airborne excellence, the use of ML and AI, gaseous contaminants, AI (AI)-based particulate matter predicting, NO_2 and SO_2 predicting, ozone concentration predicting, Early cautionary systems, AI and ML-based diagnosis of chronic respiratory disorders, Emissions of greenhouse gases, the effects of air toxins on the atmosphere and people, and the employment of composite and innovative AI models to forecast air toxic waste.

The selected papers for this review paper were sourced from the following international peer-reviewed journals: EMA, Journal of Hydrology, EGH (Environmental Geochemistry and Health), APR (Atmospheric Pollution Research, Predicting Journal, JCP, EAAI (Engineering Applications of Artificial Intelligence), EMA

(Environmental Modelling and Assessment), Ecological toxic waste, ESPR, Atmospheric Research, Aerosol and Airborne excellence Research, EMS, Remote Sensing of Atmosphere, Applied Soft Computing, CIN (Computational Intelligence and Neuroscience), Diagnostics, Ecological Science and Policy, Cancer in Lung, IJMS (Environmental Modelling and Assessment), JBD (Journal of Big data), JAIHC (Journal of Ambient Intelligence and Humanized Computing), JMIR (Journal of Medical Internet Research), JMS (Journal of Marine Systems), ecofriendly. The process of finding pertinent publication strategies was iterated until the citation trails vanished. To locate other references, the orientation lists of the chosen study publications were also examined. We took great care to ensure that the title, abstract, keywords, and conclusion of every pertinent article matched the purpose of the works assessment. Second, the substance of the chosen articles was examined to determine how applicable they were. It was decided to extend the review term from 2015 to 2022. Only English-language papers were included in the review; research articles on AI and air toxic waste predicting, AI and HH, AI and weather variation, and AI were the only ones covered; only papers published in peer-reviewed journals were included in the review; all further journals, including session identifications and intelligences were removed.

26.3 Possible Use of AI in Air Toxic Waste Reduction

Although AI faced many challenges in the twentieth century, the development of the cyberspace in the 21st period is enabling AI to realize its complete possible, which is only limited by the uncharted territory of AI itself. Almost every scientific field is interested in AI, which is one of the most remarkable growths in science today (Bai et al. 2019a). AI technologies could be applied to toxic waste control and airborne excellence observing. Numerous research has shown that AI expertise could support in the drop of ecological toxic waste. In the ecological sector, AI is being used more and more to model complex ecological issues, especially airborne excellence management (Sun and Sun 2017).

Among the most sophisticated technologies, AI is used in nearly each trade. So, the best alternate for ecological observing is to use AI, and its predicted value is the most accurate (Liu et al. 2022). For early cautionary systems, airborne excellence prediction, and ecological quality observing, most researchers and authors today employ sophisticated and complex algorithms in its place of basic ML and AI methods. Interpolation is a powerful tool for following the present toxic waste scenario and provides a rapid and accurate way to identify toxic waste hotspot locations (Masood and Ahmad 2021). Two approaches are looked for modelling city airborne toxic waste by means of exclamation (Yang and Wang 2017).

26.3.1 Different AI Methods for Airborne Excellence Predicting and Observing Systems

Improved observing data enables more accurate airborne excellence predicting, and an airborne excellence observing scheme establishes the foundation for airborne excellence predicting and tracing back. By giving early cautionary of dangerous air toxic waste, airborne excellence predicting is a useful strategy for preserving human health (Zaman et al. 2017). Precise airborne excellence predicting is crucial for both the atmosphere and individuals coping with actual air toxic waste. In response to air toxic waste advisories, cities and citizens might reduce traffic, limit outdoor activities, and close down businesses (Wang et al. 2017). Early cautionary systems for toxic waste are current in many nations. Because of this, air toxic waste predicting ought to be prioritized by all governments as the cornerstone of toxic waste cautionary and switch patterns (Bai et al. 2018). Several predicting models have been proposed (Mo et al. 2019), primarily for toxin concentrations. Predicting models can be categorized into 3 categories based on their concepts: ML models, numerical predicting models, and numerical predicting models (Mo et al. 2019). Due to their ease of use, numerical models are sometimes used in airborne excellence predicting; however, they are only able to predict future toxin concentrations by analyzing the relationship between past data's toxin concentration and climatic parameters, which requires knowledge of the sources of toxic waste (Mo et al. 2019). General numerical models include the Markov model, depressing model (DM), The model of autoregressive unified affecting typical (MAUAT) (Li and Zhu 2018), and model of numerous direct reversion (NDR) (Mo et al. 2019). One aspect of AI is ML. Regression and classification issues benefit greatly from ML's strong resilience and mistake tolerance, which makes it some of the most successful methods for toxin predicting. It was claimed that a variety of AI algorithms were applied to prediction air toxic waste and create early-cautionary and observing systems for airborne excellence (Wang et al. 2018). To predict airborne excellence, numerous AI systems have been proposed and implemented. Large volumes of historical toxin data and temporal parameters are needed to illustrate AI algorithms for predicting different degrees of air toxic waste (Bai et al. 2018). Weather variation and urbanization are driving an increasing number of airborne excellence and toxic waste assessments based on AI.

A new revision found that AI-augmented representations perform better at predicting univariate period sequence facts. Forecasted airborne excellence using the MAUAT numerical model. RSV and ANN, two ML algorithms, were contrasted with MAUAT. Prediction quality significantly improved, according to the findings (Xu et al. 2017). It has been shown that ANN efforts particularly well for tasks that get more complex. A sophisticated method that has been effectively applied to forecast air toxic waste is also used by ANN models (Li et al. 2018). By collecting crucial data from daily accessible climatological parameters, Elangasinghe et al. (Ahmadi et al. 2019) developed an artificial neural net-effort air toxin forecast perfect that is able to completely detention the period disparity of toxin attentions under specified conditions.

A double-layered ESTM neural net-effort was used by Pardo et al. (Bai et al. 2019b) to forecast the Madrid airborne excellence model. While deeper ESTM net-efforts need more time and money to process, they improve prediction accuracy (Cheng et al. 2019). A composite model based on Kalman filtering and ESTM was proposed by Song et al. (Kamińska 2019) to evaluation the attentiveness of different airborne excellence components (Masood and Ahmad 2020). A composite model for PM2.5 airborne excellence predicting that combines ESTM and chart convolutional net-efforts is proposed by Qi et al. (2019) (GC-ESTM). Historical climatological parameters, geographic features, and other time-series attributes are included in the time–space mapping data.

Wen et al. currented the spacio-temporal neural net-effort extension for convolution-extended temporary recollection (NNECTR) model (Masood and Ahmad 2021). In adding, the utilization of climatological and particular figures is employed to improve forecast results. Nevertheless, the temporal features model frame effort and the weather model frame effort require an enormous quantity of information archives, which meaningfully raises the temporal difficulty of the planned technique. An original predicting prototypical for $PM_{2.5}$ concentration was created by Zeng et al. (Masood and Ahmad 2020) by combining the Enhanced stationary wavelet transformation (ESWT) (Liu et al. 2021; Cheng et al. 2019) with a Nested ESTM (N-ESTM) neuronal net-effort. The proposed method combines layered ESTM neural net-efforts, a deep learning technology, with a adapted Stationary Wavelet Transformation built on the AI-SWT method (Cheng et al. 2019) to increase prediction performance. A novel Fuzzy K-Medoid clustering method was created to forecast SO_2 concentration (Bai et al. 2018). In a different study, the methodologies of linear regression (LR), Regressor for decision trees (RDT), and random forest regression (RFR) are used to foresee air toxic waste. Created on the provided data set, the RFR achieved improved than the LR and RDT, according to the results. Using decision tree methods, the correlation clusters of PM10 and other contaminants in order to determine the air toxic waste threat (Liu et al. 2019). A tree-based ML plan to predict the PM10 concentration. In order to rise prediction accuracy and time performance, Wang and Kong's effort created airborne excellence predictive modeling based on improved algorithms based on decision trees (Sun and Huang 2022).

Feng et al.'s introduction of ANN approaches increased the day-to-day estimated $PM_{2.5}$ attentiveness prediction correctness by employing air mass trajectory analysis (Feng et al. 2015). NNGR was used to foresee the levels of $PM_{2.5}$ in three metropolitan clusters (Zhu et al. 2018). The NNGR technique has the potential to be highly dependable and useful for exact expecting of the $PM_{2.5}$ level in metropolitan areas.

Based on advanced data preprocessing technologies and AI, the writer developed a new airborne excellence early-cautionary scheme (Mo et al. 2019). The prediction model was constructed with the use of three powerful swarm intelligence methods: the effective ANN Abrupt Learning Device (ALD) (Liu et al. 2021; Mo et al. 2019), the enhanced Collaborative Experimental approach Disintegration with ductile Clatter (CEADDC) (Liu et al. 2021), and The Algorithm for Hulk amendment (AHA). The ambiguous inclusive valuation method, which offered comprehensible information on airborne excellence and suitable actions, was then used to further investigate the prediction findings (Mo et al. 2019). The recreation outcomes exposed that the CEADDC-AHA-ALD model that was suggested performed better than earlier models and that the AHA optimization strategy and the CEADDC decomposition technique were important factors in refining the correctness of neural net-effort predictions (Mo et al. 2019). This planned airborne excellence early-cautionary scheme is anticipated to be very significant in the future due to its accuracy and effectiveness (Mo et al. 2019). Amuthadevi et al. used Numerical Multilevel Deterioration, Neuro-Fuzzy, and Long-Short Term Memory and Deep Learning (LTRSDL), including Non Linear Artificial Neural Net-efforts (ANN), to investigate the current toxic waste levels. The results showed that LTRSDL performs improved than regression, Neuro-fuzzy, and ANN methods (Xu et al. 2017). In order to find topographical associations from nearby airborne excellence and climatological data, (Qi et al. 2019) created a composite prototypical that uses a graph convolutional neural net-effort and an RNN prototypical to version for chronological dependence. Similar to this, developed a composite deep-learning prototypical for air toxic waste predicting that combines an RNN prototypical for spatial sequential modelling with a convolution neural net-effort (Gan et al. 2018). But neither model has produced high determination estimates for ranges anywhere data of airborne excellence are scarce. An additional sophisticated data ambitious method for assessing toxic waste is the data-driven Granger Causality technique, which predicts airborne excellence at high spatial sequential determination with 80% exactness by utilizing information from 4 domains: transport, climatology, toxin acquaintance, and city geomorphology.

The concert variances transversely algorithms for Deep-Learning such as CNS, ACRU, BiESTM, BiACRU, ESTM, and a composite CNS-ESTM model are explored. They used CNS-ESTM to foresee the timely estimate of $PM_{2.5}$ concentration, using a spatial sequential feature by joining climatological data, historic toxin statistics, and $PM_{2.5}$ absorption in neighbouring positions. The results of the experiment demonstrate that their composite CNS-ESTM multi-variate method exceeds all the established traditional models in terms of predictive performance and produces predictions that are more accurate. The practice of composite AI methods improves the accuracy and dependability of early cautionary system growth and ecological toxic waste observing. An amount of AI models and ML tools for predicting ecological contamination are shown in Table 26.2, along with performance standards (Wu and Lin 2019). The researcher's cast-off association examination and artificial neural net-efforts (ANNs), which included ripple ANNs (RANNs), to ascertain the direct and indirect associations among the air toxic waste index (ATWI) and climatological factors. Though the RANN prototypical achieved well during the predicting phase than the ANN model, both copies positively duplicated the ATWIs. Tierce composite replicas—RANN, RMAUAT, and RRAS—were industrialized to prediction the 2016 $PM_{2.5}$. The authors looked into how the weather affected the COVID-19 lockdown, what restrictions were in place, and how PM_2 and NO_2 levels decreased. To calculate the consequence of respectively aspect on $PM_{2.5}$ and NO_2, five widespread preservative copies with climatological constraints and multi sequential variations were employed. GAMs are a useful technique for estimating how human activity may affect airborne excellence because they blend strong interpretability with fitting performance (Zhu et al. 2019b).

To break point-segment methods to observe how the COVID19 lockdown pretentious trends in airborne excellence throughout the United Kingdom. Their findings revealed that while O_3 concentration rose 20% during the lockdown, NO and NO_2 concentrations near roadside decreased by up to 32–50%; additionally, PM_{10} and $PM_{2.5}$ absorptions enlarged at all areas positions. Climate, cyclic, and tendencies remained occupied into consideration as examined the worldwide valuation of non-linear associations among atmospheric procedures and everyday milled level O_3, PM_{10}, NO_2 and $PM_{2.5}$ remarks remained apprehended in urban and toxin exact EIBoost copies for over 600 metropolises. EIBoost is an Exciting Incline Boosting technique founded on decision making structure. EIBoost is a mechanism adaptation that uses gradient boosting to do additive optimization in functional space. Minor realms didn't meet the EIBoost copies evaluation criteria for a multi-prototypical collaborative prediction, which led to greater experiential disparities, as the author pointed out in the restriction (Liu et al. 2020). By Multiple Linear Regression (NDT) copies with cyclical and climatological forecasters shown a situation learning at city circulation places to estimation variations in air toxic waste stages as a outcome of the COVID19 lockdown. The QQ Plotting upright alteration improves classical routine and foresees simple state. Large-scale data collection on a variety of variables, including temperature and humidity, is necessary for meaningful weather science. However, managing such vast data sets across multiple geographic locations can be difficult due to dataset unavailability, climatic conditions, and seasonal fluctuations. The composite model, which combines the ensemble approach, spatial–temporal approaches, and data decomposition, outpaces the further in standings of correctness and predictive power. The composite models performed recovering than supplementary copies in terms of toxic waste forecasts.

Large-scale vast longitudinal and sequential information in a change of plans, calculation efficacy, give in procedures and determination streams to version for significant features of longitudinal procedures, like longitudinal nonstationary, and scalability to prototypical added ecological experiences crosswise diverse physical ranges are all advantages that geoAI technologies can offer ecological epidemiology. When large amounts of information are shared with wide physical study extents, geoAI methods can help overcome challenges related to ecological acquaintance modeling, such as calculation dispensation and period inadequacies, as fine as information connected restrictions that impact temporal and longitudinal determination. Several studies have shown that AI technology outperforms traditional numerical methods in terms of prediction accuracy. Among their many significant benefits are the ability to handle enormous amounts of data sets, discover all potential interactions between predictor variables, and implicitly recognize complex nonlinear relationships between dependent and independent components. Conventional AI performs better than numerical methods, but not as well as the composite model. Since ANN methods are easy to develop and have a high degree of accuracy, many

Table 26.2 A range of AI methods, ML algorithms, and performance assessment standards for air toxic waste prediction

Parameter inputs for the AI predicting model	Proposed AI model	Parameter inputs of output model	Comparative model	Toxins examined	Data resolution	References
Historical concentration of $PM_{2.5}$, CO, O_2, NO_2 & PM_{10}	CMCED	NO_2, SO_2, PM_{10}, $PM_{2.5}$, CO, O_2	RAS, MAUAT	NO_2, SO_2	Regularly	Dotse et al. (2018)
Spatial and temporal features toxin data	MLPN	PM_{10}	ANN	PM_{10}	Hourly	Zhu et al. (2018)
Routine AQI	CMCED-AM	Airborne excellence	NNGR-AM-ALD		Regularly	Gan et al. (2018)
Historical toxin concentration of NO_2, SO_2, PM_{10}, $PM_{2.5}$, CO, O_2	CEADDC-ALD	NO_2, SO_2, CO, O_2, $PM_{2.5}$, PM_{10}	CMCED-AM-ALD	PM_{10}, NO_2	Regularly	Wu and Lin (2019)
Outdoor toxin	MLPN	Pm_{10}		Pm_{10}	Hourly	Wu and Lin (2019)
Toxin concentration data	PCA-CH-MLLS Composite Model	$PM_{2.5}$	NNGRMLLS	$PM_{2.5}$	Regularly	Masood and Ahmad (2021)
Regularly AQI	Novel composite model CMCED-AM	AQI	NMD-DE-ALD, CMCED-AM-ALD, CMCED-ALD, CMCED-DE-ALD	Airborne excellence	Weekly	Gan et al. (2018)
NO_2, SO_2, CO, PM_{10}	CMCED-MLLS	NO_2, SO_2, CO, PM_{10}	MLLS-ETMN, CMCED-MLLS	SO_2, NO_2, CO, O_2, PM_{10}, $PM_{2.5}$	Regularly, Hourly	Wang et al. (2018)
Regularly $PM_{2.5}$, PM_{10}, SO_2	Novel composite model CMCED-AGWO	$PM_{2.5}$, PM_{10}, SO_2	CMCED-PSO, ETMN-GWO	$PM_{2.5}$, PM_{10}, SO_2	Hourly	Sun and Sun (2017)

(continued)

Table 26.2 (continued)

Parameter inputs for the AI predicting model	Proposed AI model	Parameter inputs of output model	Comparative model	Toxins examined	Data resolution	References
Data from Airborne excellence observing net-effort		O_3, CO, NO_2, SO_2, NO	ANN	O_3, CO, NO_2, SO_2, NO	Short term	Yang and Wang (2017)
Climatological data and regular PM_{10} concentration data	Composite model	PM_{10}	NNBE	PM_{10}	Regularly	Dotse et al. (2018)
Period sequences of $PM_{2.5}$	MLLS	$PM_{2.5}$	CMCED, LSTN, OPSGSA	$PM_{2.5}$	Hourly	Gan et al. (2018)
Toxin concentration data	CEADDC, ALD	NO_2, CO, PM_{10}, $PM_{2.5}$, O_3	MAUAT, NNGR, PSO-ALD, ALD, EEMD, ALD	CO_2, SO_2, PM, O_3	Regularly	Wang et al. (2018)
$PM_{2.5}$ concentration data	WNNP	$PM_{2.5}$	NNBE	$PM_{2.5}$	Long term	Liu et al. (2022)
Toxin data and temporal data	MAUAT	SO_2, $PM_{2.5}$, NO_2, PM_{10},		$PM_{2.5}$	Regularly	Zhang et al. (2018)
Consumption of fossils fuels	ANN	CO_2		CO_2	Short term	Yang and Wang (2017)
Climatological variables and $PM_{2.5}$	ETMN-ESTM	$PM_{2.5}$	ESTM	$PM_{2.5}$	Hourly	Bai et al. (2019a)
Toxin concentration	Novel composite AI model	$PM_{2.5}$		$PM_{2.5}$	Regularly	Liu et al. (2020)
Climatological and toxin parameter	RAS, ANN	$PM_{2.5}$		$PM_{2.5}$	Short term	Wang et al. (2018)

(continued)

Table 26.2 (continued)

Parameter inputs for the AI predicting model	Proposed AI model	Parameter inputs of output model	Comparative model	Toxins examined	Data resolution	References
Climatological, historical airborne excellence data	CNS, ESTM		MLPN, LSTN	$PM_{2.5}$	Hourly	Xu et al. (2017)
Toxin data time series		NO_2, O_3, SO_2, CO	SER	NO_2, O_3, SO_2, CO	Short term, Hourly	Li et al. (2018)

Expand appreciations: AI—Artificial Intelligence; EMCED—Empirical mode for complementary ensembles Decomposition; MLPN—Multi-Layer Perceptron Network; AM—Alternating mode; CEADDC—Collaborative Experimental approach Disintegration with ductile Clatter; ALD—Abrupt Learning Device; CH—Cuckoo hunt; MLLS—Machine learning with least squares; AGWO—Algorithm of Grey Wolf Optimizer; MLLS—Machine learning with least squares; WNNP—Wavelet-Based Neural Network Performance; MAUAT—Autoregressive Integrated Moving Average; ANN—Artificial Neural Net-effort; EEMD—Extended Temporary Memory neural net-effort; ESTM—Extended Short-Term Memory; RAS—Route Appliance Support; CNS—convolutional neural systems; LSTN—Least Square Temporal Network; NNBE—Neural Network Backpropagation Effort; OPSGSA—Optimization of particle swarms and gravitational search algorithm; PSO—Optimization of particle swarm; NNGR—Neural Networks using Generalized Regression researchers reportedly use them alone or in combination with other methods, such as a composite approach that combines ANN and fuzzy neural net-effort (FNN). It can be challenging to decide which ML technique is best for a given task because there are so many to choose from. Stated differently, the predicting prototypical will function well unknown extra variables are investigated or if a composite classical is created and pragmatic (Bai et al. 2018). Ultimately, the composite frame effort improves forecast accuracy, stability, and air toxic waste cautionary accuracy. Composite models outperform single models in airborne excellence prediction, despite the lack of a perfect method; nonetheless, the computational cost is increased due to the complexity of model creation. Composite models with complex frame efforts can demand additional processing time due to their structure (Liu et al. 2021).

26.4 Predicting of Different Air Toxins Using AI Methods

Air toxic waste has a quantity of harmful properties on the ecosystem and hominoid fitness, as of mentioned in the Sect. 1.1. By giving early cautionary of dangerous air toxic waste, airborne excellence predicting is a useful strategy for protecting human health (Zaman et al. 2017). The most prevalent air toxins are sulphur-dioxide, carbon-dioxide, nitrogen-dioxide and particulate stuff; all of these have negative things on the atmosphere and public fitness. Because of AI, academic researchers are interested in creating more accurate and reliable air toxic waste forecast systems (Masood and Ahmad 2021). Support vector machines, deep neural net-efforts, fuzzy logic, artificial neural net-efforts, and ambiguous reason are the greatest effective intellectual analysts for air toxic waste predicting. Artificial neural net-efforts have recently made a wide range of potentials for air toxic waste prediction possible.

Ambiguous sense is widely employed as an estimate procedure in the arena of metrological airborne toxic waste predicting due to its advantages over classic AI approaches (Masood and Ahmad 2021). RAS has been widely used in conjunction with other ML methods during the past 20 years to forecast a number of air toxins, including carbon dioxide, sulphur dioxide, ozone, and particulate matter (Masood and Ahmad 2021). This section offers a quick overview of toxins and the appropriate AI technologies for predicting them, as well as an evaluation of particulate matter and vaporous toxins predicting methodologies founded on performance and applicability.

26.4.1 Predicting Ozone Concentration with AI Methods

Ozone is a type of toxic gas that, because of its strong oxidizing qualities, can be damaging to active things once it is current on the external of the planet (Liu et al. 2021). At high concentrations, it can cause irritation to the skin, eyes, and nose as well as lesions to the upper respiratory tract. High ozone absorptions damage agrarian crops, and homestead damage whitethorn worsen (Liu et al. 2021; Masood and Ahmad 2020). The writer developed a multi-toxin prediction method using ANN and VRS technology (Masood and Ahmad 2021). Four weather variable quantity and continual toxin absorptions were cast-off as inputs while developing the AI methods. The Particle Swarm Algorithm, a meta-heuristic procedure, was used to subordinate the processing complication of the planned method. The VRSPSO technique performed well in every season, according to the results. Zhou et al. (2010) estimated everyday ozone stages by means of a multiple disintegration method and intellectual procedure optimization in two seriously contaminated cities utilizing a novel twin measure collective knowledge prototypical with fault restoring. The RMSE standards of the suggested prototypical for low O_3 levels are 2.5318 and 3.2068, respectively, while the RMSE standards for in height ozone stages are 2.8450 and 3.8701. The playwrights claim that the novel prototypical they have proposed performs better than the comparative models. A fuzzy binary relation-based method was used by Chattopadhyay et al. (2021) to examine air toxic waste done India, during the upright wet season to season transition period.

According to research, nitrogen dioxide and ozone had an impact on the general AEK finished the contaminated learning region from the upright wet season through the wintertime. The authors used the SANI prototypical and the Regression using semi-experimental methods (RSEM) to predict the absorptions of sulphur dioxide, carbon dioxide, nitrogen dioxide, and ozone. When it comes to predicting time series data, the SANI performs better than regression models built on the routine parameters of both copies. To predict the ozone side by side in the municipal sector, Mishra and Goyal employed three AI-based modelling methods: Neuro-Fuzzy, ANN, and NDT (Zhou et al. 2010). Furthermore, it was demonstrated that the NF prototypical accomplished better in relations of predicting accuracy than the ANN and NDT models. ANN and NDT models were used to assess the predictions of nitrogen dioxide, ozone, nitrous oxide, carbon monoxide, and sulphur dioxide concentration levels by Alimissis et al. (2018). The study's conclusions indicate that here is adequate numerical signal to conclude that the ANN prototypical predicts outcomes more accurately than the NDT method. Additionally, the authors used numerous direct deteriorations and genetic factor appearance program design to identify 2 experimental correlations for OZONE concentration prediction. While NDT has an RMSE worth of 0.60 and an association constant of 21.20, GEP has an RMSE worth of 13.52 and a correlation value of 0.82. The results show that GEP is a more reasonable and trustworthy method for determining ozone concentrations.

26.4.2 AI Methods for Predicting CO

In the human body, hemoglobin and carbon dioxide combine to form carboxyhemoglobin, which lowers the blood's ability to carry oxygen and is potentially fatal in high doses quickly after exposure. Newborns, the ageing, and individuals with mood/lung situations are particularly vulnerable to CO poisoning. Therefore, predicting atmospheric CO levels is essential for early-cautionary system implementation. Yeganeh et al. (Zhou et al. 2010) assessed the effectiveness of an original RAS created prediction model unified with fractional minimum quadrangles (FMQ) to guesstimate CO absorptions equal. Their research indicates that the FMQ-RAS collaborative approach outperforms other methods and crops good results (Zhou et al. 2010). Likewise, the writer created an appropriate technique to gauge the degree of ambiguity in Vector regression with support (VRS), artificial neural net-efforts (ANN), and SANI for CO predicting. Their outcomes displayed that the range of acceptable ambiguity in CO estimate using VRS models was found. Jain and Khare conducted a significant study in which they used a neuro ambiguous prototypical to forecast CO for 2 distinct places in Capital Delhi. 10 variables remained taken into account: the number of vehicles, temperature, mixing height, mist protection, air compression, moisture, sunshine periods, and breeze way and rapidity. The modelling consequences displayed that the Neuro ambiguous prototypical valued CO at multifaceted city points accurately, with a directory of arrangement (DA) in the 0.87–0.92 range (Zhou et al. 2010). 1 day in advance predicting of everyday CO attentiveness points using ANN and SANI copies. The ANN prototypical foreseen everyday CO absorptions extra correctly than the SANI prototypical, created on the uncertainty investigation. Based on the outcomes of the previous studies, every predicting model has a different capacity to predict the level of CO. Predicting models powered by AI achieve well and additional precise. For CO predicting, the ANN, VRS, and ambiguous sense predicting copies would be the most effective and dependable options overall.

26.4.3 AI-Based Predicting of Carbon Dioxide

Because of its heat-trapping properties, sulphur dioxide is the main contributor to worldwide warming and needs to be reduced or removed. Significant rises in Sulphur Dioxide (5000 ppm) could be harmful to human health. In addition to causing respiratory and cardiovascular disorders, carbon dioxide also decreases the pH of the serum, leading to acidosis. It is therefore essential to create a suitable sulphur dioxide emission predicting model. Major toxins such as hydrocarbons in total (HCT), (ozone) O3, NO (nitrous oxide), SO_2 (sulphur dioxide), CH_4 (methane), CO (carbon monitrous), and NO_2 (nitrogen dioxide) were predicted using an artificial neural net-effort (ANN). In their study, working artificial neural net-efforts to estimate CO secretions in 5 Central Easterly homelands. Using a GDHM (General Data Handling Methods) ANN, CO secretions—the greatest significant GG (greenhouse gas)—are determined for 5 central Easterly homelands. In contrast, the GDHM model's R2 values and average absolute relative error, which recurrent the model's accuracy in prominent carbon dioxide secretions, are 0.9998 and 2.3%, respectively. The results demonstrate that ANN approaches are sufficiently accurate to estimate carbon dioxide emissions. Using real-world data, Norhayati and Rashid (2018) estimated the Carbon Dioxide secretions from a scientific unused incinerator herbal by the

SANI model. They found that the approach forecasts carbon emissions accurately. To estimate Bahrain's area carbon dioxide emissions, employed a variety of methods, such as method of Gaussian process regression and holt, a nitrous oxide non-linear auto-regressive prototypical for foreseeing carbon dioxide emissions. With an RMSE of 0.206, the neural net-effort prototypical has the deepest RMSE of all the copies. The research showed that the neural net-effort time series non-linear auto-regressive prototypical accomplished well for Carbon dioxide secretions predicting based on the RMSE values. In recent study, Wei et al. (Lin et al. 2020) built a prototypical created on aspect investigation and an updated Abrupt Learning Device for evaluating carbon emission intensity. The author claims that the planned prototypical exactly foresees carbon secretion concentration.

In several domains, the neural net-effort has proven to be a dependable estimate and foreseeing method overall. Multi-layer perceptron artificial neural net-effort (ANN) was swiftly embraced in time series predicting because to its approximation capabilities, and it was introduced as a way to handle categorization difficulties. One of the greatest popular methods for predicting time sequence, such carbon dioxide secretions, is the neural net-effort based methodology. Neural net-effort based methods for predicting models are anticipated to effort well even in non-linear links.

26.4.4 AI-Based Predicting of Nitrogen Dioxide and Sulphur Dioxide

Fossil fuel combustion in industry and transportation accounts for most nitrogen dioxide emissions (Najjar 2011). Caustic shower is similarly believed to be significantly influenced by Nitrogen Dioxide. In addition, Nitrogen Dioxide exposure has negative impacts on ecosystems and human persons, which raises the occurrence of lung diseases. Caustic shower is mostly produced by two main air toxins: sulphur dioxide and nitrogen dioxide. Buildings, vegetation, water, and people's health can all be impacted by acid rain. Therefore, creating a reliable model for predicting and cautionary of sulphur dioxide and nitrogen dioxide is essential. An ANN method was planned to estimate continual quantities of carbon dioxide, nitrogen dioxide, and ozone. Their results showed that the nitrogen dioxide based ANN prototypical had a low R value of 0.45 and a large standard deviation of 0.614. In order to estimate the concentrations of PM and sulphur dioxide, developed composite prediction models based on a meta heuristic method known as the aged Casanova optimizer. Likewise, author looked into how well NDT and PEA-based ANN models estimated nitrogen dioxide concentrations (Masood and Ahmad 2021). Data on airborne excellence and meteorology were used in the growth and validation of the ANN approaches (Masood and Ahmad 2021). The NDT procedures based on association constant, which had short principles of 0.80, are less accurate than the PEA-ANN approach. Wang et al. planned 2 separate composite adaptive predicting copies for one step forward estimate of sulphur dioxide and PM10, and their results showed that combining ANN and RAS methods provided better airborne excellence predicting concert than only numerical methods. A recurring neural net-effort predictor was planned by to estimate the everyday maximum attentiveness of ozone, sulphur dioxide, carbon dioxide, PM10, and nitrogen dioxide in Italy. The writers claim that the ALDan prototypical yields lowest MSE, MAE, and RMSE values.

The usefulness of multiple linear regression (NDT) and multilayer perceptron (MP) for foreseeing the amount of sulphur dioxide in Tehran's air was investigated. They employed the climatological restriction, period restriction, city lime interplanetary data, and city circulation information as participation restrictions for the prediction of sulphur dioxide everyday attentiveness. The MP prototypical had RMSE and R2 standards of 0.40 and 0.8, whereas the NDT approach had RMSE and R2 values of 6.024 and 0.707, separately. The results of the investigation showed that here was enough numerical indication to imply that the ANN prototypical was more predictive than the NDT prototypical. A 2-step composite model constructed on EMCED (empirical mode for complementary ensembles decomposition), VRS (vector regression for support), AGWO (algorithm of grey wolf optimizer), and ACS (algorithm of cuckoo search) was recently developed by Zhu et al. (2019b) for sulphur dioxide and nitrogen dioxide predicting in 4 metropolises in central China. In the method selection and assessment process, concert assessment system of measurement including RMSE, MAE and MAPE are cast-off to compare the proposed model with other established methodologies. The authors claim that the most accurate model for predicting sulphur dioxide and nitrogen dioxide was the innovative composite prototypical EMCED-ACS-AGWO-VRS, which employed VRS-ACS for small regularity information and VRS-AGWO for large regularity information. An original method for predicting the everyday regular of the Tianjin area nitrogen dioxide attentiveness estimate utilizing a separate ripple convert and a TEM net-effort. Significant historical data, weather data, and five main air toxins were selected as input parameters to produce an effective nitrogen dioxide attentiveness estimate for the following daytime. The value of MAPE was utilized to assess the estimate concert of the proposed AI approaches, which were valued to be ACRU, VRS, and only TEM prototypical. In comparison to the only TEM prototypical, the MAPE declined to 11.57% from 17.85%, but the RMSE and MAE enlarged to 5.9290 and 4.3376, individually. The author claims that the recommended model structure is more appropriate for estimating Tianjin's nitrogen dioxide concentration.

26.4.5 Predicting Particulate Matter with AI Methods

A medical study claims that PM harms human DNA, as well as the respiratory, cardiovascular, neurological, and immune systems, to differing degrees. Therefore, it is essential to create an effective PM predicting and cautionary prototypical in order to inform people around the state of the atmosphere in the coming hours. An algorithm of Multi-layer Perceptron for PM_{10} concentration predicting (Masood and Ahmad 2021). The IA, RMSE, R2, and MAE standards remained, in that order, 0.77,

25.02, 0.38, and 19.01 (Masood and Ahmad 2021). The outcomes demonstrated that the ANN might effectively predict PM_{10} absorptions and do so with accuracy. The innovative methods for determining $PM_{2.5}$ absorptions in China on 2019. The 3 diverse composite algorithm copies be situated provided by the learning: WANN, WRAS, and W-MAUAT (Masood and Ahmad 2021). The estimate abilities of the planned composite copies remained contrasted with those of independent methods such as RAS, MAUAT, and ANN. The selected AI methods' predicting prediction was evaluated using concert assessment principles including RSME and MAE. Simulation findings showed that composite copies outperformed separate AI methods in relations of precision accuracy and prediction capabilities. A wavelet-AI composite prototypical was established to estimate PM_{10} toxic waste absorptions in the short and medium term. The results of the learning showed that wavelet adaptive neuro ambiguous implication scheme and WRAS methods beat other copies in continual and daily predicting, with correctness levels of roughly 96% and 99%, individually. Wave-let transform-related AI methods outperform other only AI methods in relations of performance, and they also play a major part in improving the correctness of other AI methods, allowing them to forecast PM_{10}. In a different study, the writer developed a original composite model that accurately predicted outdoor $PM_{2.5}$ levels and was utilized to evaluate the effects on health and financial losses.

The proposed approach, according to the authors, may be used in other systems, such as health challenges, in addition to providing early cautionary information. Moreover, it ensemble ML methods such as EIBoost-RF-MAUAT to anticipate and foresee $PM_{2.5}$ absorptions. The results suggest that $PM_{2.5}$ absorptions can be predicted using composite models. The created models demonstrate the efficiency of ML methods in foreseeing particulate matter concentrations, suggesting that they might be functional to the estimate of atmospheric-toxic waste. An ascendable, circulated Decision tree with gradient enhancement (DTGE) ML system is called extreme boosting (EIBoost). The EIBoost ML technique is more effective than the WRF-Chem model in predicting PM2.5 attentiveness, rendering to the authors' investigation of its use in the Shanghai case study. A hereditarily enhanced arbitrary forestry—posterior spread neural net-efforts approach for predicting everyday PM10 in Southeast Asia, was currented by Dotse et al. (2018) in a different learning. The author suggests using the proposed method for controlling and observing haze conditions in real time. In the meantime, Zhu et al. (2018) have currented the EMCED-OPSGSA-VRS-NNGR model, a composite predicting system, for everyday $PM_{2.5}$ concentration. The author claimed that the unique composite prototypical was effective in predicting $PM_{2.5}$ and might also be estimate other toxins based on the testing results. The writer was the initial to optimize for continual $PM_{2.5}$ attentiveness estimate using a mix of DWP (decomposition with wavelet packets) and EMCED (Gan et al. 2018). The DWP-Bi-TEM -SAE-RAA-II is a composite multiple resolve multiple detached collaborative prototypical created on Wavelet-packet disintegration, Bi-directional extended Provisional retention, Genetic algorithm II for non-dominated sorting., and Loaded auto encoder that Liu et al. (2021) proposed for predicting everyday PM attentiveness levels. The writers claim that the recommended model fared well than the added benchmark copies.

Bai et al. developed a predictive collaborative observed method breakdown prototypical using an Extended Temporary Memory neural net-effort (ETMN) for Continual $PM_{2.5}$ attentiveness estimate (Bai et al. 2019a). According to the results, E-TEM methods outperformed TEM and individual fodder advancing neural net-efforts in terms of RMSE, MAPE, and correlation coefficient (Bai et al. 2019a). For everyday attentiveness of $PM_{2.5}$ level estimates, Sun et al. in China created an original composite technique built on principle element analysis (PEA) and MLLS supplemented by Sun and Sun (2017). According to the findings, the recommended strategy performs better in $PM_{2.5}$ concentration forecasts than MLLS prototypical and a generic deterioration neural net-effort method. Thus, the model that the author proposes has the possible to be useful in airborne excellence prediction methods. Another intriguing study by Liu et al. (Liu et al. 2019) included the growth of a novel composite method for $PM_{2.5}$ concentrations known as DWP-PSO-BP-Adaboost. An algorithm of Adaboost and PSO considerably improved predicting exactness, the DWP-PSO-BP-Adaboost prototypical surpassed all others in multiple steps predicting, and the DWP was effective in improving prediction concert (Liu et al. 2019).

These results suggest that composite models could be useful in PM concentration prediction. In previous studies, the majority of academics preferred AI models over numerical models due to their accuracy, time series, and nonlinearity. The majority of studies suggested using either one or two composite models to predict particle matter. Furthermore, composite models that combine information disintegration and collaborative methods are highly desirable and reliable for better particulate matter predicting. The consequences of the investigation showed that there was adequate numerical indication to support the suggestion that the ANN model was better than the NDT prototypical at predicting nitrogen dioxide and sulphur dioxide toxic waste. The finest and most dependable options for carbon monoxide predicting would be the VRS, ANN and ambiguous logic predicting copies, according to earlier research. In many different disciplines, neural net-efforts have confirmed to be a dependable system for predicting and estimate. Multi-layer perceptron artificial neural net-efforts (ANNs) were extensively cast-off in period sequence foreseeing and introduced as a way to handle categorization difficulties due to its approximation capacity. For predicting period sequence like carbon dioxide emissions, one of the greatest popular methods is the neural net-effort-based methodology. Subsequently ANN intensive predicting copies are quick and reserve effectual once it originates to everyday projections, the emission inventories have less uncertainty. In relations of estimate accurateness, the composite prototypical also achieves well than individual AI methods. It was discovered that composite AI methods were preferred and ideal for creating early-cautionary systems and airborne excellence observing.

26.5 Human Health and AI

AI has advanced greatly in the medical field. AI has been essential in recent times in enhancing the healthcare system's ability to identify illnesses like tumor, impurities, and the creation of surgical machines and accurate diagnostic tools. AI is particularly useful in the following areas: early illness diagnosis and treatment; cancer detection; control of chronic conditions; drug growth; robot-assisted surgery; delivery of health-care; and scientific research and testing. Recent significant advancements in AI have given hope for the field of disease diagnosis. Many wearable apps and devices use AI methods to detect vital signs and health issues in the body; they can predict the possibility of a health issue before it manifests itself. The main focus of this section is on using ML and AI methods to forecast chronic pulmonary airway issues. The impacts of air toxic waste on COVID19 poison and mortality were briefly conversed, as well as the effect and repercussions of air toxic waste on humanoid wellbeing. The application of AI in the health care scheme was then the main focus, with a focus on observing admissions for cardiac illnesses and identifying respiratory problems. The different AI and ML methods for observing and managing acute aggravating circumstances, hospital stays, and death rates. In general, it highlights how air toxic waste affects human health and how crucial AI technologies are to the healthcare system; this section offers a quick overview of these topics.

26.5.1 AI Tools for Respiratory Disease Prognosis

Air toxic waste is a worldwide hazard to commons fitness and well-being outstanding to its significant and harmful effects on humanoid well-being. Asthma and other airborne disorders are brought on by toxic waste, and they have a serious negative effect on people. Public safety can be improved with the help of an accurate assessment of air toxic waste. Chronic airway illnesses are common, especially in developing countries, and are typified by inflammation and restriction of airflow. The two conditions with the highest death rates and economic effects worldwide are asthma and prolonged disruptive pulmonic ailment (PDPA). The lungs are the initial organs to be affected by any ecological toxins inhaled; these toxins cause irritation to the lungs, increase the threat of breathing contaminations, and constrict the airlines. There was a significant positive link seen in 122 Chinese cities among recently established COVID19 gears and PM_{10}, $PM_{2.5}$, nitrogen dioxide, and ozone. A thorough nationwide search revealed a strong association among the likelihood of COVID19 fatalities and chronic $PM_{2.5}$ exposure; this learning also showed that long-standing disclosure to air toxic waste meaningfully increases the COVID19 death ratio, with an 8% increase in COVID19 casualty rates being linked to a per $\mu g/m^3$ increases in $PM_{2.5}$ exposure. There is a substantial overtone among $PM_{2.5}$ levels and COVID19 morbidity, according to a study that collected informations from 25 Indian cities. Another recent study supported the theory that SARS-CoV-2 death in

England is influenced by air toxic waste. In three French cities, a direct correlation between COVID19 death and air toxic waste ($PM_{2.5}$ and PM_{10}) was found. In Italy, it was discovered that there was a confident association among COVID19 toxicities and crushed flat OZONE, AEK, and PM. Furthermore, a strong positive correlation was found between Wuhan, China's everyday COVID19 demise toll and the airborne excellence key (AEK), namely $PM_{2.5}$. A great ambient carbon dioxide attentiveness raises the risk of SARS-CoV-2 broadcast. According to another study, a significant portion of COVID19 bags existed found in the most contaminated areas, infected people required ICU admission, and the death rate in the contaminated areas was double as great as in the additional areas.

Comparable patterns have observed in the Czech Republic, wherever a higher percentage of COVID19 analyses have complete in extremely polluted parts of Prague. An account of continuing $PM_{2.5}$ experience was linked to higher COVID19 dissemination amounts in Peru. In general, hospitalization and COVID19 related mortality are more likely as a result of air toxic waste. AI has garnered significant attention in the ecological and medical sciences in recent years as the supreme extensively castoff tools instrument for reducing and regulating the negative impacts of numerous air toxins (Mo et al. 2019). In order to identify, track, and treat disorders associated with air toxic waste, frequent investigators must employ AI methods in health-care resolution creation instruments. Using datasets of CT-scan pictures, created a deep-learning built method to mimic the growth of tumour cubicles in the respiratory. An innovative strategy for detecting cardiac conditions was proposed utilizing an ensemble approach based on support vector machines and fuzzy logic. The relationship investigated between hospitalizations for cardio-pulmonary conditions and ecological toxic waste. They used AI methods to forecast hospitalizations for cardiorespiratory conditions. They introduced and contrasted the improved Temporary Extended Memory (ITEM) model with other AI methods, including TEM, DL, and Autoregressive models using vectors (VAR). The proposed ITEM model outperformed the TEM and additional methods in the inquiry in recognizing and predicting monthly hospitalization forms. The link between outdoor pulmonary sickness admissions and every week hospitalization attributable to pulmonary exposure was examined by the author using the VRS algorithm. The results show a high correlation between routine respiratory hospitalizations, duration, and economic burden due to air toxic waste, and that VRS is a reliable predictor of daily admissions for pulmonary illnesses. Consequently, air toxic waste-based AI systems have the potential to accurately predict cardiovascular-related hospitalizations. Acute exacerbation of long-term respiratory conditions accelerates the progression of the illness, reduces superiority of lifetime, and raises death danger. Ecological exposure and response assessments indicate that air toxic waste raises the risk of severe exacerbation of continuing respiratory conditions.

PDPA exacerbation by information's since the Domestic Clinic and Ambulatory Health-Care Review. They compared four ML founded copies (Arbitrary forestry, Increasing, deep-neural net-effort, and Lariat reversion) to out-dated logistic reversion by predictors that were usually nearby, such as triage information's. Here In an effort to reduce training errors, the ensemble modelling technique known as "boosting" builds a robust classifier since a large amount of fragile classifiers. Adaptative boosting method, or AdaBoost, is some of the greatest well known improving algorithms and was among the firstly of its nice. EIBoost, Incline Boost, and Brown-Boost are supplementary algorithms that demonstrate increasing strategies. It was evident from the consequences that ML prototypes performed improved. ANN must castoff to display and achieve acute aggravating circumstances as well as hospital admissions. In a unit learning of 31,432 mature asthma persevering, a period searching prediction prototypical founded on an ANN that exploited scientific information in the specific period edge to foresee breathing attacks was described. 26 scenarios for severe PDPA exacerbation using accepted numerical methods including logistic reversion examination and Cox reversion examination. ANN, such as Multi-layer Receptors (MP), Echo-state Netefforts (EN), and Extreme-Learning Machines (ALD), were utilized by Polezer et al. (2018) to measure the adverse influence of atmospheric air toxic waste on personal well-being. The possessions of temperature, comparative moistness, and $PM_{2.5}$ on respiratory hospitalizations were examined in three well-known ANN. Using numerical regression models, the special effects of airway toxins on humanoid well-being have been thoroughly investigated. Other data processing methods should be investigated if linear numerical regression performs poorly due to insufficient datasets. The results suggest that artificial neural net-efforts (ANN) could be a more accurate way to identify the effects on lung well-being than numerical reversion prototypes, and that it could be particularly useful in situations where there is enough data. It was determined that AI might be used to replicate a doctor's cognitive functions in order to provide an automated PFT interpretation.

A ML method diagnose the most common obstructive illnesses with an overall accuracy of 68%. ANN may provide a helpful alternative method because, despite their ease of construction, numerical regression models are typically insufficient for evaluating the influence of air toxic waste on humanoid well-being. Meanwhile ML built approaches presented capacity in enlightening forecast skills, numerous revisions must cast-off ML procedures to delight PDPA aggravating circumstances of airline illnesses; however, only an insufficient identification must issue on AI usage in analysing the probable fitness properties of airborne toxic waste; though, AI/ML approaches can't however completely substitute doctors in the analysis and handling of unadorned breathing ailments. The usage of AI in the medical area is severely limited because of difficulties in obtaining sufficient data, trials, and training, even if ML methods are constantly increasing. Without a question, AI will have a significant impact on future health-care options.

26.6 Weather Variation Prediction Using AI and ML Methods

The harm that weather variation and worldwide warming pose to long-term growth has made them contentious issues in the international community. The overuse of natural resources, coupled with the combustion of fossil fuels and non-renewable energy sources, currents significant ecological hazards, including scarcity of potable water, deforestation, and worldwide warming. Since human activity causes the weather to variation, it becomes increasingly important to exactly foresee meteorological conditions measures that may have an influence on our own surroundings and well-being. As a result of anthropogenic weather variation, intense oceanic heating events, also known as Aquatic Warmness Sprays, are often occurring added commonly and to a larger extent (Hobday et al. 2016). The increasing worldwide temperatures lead heat-waves to become more powerful, prolonged, and frequent. Exposure to prolonged heat-waves can harm the atmosphere, people's health, crops, and livestock. More attention has been paid to land-based heat-waves than to marine heat-waves (MHWs), which cause significant financial burdens and detrimental impacts on marine life and ecosystems. Acute heat-related exposure of marine life and ecosystems may result in irreversible damage or even the near extinction of species, as seen in the case of widespread bleaching of coral reefs, die-offs in biodiversity, and financial losses in the aquaculture and fisheries industries. Deep learning, a type of AI, can also be used to forecast extreme weather events and the atmosphere. Owing to their capacity to incorporate a greater amount of information and knob the challenges posed by real-life weather, information handing out for metrological and marine changing aspects, as fit as marine and metrological biological chemistry, into their calculations; this enhances the precision of meteorological conditions and meteorological conditions demonstrating, rendering it added beneficial to decision makers (Lee et al. 1894). Then, by expanding on the relationships between weather and observed weather patterns, AI (AI) can offer more accurate alerts for impending weather phenomena, including extreme catastrophes. This section usually covers a few previous efforts on heat-wave predicting using AI methods. One of AI's most evident uses for weather variation is to improve weather modelling and predicting.

A number of very good weather prediction approaches have demonstrated truncated talent in current decades outstanding to differences in their forecast skills caused by weather variation. Foreseeing slightly climatological adjustable can be stimulating, smooth with the greatest advanced ML algorithms. Conversely, fresh disasters are intimately related with climate difference, and greatest prototypes have underperformed as a result of their inability to adequately account for new causes of expectedness that require emerged as a consequence of new climate difference. Many Machine Learning procedures have been used to forecast numerous natural events in recent years due to their ability to comprehend complex multivariate processes. Only a few studies have looked at the potential of marine shallow illness estimates to foresee MHW. Any of the earliest efforts to estimate MHWs. The study evaluated the likelihood of 4 MHW actions occurring in the System of California Current (SCC) among 2014 and 2016 by 8 diverse joined climate prototypes. The effectiveness of ML algorithms, including the Deep-Neural Net-effort prototypical, AdaBoost collaborative, and TEM, in predicting everyday marine shallow infection (MSI) has been demonstrated. The ability of the EIBoost, Gene (RF), long-temporary memory (TM), multi-layer perceptron (MP), and generalized additive models (GAMs) to forecast SST over a period of 562 days was examined by Wolff et al. Khan et al. developed a reliable heat wave forecast system for Pakistan based on weather variation using ML methods such as ANN, RAS, and arbitrary plantations. With an R2 of 0.86 and an OMSFN (origin mean square fault normalised) of 35%, RAS algorithms performed better in foreseeing HWD than other ML models during the validation period. The Earth System Simulator and Australian Community Weather: Seasonal Version 1 (ESSACW-S1) marine air prototypical should be intelligent to predict the 2021 maritime heat-waves in the Excessive Obstruction Ridge on an act periodic measure.

The model could predict the start of MHWs one week ahead of time with accuracy, but it could not predict when the MHW would terminate. The authors of the latest study also looked into the use of ESSACW-S2 for regular MHW predictions, which had a 40% success level when projected four months ahead of time. The authors emphasized the need of developing seasonal forecasts that are more accurate. Over the last ten existences, a proportion of investigation has been prepared in an effort to predict SST. These studies have used a variety of deep learning buildings, such as CNSs and convolutional long-term convolutional memory (CLCM) net-efforts, as well as ocean-weather models based on physical equations. The results showed that TEM under-performed in comparison to other models due to its inability to identify the high-frequency fluctuations in the input dataset; in oceanography, RF has been shown to be a dependable technique for modelling time series data. Out of all the models, the AF (Arbitrary Forestry) and RAS procedures gave the greatest results; however, the ensemble decomposition composite models gave even better results.

26.7 Conclusion

Airborne contamination positions substantial challenges to both humanoid wellbeing and the environment in the twenty-first century, necessitating effective mitigation strategies. Utilizing airborne excellence estimating and prediction, through premature cautioning systems, is crucial for minimizing human exposure to pollution. Recent years have witnessed the rise of artificial intelligence (AI) technologies in environmental pollution control, offering efficient alternatives to address the difficulties of real-life ecological apprehensions. This broadside examines various AI skills and ML algorithms used in ecological pollution estimating and premature cautioning schemes. It starts by revising the foundations and possessions of harmful airborne impurities, along with existing pollution forecasting tools. The discussion encompasses the application of AI techniques in forecasting major impurities and their effects on humanoid wellbeing and temperature estimate. AI-based techniques have established significant latent for in height correctness pollution estimate, overcoming limitations of traditional physical models such as computational intensity and data dependency. ANN, SVM, ambiguous sense, and hybrid models have emerged as effective tools for pollution forecasting, offering improved accuracy and fault tolerance. Deep-learning copies like convolutional neural systems (CNS) and long temporary remembrance systems (LTRS) show promise in enhancing forecast accuracy, particularly in densely populated areas where pollution concentration poses severe health risks. However, selecting the most suitable forecasting method depends on various factors, including location, pollutants, and dataset size. While AI technology offers rapid and reliable responses, challenges such as dataset size and computational complexity persist. Future research should explore AI methodologies for forecasting water and terrestrial smog, in calculation to addressing contests related to weather forecasting in need of secretions. Association among investigators, representatives, and technologists is key for developing robust forecasting systems to combat air pollution effectively.

References

Aayush K, Vishal D, Hammad N, Manu K (2020) Application of AI in curbing air toxic waste: the case of India. Asian J Manag 11:285–290

Ahmadi MH, Jashnani H, Chau K-W, Kumar R, Rosen MA (2019) Carbon dioxide emissions prediction of five Middle Eastern countries using artificial neural networks. Energy Sources Part A Recov Util Environ Eff

Alimissis A, Philippopoulos K, Tzanis C, Deligiorgi D (2018) Spatial estimation of urban air pollution with the use of artificial neural network models. Atmos Environ 191:205–213

Asha P, Natrayan L, Geetha B, Beulah JR, Sumathy R, Varalakshmi G, Neelakandan S (2022) IoT enabled ecological toxicology for air toxic waste observing using AI methods. Environ Res 205:112574

Bai L, Wang J, Ma X, Lu H (2018) Air toxic waste forecasts: an overview. Int J Environ Res Public Health 15:780

Bai Y, Zeng B, Li C, Zhang J (2019a) An ensemble extended short-term memory neural neteffort for hourly $PM_{2.5}$ concentration predicting. Chemosphere 222:286–294

Bai Y, Li Y, Zeng B, Li C, Zhang J (2019) Hourly PM2.5 concentration forecast using stacked autoencoder model with emphasis on seasonality. J Clean Prod 224:739–750

Cetin M (2016) A variation in the amount of CO_2 at the center of the examination halls: case study of Turkey. Stud Ethno-Med 10:146–155

Cetin M, Onac AK, Sevik H, Sen B (2019) Temporal and regional variation of some air toxic waste parameters in Bursa. Air Qual Atmos Health 12:311–316

Chaloulakou A, Saisana M, Spyrellis N (2003) Comparative assessment of neural netefforts and regression models for predicting summertime ozone in Athens. Sci Total Environ 313:1–13

Chattopadhyay G, Chattopadhyay S, Midya SK (2021) Fuzzy binary relation based elucidation of air quality over a highly polluted urban region of India. Earth Sci Inform 14:1625–1631

Chen X, Yin L, Fan Y, Song L, Ji T, Liu Y, Tian J, Zheng W (2020) Temporal evolution characteristics of $PM_{2.5}$ concentration based on continuous wavelet transform. Sci Total Environ 699:134244

Cheng Y, Zhang H, Liu Z, Chen L, Wang P (2019) Hybrid algorithm for short-term forecasting of $PM_{2.5}$ in China. Atmos Environ 200:264–279

Dotse S-Q, Petra MI, Dagar L, De Silva LC (2018) Application of computational intelligence methods to forecast daily PM_{10} exceedances in Brunei Darussalam. Atmos Pollut Res 9:358–368

Elsunousi AAM, Sevik H, Cetin M, Ozel HB, Ozel HU (2021) Periodical and regional variation of particulate matter and CO_2 concentration in Misurata. Environ Monit Assess 193:707

Feng X, Li Q, Zhu Y, Hou J, Jin L, Wang J (2015) Artificial neural netefforts predicting of $PM_{2.5}$ toxic waste using air mass trajectory based geographic model and wavelet transformation. Atmos Environ 107:118–128

Fernández JD, Vico F (2013) AI methods in algorithmic composition: a comprehensive survey. J Artif Intell Res 48:513–582

Gan K, Sun S, Wang S, Wei Y (2018) A secondary-decomposition-ensemble learning paradigm for predicting $PM_{2.5}$ concentration. Atmos Pollut Res 9:989–999

Gu Y, Li B, Meng Q (2022) Composite interpretable predictive ML model for air toxic waste prediction. Neurocomputing 468:123–136

Han X, Liu Y, Gao H, Ma J, Mao X, Wang Y, Ma X (2017) Predicting $PM_{2.5}$ induced male lung cancer morbidity in China using satellite retrieved $PM_{2.5}$ and spatial analysis. Sci Total Environ 607:1009–1017

Hobday AJ, Alexander LV, Perkins SE, Smale DA, Straub SC, Oliver EC, Benthuysen JA, Burrows MT, Donat MG, Feng M (2016) A hierarchical approach to defining marine heatwaves. Prog Oceanogr 141:227–238

Jerrett M, Arain A, Kanaroglou P, Beckerman B, Potoglou D, Sahsuvaroglu T, Morrison J, Giovis C (2005) A review and evaluation of intraurban air toxic waste exposure models. J Expo Sci Environ Epidemiol 15:185–204

Kamińska, J.A. A random forest partition model for predicting NO_2 concentrations from traffic flow and meteorological conditions. Sci Total Environ 651:475–483

Kampa M, Castanas E (2008) Human health effects of air toxic waste. Environ Pollut 151:362–367

Kemp AC, Horton BP, Donnelly JP, Mann ME, Vermeer M, Rahmstorf S (2011) Weather related sea-level variations over the past two millennia. Proc Natl Acad Sci USA 108:11017–11022

Lee D, Kang S, Shin J (1894) Using deep learning methods to forecast ecological consumption level. Sustainability 2017:9

Li T, Li X, Wang L, Ren Y, Zhang T, Yu M (2018) Multi-model ensemble forecast method of $PM_{2.5}$ concentration based on wavelet neural networks. In: Proceedings of the 2018 1st international cognitive cities conference (IC3), Okinawa, Japan, pp 81–86

Li C, Zhu Z (2018) Research and application of a novel hybrid air quality early-warning system: a case study in China. Sci Total Environ 626:1421–1438

Lin S, Wei D, Sun Y, Chen K, Yang L, Liu B, Huang Q, Paoliello MMB, Li H, Wu S (2020) Region-specific air pollutants and meteorological parameters influence COVID-19: a study from mainland China. Ecotoxicol Environ Saf 204:111035

Liu H, Yang R (2021) A spatial multi-resolution multi-objective data-driven ensemble model for multi-step airborne excellence index predicting based on real-time decomposition. Comput Ind 125:103387

Liu H, Yan G, Duan Z, Chen C (2021) Intelligent modeling strategies for predicting airborne excellence time series: a review. Appl Soft Comput 102:106957

Liu X, Lu D, Zhang A, Liu Q, Jiang G (2022) Data-driven ML in ecological toxic waste: gains and problems. Environ Sci Technol 56:2124–2133

Liu H, Jin K, Duan Z (2019) Air PM2.5 concentration multi-step predicting using a new composite modeling method: comparing cases for four cities in China. Atmos Pollut Res 10:1588–1600

Liu H, Duan Z, Chen C (2020) A composite multi-resolution multi-objective ensemble model and its application for predicting of daily $PM_{2.5}$ concentrations. Inf Sci 516:266–292

Mannucci PM, Franchini M (2017) Health effects of ambient air toxic waste in developing countries. Int J Environ Res Public Health 14:1048

Masood A, Ahmad K (2021) A review on emerging AI (AI) methods for air toxic waste predicting: fundamentals, application and performance. J Clean Prod 322:129072

Masood A, Ahmad K (2020) A model for particulate matter (PM2.5) prediction for Delhi based on machine learning approaches. Procedia Comput Sci 167:2101–2110

Mo X, Zhang L, Li H, Qu Z (2019) A novel airborne excellence early-cautionary system based on AI. Int J Environ Res Public Health 16:3505

Najjar YS (2011) Gaseous toxins formation and their harmful effects on health and atmosphere. Innov Energy Policies 1:1–8

Norhayati I, Rashid M (2018) Adaptive neuro-fuzzy prediction of carbon monoxide emission from a clinical waste incineration plant. Neural Comput Appl 30:3049–3061

Ong BT, Sugiura K, Zettsu K (2016) Dynamically pre-trained deep recurrent neural netefforts using ecological observing data for predicting $PM_{2.5}$. Neural Comput Appl 27:1553–1566

Polezer G, Tadano YS, Siqueira HV, Godoi AF, Yamamoto CI, de André PA, Pauliquevis T, de Fatima Andrade M, Oliveira A, Saldiva PH (2018) Assessing the impact of $PM_{2.5}$ on respiratory disease using artificial neural networks. Environ Pollut 235:394–403

Qi Y, Li Q, Karimian H, Liu D (2019) A hybrid model for spatiotemporal forecasting of $PM_{2.5}$ based on graph convolutional neural network and long short-term memory. Sci Total Environ 664:1–10

Stein AL (2020) AI and weather variation. Yale J Regul 37:890

Sun W, Sun J (2017) Daily $PM_{2.5}$ concentration prediction based on principal component analysis and MLLS optimized by cuckoo search algorithm. J Environ Manag 188:144–152

Sun W, Huang C (2022) Predictions of carbon emission intensity based on factor analysis and an improved abrupt learning device from the perspective of carbon emission efficiency. J Clean Prod 338:130414

Wang D, Wei S, Luo H, Yue C, Grunder O (2017) A novel hybrid model for air quality index forecasting based on two-phase decomposition technique and modified extreme learning machine. Sci Total Environ 580:719–733

Wang C, Ye Z, Yu Y, Gong W (2018) Estimation of bus emission models for different fuel types of buses under real conditions. Sci Total Environ 640:965–972

Wu Q, Lin H (2019) Daily urban airborne excellence index predicting based on variational mode decomposition, sample entropy and ESTM neural neteffort. Sustain Cities Soc 50:101657

Xu Y, Du P, Wang J (2017) Research and application of a hybrid model based on dynamic fuzzy synthetic evaluation for establishing air quality forecasting and early warning system: a case study in China. Environ Pollut 223:435–448

Yang Z, Wang J (2017) A new air quality monitoring and early warning system: air quality assessment and air pollutant concentration prediction. Environ Res 158:105–117

Zaman NAFK, Kanniah KD, Kaskaoutis DG (2017) Estimating particulate matter using satellite based aerosol optical depth and meteorological variables in Malaysia. Atmos Res 193:142–162

Zhang L, Lin J, Qiu R, Hu X, Zhang H, Chen Q, Tan H, Lin D, Wang J (2018) Trend analysis and forecast of $PM_{2.5}$ in Fuzhou, China using the MAUAT model. Ecol Indic 95:702–710

Zhou Y, Zhao X, Lin K-P, Wang C-H, Li L (2019) A Gaussian process mixture model-based hard-cut iterative learning algorithm for airborne excellence prediction. Appl Soft Comput 85:105789

Zhou J-H, Zhao J-G, Li P (2010) Study on gray numerical model of air toxic waste in wuan city. In: Proceedings of the 2010 international conference on challenges in ecological science and computer engineering, Wuhan, China, pp 321–323

Zhu F, Ding R, Lei R, Cheng H, Liu J, Shen C, Zhang C, Xu Y, Xiao C, Li X (2019a) The short-term effects of air toxic waste on respiratory diseases and lung cancer mortality in Hefei: a time-series analysis. Respir Med 146:57–65

Zhu S, Qiu X, Yin Y, Fang M, Liu X, Zhao X, Shi Y (2019b) Two-step-composite model based on data preprocessing and intelligent optimization algorithms (CS and GWO) for NO_2 and SO_2 predicting. Atmos Pollut Res 10:1326–1335

Zhu S, Lian X, Wei L, Che J, Shen X, Yang L, Qiu X, Liu X, Gao W, Ren X (2018) $PM_{2.5}$ predicting using RSV with OPSGSA algorithm based on CMCED, NNGRand GCA considering climatological factors. Atmos Environ 183:20–32

Chapter 27
Systematic Review on Role of Microorganisms in Bioremediation

Gopi Devarajan, Suganya Ilango, Aishwarya Vetrivel, and T. G. Nithya

Abstract Worldwide, environmental contamination and its remediation are major problems. Heavy metals, dyes, hydrocarbons, pesticides, and other pollutants are some of the main causes of environmental deterioration. Eliminating residual contaminants is also one of the toughest challenges. Since bioremediation restores the damaged site to its previous state, it is one of the most successful methods for lowering environmental toxins. Through the action of bacteria, fungus, and plants, it degrades, removes, alters, immobilizes, or detoxifies numerous chemicals and physical wastes from the environment. Microorganisms' natural biodegradation of complex organic compounds implies that they can be reduced to more basic inorganic forms. Bioremediation, is a technology developed by humans that employs microorganisms to reduce pollutants, methods such as bioaugmentation, biostimulation, bioventing, and bio attenuation are some of the types of bioremediation used to increase the capacity of microorganisms to degrade pollutants. These organisms will use the pollutants as a source of energy. Using symbiotic microorganisms, genetically engineered microorganisms have been used to improve pollutant cleanup and ensure safe biodegradation. Hence, this review focuses on the most recent developments in bioremediation techniques, the mechanisms by which microorganisms degrade various pollutants, and the possible uses of bioremediation as a worldwide pollution mitigation strategy.

Keywords Bioremediation · Microorganisms · Emerging technology · Pollutants · Bioremediation techniques

G. Devarajan (✉)
Department of Economics, Faculty of Science and Humanities, SRM Institute of Science and Technology, Kattankulathur, Chennai, Tamil Nadu 603203, India

S. Ilango · T. G. Nithya
Department of Biochemistry, Faculty of Science and Humanities, SRM Institute of Science and Technology, Kattankulathur, Chennai, Tamil Nadu 603203, India

A. Vetrivel
Department of Biotechnology, Dr. G.R. Damodaran College of Science, Coimbatore, Tamil Nadu 641014, India

© The Author(s), under exclusive license to Springer Nature Switzerland AG 2025
A. Jeyaseelan and K. Murugasen (eds.), *Sustainable Environmental Remediation: Avenues in Nano and Biotechnology*, Environmental Science and Engineering,
https://doi.org/10.1007/978-3-031-78483-5_27

27.1 Introduction

Environmental contamination has recently emerged as one of the world's most sensitive topics. The rapid growth of the human population, urbanization, hazardous farming practices, and industrialization all had a part in the vast environmental pollution that resulted from the discharge of various waste products. Pollutants found in sewage, industrial effluent, municipal and household solid waste include heavy metals Cd, Ld, Mercury, Cu, and Pb; organic chemicals and volatile organic compounds like Benzene, Toluene, Xylenes, Dichloromethane; and inorganic chemicals like NH_3, N_2, P, NO_3, and NO_2. Chlorine-containing compounds, dyes such as anthraquinone, vat dye, triphenylmethane, Azodye and greenhouse gas emissions, polymers, nuclear waste, petroleum hydrocarbons, paints, disinfectants, polychlorinated biphenyls, and other biphenyls, lubricants and detergents, contaminate agriculture soil, ground water, rivers and air in general terrestrial and aquatic ecosystem.

Natural biodegradation by microorganisms suggests that complex organic materials can be broken down into simpler inorganic forms. While bioremediation is a human-engineered method that uses microorganisms to lessen pollution. Natural attenuation, biostimulation, bioaugmentation are the methods used to boost the ability of microorganisms to fight against pollutants. Pollutants will be used as a source of energy by these microorganisms. Bacteria, fungus, algae, and protozoa are some of the potential microorganisms that can breakdown environmental pollutants. There are two methods for carrying out the bioremediation process: ex situ and in situ bioremediation.

To treat and manage these wastes, many processes—physical, chemical, and biological methods or in combination—are available. Some uneconomical physicochemical approaches, are gradually losing momentum due to intrinsic issues with secondary contamination and nonsustainable contaminant control.

Although they have been used for a long time, both chemical and physical approaches have disadvantages. For example, the chemical method requires specialized knowledge and equipment, and the physical method is expensive. This has increased demand for biological remediation (bioremediation), a superior substitute. The most economical, environmentally friendly, and efficient method of changing pollutants is bioremediation. Both microbes and plants can be employed in biological remediation; however, microbes are preferred since they are easier to control and require less time to grow than plants.

The goal of this review is to represent the current trend of microorganism application and role in bioremediation and to give important background for efficient evaluation of bioremediation processes.

27.2 Conclusion

Microbiologists face a difficult problem when trying to develop a good bioremediation approach since it requires a thorough understanding of degradative microbial communities. As a result, proteomics, metabolomics, and genome analysis have become crucial tools for discovering all of the microbial communities that have not yet been identified but are capable of decomposing pollutants and for identifying the many metabolites that organisms produce in reaction to stress. Bioremediation techniques have been found to be significantly more successful and efficient than conventional remediation approaches, and they also help to maintain the biological balance of the ecosystem. It will be necessary to evaluate in situ sediment bioremediation on a large scale by increasing laboratory or small-scale research and figuring out the financial and environmental implications in order to reach a conclusion. They are therefore crucial components for creating eco-friendly and long-lasting bioremediation treatments for contaminated locations.

The manufacturer's authorised representative in the EU is Springer Nature Customer Service Centre GmbH, Europaplatz 3, 69115 Heidelberg, Germany. If you have any concerns regarding our products, please contact ProductSafety@springernature.com

Printed and bound by CPI Group (UK) Ltd, Croydon, CR0 4YY
01/10/2026
02225965-0017